Probability and Mathematical Statistics (Continued)

ROHATGI • An Introduction to Probability Theory and Mathematical Statistics

SCHEFFE • The Analysis of Variance

SEBER • Linear Regression Analysis

SERFLING • Approximation Theorems of Mathematical Statistics

WILKS • Mathematical Statistics

WILLIAMS • Diffusions, Markov Processes, and Martingales, Volume I: Foundations

ZACKS • The Theory of Statistical Inference

Applied Probability and Statistics

ANDERSON, AUQUIER, HAUCK, OAKES, VANDAELE, and WEISBERG • Statistical Methods in Comparative Studies

BAILEY • The Elements of Stochastic Processes with Applications to the Natural Sciences

BAILEY • Mathematics, Statistics and Systems for Health

BARNETT and LEWIS • Outliers in Statistical Data

BARTHOLEMEW • Stochastic Models for Social Processes, *Second Edition*

BARTHOLOMEW and FORBES • Statistical Techniques for Manpower Planning

BECK and ARNOLD • Parameter Estimation in Engineering and Science

BELSLEY, KUH, and WELSCH • Regression Diagnostics: Identifying Influential Data and Sources of Collinearity

BENNETT and FRANKLIN • Statistical Analysis in Chemistry and the Chemical Industry

BHAT • Elements of Applied Stochastic Processes

BLOOMFIELD • Fourier Analysis of Time Series: An Introduction

BOX • R. A. Fisher, The Life of a Scientist

BOX and DRAPER • Evolutionary Operation: A Statistical Method for Process Improvement

BOX, HUNTER, and HUNTER • Statistics for Experimenters: An Introduction to Design, Data Analysis, and Model Building

BROWN and HOLLANDER • Statistics: A Biomedical Introduction

BROWNLEE • Statistical Theory and Methodology in Science and Engineering, *Second Edition*

BURY • Statistical Models in Applied Science

CHAMBERS • Computational Methods for Data Analysis

CHATTERJEE and PRICE • Regression Analysis by Example

CHERNOFF and MOSES • Elementary Decision Theory

CHOW • Analysis and Control of Dynamic Economic Systems

CLELLAND, deCANI, and BROWN • Basic Statistics with Business Applications, *Second Edition*

COCHRAN • Sampling Techniques, *Third Edition*

COCHRAN and COX • Experimental Designs, *Second Edition*

CONOVER • Practical Nonparametric Statistics, *Second Edition*

COX • Planning of Experiments

DANIEL • Biostatistics: A Foundation for Analysis in the Health Sciences, *Second Edition*

DANIEL • Applications of Statistics to Industrial Experimentation

DANIEL and WOOD • Fitting Equations to Data: Computer Analysis of Multifactor Data, *Second Editon*

DAVID • Order Statistics

DEMING • Sample Design in Business Research

continued on back

FUNDAMENTALS OF QUEUEING THEORY

A WILEY PUBLICATION IN APPLIED STATISTICS

FUNDAMENTALS OF QUEUEING THEORY

Donald Gross

Professor of Operations Research
The George Washington University

Carl M. Harris

Professor of Industrial Engineering and Operations Research
Syracuse University

JOHN WILEY & SONS

New York   Chichester   Brisbane   Toronto

Library of Congress Cataloging in Publication Data:

Gross, Donald.
    Fundamentals of queueing theory.

    (Wiley series in probability and mathematical statistics)
    1. Queueing theory. I. Harris, Carl M., 1940– joint author. II. Title.
T57.9.G76   519.8'2   73-20084
ISBN 0-471-32812-X

Printed in the United States of America

10 9 8 7 6

To
[2(Alice) + Paul]

PREFACE

This book is the outgrowth of our experiences in teaching a one-semester course in queueing theory, which serves as an elective in a Master of Science operations research program offered by the Department of Operations Research, School of Engineering and Applied Science of The George Washington University. We believe there has been a great gap in books available which could serve as a text for such a course. In writing this book, we have expanded, modified, and revised our course notes several times, and also incorporated, where applicable, our experiences in the research and application of queueing theory.

COURSE COVERAGE

This book can be used as a text for a one or two semester, or a one or two quarter course, such as the one described above, namely, a graduate-level elective course in an operations research, systems analysis, or industrial engineering program of an engineering school or a quantitatively oriented program of a business school. The chart below is our recommendation for the material to include in courses of varying duration. Naturally, the individual instructor should feel free to make adjustments depending on his or her personal interests and the background of the students.

In addition, we feel that this book can also be extremely valuable for analysts working in industry or government who may wish to fill in some theoretical gaps in the development of queueing models. The last two chapters on simulation and application, respectively, might be of particular interest to practicing analysts. In toto, we think that the book will serve as a comprehensive reference in the area of queueing theory. It has a degree of mathematical sophistication which will allow the understanding and use

Recommended Coverage

Course Length	Text Material
One Quarter	Chapters 1, 2 (through 2.6.1), 3 (omit 3.2.1 and 3.5.2), 4 (through 4.4.1), 5 (5.1.1 only)
One Semester	Chapters 1, 2, 3, 4, 5 (5.1.1, 5.1.2, 5.3.1), 6 (6.1 only), 7 (7.2 only)
Two Quarters	Chapters 1, 2, 3, 4, 5 (5.1.1, 5.1.2, 5.1.10, 5.2.2, 5.3), 6 (6.1 only), 7 (7.2 only), 8, 9
Two Semesters	Entire Text

of most of the important queueing models, but is not so mathematically abstract and imposing as to prevent all but mathematicians from understanding the material.

PREREQUISITE KNOWLEDGE

The background assumed on the part of the reader is a knowledge of undergraduate differential and integral calculus and of elements of differential equations, and the experience of a probability and statistics course based upon a calculus foundation. Some knowledge of special "higher" mathematical topics such as transforms, difference equations, and Markov processes would be helpful for the more advanced material, although certainly not necessary, since appendices on these subjects are included. Thus the undergraduate background of most engineering, physical science, and mathematics majors, as well as *some* economics, business administration, and social science majors, would be adequate.

PHILOSOPHY AND ORGANIZATION

The general plan of the book is to present, in detail, the basic material in the early chapters. For example, in Chapter 1 we spend a sizable amount of time on deterministic queues. In addition to adding this topic for completeness, we believe it serves to familiarize the reader in a graphic way with the actual "workings" of a queue, such as, how customers arrive and depart and how the line builds up and depletes.

In Chapter 2 we develop the basic exponential-interarrival-time exponential-service-time single-channel model $(M/M/1)$ and then in Chapter 3 derive the general birth–death equations, showing that the $M/M/1$ model falls out as a special case. We purposely do this partial repetition as a pedagogical tool, since we feel that $M/M/1$ more naturally follows from the Poisson process than does the birth–death process. Furthermore, the simplicity of $M/M/1$ allows us to illustrate early in the text the three methods of solving difference equations, namely, iteration, generating functions, and linear difference methods (the latter sometimes referred to as the operator or characteristic-equation approach).

We introduce only touches of transient analyses to show their complexity, but give the reader a feel for the speed of convergence to steady state by observing in detail the very simple exponential-exponential single-channel no-waiting-room $(M/M/1/1)$ case.

We also provide an extensive discussion of the concepts of steady state and ergodicity, topics which often have been glossed over or completely ignored. All this is done in Chapter 2, so that the student, after reading the first two chapters, should have a solid feeling for the nature and types of analyses involved in basic queueing models.

The general approach used in this text (except where other types of analyses are indicated) is the rather formal one of first setting-up difference equations, and then obtaining the differential-difference equations and finally the steady-state difference equations which then lead to the steady-state probabilities. We do, however, mention briefly (Problem 2.1) the method of detailed (or stochastic) balance which yields the steady-state probabilities directly, but our personal taste dictated the more formal procedure.

While the book is titled "Fundamentals," some of the material in Chapters 5 and 6 (especially Chapter 6) is quite advanced. We do not go into great detail for this advanced material, but present results, sketch derivations, and reference the literature freely, so that the interested reader can gain greater depth by going to the references if he or she so desires. The first six chapters deal with the development of analytical queueing models while the last three deal with application and implementation.

ACKNOWLEDGMENTS

Many colleagues and students have contributed to this book through their encouragement and day-to-day discussions on the topics in general and portions of the text in particular. We are especially thankful for the help provided by Professors George S. Fishman, Irwin Greenberg, N. U.

Prabhu, and Nozer D. Singpurwalla, and by Dr. Daniel P. Heyman. We are indebted to the administration of The George Washington University, to Dr. Harold Liebowitz, Dean of Engineering and Applied Science, and to Professor William H. Marlow, Chairman, Department of Operations Research and Director, Institute for Management Science and Engineering, for their support of this effort. We are also most grateful to the Office of Naval Research for support of portions of this work under Contract N00014-67-A-0214-0001. Special thanks are due Mr. Ross Tomlinson, Mrs. Bettie Taggart, Mrs. B. J. Walker, Miss Mary L. Vincent and Miss Robin Meader for their invaluable aid in computer programming, typing, and proofreading and to Mr. M. Morse for his fine job on the illustrations. Research Assistants T. R. Thiagarajan and K. L. Chhabra provided help in the area of computer programming and calculations. In addition, thanks go to our many students, particularly F. Al-Khayyal, H. D. Kahn, B. Rider, and S. S. Shanker for finding typographical (and other) errors, and providing solutions for many of the problems. We also wish to express sincere gratitude to our many teachers through the years of our education, particularly our advisors, Professors Robert E. Bechhofer, Clifford W. Marshall, Andrew Schultz, Jr., and Lionel Weiss.

Finally, we wish to express our sincere appreciation to our wives for their constant encouragement and reminders that we should complete our task forthwith, so that we might devote more time to our families, and to our parents, Frank and Marion Gross, and Benjamin and Marion Harris, without whose encouragement to complete our education this work would never have been undertaken.

DONALD GROSS
CARL M. HARRIS

Washington, D.C.
January 1974

CONTENTS

FUNDAMENTALS OF
QUEUEING THEORY

Chapter 1
INTRODUCTION

All of us have experienced the annoyance of having to wait in line. Unfortunately, this phenomenon is becoming more and more prevalent in our increasingly congested and urbanized society. We wait in line in our cars in traffic jams or at toll booths; we wait in line at supermarkets to check out; we wait in line in barber shops or beauty parlors; we wait in line at post offices (especially around Christmas time); and so on, ad infinitum. We, as customers, do not generally like these waits, and the managers of the establishments at which we wait also do not like us to wait, since it may cost them business. Why then is there waiting? The answer is relatively simple. There is more demand for service than there is facility for service available. Why is this so? There may be many reasons; for example, there may be a shortage of available servers; it may be infeasible economically for a business to provide the level of service necessary to prevent waiting; or there may be a space limit to the amount of service that can be provided. Generally these limitations can be removed with the expenditure of capital, and to know how much service should then be made available, one would need to know answers to such questions as, "How long must a customer wait?" and "How many people will form in the line?" Queueing theory attempts (and in many cases succeeds) to answer these questions through detailed mathematical analysis. The word "queue" is in more common usage in Great Britain than in the United States, but it is rapidly gaining acceptance in this country, although it must be admitted that it is just as displeasing to spend time in a queue as in a waiting line.

1.1 DESCRIPTION OF THE QUEUEING PROBLEM

A queueing system can be described as customers arriving for service, waiting for service if it is not immediate, and if having waited for service, leaving the system after being served.[1] Such a basic system can be

---

[1] The term "customer" is used in a general sense and does not imply necessarily a human customer. For example, a customer could be a ball bearing waiting to be polished, an airplane waiting in line to take off, or a computer program waiting to be run on a time-shared basis.

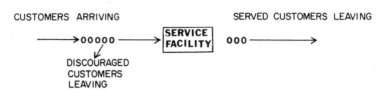

Fig. 1.1   Schematic diagram of a queueing process.

schematically shown as in Figure 1.1. Although any queueing system may be diagrammed in this manner, it should be rather clear that a reasonably accurate representation of such a system would require a detailed characterization of the underlying processes. The primary categories of such a characterization are taken up in detail in the next section.

1.2   CHARACTERISTICS OF QUEUEING PROCESSES

Six basic characteristics of queueing processes are explored in this section. These are as follows:

(i)   Arrival pattern of customers.
(ii)   Service pattern of servers.
(iii)   Queue discipline.
(iv)   System capacity.
(v)   Number of service channels.
(vi)   Number of service stages.

In most cases, these six basic characteristics provide an adequate description of a queueing system.

1.2.1   Arrival Pattern of Customers

The arrival pattern or *input* to a queueing system is often measured in terms of the average number of arrivals per some unit of time (mean arrival rate) or by the average time between successive arrivals (mean interarrival time). Since these quantities are clearly related, either one of these measures suffices in describing the system input. In the event that the stream of input is deterministic (completely known and thus void of uncertainty), then the arrival pattern is fully determined by either the mean arrival rate or the mean interarrival time. On the other hand, if there is uncertainty in the arrival pattern (often referred to as random, probabilistic, or stochastic), then these mean values provide only measures of central tendency for the input process, and further characterization is required in the form of the probability distribution associated with this random process.

Another factor of interest concerning the input process is the possibility that arrivals come in batches instead of one at a time. In the event that more than one arrival can enter the system simultaneously (entering the system does not necessarily mean entering into service, but may instead require joining the line when immediate service is not available), the input is said to occur in *bulk* or *batches*. In the bulk-arrival situation, not only may the time between successive arrivals of the batches be probabilistic, but also the number of customers in a batch (the batch size).

It is also necessary to know the reaction of a customer upon entering the system. A customer may decide to wait no matter how long the queue becomes, or if the queue is too long to suit him, may decide not to enter it. If a customer decides not to enter the queue upon arrival, he is said to have *balked*. On the other hand, a customer may enter the queue, but after a time lose patience and decide to leave. In this case he is said to have *reneged*. In the event that there are two or more parallel waiting lines, customers may switch from one to another, that is, *jockey* for position. These three situations are all examples of queues with *impatient customers*.

One final factor to be considered regarding the arrival pattern is the manner in which the pattern changes with time. An arrival pattern that does not change with time (i.e., the form and the values of the parameters of the probability distribution describing the input process are time-independent) is called a *stationary* arrival pattern. One that is not time-independent in the sense described above is called *nonstationary*.

1.2.2 Service Patterns of Servers

Much of the discussion above concerning the arrival pattern is appropriate in discussing service patterns. For example, service patterns can also be described by a rate (number of customers served per some unit of time) or as a time (time required to service a customer). One important difference exists, however, between service and arrivals. When one speaks about service rate or service time, these terms are conditioned on the fact that the system is not empty; that is, there is someone in the system requiring service. If the system is empty, the service facility is idle. Service may also be deterministic or probabilistic; hence in the latter case the probability distributions associated with service are conditional, based on a nonempty system.

Service may also be single or batch. One generally thinks of one customer being served at a time by a given server, but there are many situations where customers may be served simultaneously by the same server, such as a computer with parallel processing, sightseers on a guided tour, or people boarding a train.

The service rate may depend on the number of customers waiting for service. A server may work faster if he sees that the queue is building up or, conversely, he may get flustered and become less efficient. The situation in which service depends on the number of customers waiting is referred to as *state-dependent* service. Although this term was not used in discussing arrival patterns, the problems of customer impatience can be looked upon as ones of state-dependent arrivals, since the arrival behavior depends on the amount of congestion in the system.

Service, as was the case for arrivals, can be stationary or nonstationary with respect to time. For example, learning may take place on the part of the server so that he becomes more efficient as he gains experience. The dependence on time is not to be confused with dependence on state. The former does not depend on the number of customers in the system, but rather on how long it has been in operation. The latter does not depend on how long the system has been in operation, but only on the state of the system at a given time, that is, on how many customers are currently in the system. Of course, any queueing system could be both nonstationary and state dependent.

Even if the service rate is high, it is very likely that some customers will be delayed by waiting in the line. In general, customers arrive and depart at irregular intervals; hence the queue length will assume no definitive pattern unless arrivals and service are deterministic. Thus it follows that a probability distribution for queue lengths would be the result of two separate processes—arrivals and services—which are generally, though not universally, assumed mutually independent.

1.2.3 Queue Discipline

Queue discipline refers to the manner by which customers are selected for service when a queue has formed. The most common discipline that can be observed in everyday life is first come, first served, or first in, first out (FIFO), as it is sometimes called. However, this is certainly not the only possible queue discipline. Some others in common usage are last in, first out (LIFO), which is applicable to many inventory systems when there is no obsolescence of stored units as it is easier to reach the nearest items which are the last in; selection for service in random order independent of the time of arrival to the queue (SIRO); and a variety of *priority* schemes, where customers are given priorities upon entering the system, the ones with higher priorities to be selected for service ahead of those with lower priorities, regardless of their time of arrival to the system.

There are two general situations in priority disciplines. In the first, which is called *preemptive*, the customer with the highest priority is allowed to enter service immediately even if a customer with lower priority is already in service when the higher priority customer enters the system; that is, the lower priority customer in service is preempted, his service stopped, to be resumed again after the higher priority customer is served. There are two possible additional variations: the preempted customer's service when resumed can either continue from the point of preemption or start anew. In the second general priority situation, called the nonpreemptive case, the highest priority customer goes to the head of the queue but cannot get into service until the customer presently in service is completed, even though this customer has a lower priority.

The number of priority classes can be any number greater than one, and if there can be more than a single customer in any given priority class in the system simultaneously, then the discipline of selecting customers within the same priority class must also be specified and can be any of those mentioned previously. As an example, suppose messages into a military communications center can have one of four priorities: routine, priority, operational immediate, and emergency. If there are several messages in the center in the same priority class, they are taken in order of time received. This amounts to a FIFO discipline within each priority class. Regulations instruct the message centers to preempt whenever a higher priority message comes in. In practice, however, they are likely (depending on the person in charge of the center) to preempt only when emergency messages come in and not to preempt for the lower priorities. Thus it is possible to simultaneously have preemptive and nonpreemptive disciplines. The priority situation can be further complicated by a decision rule to the effect that preemption will occur if service is not too far along, but not if the customer in service is near completion. This is a common practice in manual transmission message centers, where if a message is just started and is long, it can be preempted; otherwise, its transmission will be completed even though a higher priority message has come in.

1.2.4 System Capacity

In some queueing processes there is a physical limitation to the amount of waiting room, so that when the line reaches a certain length, no further customers are allowed to enter until space becomes available by a service completion. These are referred to as finite queueing situations; that is, there is a finite limit to the maximum queue size. A queue with limited waiting room can be viewed as one with forced balking where a customer is forced to balk if he arrives at a time when queue size is at its limit. This is a simple case of balking, since it is known exactly under what circumstances arriving customers must balk.

1.2.5 Number of Service Channels

The number of service channels refers to the number of parallel service stations which can service customers simultaneously. Figure 1.1 depicts a single-channel system, while Figure 1.2 shows two variations of multiple-channel systems. The two multichannel systems differ in that the first has a single queue, while the second allows a queue for each channel. A barber shop with many chairs is an example of the first type of multichannel system (assuming no customer is waiting for any particular barber), while a supermarket fits the second description. It is generally assumed that the service mechanisms of parallel channels operate independently of each other.

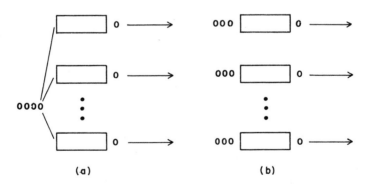

Fig. 1.2 Multichannel queueing system.

1.2.6 Stages of Service

A queueing system may have only a single stage of service such as the barber shop and supermarket examples, or it may have several stages. An example of a multistage queueing system would be a physical examination

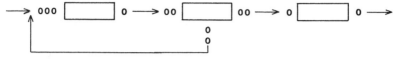

Fig. 1.3 Multistage queueing system.

procedure, where each patient must proceed through several stages, such as medical history; ear, nose, and throat examination; blood tests; electrocardiogram; eye examination; and so on. In some multistage queueing processes recycling may occur. Recycling is common in manufacturing processes where quality control inspections are performed after certain stages, and parts that do not meet quality standards are sent back for reprocessing. A multistage queueing system with some recycling (or feedback as it is sometimes called) is depicted in Figure 1.3.

The six characteristics of queueing systems discussed in this section are generally sufficient to describe completely a process under study. One can see from the discussion thus far that there exists a wide variety of queueing systems that can be encountered. Before performing any mathematical analyses, however, it is absolutely necessary to describe adequately the process being modeled. Knowledge of the aforementioned six characteristics is essential in this task.

It is extremely important to use the correct model or at least the model that best describes the real situation being studied. A great deal of thought is often required in this model-selection procedure. For example, let us reconsider the supermarket mentioned previously. Suppose there are c checkout counters. If customers were to choose a checkout counter on a purely random basis (without regard to the queue length in front of each counter) and never switch lines (no jockeying), then we truly have c independent single-channel models. If, on the other hand, there is a single waiting line and when a checker becomes idle, the customer at the head of the line (or with the lowest number if numbers are given out) enters service, we have a c-channel model. Neither, of course, is generally the case in most supermarkets. What usually happens is that queues form in front of each counter, but new customers enter the queue which is the shortest (or has shopping carts which are lightly loaded). Also, there is a great deal of jockeying between lines. Now the question becomes which choice of models (c independent single-channels or a single c-channel) is more appropriate? If there were complete jockeying, the single c-channel model would be quite appropriate, since even though in reality there are c lines, there is little difference, when jockeying is present, between these two cases. This is so because no servers will be idle as long as customers are waiting for service, which would not be the case with c truly independent single-channels. As jockeying is relatively easy to accomplish in super-markets, the c-channel model would be the more appropriate and realistic model rather than the c-single-channels model, which one might have been tempted to choose initially prior to giving much thought to the process. Thus it is important not to jump to hasty conclusions but to select carefully the most appropriate model.

1.3   MEASURES OF EFFECTIVENESS

Up to now the concentration has been on the physical description of queueing processes. What, then, might one like to know about the effectiveness of a queueing system? Generally there are three system responses of interest. These are (1) some measure of the waiting time that a customer might be forced to endure; (2) an indication of the manner in which customers may accumulate; and (3) a measure of the idle time of the servers. Since most queueing systems have stochastic elements, these measures are often random variables and their probability distributions, or at the very least their expected values, are desired to be found.

The task of the queueing analyst is one of two things. He must either determine the values of appropriate measures of effectiveness for a given process or he must design an "optimal" (according to some criterion) system. To do the former, he must relate waiting delays, queue lengths, and such, to the given properties of the input stream and the service proce-dures. On the other hand, for the design of a system the analyst would probably want to balance customer waiting time against the idle time of servers according to some inherent cost structure. If the cost of waiting and idle service can be obtained directly, they could be used to determine the optimum number of channels to maintain and the service rates at which to operate these channels. Also, to design the waiting facility it is necessary to have information regarding the possible size of the queue to plan for waiting room. There may also be a space cost which should be considered along with customer-waiting and idle-server costs to obtain the optimal system design. In any case, the analyst would strive to solve his problem by analytical means; however, if these fail, he must resort to simulation.

1.4   NOTATION

As a shorthand in describing a queueing process, a notation has evolved, due for the most part to Kendall (1953), which is now rather standard throughout the queueing literature. A queueing process is described by a series of symbols and slashes such as $A/B/X/Y/Z$, where A indicates in some way the interarrival-time distribution, B the service pattern as described by the probability distribution for service time, X the number of parallel service channels, Y the restriction on system capacity, and Z the queue discipline.[2] Some standard symbols for these characteristics are presented in Table 1.1. For example, the notation $M/D/2/\infty/\text{FIFO}$ indicates a queueing process with exponential interarrival times, deterministic service times, two parallel servers, no restriction on the maximum number allowed in the system, and first-in, first-out queue discipline.

TABLE 1.1 QUEUEING NOTATION[a]

Characteristics	Symbol	Explanations
Interarrival-time distribution (A)	M	Exponential
	D	Deterministic
	E_k	Erlang type k $(k = 1, 2, \ldots)$
	GI	General independent
Service-time distribution (B)	M	Exponential
	D	Deterministic
	E_k	Erlang type k
	G	General
Number of parallel servers (X)	$1, 2, \ldots, \infty$	
Restriction on system capacity (Y)	$1, 2, \ldots, \infty$	
Queue discipline (Z)	FIFO	First in, first out
	LIFO	Last in, first out
	SIRO	Service in random order
	PRI	Priority
	GD	General discipline

[a]This notation is consistent with that recommended in the *Queuing Standardization Conference Report*, May 11, 1971, except for our use of *GI* for general arrivals rather than the recommended *G*, since the former is more common in the existing literature, and FIFO, LIFO, SIRO, and PRI in place of FCFS, LCFS, RSS, and PR, respectively, as the latter are not pronounceable in English.

[2]Note that Appendix 1 contains a dictionary of symbols used throughout this text.

In many situations only the first three symbols are used. Current practice is to omit the service-capacity symbol if no restriction is imposed ($Y = \infty$) and to omit the queue discipline if it is first in, first out (Z = FIFO). Thus $M/D/2$ would be used to represent a queueing system with exponential input, deterministic service, two servers, no limit on system capacity, and first-come, first-served discipline.

The symbols in Table 1.1 are, for the most part, self-explanatory; however, a few require further comment. The GI and G symbols represent general probability distributions; that is, no assumption is made as to the precise form of the distribution. Results in these cases are applicable to any probability distribution. General independent requires that interarrival times be independent and identically distributed random variables. General service also requires that service times be independent as well as identically distributed, but the convention developed by Kendall dropped the word independent when referring to service, probably to allow more freedom in notation and not to force the same letter(s) to be used twice. It may also appear strange that the symbol M is used for exponential. The use of the symbol E, as one might expect, would be too confusing with E_k, which is used for the type k Erlang distribution (a gamma with an integral number of degrees of freedom). So M is used instead, where the M comes from the Markovian property of the exponential, which is developed in some detail in Section 1.9.

The reader may have noticed that the list of symbols is not complete. For example, there is no indication of a symbol to represent bulk arrivals, nothing to represent series queues, no symbol to denote any state dependence, and so on. If a suitable notation does exist for any previously unmentioned model, it is indicated when that particular model is brought up in the text. However, there still remain models for which no symbolism has either been developed or accepted as standard, and this is generally true for those models less frequently analyzed in the literature.

1.5   HISTORY

Queueing theory was developed to provide models to predict behavior of systems that attempt to provide service for randomly arising demands, and not unnaturally, then, the earliest problems studied were those of telephone traffic congestion. The pioneer investigator was the Danish mathematician A. K. Erlang, who, in 1909, published *The Theory of Probabilities and Telephone Conversations*. In later works he observed that a telephone system was generally characterized by either (1) Poisson input, exponential holding (service) times, and multiple channels (servers), or (2)

Poisson input, constant holding times, and a single channel. Erlang was also responsible for the notion of stationary equilibrium, for the introduction of the so-called balance-of-state equations, and for the first consideration of the optimization of a queueing system.

Work on the application of the theory to telephony continued after Erlang. In 1927 Molina published his *Application of the Theory of Probability to Telephone Trunking Problems*, which was followed 1 year later by Thornton Fry's *Probability and Its Engineering Uses*, which expanded much of Erlang's earlier work. In the early 1930s, Felix Pollaczek did some further pioneering work on Poisson input, arbitrary output, and single- and multiple-channel problems. Additional work was done at that time in Russia by Kolmogorov and Khintchine, in France by Crommelin, and in Sweden by Palm. The early work in queueing theory picked up momentum rather slowly, but the trend has now changed and there has been a flood of papers in the area; so much so, in fact, that each monthly issue of the *Mathematical Reviews* usually contains a dozen or more queueing papers. Similar situations are being experienced by management science and operations research abstracting services. However, much of what has been done recently has been often a minor variation on some established theme, although frequently much substantive work has appeared. Some representatives of the more recent contributions are works by D. R. Lindley on integral equations, N. T. J. Bailey, and W. Lederman and G. E. Reuter on time-dependent solutions, L. Takács on waiting time, D. R. Cox on supplementary variables, D. G. Kendall on imbedded chains, D. G. Champernowne on the use of random walks, and S. Karlin and J. L. McGregor on birth-death processes. Some additional researchers who have been very productive and have provided significant works to the literature are V. E. Beneš, U. N. Bhat, R. W. Conway, D. P. Gaver, J. D. C. Little, W. L. Maxwell, P. M. Morse, M. F. Neuts, N. U. Prabhu, E. Reich, T. L. Saaty, and R. Syski.

Queueing theory originated as a very practical subject, but regrettably most of the recent literature has been of little direct practical value. It is now imperative for queueing theorists to once again become concerned about the application of the sophisticated theory that has mostly arisen in these past 20 years. It is clear that the emphasis in the literature on the exact solution of queueing problems with clever mathematical tricks must become secondary to model building and the direct use of these techniques in management decision-making. Most real problems do not correspond exactly to a mathematical model, but very little of the literature deals with approximate solutions, sensitivity analyses, and the like. The development of the practice of queueing theory must not be restricted by a lack of closed-form solutions. A problem solver cannot become frustrated with transform solutions, and he must learn to put to work the developed theory. These points should be kept in mind by the reader throughout this text.

1.6   COMMON AREAS OF APPLICATION

There are many well-known, common, but nontrivial, areas of application for the theory of queueing. Let us just bring up a few of the more widely known ones here. The first to be mentioned is, in fact, the first area of application, that is, telephone conversations. The first work of any great significance that dealt with a real waiting-line situation was the previously mentioned work of A. K. Erlang begun in 1909. He was an employee of the Danish Telephone Company in Copenhagen, and his work was the application of established techniques of probability to the problem of determining the optimum number of telephone lines to handle prescribed call frequencies. To look at telephone conversations as a waiting line, Erlang had to first define what would be meant by arrivals and servers, and in fact, what would be the waiting line itself. For telephone calls, as in many other examples, it is not at all immediately obvious which is which. With a bit of thought, however, it should become clear that the arrivals are just the calls, that the sequence of calls in time would form the input stream, and that call duration is the service time. The frequency of these calls might be difficult to measure, because sometimes calls may find a busy signal and it may not be known whether these were legitimate calls. But in the end, the actual number of cables that a telephone company uses is determined by an application of these kinds of results. In particular, it might be desired that busy signals occur only 5% of the time. So one would find the number of channels that would serve this multiple-channel queueing system and leave a customer waiting 5% of the time on the average. The servers are therefore the circuits or the telephone lines themselves, and the queue (waiting line) is the collection of uncompleted calls. Now in the case of ordinary calls, a call is placed and is either completed or the caller hears a busy signal and hangs up. This can be construed as a situation in which there is no waiting room, with customers who find busy signals forced to reenter the input stream. On the other hand, these same callers might be considered to wait until they succeed in finding a free line.

Telephony remained the principal application of the theory through about 1950. But since then, numerous other applications have been found.

For illustrative purposes, a second interesting application is the landing of aircraft [see, for example, Galliher and Wheeler (1958), Rosenshine (1967) and (1968)]. This does by no means imply that this application was chronologically second to telephony with no others in between. Rather, aircraft-landing problems have many of the ingredients necessary for the appropriate application of queueing techniques. For example, it seems clear that the customers are the airplanes, that the servers are the runways, and that the all-too-familiar stack represents the waiting line. The ability of the analyst to distinguish these and other basic characteristics is essential before work can begin on the solution of any problem. Work has recently been done on the use of some new results on state-dependent queues (i.e., queues in which the service rate is a function of the number of waiting customers) to landing problems. It seems reasonable to assume here that the server works faster to reduce a growing line. In other words, the service rate may not be a constant, but rather may increase when the stack gets longer in an effort to reduce its size. On the other hand, is it not possible that servers get into each other's way and that instead the service rate decreases with increasing queue size? The larger the stack becomes, the more difficult it might be to service any additional arrivals.

There are many other important applications of the theory, most of which have been well-documented in the literature of probability, operations research, and management science. Some of the more prominent of these are machine repair, toll booths, taxi stands, inventory control, the loading and unloading of ships, scheduling patients in hospital clinics, production flow, and applications in the computer field with respect to program-scheduling, time-sharing, and system design.

1.7   DETERMINISTIC QUEUEING MODELS

The conceptually simplest class of queueing problems are those for which probability distributions are not necessary to describe arrival and service patterns. Instead, the units of input arrive at known points in time and service times are fixed constants. A queueing model which falls into this class is said to be deterministic, since there are no probability distributions associated in any way with the problem.

Consider then the elementary case of a constant rate of arrivals to a single channel which possesses a constant service rate. These regularly spaced arrivals are to be serviced first come, first served (FIFO). Let it also be assumed that at time $t = 0$ there are no customers waiting and that the channel is empty. Let λ be defined as the number of arrivals per unit time, and $1/\lambda$ then will be the constant time between successive arrivals.[3]

---

[3]The particular unit of time (minutes, hours, etc.) is up to the choice of the analyst. However, consistency must be adhered to once the unit is chosen so that the same basic unit is used throughout the analysis.

Similarily, if μ is to be the rate of service in terms of completions per unit time when the server is busy, then $1/\mu$ is the constant service time. We first consider the case where $\lambda > \mu$ $(1/\lambda < 1/\mu)$; that is, the arrival rate is greater than the service rate. In this situation, the queue length would keep increasing and grow beyond any bound. Each successive customer would wait longer than his predecessor for service until eventually customers would be waiting forever. To prevent this, forced balking is imposed on customers whenever the number in the system gets to a certain size. This can be viewed as limiting the amount of waiting room (i.e., imposing a finite system-capacity constraint). The limit on the system's size will be set to be $K-1$, so that if a would-be customer to the system would increase the total number in the system to K, he is refused entry. Thus the system size will be at most $K-1$. We would like to calculate for this $D/D/1/K-1$ model the number in the system at time t, say $n(t)$, and the time the nth arriving customer must wait in the queue to obtain service, say $W_q^{(n)}$.[4]

Under the assumption that as soon as a service is completed another is begun, the number in the system (including the customer in service) at time t is determined by the equation

$$n(t) = \{\text{number of arrivals in the interval } (0, t]\}$$

$$- \{\text{number of services completed in } (0, t]\} \qquad (1.1)$$

$$= \left[\frac{t}{1/\lambda}\right] - \left[\frac{t - 1/\lambda}{1/\mu}\right]$$

$$= [\lambda t] - \left[\mu t - \frac{\mu}{\lambda}\right], \qquad (1.2)$$

where $[x], x \geqslant 0$, is the greatest integer $\leqslant x$ and where $n(0) = 0$.

Equation (1.2) can be demonstrated by the diagrams of Figures 1.4a and b. Suppose $1/\lambda = 3$, $1/\mu = 5$, and $t = 14$. Then $[14/3] = 4$ arrivals have come and $[(14-3)/5] = 2$ services have been completed, so that $n(14) = 2$.

Equation (1.2) is valid only up until the first balk occurs. If we let t_i be

[4]It should be pointed out that there can be two waiting times of interest—the time spent waiting for service which we define as $W_q^{(n)}$ and the time spent in the system which we shall henceforth call $W^{(n)}$. The reader is cautioned that various authors are not consistent and sometimes either one is referred to as waiting time. In this text we differentiate the two with the subscript q.

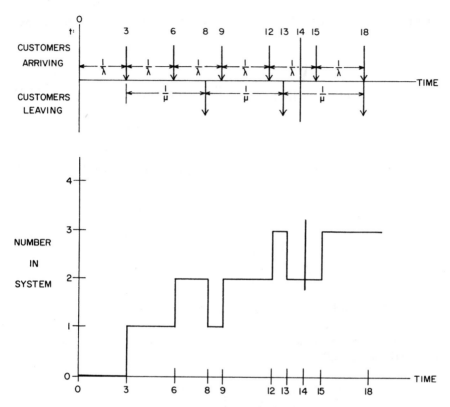

Fig. 1.4   Deterministic queueing process. (a) Arrivals and departures versus time; (b) number in system versus time.

the time until the first balk, then we have from (1.2) that

$$n(t_i) = K = [\lambda t_i] - \left[\mu t_i - \frac{\mu}{\lambda} \right], \tag{1.3}$$

where t_i is the smallest real number satisfying (1.3). Now customers arriving between t_i and the time until the next service is completed balk and $n(t)$ remains at $K-1$. Then $n(t)$ would drop to $K-2$ when the next service is completed unless an arrival comes at the same time. If this is not the case, $n(t)$ goes back up to $K-1$ when the next arrival comes. So $n(t)$ will never go below $K-2$, since we are assuming $1/\lambda < 1/\mu$ and an arrival comes before another service is completed; that is, at most one service can be completed between successive arrivals. Only when $1/\mu$ is a multiple of $1/\lambda$ will we never have the situation where $n(t)$ drops to $K-2$ after the first balk occurs. Thus if $1/\mu = m(1/\lambda)$, where m is an integer > 0, we have

$$n(t) = \begin{cases} 0 & (t < 1/\lambda) \\ [\lambda t] - \left[\mu t - \dfrac{\mu}{\lambda}\right] & (1/\lambda \leqslant t < t_i), \\ K - 1 & (t \geqslant t_i) \end{cases} \qquad (1.4)$$

where t_i is found from (1.3).

Consider the following example where $1/\mu = 8$ and $1/\lambda = 4$. From (1.4) we have

$$n(t) = \begin{cases} 0 & (t < 4) \\ \left[\dfrac{t}{4}\right] - \left[\dfrac{t}{8} - \dfrac{1}{2}\right] & (4 \leqslant t < t_i), \\ K - 1 & (t \geqslant t_i) \end{cases}$$

where t_i is found from (1.3). Figure 1.5 is a depiction of this process over time.

Let us now set K equal to 5. To then find t_i, we desire the smallest real value satisfying (1.3), that is,

$$5 = \left[\frac{t_i}{4}\right] - \left[\frac{t_i}{8} - \frac{1}{2}\right] = \left[\frac{t_i}{4}\right] - \left[\frac{t_i - 4}{8}\right].$$

Since t_i is a time of arrival it must be divisible by 4. By trial and error we can find t_i to be 32, which implies a balk by the eighth arrival. Of course,

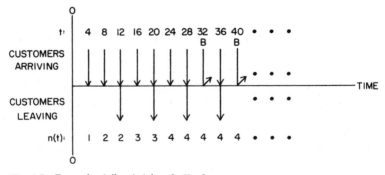

Fig. 1.5 Example: $1/\lambda = 4$, $1/\mu = 8$, $K = 5$.

we could find t_i directly from Figure 1.5, however Equation (1.3) yields t_i without resorting to graphing the process.

To find the waiting times in queue until service begins, we observe that the waits, $W_q^{(n)}$ and $W_q^{(n+1)}$, of two successive customers in any single-channel queue (deterministic or otherwise) are related by the simple recurrence relation[5]

$$W_q^{(n+1)} = \begin{cases} W_q^{(n)} + S^{(n)} - T^{(n)} & \left(W_q^{(n)} + S^{(n)} - T^{(n)} > 0 \right) \\ 0 & \left(W_q^{(n)} + S^{(n)} - T^{(n)} \leqslant 0 \right) \end{cases}, \quad (1.5)$$

where $S^{(n)}$ is the service time of the nth customer and $T^{(n)}$ is the interarrival time between the nth and $(n+1)$st customers. This can be seen by a simple diagram as shown in Figure 1.6.

For the problem at hand, we shall use the foregoing approach to secure a formula for $n < 8$. For $n \geqslant 8$, it should be clear that the wait $W_q^{(n)}$ is always $(K-2)8 = 24$ since after the first balk each arrival finds $K-2$ ahead of him, each requiring a service of $S^{(n)} = 8$. This also checks with the recursive relation of (1.5), since now $W_q^{(n+1)} = W_q^{(n)} + 8 - 8 = W_q^{(n)}$. For $n < 8$, we have $S^{(n)} = 1/\mu = 8$ and $T^{(n)} = 1/\lambda = 4$, for all n. Hence $W_q^{(n+1)} = W_q^{(n)} + 4$, or the first finite difference $\Delta W_q^{(n)} \equiv W_q^{(n+1)} - W_q^{(n)}$ is 4. This simple difference equation has the solution $W_q^{(n)} = 4n + C$, where C is an arbitrary constant. To find this constant, we employ the boundary condition that $W_q^{(1)} = 0$, which gives $0 = 4 + C$, or $C = -4$. Thus[6]

$$W_q^{(n)} = \begin{cases} 4(n-1) & (n < 8) \\ 24 & (n \geqslant 8) \end{cases}.$$

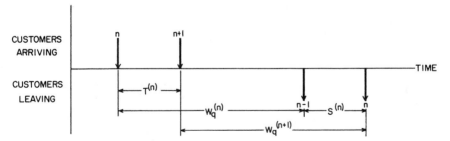

Fig. 1.6 Successive waiting times.

[5]This is an important general relation which is also utilized in later portions of the text.

[6]Strictly speaking, since n is an integer, we should write $(n = 0, 1, 2, \ldots, 7)$; however, whenever it is clear from the context, we take the liberty of using the shorter $(n < 8)$ notation.

In general, whenever $1/\mu = m(1/\lambda)$, for $n < \lambda t_i$, $S^{(n)} = 1/\mu$, and $T^{(n)} = 1/\lambda$, we find that $W_q^{(n+1)} = W_q^{(n)} + (1/\mu - 1/\lambda)$ and thus that $\Delta W_q^{(n)} = 1/\mu - 1/\lambda$. The solution to this difference equation is, as before, $W_q^{(n)} = (1/\mu - 1/\lambda)n + C$. From the boundary condition $W_q^{(1)} = 0$, we obtain $C = -(1/\mu - 1/\lambda)$. Thus we can write

$$
W_q^{(n)} = \begin{cases} \left(\dfrac{1}{\mu} - \dfrac{1}{\lambda}\right)(n-1) & (n < \lambda t_i) \\[2mm] (K-2)\dfrac{1}{\mu} & (n \geqslant \lambda t_i) \end{cases}. \tag{1.6}
$$

In the case where $1/\mu \neq m(1/\lambda)$ (i.e., the interarrival time is not a multiple of the service time), things are more complicated. Consider the example where $1/\lambda = 4$, $1/\mu = 6$, and $K = 5$. Equations (1.2) and (1.3) still hold, so from (1.3) we find t_i to be 44. A graph of this system is shown in Figure 1.7.

The complication sets in because, after the first balk, $n(t)$ occasionally drops to $K - 2 = 3$. In fact, at a time 2 after each balk it drops to $K - 2$ until the next arrival comes at a time 2 later. Thus $n(t)$ is given by

$$
n(t) = \begin{cases} 0 & (t < 1/\lambda = 4) \\[2mm] \left[\dfrac{t}{4}\right] - \left[\dfrac{t-4}{6}\right] & (4 \leqslant t < t_i = 44) \\[2mm] 4 & (44 + 12k \leqslant t < 46 + 12k; \quad 48 + 12k \leqslant t < 58 + 12k) \\[2mm] 3 & (46 + 12k \leqslant t < 48 + 12k; \quad k = 0, 1, 2, \ldots) \end{cases}
$$

The formula for the waiting time $W_q^{(n)}$ given by (1.6) is still valid over the range of values $n < \lambda t_i$, but we can see from the following figure that there are two possible waits for all customers beyond $n = 10$. In particular,

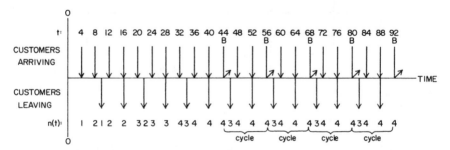

Fig. 1.7 Example: $1/\lambda = 4$, $1/\mu = 6$, $K = 5$.

the odd-numbered joined arrivals after 10 wait 16, while the even-numbered ones have waits of 18.

Thus we see that an expression for the waiting time of the nth customer is

$$
W_q^{(n)} = \begin{cases} 2(n-1) & (n<11) \\ 16 & (n \geqslant 11, \quad n \text{ odd}) \\ 18 & (n \geqslant 11, \quad n \text{ even}) \end{cases}
$$

While it is relatively straightforward to work out $n(t)$ and $W_q^{(n)}$ for a specific example such as this, it is far more difficult if, indeed, at all possible to get general equations such as (1.4) and (1.6). However, one can determine $n(t)$ and $W_q^{(n)}$ for any specific case by graphing the system and analyzing the graph as was done here. The analysis is more cumbersome as K gets larger since more graphing is required. The analysis also gets more difficult as the least common multiple of $1/\lambda$ and $1/\mu$ gets larger.

In the prior example, the common multiple of $1/\lambda$ and $1/\mu$ was 12. As this gets larger, the *steady-state* $(t > t_i)$ characteristics become less "obvious"; that is, the cycles are longer and more complex. To illustrate this we present the following example. Let $1/\lambda = 3$, $1/\mu = 7$, and $K = 4$. The diagram of this example is shown in Figure 1.8. The reader can verify from Figure 1.8 that $n(t)$ and $W_q^{(n)}$ are given as follows:

$$
n(t) = \begin{cases} 0 & (0 \leqslant t < 1/\lambda = 3) \\ \left[\dfrac{t}{3}\right] - \left[\dfrac{t-3}{7}\right] & (3 \leqslant t < t_i = 15) \\ 3 & (15 + 21k \leqslant t < 17 + 21k; \quad 18 + 21k \leqslant t < 31 + 21k; \\ & \quad 33 + 21k \leqslant t < 36 + 21k) \\ 2 & (17 + 21k \leqslant t < 18 + 21k; \quad 31 + 21k \leqslant t < 33 + 21k; \\ & \quad k = 0, 1, 2, \dots) \end{cases}
$$

$$
W_q^{(n)} = \begin{cases} 4(n-1) & (n < \lambda t_i = 5) \\ 13 & (n = 5 + 3k) \\ 14 & (n = 6 + 3k) \\ 12 & (n = 7 + 3k; \quad k = 0, 1, 2, \dots) \end{cases}
$$

This example is somewhat more complicated than the last due to the least common multiple between $1/\lambda$ and $1/\mu$ being 21. The reader may have

20

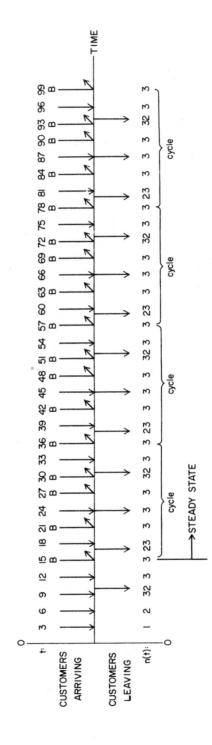

Fig. 1.8 Example: $1/\lambda = 3$, $1/\mu = 7$, $K = 4$.

noticed that the cycle length after reaching steady state (time of first customer balk) is equal to the least common multiple.

Thus far we have considered cases only where $1/\lambda < 1/\mu$; when $1/\lambda \geqslant 1/\mu$, we have a very simple situation since there is never more than one in the system. Hence $n(t)$ is either one or zero and $W_q^{(n)} = 0$, for all n. It is sometimes of interest to consider such a situation as this if we start the process with M customers in the system. We are interested in the system until it reaches its steady state of $n(t) = 1$ or 0; that is, we are interested in the *transient* characteristics of the process. Let t_i now represent the time at which steady state is reached. We consider the case when $1/\lambda > 1/\mu$, for the case where $1/\lambda = 1/\mu$ is trivial in that $t_i = 0$, since $n(t) = M$ for all t, and if the queue discipline is first come, first served, $W_q^{(n)} = (M-1)/\mu$ for all n. Now when $1/\lambda > 1/\mu$, t_i is the smallest value for which $n(t_i) = 0$. For $t > t_i$, $n(t) = 0$ when $t_i \leqslant t < t_1$ and $t_k + 1/\mu \leqslant t < t_k + 1/\lambda$ and $n(t) = 1$ when $t_k \leqslant t < t_k + 1/\mu$, where $t_k = [\lambda t_i]/\lambda + k/\lambda$ and $k = 1, 2, \ldots$. For $t < t_i$, $n(t) = M - \{[t/(1/\mu)] - [t/(1/\lambda)]\}$, the number of completed services in time t minus the number of new arrivals in time t subtracted from M. To find t_i we have

$$M = [\mu t_i] - [\lambda t_i]. \tag{1.7}$$

(We suggest the reader diagram this procedure as an aid to following the above development.) Summarizing, we obtain

$$n(t) = \begin{cases} M - \{[\mu t] - [\lambda t]\} & (0 \leqslant t < t_i) \\ 0 & (t_i \leqslant t < t_1; \; t_k + 1/\mu \leqslant t < t_k + 1/\lambda) \\ 1 & (t_k \leqslant t < t_k + 1/\mu; \; t_k = [\lambda t_i]/\lambda + k/\lambda; \\ & \qquad k = 1, 2, \ldots) \end{cases}$$

Consider the example for which $1/\lambda = 3$, $1/\mu = 1$, and $M = 7$. Figure 1.9 shows this process. From the aforementioned we have

$$n(t) = \begin{cases} 7 - [t] + \left[\dfrac{t}{3}\right] & (0 \leqslant t < t_i) \\ 0 & (t_i \leqslant t < t_1; \; t_k + 1 \leqslant t < t_k + 3) \\ 1 & (t_k \leqslant t < t_k + 1; \; t_k = 3[t_i/3] + 3k) \end{cases}$$

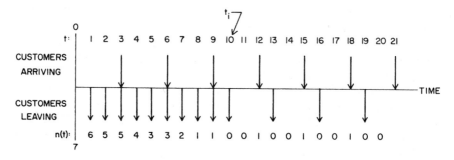

Fig. 1.9   Example: $1/\lambda=3$, $1/\mu=1$, $M=7$.

From (1.7) we find that

$$7=[t_i]-\left[\frac{t_i}{3}\right],$$

which yields $t_i=10$. Thus

$$n(t)=\begin{cases} 7-[t]+\left[\dfrac{t}{3}\right] & (0\leqslant t<10) \\[2mm] 0 & (10\leqslant t<12;\,10+3k\leqslant t<12+3k) \\[2mm] 1 & (9+3k\leqslant t<10+3k) \end{cases}.$$

To find the waiting time, we must make some assumption as to the order of service for the initial M customers. If we assume random selection of the initial M and thereafter, first come, first served, we can calculate the waiting time as follows. For the initial M customers, one has zero wait, one has a wait of $1/\mu$, one has a wait of $2/\mu$, and so on, the last having a wait of $(M-1)/\mu$. On the average, the wait would be

$$\left(\frac{1}{M}\right)\left(\frac{1}{\mu}\right)\sum_{i=1}^{M-1}i=\frac{(M-1)M}{2}\left(\frac{1}{M\mu}\right)=\frac{M-1}{2\mu}.$$

To serve the initial M customers takes $M(1/\mu)$. Thus the first new customer must wait $M(1/\mu)-1/\lambda$ before getting service. The second must wait $(M+1)(1/\mu)-2/\lambda$, the third, $(M+2)(1/\mu)-3/\lambda$, and so on until customer $[\lambda t_i]$, who enters service immediately. Note that we again have

$$W_q^{(n+1)}=W_q^{(n)}+S^{(n)}-T^{(n)}=W_q^{(n)}+\frac{1}{\mu}-\frac{1}{\lambda},$$

the recursive relation of (1.5). Hence $W_q^{(n)}$ can be given as

$$
W_q^{(n)} = \begin{cases} \dfrac{M-1}{2\mu}, & \text{average wait for initial } M \text{ customers} \\[2mm] (M+n-1)\dfrac{1}{\mu} - \dfrac{n}{\lambda} & (n=1,2,\dots,[\lambda t_i]) \\[2mm] 0 & (n>[\lambda t_i]) \end{cases}
$$

$$(1.8)$$

or more specifically for this example,

$$
W_q^{(n)} = \begin{cases} \dfrac{3}{\mu}, & \text{average for initial 7 customers} \\[2mm] \dfrac{6+n}{\mu} - 3n & (1 \leqslant n \leqslant 3) \\[2mm] 0 & (n>3) \end{cases}
$$

In this section we have treated some deterministic queueing models to illustrate the approach toward solution, as well as the nature of general queueing processes. Although one could complicate these types of models even further, for example, by allowing periodic balking or instituting a priority queue discipline, exact solutions are always possible by graphing since all factors are deterministic. This is not the case for the nondeterministic models and exact solutions become exceedingly more difficult.

1.8   POISSON PROCESS AND THE EXPONENTIAL DISTRIBUTION

The most common stochastic queueing models assume that interarrival times and service times obey the exponential distribution or, equivalently, that the arrival rate and service rate follow a Poisson distribution. In this section we will derive the Poisson distribution and show that assuming the number of occurrences in some time interval to be a Poisson random variable is equivalent to assuming the time between successive occurrences to be an exponentially distributed random variable.

We consider an arrival process $\{N(t),\ t \geqslant 0\}$, where $N(t)$ denotes the total number of arrivals up to t, with $N(0)=0$, and which satisfies the following three assumptions:

(i) The probability that an arrival occurs between time t and time $t+\Delta t$ is equal to $\lambda \Delta t + o(\Delta t)$. We write this as $\Pr\{\text{arrival occurs between } t \text{ and } t+\Delta t\} = \lambda \Delta t + o(\Delta t)$, where λ is a constant independent of $N(t)$, Δt is an incremental element, and $o(\Delta t)$ denotes a quantity that becomes negligible when compared to Δt as $\Delta t \to 0$; that is,

$$\lim_{\Delta t \to 0} \frac{o(\Delta t)}{\Delta t} = 0;$$

(ii) Pr{more than one arrival between t and $t + \Delta t$} $= o(\Delta t)$;

(iii) The numbers of arrivals in nonoverlapping intervals are statistically independent; that is, the process has independent increments.

We wish to calculate $p_n(t)$, the probability of n arrivals in a time interval of length t, n being an integer ≥ 0. We will do this by first developing differential-difference equations for the arrival process as follows:[7]

$$p_n(t + \Delta t) = \Pr\{n \text{ arrivals in } t \text{ and zero in } \Delta t\}$$

$$+ \Pr\{n - 1 \text{ arrivals in } t \text{ and one in } \Delta t\}$$

$$+ \Pr\{n - 2 \text{ arrivals in } t \text{ and two in } \Delta t\}$$

$$+ \dots + \Pr\{0 \text{ arrivals in } t \text{ and } n \text{ in } \Delta t\} \qquad (n \geq 1). \qquad (1.9)$$

Using assumptions (i), (ii), and (iii), Equation (1.9) becomes

$$p_n(t + \Delta t) = p_n(t)[1 - \lambda\Delta t - o(\Delta t)] + p_{n-1}(t)[\lambda\Delta t + o(\Delta t)] + o(\Delta t), \quad (1.10)$$

where the last term, $o(\Delta t)$, represents the terms

$$\Pr\{n - j \text{ arrivals in } t \text{ and } j \text{ in } \Delta t; \quad 2 \leq j \leq n\}.$$

For the case $n = 0$, we have

$$p_0(t + \Delta t) = p_0(t)[1 - \lambda\Delta t - o(\Delta t)]. \qquad (1.11)$$

Rewriting (1.10) and (1.11) and combining all $o(\Delta t)$ terms, we have

$$p_0(t + \Delta t) - p_0(t) = -\lambda\Delta t p_0(t) + o(\Delta t) \qquad (1.12)$$

and

$$p_n(t + \Delta t) - p_n(t) = -\lambda\Delta t p_n(t) + \lambda\Delta t p_{n-1}(t) + o(\Delta t) \qquad (n \geq 1). \quad (1.13)$$

We divide (1.12) and (1.13) by Δt, take the limit as $\Delta t \to 0$, and obtain the

---

[7]Many mathematicians argue that these types of equations are incorrectly referred to and that the correct term for them is difference-differential. The term differential-difference, they claim, is appropriate only when differencing and differentiating occur on the same variable. One finds throughout the queueing literature both these terms in use to describe the types of equations above, with the "incorrect" term being the more common. Recognizing this fact, we sacrifice mathematical precision for greater consistency with the literature and continue to refer to equations such as these as differential-difference equations.

following differential-difference equations

$$
\begin{cases}
\lim_{\Delta t \to 0} \left[\dfrac{p_0(t+\Delta t) - p_0(t)}{\Delta t} = -\lambda p_0(t) + \dfrac{o(\Delta t)}{\Delta t} \right] \\[2em]
\lim_{\Delta t \to 0} \left[\dfrac{p_n(t+\Delta t) - p_n(t)}{\Delta t} = -\lambda p_n(t) + \lambda p_{n-1}(t) + \dfrac{o(\Delta t)}{\Delta t} \right] \quad (n \geqslant 1),
\end{cases}
$$

which reduce to

$$
\frac{dp_0(t)}{dt} = -\lambda p_0(t) \tag{1.14}
$$

and

$$
\frac{dp_n(t)}{dt} = -\lambda p_n(t) + \lambda p_{n-1}(t) \quad (n \geqslant 1). \tag{1.15}
$$

We now have an infinite set of linear, first order, ordinary differential equations to solve. Recall that a linear, first order, differential equation of the form

$$
\frac{dy(x)}{dx} + \Phi(x)y(x) = \Psi(x) \tag{1.16}
$$

has the solution

$$
y(x) = Ce^{-\int \Phi(x)\,dx} + e^{-\int \Phi(x)\,dx} \int e^{\int \Phi(x)\,dx} \Psi(x)\,dx, \tag{1.17}
$$

where C is a constant to be determined by boundary conditions (see Appendix 2). We will, therefore, use (1.17) to sequentially solve the infinite set of equations given by (1.14) and (1.15), utilizing mathematical induction to arrive at the final answer. It is left as an exercise (see Problem 1.10) to show that applying (1.17) to (1.14) and (1.15) for $n = 0$, 1, and 2, and using the boundary condition $p_n(0) = 0$ for $n > 0$ and $p_0(0) = 1$ yields

$$
\begin{cases}
p_0(t) = e^{-\lambda t} \\[1em]
p_1(t) = \lambda t e^{-\lambda t} \\[1em]
p_2(t) = \dfrac{(\lambda t)^2}{2} e^{-\lambda t} \\[1.5em]
p_3(t) = \dfrac{(\lambda t)^3}{3!} e^{-\lambda t}
\end{cases} \tag{1.18}
$$

From (1.18) we conjecture the general formula to be

$$p_n(t) = \frac{(\lambda t)^n}{n!} e^{-\lambda t} \qquad (n \geq 0).$$ (1.19)

We will prove this by induction. Using (1.15) for $n+1$, we have

$$\frac{dp_{n+1}(t)}{dt} + \lambda p_{n+1}(t) = \lambda p_n(t).$$

Let $\Phi(t) = \lambda$ and $\Psi(t) = \lambda p_n(t) = \frac{\lambda(\lambda t)^n}{n!} e^{-\lambda t}$; hence, from (1.19), (1.16), and (1.17), we obtain

$$p_{n+1}(t) = Ce^{-\int \lambda \, dt} + e^{-\int \lambda \, dt} \int e^{\int \lambda \, dt} \frac{\lambda^{n+1} t^n}{n!} e^{-\lambda t} \, dt$$

$$= Ce^{-\lambda t} + e^{-\lambda t} \frac{\lambda^{n+1} t^{n+1}}{(n+1)n!}.$$

Use of the boundary condition $p_{n+1}(0) = 0$ yields $C = 0$ and gives

$$p_{n+1}(t) = \frac{(\lambda t)^{n+1} e^{-\lambda t}}{(n+1)!}.$$

Thus assuming (1.19) holds for n, we have shown it also for $n+1$. We have proven it holds for $n = 0, 1, 2,$ and 3 in (1.18); thus the proof by induction is complete. (Note that the proof by induction requires our showing (1.19) holds only for $n = 0$; we proved it also holds specifically for $n = 1, 2,$ and 3 to aid in conjecturing the general formula.) Equation (1.19) is the well-known formula for the Poisson probability distribution with mean λt. Thus if we consider the random variable defined as the number of arrivals to a system in a time t, this random variable has the Poisson distribution given by (1.19) with a mean of λt arrivals, or a mean arrival rate of λ.

Poisson processes have a number of interesting additional properties. One of the most important is the fact that the numbers of occurrences in intervals of equal width are identically distributed (stationary increments). In particular, for $t > s$, $N(t) - N(s)$ is identically distributed as $N(t+h) - N(s+h)$, with frequency function

$$p_n(t-s) = e^{-\lambda(t-s)} \frac{[\lambda(t-s)]^n}{n!}.$$

This can easily be seen by the following argument. Since the Poisson has independent increments [assumption iii], there is no loss of generality if $N(s)$ and $N(s+h)$ are assumed to be zero. Then if the Poisson derivation is carried out for both $N(t)$ and $N(t+h)$ under assumptions i, ii, and iii, the foregoing formula results for each.

We now show that if the arrival process follows the Poisson distribution, an associated random variable defined as the time between successive arrivals (interarrival time) follows the exponential distribution. Let T be the random variable "time between successive arrivals"; then

$$\Pr\{T \geqslant t\} = \Pr\{\text{zero arrivals in time } t\} = p_0(t) = e^{-\lambda t}.$$

Letting $A(t)$ represent the cumulative distribution function of T, we have

$$A(t) = \Pr\{T \leqslant t\} = 1 - e^{-\lambda t}.$$

The density function, $a(t)$, then is given by

$$a(t) = \frac{dA(t)}{dt} = \lambda e^{-\lambda t}.$$

Thus T has the exponential distribution with mean $1/\lambda$. We would intuitively expect the *mean* time between arrivals to be $1/\lambda$ if the *mean* arrival rate is λ. Our analysis substantiates this. It can also be shown that if the interarrival times are independent and have the same exponential distribution, then the arrival rate follows the Poisson distribution, and a proof of this assertion follows accordingly.

To begin, let the cumulative distribution function (CDF) of the arrival counting process, $\Pr\{N(t) \leqslant n\}$, be denoted by $P_n(t)$. Then it follows that

$$p_n(t) = \Pr\{N(t) = n\}$$

$$= P_n(t) - P_{n-1}(t).$$

But

$$P_n(t) = \Pr\{(\text{sum of } n+1 \text{ interarrival times}) > t\}.$$

However, the sum of independent and identical exponential random variables has an Erlang distribution (which is a special type of gamma distribution)[8]; hence

$$P_n(t) = \int_t^\infty \frac{\lambda(\lambda x)^n}{n!} e^{-\lambda x} dx. \tag{1.20}$$

[8]See Problem 2.5b and Table A5.3 of Appendix 5.

The transformation of variables $u = x - t$ gives

$$P_n(t) = \int_0^\infty \frac{\lambda^{n+1}(u+t)^n}{n!} e^{-\lambda t} e^{-\lambda u} du$$

$$= \int_0^\infty \frac{\lambda^{n+1} e^{-\lambda t} e^{-\lambda u}}{n!} \sum_{i=0}^n u^{n-i} t^i \frac{n!}{(n-i)! i!} du,$$

from the binomial theorem. The $\sum$ and $\int$ may be switched to give

$$P_n(t) = \sum_{i=0}^n \frac{\lambda^{n+1} e^{-\lambda t} t^i}{(n-i)! i!} \int_0^\infty e^{-\lambda u} u^{n-i} du.$$

But $\int_0^\infty e^{-u} u^{n-i} du$ is the well-known gamma function denoted by $\Gamma(n - i + 1)$ which equals $(n - i)!$. So with the proper change of variables to account for λ in the $e^{-\lambda u}$ term of the integral we get

$$P_n(t) = \sum_{i=0}^n \frac{(\lambda t)^i e^{-\lambda t}}{i!},$$

which is clearly recognizable as the CDF of the Poisson process.

The Poisson/exponential arrival process derived here is sometimes referred to as completely random arrivals. Although the reader might think that completely random would allude to some sort of haphazard arrival process or a uniform distribution for interarrival times, when encountered in queueing literature it specifically refers to the Poisson arrival rate-exponential interarrival time pattern. This can be explained in light of the following characteristic of a Poisson process. Given that k arrivals have occurred in an interval $[0, T]$, the k times $\tau_1 < \tau_2 < \ldots < \tau_k$ at which the arrivals occurred are distributed as the order statistics of k uniform random variables on $[0, T]$. Note that it is not interarrival times, but rather the times at which arrivals occurred, that are uniformly distributed. This can be proven as follows.

$$f_{\tau_1, \tau_2, \ldots, \tau_k}(t_1, t_2, \ldots, t_k | k \text{ arrivals in } \{0, T\}) dt_1 dt_2 \ldots dt_k \equiv f_\tau(t | k) dt$$

$$\doteq \Pr\{t_1 \leqslant \tau_1 \leqslant t_1 + dt_1, \ldots, t_k \leqslant \tau_k \leqslant t_k + dt_k | k \text{ arrivals in } [0, T]\}.$$

Using the definition of conditional probability gives

$$f_\tau(t|k)\,dt = \frac{\Pr\{t_1 \leqslant \tau_1 \leqslant t_1 + dt_1, \ldots, t_k \leqslant \tau_k \leqslant t_k + dt_k \text{ and } k \text{ arrivals in } [0,T]\}}{\Pr\{k \text{ arrivals in } [0,T]\}}.$$

The numerator of the last term above can be found by making direct use of the Poisson density function and its properties, since we wish to find the probability that exactly one event occurs in each of the k time intervals, $(t_i, t_i + dt_i)$, and no events occur elsewhere, that is, in $T - dt_1 - dt_2 - \cdots - dt_k$. Therefore, since the probability of k occurrences in a time t is $(\lambda t)^k e^{-\lambda t}/k!$, we have

$$f_\tau(t|k)\,dt \doteq \frac{\lambda\,dt_1 e^{-\lambda dt_1}\lambda\,dt_2 e^{-\lambda dt_2}\cdots\lambda\,dt_k e^{-\lambda dt_k}e^{-\lambda(T - dt_1 - dt_2 - \ldots - dt_k)}}{(\lambda T)^k e^{-\lambda T}/k!},$$

which reduces to

$$f_\tau(t|k)\,dt \doteq \frac{\lambda^k dt_1 dt_2 \cdots dt_k e^{-\lambda T}}{(\lambda T)^k e^{-\lambda T}/k!} = \frac{k!}{T^k}dt_1 dt_2 \ldots dt_k.$$

Hence

$$f_{\tau_1 \tau_2 \ldots \tau_k}(t_1, t_2, \ldots, t_k | k \text{ arrivals in } [0,T]) = \frac{k!}{T^k}, \tag{1.21}$$

which is identical to the joint density of the order statistics of k random variables uniform on $[0, T]$.

Making similar assumptions as done above for arrivals, one could utilize the same type of process to describe the service pattern. If we change the three assumptions in the beginning of this section slightly by using the word service instead of arrival and condition the probability statements by requiring the system to be nonempty, we would obtain a Poisson service rate or an exponential service-time distribution for describing the service pattern. In the following section, we prove an important property of the exponential distribution which aids in a relatively simple analysis of queueing problems when arrival and service patterns exhibit the Poisson/ exponential characteristics as derived in this section.

1.9 MARKOVIAN PROPERTY OF THE EXPONENTIAL DISTRIBUTION

We will now prove the Markovian or, as it is sometimes called, the memorylessness property, of the exponential distribution. To explain this property in words, suppose service times were exponentially distributed. This property states that the probability that a customer currently in service is completed at some future time t is independent of how long he has already been in service. One can readily see why the term memoryless is applied to this property. Thus what we wish to prove is that

$$\Pr\{T \leqslant t_1 | T \geqslant t_0\} = \Pr\{0 \leqslant T \leqslant t_1 - t_0\}. \tag{1.22}$$

The proof is relatively straightforward and proceeds as follows. From the definition of conditional probability we have

$$\Pr\{T \leqslant t_1 | T \geqslant t_0\} = \frac{\Pr\{(T \leqslant t_1) \text{ and } (T \geqslant t_0)\}}{\Pr\{T \geqslant t_0\}}$$

$$= \frac{\displaystyle\int_{t_0}^{t_1} \lambda e^{-\lambda t} dt}{\displaystyle\int_{t_0}^{\infty} \lambda e^{-\lambda t} dt}$$

$$= -\frac{e^{-\lambda t_1} - e^{-\lambda t_0}}{e^{-\lambda t_0}}$$

$$= 1 - e^{-\lambda(t_1 - t_0)} = \Pr\{0 \leqslant T \leqslant t_1 - t_0\}.$$

It is also true that the exponential distribution is the only continuous distribution which exhibits this memoryless property. The proof of this assertion rests on the fact that the only continuous function solution of the equation

$$g(s + t) = g(s) + g(t)$$

is the linear form

$$g(y) = Cy, \tag{1.23}$$

where C is an arbitrary constant. This rather intuitive result turns out not to be a trivial matter to prove. However, the proof is well-documented in the literature [for example, see Parzen (1962)] and, for the purposes of this text, this additional detail is not necessary. Then, under the assumption of the foregoing result, we proceed as follows. We wish to show that if (1.22) holds, that is, if

$$\Pr\{T \leqslant t_1 | T \geqslant t_0\} = \Pr\{0 \leqslant T \leqslant t_1 - t_0\},$$

then

$$\Pr\{T \leqslant t\} = F(t) = 1 - e^{Ct}.$$

Now subtract both sides of (1.22) from 1 and denote the complementary CDF by $\tilde{F}$. Thus

$$\tilde{F}(t_1 | T \geqslant t_0) = \tilde{F}(t_1 - t_0). \tag{1.24}$$

From the laws of conditional probability, we can write (1.24) as

$$\frac{\tilde{F}(t_1 \text{ and } T \geqslant t_0)}{\tilde{F}(t_0)} = \frac{\tilde{F}(t_1)}{\tilde{F}(t_0)} = \tilde{F}(t_1 - t_0),$$

or

$$\tilde{F}(t_1) = \tilde{F}(t_0)\tilde{F}(t_1 - t_0).$$

Letting $t = t_1 - t_0$ yields

$$\tilde{F}(t + t_0) = \tilde{F}(t_0)\tilde{F}(t),$$

and when natural logarithms are taken of both sides, it is found that

$$\ln \tilde{F}(t + t_0) = \ln \tilde{F}(t_0) + \ln \tilde{F}(t).$$

We then find from (1.23) that

$$\ln \tilde{F}(t) = Ct,$$

or

$$\tilde{F}(t) = e^{Ct}.$$

Thus

$$F(t) = 1 - e^{Ct}.$$

There are many possible and well-known generalizations of the Poisson/exponential process, most of which have rather obvious applications to queues and are taken up in greater detail later in the text.

The most obvious of the generalizations is a truncation of the infinite domain, that is, the omission of some of the nonnegative integers from the range of possible values. This is done whenever the removed values are either theoretically meaningless or practically unobservable. An example of this occurs in the $M/M/c/c$ queue and gives rise to *Erlang's loss formula*.

The only change to be made here, with caution, is the rescaling of the respective probabilities since the Poisson terms no longer sum to one.[9]

Another generalization arises if we go back to the axiomatic derivation and no longer permit λ to be a constant independent of time. If instead the functional relationship is denoted by $\lambda(t)$, then the probability of one occurrence in a small time increment is rewritten as $\lambda(t)\Delta t + o(\Delta t)$, and it turns out that the resulting distribution of the counting process is the so-called nonhomogeneous Poisson given by

$$p_n(t) = e^{-\int_0^t \lambda(s)\,ds} \frac{\left[\int_0^t \lambda(s)\,ds\right]^n}{n!} \qquad (n \geqslant 0).$$

A third, and very common, generalization occurs when one relaxes the Poisson assumption that more than one occurrence in Δt has probability $o(\Delta t)$. Instead, let

$$\Pr\{i \text{ occurrences in } (t, t+\Delta t)\} = \lambda_i \Delta t + o(\Delta t) \qquad (i = 1, 2, \ldots, n)$$

with

$$\sum_{i=1}^n \lambda_i = \lambda.$$

It should be immediately clear that this is equivalent to now allowing the event of i simultaneous occurrences in Δt with probability $\lambda_i \Delta t + o(\Delta t)$, and each individual stream of occurrences of the same batch size (i) itself forms a Poisson process. If these substreams are denoted by $N_i(t)$, then it should also be clear that the total process is

$$N(t) = \sum_i i N_i(t),$$

with probability function

$$p_n(t) = \Pr\{n \text{ occurrences in } [0, t]\}$$

$$= \sum_{i=0}^n e^{-\lambda t} \frac{(\lambda t)^i}{i!} c_n^{(i)} \qquad (c_0^{(0)} \equiv 1),$$

where $c_n^{(i)}$ is the probability that i occurrences give a grand total of n (i.e., the probability associated with the i fold convolution of the batch-size

[9] This is taken up in greater detail in Chapter 3, Section 3.4.

probabilities $\{\lambda_i/\lambda\}$). The process $N(t)$ is known as the *multiple Poisson* and clearly also has the stationary and independent increment properties.

The foregoing probability function $p_n(t)$ has an alternative derivation as a compound distribution since it admits of a random sum interpretation as follows.[10] Consider the process $N(t)$ to be defined by

$$N(t) = \sum_{n=1}^{M(t)} Y_n,$$

where $M(t)$ is a regular Poisson process and $\{Y_n\}$ is a sequence of independent and identically distributed (IID) discrete random variables with probabilities

$$c_j = \Pr\{Y_n = j\} \qquad \text{(for all } n\text{)}$$

$$= \frac{\lambda_j}{\sum_j \lambda_j};$$

that is, occurrences happen according to a Poisson process $\{M(t)\}$ but are not necessarily singlets in that their size is j with probability c_j. Then, by the laws of probability,

$$\Pr\{N(t) = m\} = \sum_{k=0}^{m} \left[\Pr\{M(t) = k\} \cdot \Pr\left\{\sum_{n=1}^{k} Y_n = m\right\} \right]$$

$$= \sum_{k=0}^{m} e^{-\lambda t} \frac{(\lambda t)^k}{k!} c_m^{(k)}.$$

In other words, the compound approach looks at the process as one Poisson stream with a randomly varying batch size, whereas the multiple approach looks at the process as the sum of Poisson streams, each with a constant batch size.

Poisson/exponential streams are special cases of a larger class of problems called renewal processes. An ordinary renewal process arises from any sequence of nonnegative IID random variables. Many of the properties that we have derived for Poisson/exponential sequences can also be derived in a renewal context. Some other results will be needed later in the text, particularly when the input is an arbitrary *GI* stream, but these will be derived as needed by direct probabilistic arguments. The reader par-

---

[10] The resultant distribution is therefore also referred to as *compound Poisson*.

ticularly interested in renewal theory is referred to Chapter 3 of Ross (1970).

In subsequent chapters of the book, the Poisson process and its associated characteristics will play a key role in the development of many queueing models. This is true not only because of the many mathematically agreeable properties of the Poisson/exponential but also because many real-life situations in fact do obey the appropriate requirements. Though it may seem at first glance that the demands of exponential interoccurrence times are rather stringent, this is not the case. A strong argument, for example, in favor of exponential inputs is the one that often occurs in the context of reliability. It is the result of the fact that the limit of a binomial distribution is Poisson[11] [for example, see Parzen (1960)], which says that if a mechanism consists of many parts, each of which can fail with only a small probability, and if the failures for the different parts are mutually independent, then the total flow of failures can be considered Poisson. There is also the additional argument that comes out of information theory. It is that the exponential distribution is the one that provides the least information, where information content or negative entropy of the distribution $f(x)$ is defined as $\int_0^\infty f(x)\log f(x)dx$. It can easily be shown that the exponential distribution has least information or highest entropy, and is therefore the most random law that can be used and thus certainly provides a reasonably conservative approach.

PROBLEMS

1.1. Discuss the following queueing situations in terms of the basic characteristics given in Section 1.2.

(a) Aircraft landing at an airport.
(b) Supermarket checkout procedures.
(c) Post office customer windows.
(d) Toll booth on a bridge.
(e) Gasoline station with several pump islands.
(f) Automatic car wash facility.
(g) Telephone calls coming into a switchboard.
(h) Appointment patients coming into a doctor's office.
(i) Tourists wishing a guided tour of the White House.
(j) Electronic components on an assembly line consisting of three operations and one inspection at end of line.
(k) Water being stored in a reservoir.

---

[11] The reader is asked to prove this assertion as Problem 1.12.

1.2. Give three examples of a queueing situation other than those in Problem 1.1 and discuss in terms of the basic characteristics of Section 1.2.

1.3. The following observations have been made regarding the time between successive arrivals to a single-server, FIFO queueing process, and the actual time to serve them (excluding any waiting time).

Customer Number	Interarrival Time	Service Time
1	—	3
2	9	7
3	6	9
4	4	9
5	7	10
6	9	4
7	5	8
8	8	5
9	4	5
10	10	3
11	6	6
12	12	3
13	6	5
14	8	4
15	9	9
16	5	9
17	7	8
18	8	6
19	8	8
20	7	3

Calculate the average time a customer must wait before being served, the average waiting time including service, the average waiting time including service of those customers who had to wait for service (exclude those who were immediately taken into service), the average length of the queue, the average length of the queue when there was a queue (exclude the situation when no queue was present), the average number in the system, and the idle time of the server. Discuss how management might make use of these measures. (*Hint*: Develop a table with the following headings: Customer Number, Time of Arrival, Time Into Service, Service Time, Time Customer Leaves System, Waiting Time for Service, Total Time Spent in System. From this table the average waiting time and system time can be calculated. Also from this table develop a second table showing times at which the system changes state (either an arrival came or a served customer left), the queue size, and number in system between state changes. Then use this new table to compute the average queue lengths and system size, and the mean idle time.)

1.4. Items arrive at an initially unoccupied inspection station at a uniform rate of one every 5 min. With the time of the first arrival set equal to 0, the chronological times for inspection completion of the first ten items were observed to be 7, 9, 13, 24, 30, 33, 34, 39, 45, and 50, respectively. By manual simulation of the operation for 50 min, using these data, develop the following sample results:

    (a) Mean number in system.
    (b) Percentage idle time experienced.

1.5. Find $n(t)$, t_i, and $W_q^{(n)}$ for the following deterministic queueing systems:

    (a) $1/\lambda = 2\frac{1}{3}$, $1/\mu = 4\frac{2}{3}$, $K = 7$.
    (b) $1/\lambda = 2\frac{1}{2}$, $1/\mu = 3\frac{1}{3}$, $K = 6$.

1.6. Find $n(t)$, t_i, and $W_q^{(n)}$ for the deterministic queueing system where

$$\frac{1}{\lambda} = 5, \qquad \frac{1}{\mu} = 2\frac{2}{3}, \qquad M = 15.$$

1.7. Show that $t_i = 44$ for the second example in the text.

1.8. Solve the same problem as in Problem 1.5a by finding the first customer who tries to enter the system and finds that his system waiting time will be greater than $(K-1)/\mu$. (Why does this work?)

1.9. Solve the following differential equations:

    (a)
$$\frac{dy}{dx} + xy = 0; \qquad y(0) = 3.$$

    (b)
$$\frac{dy}{dx} + \left(\frac{2x}{x^2+1}\right)y = x; \qquad y(1) = \frac{1}{2}.$$

1.10. Derive Equations (1.18) of Section 1.7 by the sequential use of Equation (1.17).

1.11. Given the frequency function found for the Poisson in Equation (1.19), find its moment generating function, say $M_{N(t)}(\theta)$, that is, the expected value of $e^{\theta N(t)}$ (see Appendix 3, Section A3.5). Check your answer with Table A5.3 (Appendix 5) and then use it to show that the mean and variance are λt.

1.12. Derive the Poisson process by using the third assumption that the numbers of arrivals in nonoverlapping intervals are statistically independent and then applying the binomial distribution.

1.13. By the use of the argument of Section 1.8, find the distribution of the counting process associated with IID Erlang interoccurrence times.

1.14. Assume that arrivals can occur singly or in batches of two, with the batch size following the probability distribution

$$f(1) = p, \qquad f(2) = 1 - p \qquad (0 < p < 1)$$

and with the time between successive batches following the exponential probability distribution

$$a(t) = \lambda e^{-\lambda t} \qquad (t \geqslant 0).$$

Show that the probability distribution for the number of arrivals in time t is the compound Poisson distribution given by

$$p_n(t) = e^{-\lambda t} \sum_{k=0}^{[n/2]} \frac{p^{n-2k}(1-p)^k (\lambda t)^{n-k}}{(n-2k)!k!},$$

where $[n/2]$ is the greatest integer $\leqslant n/2$.

Chapter 2

SINGLE-CHANNEL EXPONENTIAL QUEUEING MODELS

In this chapter we study the simplest probabilistic queueing models which can be treated analytically, namely, single-channel models with exponential interarrival times, exponential service times, and a FIFO queue discipline. To begin, we would like to find the state probabilities, $p_n(t) = \Pr\{n$ in the system at time $t\}$. Most often, finding $p_n(t)$ is quite a difficult task (as shall be seen in Section 2.6 even for this simplest of queueing processes), so we first look for the "steady-state" probabilities $p_n = \lim_{t\to\infty} p_n(t)$. That is to say, we hope that the system state eventually becomes essentially independent of the initial state so that no matter what time we query the system, the *probability* of finding n remains constant. Fortunately, frequently in practice, the steady-state characteristics are of main interest anyway.

Exactly what is the steady state? The idea of a steady state is related to the concept of ergodicity, which deals with the problem of determining measures of a stochastic process $X(t)$ from a single realization.[1] That is to say, $X(t)$ is ergodic in the most general sense if, with probability one, all its "measures" can be determined or well approximated from a single realization, $X_0(t)$, of the process. Since statistical parameters of the process are usually expressed as time averages, this is often stated as $X(t)$ is ergodic if time averages equal ensemble averages, that is, expected values (see Figure 2.1). One is usually not interested in all, but rather in certain parameters of a process. We can then define ergodicity with respect to these parameters, and a process might thus be ergodic for certain parameters, but not for others.

Mathematically now, the time average of the square of a realization of a

---

[1]A stochastic process $X(t)$ is a collection of parametrized random variables where the parameter t runs through a given set T. For our purposes, t represents time and $X(t)$ the state of the system at time t (see Appendix 4).

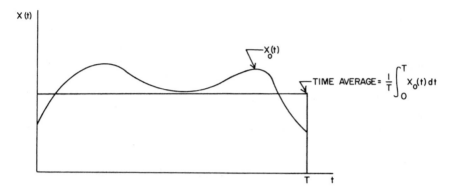

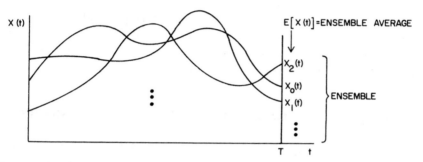

Fig. 2.1 Ergodicity.

process, for example, would be written as

$$N_T = \frac{1}{T} \int_0^T [X_0(t)]^2 \, dt.$$

This is then the second moment of the sample function $X_0(t)$. The ensemble average at time t would be denoted by

$$E\big[\{X(t)\}^2\big] = \lim_{n \to \infty} \frac{\displaystyle\sum_{i=0}^n \{X_i(t)\}^2}{n} = m_2(t),$$

and is the population second moment. To then say that the process is ergodic with respect to its second moment is equivalent to requiring that

$$\lim_{T \to \infty} N_T = \lim_{t \to \infty} m_2(t) = \lim_{T \to \infty} \frac{1}{T} \int_0^T m_2(t) \, dt = N.$$

We thus say that a process possesses a steady state if this ergodic property holds for all its moments (ergodic in distribution function), and thus that each moment possesses a limit (becomes free of "time"). In practice, we reach a steady state if "time is no longer of the essence" and all transient components have been wiped out. Clearly, processes need not be ergodic with respect to any specific number of moments, and the conditions under which any moment is ergodic are the subject of a large class of theorems commonly referred to as ergodic theorems. It is generally not necessary to appeal to these theorems to determine the conditions under which queueing processes are ergodic, since they usually fall out from other considerations. If, however, the reader is interested in a further look into this area, he is referred to Parzen (1962).

So the steady-state probability p_n can be interpreted as the probability of finding n customers in the system at an arbitrary point of time after the process has reached statistical equilibrium. It is not true that all systems may reach steady state; that is, $\lim_{t \to \infty} p_n(t)$ may not yield a true probability distribution. The analysis, however, will point this out.

2.1 STEADY-STATE SOLUTION FOR THE $M/M/1$ MODEL

We first start the analysis by attempting to find the state probabilities $\{p_n(t)\}$ for the single-channel Poisson input and exponential service model in which there is no limit on system capacity and customers are served on a first-in, first-out basis. The procedure utilized here illustrates the methodology for solving a wide range of models. The analysis can be looked at as consisting of four steps as follows:

Step 1.   Difference equations for $p_n(t)$;
Step 2.   Differential-difference equations for $p_n(t)$;
Step 3.   Solution for $p_n(t)$–transient solution;
Step 4.   Limiting solution for p_n–steady state.

In this section we go through steps 1, 2, and 4. Step 3 is considered separately in Section 2.6.

Before proceeding with step 1, let us first consider the arrival and service mechanisms. The density functions for the interarrival times and service times are given, respectively, as

$$\begin{cases} a(t) = \lambda e^{-\lambda t} \\ b(t) = \mu e^{-\mu t} \end{cases},$$

where $1/\lambda$ then is the mean interarrival time and $1/\mu$ the mean service time. Interarrival times, as well as service times, are assumed to be statistically independent. Since interarrival and service times are exponential, and arrival and conditional service rates Poisson, we have

$$
\begin{cases}
\Pr\{\text{an arrival occurs in an infinitesimal interval of length } \Delta t\} \\
\qquad\qquad\qquad\qquad\qquad\qquad\qquad\qquad\qquad = \lambda\Delta t + o(\Delta t) \\[6pt]
\Pr\{\text{more than one arrival occurs in } \Delta t\} = o(\Delta t) \\[6pt]
\Pr\{\text{a service completion in } \Delta t | \text{ system not empty}\} = \mu\Delta t + o(\Delta t) \\[6pt]
\Pr\{\text{more than one service completion in } \Delta t | \text{ more than one in system}\} \\
\qquad\qquad\qquad\qquad\qquad\qquad\qquad\qquad\qquad = o(\Delta t).
\end{cases}
$$

We have, then, a process with arrivals and departures occurring randomly over time, with the probability mechanism just described. Arrivals can be considered as "births" to the system, since if the system is in state E_n (we consider system state as the number in the system, so that E_n indicates that n are in the system) and an arrival occurs, the state is changed to E_{n+1}. On the other hand, a departure occurring while the system is in state E_n sends the system down one to E_{n-1} and can be looked upon as a "death." This type of process is often referred to as a *birth-death* process, and the analysis we will perform to obtain the difference equations is quite a typical approach. More detail on general birth-death processes is presented in Chapter 3.

2.1.1 Step 1—Difference Equations for $p_n(t)$

To write the difference equation for $p_n(t)$ we first consider how the system can get to state E_n at time $t+\Delta t$. To be in state E_n at time $t+\Delta t$, the system could have been in state E_n at time t and during Δt have j arrivals and j service completions, or have been at state E_{n+j} at time t and during Δt have $j+k$ services completed and k arrivals. Also, the system could have been in state E_{n-j} at time t and during Δt have $j+k$ arrivals and k service completions. We know, however, that the probability of more than one arrival in Δt or more than one service completion in Δt is given by $o(\Delta t)$. Thus we need consider explicitly the cases where, at most, only one arrival and/or one service completion occurs during Δt. The system can be in state E_n at time t and have no arrivals or service completions in Δt or be in state E_{n-1} at time t and have, during Δt, one arrival and no service completions, or finally, the system can be in state E_{n+1} at time t and have, during Δt, one service completion and no arrivals. We assume for the time being that $n \geqslant 1$. We can write, therefore, that

$\Pr\{$ system is in E_n at time $t+\Delta t \} = \Pr\{$ system is in E_n at time t

and no arrivals occur during Δt and no service is completed during $\Delta t \}$

$+ \Pr\{$ system is in E_n at time t and one arrival occurs during Δt

and one service is completed during $\Delta t \}$

$+ \Pr\{$ system is in E_{n+1} at time t and one service completion

occurs during Δt and no arrivals occur during $\Delta t \}$

$+ \Pr\{$ system is in E_{n-1} at time t and one arrival occurs during

Δt and no service completions occur during $\Delta t \} + o(\Delta t)$.

Since arrivals and service are both independent of each other and the state at t, it is seen that

$p_n(t+\Delta t) = p_n(t) \cdot \Pr\{$ no arrivals in $\Delta t \} \cdot \Pr\{$ no service completions in $\Delta t \}$

$+ p_n(t) \cdot \Pr\{$ one arrival in $\Delta t \} \cdot \Pr\{$ one service in $\Delta t \}$

$+ p_{n+1}(t) \cdot \Pr\{$ one service completed in $\Delta t \} \cdot \Pr\{$ no arrivals in $\Delta t \}$

$+ p_{n-1}(t) \cdot \Pr\{$ one arrival in $\Delta t \} \cdot \Pr\{$ no service completions in $\Delta t \}$

$+ o(\Delta t) \qquad (n \geqslant 1)$.

Because interarrival and service times are distributed exponentially, it is not necessary to know how long since the last arrival or how long the current customer in service has been there. We are thus able to write that

$$p_n(t+\Delta t) = p_n(t)[1-\lambda\Delta t - o(\Delta t)][1-\mu\Delta t - o(\Delta t)]$$

$$+ p_n(t)[\lambda\Delta t + o(\Delta t)][\mu\Delta t + o(\Delta t)]$$

$$+ p_{n+1}(t)[\mu\Delta t + o(\Delta t)][1-\lambda\Delta t + o(\Delta t)]$$

$$+ p_{n-1}(t)[\lambda\Delta t + o(\Delta t)][1-\mu\Delta t + o(\Delta t)]$$

$$+ o(\Delta t). \qquad (2.1)$$

Combining all o(Δt) terms and realizing terms with $(\Delta t)^2$ are also o(Δt), we can write

$$p_n(t+\Delta t) = p_n(t)[1-\lambda\Delta t - \mu\Delta t] + p_{n+1}(t)[\mu\Delta t]$$

$$+ p_{n-1}(t)[\lambda\Delta t] + o(\Delta t) \qquad (n \geq 1). \qquad (2.2)$$

Since p_{n-1} does not exist for $n=0$, (2.2) is invalid for $n=0$ and that state must be considered separately. The system can be in E_0 at $t+\Delta t$ if it were in E_0 at t and no arrivals came during Δt (no service is possible since the system is empty), or the system can be in E_1 at time t and have no arrivals but one service completion in Δt. If the system is in any higher state at time t, multiple services would be required during Δt and the probability of this is o(Δt). Thus

$$p_0(t+\Delta t) = p_0(t)[1-\lambda\Delta t + o(\Delta t)]$$

$$+ p_1(t)[1-\lambda\Delta t + o(\Delta t)][\mu\Delta t + o(\Delta t)] + o(\Delta t)$$

$$= p_0(t)[1-\lambda\Delta t] + p_1(t)[\mu\Delta t] + o(\Delta t). \qquad (2.3)$$

The difference equations, then, for the $M/M/1$ model are given as

$$\left\{ \begin{array}{l} p_n(t+\Delta t) = p_n(t)[1-\lambda\Delta t - \mu\Delta t] + p_{n+1}(t)[\mu\Delta t] \\[2mm] \qquad\qquad + p_{n-1}(t)[\lambda\Delta t] + o(\Delta t) \qquad (n \geq 1) \\[2mm] p_0(t+\Delta t) = p_0(t)[1-\lambda\Delta t] + p_1(t)[\mu\Delta t] + o(\Delta t) \end{array} \right. \qquad (2.4)$$

We note that these difference equations are with respect to both t and n, that is, time and system state.

2.1.2 Step 2—Differential-Difference Equations

Equation (2.4) may be rewritten slightly to yield

$$\left\{ \begin{array}{l} p_n(t+\Delta t) - p_n(t) = -(\lambda+\mu)\Delta t p_n(t) + \mu\Delta t p_{n+1}(t) \\[2mm] \qquad\qquad + \lambda\Delta t p_{n-1}(t) + o(\Delta t) \qquad (n \geq 1) \\[2mm] p_0(t+\Delta t) - p_0(t) = -\lambda\Delta t p_0(t) + \mu\Delta t p_1(t) + o(\Delta t) \end{array} \right.$$

If we divide through by Δt and take the limit as $\Delta t \rightarrow 0$, it is found that

$$
\begin{cases}
\dfrac{dp_n(t)}{dt} = -(\lambda + \mu)p_n(t) + \mu p_{n+1}(t) \\[2mm]
\qquad\qquad + \lambda p_{n-1}(t) \qquad (n \geqslant 1) \\[2mm]
\dfrac{dp_0(t)}{dt} = -\lambda p_0(t) + \mu p_1(t)
\end{cases}
\qquad (2.5)
$$

The equations of (2.5) are differential-difference equations, differential equations in t, and difference equations in n.

2.1.3 Step 3—Solution of the Differential Equations for $p_n(t)$

Since this step is generally quite cumbersome and for some models impossible, the solution of (2.5) is treated separately in Section 2.6.

2.1.4 Step 4—Steady-State Solution for p_n

To get the steady-state solution for p_n, the probability of n customers in the system at an arbitrary point of time after steady state is reached, we take the limit as $t \rightarrow \infty$ of (2.5). Let it be assumed for the time being that the steady-state solution exists. When $p_n(t)$ is independent of time, as it is in the steady state, $dp_n(t)/dt$ is zero, and (2.5) may be written as

$$
\begin{cases}
0 = -(\lambda + \mu)p_n + \mu p_{n+1} + \lambda p_{n-1} \qquad (n \geqslant 1) \\[2mm]
0 = -\lambda p_0 + \mu p_1
\end{cases}
$$

or

$$
\begin{cases}
p_{n+1} = \dfrac{\lambda + \mu}{\mu} p_n - \dfrac{\lambda}{\mu} p_{n-1} \qquad (n \geqslant 1) \\[2mm]
p_1 = \dfrac{\lambda}{\mu} p_0
\end{cases}
\qquad (2.6)
$$

We now have to solve a set of difference equations in the one variable n. We present three procedures for doing this. The first and probably the most straightforward is an iterative procedure, the second involves generating functions, and the third involves the concept of linear operators and is analogous to methods used for differential equations. The reason for

presenting the three methods of solution is that one may be more successful than the others, depending on the particular model at hand. For the
model under study here, all three methods work equally well and we take
this opportunity to illustrate their use.

Iterative Method of Solving the Steady-State Difference Equations

for $\{p_n\}$.

Equations (2.6) can be used iteratively to obtain the following (see Problem
2.2):

$$\begin{cases} p_1 = \dfrac{\lambda}{\mu} p_0 \\[2em] p_2 = \left(\dfrac{\lambda}{\mu}\right)^2 p_0 \\[2em] p_3 = \left(\dfrac{\lambda}{\mu}\right)^3 p_0 \end{cases} \qquad (2.7)$$

At this point it is reasonable to conjecture that

$$p_n = \left(\frac{\lambda}{\mu}\right)^n p_0. \qquad (2.8)$$

We prove that this is, in fact, the case by mathematical induction. The
formula clearly holds for the first and second values, 1 and 2, of the index
n. We then assume that it holds for n and $n-1$, and show it also holds for
$n+1$ as follows. Using (2.6) and (2.8) we get

$$p_{n+1} = \frac{\lambda+\mu}{\mu}\left(\frac{\lambda}{\mu}\right)^n p_0 - \frac{\lambda}{\mu}\left(\frac{\lambda}{\mu}\right)^{n-1} p_0$$

$$= \left[\frac{\lambda^{n+1}+\mu\lambda^n}{\mu^{n+1}} - \left(\frac{\lambda}{\mu}\right)^n\right]p_0$$

$$= \left(\frac{\lambda^{n+1}+\mu\lambda^n-\mu\lambda^n}{\mu^{n+1}}\right)p_0$$

$$= \left(\frac{\lambda}{\mu}\right)^{n+1} p_0. \qquad (2.9)$$

Now invoke the principle of mathematical induction and Equation (2.8) is verified. That is, we know that the formula holds for $n=1$ and $n=2$. Hence it holds for $n=3$. Since it holds for $n=3$, it also is valid for $n=4$, and so on, *ad infinitum*.

It remains now only to obtain p_0. This can be accomplished by utilizing the boundary condition that $\sum_{n=0}^{\infty} p_n = 1$, since p_n is a probability distribution. Using (2.9),

$$1 = \sum_{n=0}^{\infty} \left(\frac{\lambda}{\mu} \right)^n p_0.$$

For the sake of notational efficiency, let us define ρ as λ/μ. The ratio ρ is often called the *utilization factor*, since it is, by definition, a measure of the average use of the service facility. It is, in fact, the expected number of arrivals per mean service time in the limit, and it is therefore often also called the traffic intensity.[2] Rewriting thus gives

$$p_0 = \frac{1}{\sum_{n=0}^{\infty} \rho^n}.$$

Now $\sum_{n=0}^{\infty} \rho^n$ is the geometric series, $1 + \rho + \rho^2 + \rho^3 + \cdots$, and converges if and only if $|\rho| < 1$. Thus for the existence of a steady-state solution, $\rho = \lambda/\mu$ must be less than 1, or equivalently, λ less than μ. This makes sense intuitively for if $\lambda > \mu$, the mean arrival rate is greater than the mean service rate and the server will get further and further behind. That is to say, system size will keep building up without limit. It is not as intuitive, however, why no steady-state solution exists when $\lambda = \mu$. One possible explanation for infinite build-up when $\lambda = \mu$ is that as the queue grows, it is more and more difficult for the server to decrease the queue because his average service rate is no higher than the average arrival rate.

Making use of the well-known expression for the sum of the terms of a geometric progression,

$$\sum_{n=0}^{\infty} \rho^n = \frac{1}{1-\rho} \qquad (\rho < 1),$$

we have

$$p_0 = 1 - \rho \qquad (\rho = \lambda/\mu < 1). \tag{2.10}$$

---

[2]Though ρ is clearly dimensionless, it is often given to be in "erlangs," in honor of the pioneering effort of A. K. Erlang.

Thus the steady-state solution is

$$\boxed{p_n = \rho^n(1-\rho) \qquad (\rho = \lambda/\mu < 1)} \ . \tag{2.11}$$

Solution of the Steady-State Difference Equations for $\{p_n\}$ by Generating Functions.

The probability generating function, $P(z) = \sum_{n=0}^{\infty} p_n z^n$ (z complex with $|z| \le 1$), can be utilized to find p_n.[3] The procedure involves finding a closed expression for $P(z)$ from Equation (2.6) and then finding the power series expansion to "pick off" the $\{p_n\}$ which are the coefficients. For some models, it is relatively easy to find a closed expression for $P(z)$, but quite difficult to find its series expansion to obtain the $\{p_n\}$. However, even if the series expansion cannot be found, $P(z)$ still provides useful information. For example, $dP(z)/dz$ evaluated at $z = 1$ gives the expected value $\sum_{n=0}^{\infty} n p_n$, which is the average number in the system. For the model under consideration here, we can completely solve for the $\{p_n\}$ using $P(z)$.

We rewrite (2.6) in terms of ρ and obtain

$$\begin{cases} p_{n+1} = (\rho+1)p_n - \rho p_{n-1} & (n \ge 1) \\ \\ p_1 = \rho p_0 \end{cases} \ . \tag{2.12}$$

When both sides of the first line of (2.12) are multiplied by z^n we find

$$p_{n+1}z^n = (\rho+1)p_n z^n - \rho p_{n-1}z^n$$

or

$$z^{-1}p_{n+1}z^{n+1} = (\rho+1)p_n z^n - \rho z p_{n-1}z^{n-1}.$$

When both sides of the foregoing equation are summed from $n = 1$ to ∞, it is found that

$$z^{-1}\sum_{n=1}^{\infty} p_{n+1}z^{n+1} = (\rho+1)\sum_{n=1}^{\infty} p_n z^n - \rho z \sum_{n=1}^{\infty} p_{n-1}z^{n-1},$$

or

$$z^{-1}\left[\sum_{n=-1}^{\infty} p_{n+1}z^{n+1} - p_1 z - p_0\right] = (\rho+1)\left[\sum_{n=0}^{\infty} p_n z^n - p_0\right] - \rho z \sum_{n=1}^{\infty} p_{n-1}z^{n-1}.$$

---

[3]The reader unfamiliar with generating functions is referred to Appendix 3, Section A3.3. We note here that the generating function is closely related to the z-transform which is given as $F(z) = \sum_{n=0}^{\infty} f(n)z^{-n}$. The z-transform of the steady-state probability distribution would be $\sum_{n=0}^{\infty} p_n z^{-n}$.

Noting that

$$\sum_{n=-1}^{\infty} p_{n+1}z^{n+1} = \sum_{n=0}^{\infty} p_n z^n = \sum_{n=1}^{\infty} p_{n-1}z^{n-1} = P(z),$$

we get

$$z^{-1}[P(z) - p_1 z - p_0] = (\rho+1)[P(z) - p_0] - \rho z P(z). \qquad (2.13)$$

From (2.12) we also have that $p_1 = \rho p_0$; hence

$$z^{-1}[P(z) - (\rho z + 1)p_0] = (\rho+1)[P(z) - p_0] - \rho z P(z).$$

Solving for $P(z)$ we finally have

$$P(z) = \frac{p_0}{1 - z\rho}. \qquad (2.14)$$

To find p_0 we use the boundary condition that $\sum_{n=0}^{\infty} p_n = 1$ in the following way. Consider $P(1)$, which can be seen to be

$$P(1) = \sum_{n=0}^{\infty} p_n 1^n$$

$$= \sum_{n=0}^{\infty} p_n = 1.$$

Thus from (2.14), we have

$$P(1) = 1 = \frac{p_0}{1 - \rho}, \qquad (2.15)$$

or

$$p_0 = 1 - \rho.$$

Because the $\{p_n\}$ are probabilities, $P(z) > 0$ for $z > 0$; hence $P(1) > 0$. From (2.15) we see that $P(1) = p_0/(1-\rho) > 0$; hence ρ must be < 1 since p_0 is a probability and is > 0. Thus

$$P(z) = \frac{1 - \rho}{1 - z\rho} \qquad (\rho < 1). \qquad (2.16)$$

It is an easy task to expand (2.16) as a power series since one merely has to utilize long division on $1/(1 - z\rho)$ or recognize it as the sum of a geometric series. So doing yields

$$\frac{1}{1 - z\rho} = 1 + z\rho + (z\rho)^2 + (z\rho)^3 + \cdots,$$

and thus the probability generating function is

$$P(z) = \sum_{n=0}^{\infty} (1-\rho)\rho^n z^n. \tag{2.17}$$

The coefficient of z^n, p_n, is given by

$$p_n = (1-\rho)\rho^n,$$

which is what was previously obtained in Equation (2.11).

Solution of the Steady-State Difference Equations for $\{p_n\}$ by
Use of Operators

If we consider a linear operator D defined on the sequence $\{a_0, a_1, a_2, \dots\}$, such that

$$Da_n = a_{n+1} \qquad \text{(for all } n\text{)},$$

then the general linear difference equation with constant coefficients

$$C_n a_n + C_{n+1} a_{n+1} + \cdots + C_{n+k} a_{n+k} = \sum_{i=n}^{n+k} C_i a_i = 0 \tag{2.18}$$

may be written as

$$\left(\sum_{i=n}^{n+k} C_i D^{i-n} \right) a_n = 0$$

since

$$D^l a_n = a_{n+l} \qquad \text{(for all } n \text{ and } l\text{)}.$$

In the event that (2.18) is of the form

$$c_2 a_{n+2} + c_1 a_{n+1} + c_0 a_n = 0, \tag{2.19}$$

then

$$(c_2 D^2 + c_1 D + c_0) a_n = 0, \tag{2.20}$$

and if the quadratic in D has the real roots r_1 and r_2, it is also true that

$$(D - r_1)(D - r_2) a_n = 0.$$

Since r_1 and r_2 are roots of Equation (2.20), $d_1 r_1^n$ and $d_2 r_2^n$ are solutions to (2.19), where d_1 and d_2 are arbitrary constants. This can be verified by substitution of $d_1 r_1^n$ and $d_2 r_2^n$ into (2.19), where $a_n = d_1 r_1^n$ or $d_2 r_2^n$. For example, letting $a_n = d_1 r_1^n$ we have, upon substitution in (2.19),

$$c_2 d_1 r_1^{n+2} + c_1 d_1 r_1^{n+1} + c_0 d_1 r_1^n = 0$$

or

$$d_1 r_1^n [c_2 r_1^2 + c_1 r_1 + c_0] = 0,$$

which is of the form of (2.20). Similarly, one can show that $d_2 r_2^n$ is a solution and hence that their sum, $d_1 r_1^n + d_2 r_2^n$, is also a solution. It can be shown in a manner similar to that used for ordinary linear differential equations that this sum is the most general solution (see Appendix 2, Section A2.2).

This approach is directly applicable to the solution of the steady-state difference equations (2.6), since these are linear difference equations of a form like (2.19). If D is defined by the operation $Dp_n = p_{n+1}$ and if (2.6) is rewritten as

$$\mu p_{n+2} - (\lambda + \mu) p_{n+1} + \lambda p_n = 0 \qquad (n = 0, 1, 2, \dots),$$

then these limiting probabilities are the solution to

$$[\mu D^2 - (\lambda + \mu) D + \lambda] p_n = 0, \tag{2.21}$$

subject to the boundary conditions that

$$p_1 = \left(\frac{\lambda}{\mu} \right) p_0$$

and

$$\sum_{n=0}^{\infty} p_n = 1.$$

This quadratic in D factors easily to give

$$[(D - 1)(\mu D - \lambda)] p_n = 0,$$

and hence

$$p_n = d_1 (1)^n + d_2 \left(\frac{\lambda}{\mu} \right)^n, \tag{2.22}$$

where d_1 and d_2 are to be found with the use of the boundary conditions. From (2.22),

$$p_1 = d_1 + d_2 \rho$$

and

$$p_2 = d_1 + d_2\rho^2.$$

Also, from (2.6), we can get the boundary conditions

$$p_1 = \frac{\lambda}{\mu}p_0 = \rho p_0,$$

and

$$p_2 = \frac{\lambda + \mu}{\mu}p_1 - \frac{\lambda}{\mu}p_0 = \left(\frac{\lambda}{\mu}\right)^2 p_0 = \rho^2 p_0.$$

Equating the two expressions for p_1 and p_2, respectively, gives two equations in the two unknowns d_1, d_2:

$$\begin{cases} d_1 + d_2\rho = p_0\rho \\ d_1 + d_2\rho^2 = p_0\rho^2. \end{cases}$$

Solving these simultaneously yields

$$d_1 = 0$$

and

$$d_2 = p_0,$$

and thus

$$p_n = \rho^n p_0.$$

Then p_0 is found as in the first method by summing p_n over all n and is

$$p_0 = 1 - \rho \qquad (\rho < 1).$$

2.2   MEASURES OF EFFECTIVENESS

The steady-state probability distribution for the system size allows us to calculate what are commonly called measures of effectiveness. Two of immediate interest are the expected number in the system and the expected number in the queue at steady state.

To derive the foregoing measures, let N represent the random variable "number of customers in the system at steady state" and L represent its expected value. We can then write

$$L = E[N] = \sum_{n=0}^{\infty} n p_n$$

$$= \sum_{n=0}^{\infty} n(1-\rho)\rho^n$$

$$= (1-\rho) \sum_{n=0}^{\infty} n\rho^n. \tag{2.23}$$

Consider $\sum_{n=0}^{\infty} n\rho^n$:

$$\sum_{n=0}^{\infty} n\rho^n = \rho + 2\rho^2 + 3\rho^3 + \ldots$$

$$= \rho(1 + 2\rho + 3\rho^2 + \ldots)$$

$$= \rho \sum_{n=1}^{\infty} n\rho^{n-1}.$$

But we observe that $\sum_{n=1}^{\infty} n\rho^{n-1}$ is simply the derivative of $\sum_{n=0}^{\infty} \rho^n$ with respect to ρ since the summation and differentiation operations may be interchanged as the functions are sufficiently well-behaved. Since $\rho < 1$,

$$\sum_{n=0}^{\infty} \rho^n = \frac{1}{1-\rho};$$

hence

$$\sum_{n=1}^{\infty} n\rho^{n-1} = \frac{d[1/(1-\rho)]}{d\rho}$$

$$= \frac{1}{(1-\rho)^2}.$$

So the expected number in the system at steady state is then

$$L = \frac{(1-\rho)\rho}{(1-\rho)^2}$$

or simply,

$$\boxed{L = \frac{\rho}{1-\rho}} \quad . \tag{2.24}$$

Equivalently, we can write

$$\boxed{L = \frac{\lambda}{\mu - \lambda}}.\tag{2.25}$$

If the random variable "number in queue in steady state" is denoted by N_q and its expected value by L_q, then we have[4]

$$L_q = E[N_q] = 0p_0 + \sum_{n=1}^{\infty} (n-1)p_n$$

$$= \sum_{n=1}^{\infty} np_n - \sum_{n=1}^{\infty} p_n$$

$$= L - (1 - p_0)$$

$$= \frac{\rho}{1-\rho} - \rho.$$

Thus the mean queue length is

$$\boxed{L_q = \frac{\rho^2}{1-\rho}},\tag{2.26}$$

or equivalently,

$$\boxed{L_q = \frac{\lambda^2}{\mu(\mu-\lambda)}}.\tag{2.27}$$

We might also be interested in the expected queue size of nonempty queues, which we denote by L_q'; that is, we wish to ignore the cases where the queue is empty. Another way of looking at this measure is to view it as the expected size of the queues which form from time to time. We can write

$$L_q' = E[N_q | N_q \neq 0]$$

$$= \sum_{n=1}^{\infty} (n-1)p_n' = \sum_{n=2}^{\infty} (n-1)p_n',$$

where p_n' is the conditional probability distribution of n in the system given the queue is not empty; that is,

$$p_n' = \Pr\{n \text{ in system} | n \geq 2\}.$$

---

[4]Note that $L_q = L - (1 - p_0)$ holds for all single-channel one-at-a-time service queues since, in the derivation, no assumptions as to the distributions of input or service are used.

From the laws of conditional probability,

$$p_n' = \frac{\Pr\{n \text{ in system and } n \geqslant 2\}}{\Pr\{n \geqslant 2\}}$$

$$= \frac{p_n}{\displaystyle\sum_{n=2}^{\infty} p_n} \qquad (n \geqslant 2)$$

$$= \frac{p_n}{1 - p_0 - p_1}$$

$$= \frac{p_n}{1 - (1 - \rho) - (1 - \rho)\rho}$$

$$= \frac{p_n}{\rho^2}.$$

The probability distribution p_n' is the probability distribution p_n normalized when the cases $n = 0$ and 1 are omitted. Thus

$$L_q' = \sum_{n=2}^{\infty} (n-1) \frac{p_n}{\rho^2}$$

$$= \frac{L - p_1 - (1 - p_0 - p_1)}{\rho^2}.$$

Hence

$$\boxed{L_q' = \frac{1}{1 - \rho}} \quad , \tag{2.28}$$

or equivalently,

$$\boxed{L_q' = \frac{\mu}{\mu - \lambda}} \quad . \tag{2.29}$$

As a side observation, it is not at all by coincidence that it turned that

$$\Pr\{n \text{ in system} \geqslant 2\} = \rho^2$$

because it can easily be established that for all n, letting N denote the random variable "steady-state number in system,"

$$\Pr\{N \geqslant n\} = \rho^n.$$

The proof is as follows:

$$\Pr\{N \geqslant n\} = \sum_{k=n}^{\infty} (1-\rho)\rho^k$$

$$= (1-\rho)\rho^n \sum_{k=n}^{\infty} \rho^{k-n}$$

$$= \frac{(1-\rho)\rho^n}{1-\rho} = \rho^n.$$

The foregoing measures of effectiveness are illustrated by the following.

Example 2.1

H. R. Cutt runs a one-man barber shop. He does not make appointments but runs his shop on a first-come, first-served basis. He finds that on Saturday mornings he is extremely busy and is contemplating hiring a part-time barber and even possibly moving to a larger building. Having obtained his Master's degree in operations research (OR) from a leading middle Atlantic university prior to embarking upon a career in barbering, he elects, before making any rash decisions, to analyze the situation.

He thus keeps careful records for a succession of Saturday mornings and finds that customers seem to arrive according to a Poisson process with a mean arrival rate of 5/hr. Because of his excellent reputation (what else would you expect from a barber with a masters in OR?) customers were always willing to wait. His data further showed that his haircutting time was exponentially distributed with an average haircut taking 10 min.

Cutt first decided to calculate the average number of customers in the shop and the average number of customers waiting for a haircut. From his data, $\lambda = 5/\text{hr}$ and $\mu = (1/10)/\text{min} = 6/\text{hr}$. This gives a ρ of 5/6. From (2.24) and (2.26) he can find $L = 5$ and $L_q = 4\frac{1}{6}$. The number waiting when there are people waiting can be found from (2.28) as $L_q' = 6$. He also is interested in the percent of time an arrival can walk right in without having to wait at all. A customer can walk right in and be served immediately if no one is in the shop. The probability of this is $p_0 = 1 - \rho = \frac{1}{6}$. Hence approximately 16.7% of the time Cutt is idle and a customer can get a haircut without waiting. Because of the Poisson process governing arrivals and the completely random phenomenon of the Poisson distribution discussed in Section 1.8, the percentage of customers that can go directly into service is also 16.7%. Thus 83.3% of the customers must wait prior to getting into the barber chair.

Cutt's waiting room has, at present, only four seats. He is interested in the probability that a customer, upon arrival, will not be able to find a seat and have to stand. This can be calculated from the $\{p_n\}$ as follows:

$$\Pr\{\text{finding no seat}\} = \Pr\{N \geqslant 5\}$$

$$= \rho^5$$

$$\doteq 0.402.$$

Thus a little over 40% of the time a customer cannot find a seat. This also implies 40% of his customers will have to stand upon arrival. Cutt is also interested in learning information concerning how long customers must wait for service. To do so he needs waiting-time measures; these are derived in the next section.

2.3 WAITING-TIME DISTRIBUTIONS

To obtain information concerning the time an arrival must spend waiting until entering service, we now proceed to derive the probability distribution for waiting time.[5] Up to now the queue discipline has had no effect on our derivations. When considering individual waiting time, however, queue discipline must be specified and we are here assuming that it is FIFO. The waiting-time random variable has an interesting property in that it is part discrete and part continuous. Waiting time is, for the most part, a continuous random variable except that there is a nonzero probability that the delay will be zero, that is, a customer entering service immediately upon arrival. So let T_q denote the random variable "time spent waiting in the queue" and $W_q(t)$ denote its cumulative probability distribution. Hence we have, from the complete randomness of the Poisson, that

$$W_q(0) = \Pr\{T_q \leqslant 0\} = \Pr\{T_q = 0\}$$

$$= \Pr\{\text{system empty at an arrival}\}$$

$$= p_0$$

$$= 1 - \rho. \tag{2.30}$$

It then remains to find $W_q(t)$ for $t > 0$.

---

[5]The time a fictitious customer would have to wait *were* he to arrive at an arbitrary point in time is called the *virtual waiting time*, and its steady-state distribution is identical to that of the actual waiting time of an arriving customer if, and only if, the input is Poisson.

Consider $W_q(t)$, the probability of a customer waiting a time less than or equal to t for service. If there are n units in the system upon arrival, in order for the customer to go into service at a time between 0 and t, all n units must have been serviced by time t. Since the service distribution is memoryless, the distribution of the time required for n completions is independent of the time of the current arrival and is the convolution of n exponential random variables, which is an Erlang type n. In addition, since the input is Poisson, the arrival points are uniformly spaced and hence the probability that an arrival finds n in the system is identical to the stationary distribution of system size. Therefore we may write that

$$W_q(t) = \Pr\{T_q \leqslant t\} = \sum_{n=1}^{\infty} [\Pr\{n \text{ completions in } \leqslant t|$$

$$\text{arrival found } n \text{ in system}\} \cdot p_n] + W_q(0)$$

$$= (1-\rho) \sum_{n=1}^{\infty} \rho^n \int_0^t \frac{\mu(\mu x)^{n-1}}{(n-1)!} e^{-\mu x} \, dx + (1-\rho)$$

$$= (1-\rho)\rho \int_0^t \mu e^{-\mu x} \sum_{n=1}^{\infty} \frac{(\mu x \rho)^{n-1}}{(n-1)!} \, dx + (1-\rho)$$

$$= \rho(1-\rho) \int_0^t \mu e^{-\mu x(1-\rho)} \, dx + (1-\rho)$$

$$= 1 - \rho e^{-\mu(1-\rho)t} \qquad (t > 0).$$

So the distribution of waiting time in queue is then

$$\boxed{W_q(t) = \begin{cases} 1-\rho & (t=0) \\ 1 - \rho e^{-\mu(1-\rho)t} & (t>0) \end{cases}}. \qquad (2.31)$$

It is left as an exercise (see Problem 2.7a) to show that while the derivative of $W_q(t)$ for $t > 0$ does not integrate to 1 over the range $(0, \infty)$ and is thus not a true density, the addition of the point $t=0$ with its probability $1-\rho$ yields a valid composite probability distribution, $w_q(t)$.

With the probability distribution of T_q we can now calculate the expected waiting time, which is denoted by W_q:[6]

$$W_q = E[T_q] = \int_0^\infty t\, dW_q(t)$$

$$= 0\left(1 - \frac{\lambda}{\mu}\right) + \int_0^\infty t\frac{\lambda}{\mu}(\mu-\lambda)e^{-(\mu-\lambda)t}\, dt$$

$$= \frac{\lambda}{\mu}\int_0^\infty t(\mu-\lambda)e^{-(\mu-\lambda)t}\, dt.$$

So

$$W_q = \frac{\lambda}{\mu(\mu-\lambda)} \qquad (2.32)$$

Also of interest would be the total time a customer had to spend in the system including service. Denote this random variable by T, its CDF by $W(t)$, its density by $w(t)$, and its expected value by W. Then it can be shown (see Problem 2.8) that

$$w(t) = (\mu-\lambda)e^{-(\mu-\lambda)t} \qquad (t > 0) \qquad (2.33)$$

$$W = E[T] = \frac{1}{\mu-\lambda} \qquad (2.34)$$

Let us now go back to Example 2.1, where W_q and W can be calculated from (2.32) and (2.34), respectively, as 50 and 60 min. Of interest also might be the expected waiting time of those customers who actually had to wait, that is, $E[T_q | T_q > 0] = W_q'$. It can be shown (see Problem 2.7b) that $W_q' = 1/(\mu-\lambda)$ also; hence those people who had to wait waited 1 hr on the average. Cutt is also interested in finding the probability that any customer will have to wait more than 1 hr for service. This can be found

---

[6]The first expression is a Riemann-Stieltjes (R-S) integral. An R-S integral, say $\int dF(x)$, allows us to work with functions which are not completely differentiable. If $F(x)$ were completely differentiable, then $\int dF(x)$ reduces to the usual Riemann integral, $\int f(x)dx$. Since $W_q(t)$ has a nonzero lump at $t=0$, the use of the R-S integral allows us to include this. See, for example, Olmsted (1959) for further discussion on R-S integrals.

from (2.31) as

$$\Pr\{T_q > 1\} = 1 - W_q(1)$$

$$= \tfrac{5}{6} e^{-1}$$

$$\doteq 0.306.$$

Thus almost 31% of his customers on a Saturday morning must wait over an hour. All these calculations, of course, assume steady state is reached. These figures are not applicable to the initial few customers who arrive soon after opening; however, the transient effects are generally "washed out" rather quickly and the steady-state calculations suffice. The transient effects are explored further in Section 2.6.

Note that the development for p_n, L, and L_q did not depend on the order in which customers are served; that is, they would also be valid for the $M/M/1/\infty/GD$ model. It turns out, as we see in the next section, that these results for W and W_q are also applicable under quite general conditions for $M/M/1/\infty/GD$. However, the CDF's, $W(t)$ and $W_q(t)$, are discipline dependent, as can be seen from the way they were developed.

2.4 RELATIONS BETWEEN EXPECTED QUEUE LENGTH AND EXPECTED WAITING TIME—LITTLE'S FORMULA

Certain relationships can be seen among some of the measures of effectiveness developed in Sections 2.2 and 2.3. These relations are not coincidental but can be substantiated on theoretical grounds. These theoretical proofs are not always simple, however, and we explore these relations in this section.

We consider first the connection between W_q and W. Comparing Equations (2.32) and (2.34) we see that

$$W = W_q + \frac{1}{\mu}. \tag{2.35}$$

This is certainly intuitive and is easily substantiated by the following argument. Let S represent the random variable "service time," and we have $T = T_q + S$. Hence $E[T] = E[T_q] + E[S]$, which directly yields Equation (2.35). In fact, $w(t)$ can be derived as the convolution of $w_q(t)$ and the service-time density, if it exists. Equation (2.35), then, is not only applicable to the $M/M/1$ model, but holds in general for any queue.

Another more subtle relationship exists between L_q and W_q. We can see from (2.27) and (2.32) that

$$L_q = \lambda W_q \qquad (2.36)$$

and also from (2.25) and (2.34) that

$$L = \lambda W. \qquad (2.37)$$

Equation (2.37) is generally referred to as Little's formula because of the work of Little (1961). These relationships can be explained intuitively as follows. Consider a customer who just arrives. On the average he steps into service after a time W_q. Suppose that when he steps into service, he looks over his shoulder and counts the number that have formed in back of him. On the average this number is L_q. It also took, on the average, $1/\lambda$ for each of the L_q to arrive, and the total time it took for the L_q arrivals to form behind him must equal his waiting time[7]; hence $L_q(1/\lambda) = W_q$. The same argument can be used on (2.37) by considering that when a customer just steps out of service and looks over his shoulder, on the average, L will be in the system and his average total waiting time is equivalent to the average time it took the L to arrive. This intuitive argument, of course, does not prove the validity of these relations in general, but it turns out that they do hold under fairly general conditions. We, however, have proved them for the $M/M/1$ model, since L, L_q, W, and W_q were derived rigorously and the relationships given by (2.36) and (2.37) are seen to hold.

The conditions under which Equation (2.37) holds are now further explored. A rather straightforward proof is possible for Poisson input to a single channel, first in, first out, regardless of the form of the service distribution, and proceeds as follows. In this case the steady-state probability that there are n in the system when a customer departs is equal to the probability that n customers arrive during the total time spent in the system by an arbitrary customer.[8] Hence from the law of total probability,

$$p_n = \int_0^\infty \Pr\{\, n \text{ arrivals during a waiting time of } T \,|\, T = t \,\} \, dW(t),$$

---

[7]The unused portion of the last interarrival time may be ignored because of the memorylessness of the exponential.

[8]We prove in Chapter 5 that this probability of n in the system at a departure point is equivalent to the probability of n in the system at any general time.

where $W(t)$ is the cumulative distribution function of the waiting time. Thus

$$p_n = \left(\frac{1}{n!}\right) \int_0^\infty (\lambda t)^n e^{-\lambda t} \, dW(t) \qquad (n \geqslant 0),$$

and

$$L = E[N] = \sum_{n=0}^\infty n p_n$$

$$= \sum_{n=1}^\infty \frac{n}{n!} \int_0^\infty (\lambda t)^n e^{-\lambda t} \, dW(t).$$

But the summation and integration can be interchanged since the appropriate functions are sufficiently well-behaved; thus

$$L = \int_0^\infty dW(t) e^{-\lambda t} \sum_{n=1}^\infty \frac{(\lambda t)^n}{(n-1)!}$$

$$= \int_0^\infty \lambda t e^{-\lambda t} \, dW(t) \sum_{n=1}^\infty \frac{(\lambda t)^{n-1}}{(n-1)!}$$

$$= \int_0^\infty \lambda t \, dW(t) e^{-\lambda t} [e^{\lambda t}]$$

$$= \lambda \int_0^\infty t \, dW(t) = \lambda E[T]$$

$$= \lambda W.$$

The proof above is valid for a Poisson input general-service FIFO queue, but these conditions are by no means necessary.[9] Jewell (1967) has supplied a general set of sufficient conditions in his paper entitled "A Simple Proof of $L = \lambda W$." His "simple" proof, however, is somewhat beyond the level of this text, and we merely list the conditions Jewell presents. They are basically as follows:

[9]The most general and current exposition on this topic is provided by Stidham (1972).

(i) For any current condition of the system, the system will become empty with probability one at some future epoch;

(ii) Whenever the system becomes empty, the arrival and waiting-time mechanisms are identically 'reset' by the next arrival;

(iii) The expectation of the system size over the first busy period is finite;

(iv) The unit-average mean wait and interarrival time taken over each busy period are both finite, where a busy period is the opposite of an idle period.

For illustration, it can be shown that the $M/M/1$ queue satisfies these assumptions since:

(i) The system will become idle with probability one because $\rho < 1$;

(ii) The arrivals are Poisson and are independent of service, and service on successive customers is independent and distributed identically;

(iii) Since $\rho < 1$, the first busy period must end; therefore the number of arrivals during this period and their waiting times are finite;

(iv) The expected number of units processed in a busy period is finite, as is the sum of their waits, and the sum of the interarrival times in the busy period is also finite.

Thus far, we have been concerned with the relationship $L = \lambda W$ given by Equation (2.37), Little's formula. We now turn our attention to the relation given by (2.36). All previous arguments remain valid if the queue wait replaces system wait everywhere and queue size replaces system size. It is necessary to obtain, therefore, only one of the set $\{L, L_q, W, W_q\}$, since if one is known, the others can be found using Equations (2.35), (2.36), and (2.37).

Taking Equations (2.36) and (2.37) together to be valid and using Equation (2.35) leads us to the additional relation

$$\frac{L}{\lambda} = \frac{L_q}{\lambda} + \frac{1}{\mu},$$

or

$$L = L_q + \frac{\lambda}{\mu}. \tag{2.38}$$

We now consider one last relationship which can be seen from Equations (2.32) and (2.25) to be

$$W_q = \frac{L}{\mu}. \tag{2.39}$$

This can also be argued intuitively, as follows. Upon arrival, on the average, a customer finds L in the system, assuming arrival points are arbitrary as they are for the Poisson case. He will enter service only after the L in front of him have been served. The expected service time of the $L-1$ in the queue is $1/\mu$. Because of the Markovian property of the exponential distribution, the customer in service upon arrival, even though he has been in service for some time, still has an expected time for completion of $1/\mu$. Thus the customer must wait, on the average, a time $L(1/\mu)$ before being able to enter service himself. This relation holds under much more limited conditions than does Little's formula. From the intuitive argument it appears that service and interarrival times must be assumed exponential. We now show that if (2.39) holds and Little's formula is valid, then $L=\rho/(1-\rho)$. Equation (2.39) states that

$$W_q = \frac{L}{\mu}.$$

But, from (2.35), we have the general relation

$$W_q = W - \frac{1}{\mu}.$$

Thus

$$\frac{L}{\mu} = W - \frac{1}{\mu}.$$

If (2.37) is valid, $W=L/\lambda$ and we get

$$\frac{L}{\mu} = \frac{L}{\lambda} - \frac{1}{\mu}$$

or

$$L\left(\frac{1}{\lambda} - \frac{1}{\mu}\right) = \frac{1}{\mu},$$

and

$$L = \frac{\lambda}{\mu - \lambda} = \frac{\rho}{1-\rho}.$$

This equation is, of course, true for the $M/M/1$ model, as shown by Equations (2.24) and (2.25). Since the sufficient conditions for the validity of Little's formula are so general, the restriction on the use of Equation (2.39), implied by the foregoing, is quite limiting. In fact, the $M/M/1$ model and its relatives are the only queueing models of practical interest which have the property given by (2.39).

We can summarize this section in Table 2.1 below.

TABLE 2.1 RELATIONS AMONG MEASURES
OF EFFECTIVENESS

Relation	Comments on Validity
$W = W_q + \dfrac{1}{\mu}$	Holds in general
$W = \dfrac{L}{\lambda}$ $W_q = \dfrac{L_q}{\lambda}$	Sufficient conditions for validity rather general, as given by Jewell and listed earlier
$L = L_q + (1 - p_0)$	Holds for all single channel one-at-a-time service queues.
$L = L_q + \dfrac{\lambda}{\mu}$	Holds whenever Little's formulas are valid
$p_0 = 1 - \dfrac{\lambda}{\mu}$	Holds for all single-channel one-at-a-time service queues in which Little's formulas are valid
$W_q = \dfrac{L}{\mu}$	Limited use—$M/M/1$ and like models

2.5 FINITE SYSTEM CAPACITY—QUEUES WITH TRUNCATION $(M/M/1/K)$

In this section we study first-come, first-served, single-channel Poisson/ exponential queueing systems with finite waiting capacity. The procedure for solution will closely follow that for the $M/M/1/\infty$ model treated in the previous sections of this chapter. Letting K denote the truncation point, that is, at most K are allowed in the system, we proceed to write the difference equations. The difference equations for the $M/M/1/\infty$ model given by (2.6) are valid for this model as long as $n < K$. We must derive an equation for $n = K$, since now, if the system is in state E_K, the probability of an arrival into the system is zero. We have then the additional difference equation

$$p_K(t+\Delta t)=p_K(t)[1-\mu\Delta t]+p_{K-1}(t)[\lambda\Delta t][1-\mu\Delta t]+o(\Delta t).$$

This gives the differential equation

$$\frac{dp_K(t)}{dt} = -\mu p_K(t)+\lambda p_{K-1}(t),$$

from which the resultant steady-state equation is

$$p_K = \frac{\lambda}{\mu}p_{K-1}.$$

The full set of steady-state difference equations, then, can be written as

$$\begin{cases} p_1 = \dfrac{\lambda}{\mu}p_0 \\[2mm] p_{n+1} = \dfrac{\lambda+\mu}{\mu}p_n - \dfrac{\lambda}{\mu}p_{n-1} \quad (1 \leqslant n \leqslant K-1) \, . \\[2mm] p_K = \dfrac{\lambda}{\mu}p_{K-1} \end{cases} \qquad (2.40)$$

From the iterative procedure used in Section 2.1 for the $M/M/1/\infty$ model, we know that

$$p_n = \left(\frac{\lambda}{\mu}\right)^n p_0 \qquad (n \leqslant K-1). \qquad (2.41)$$

We also see from the last relationship of (2.40) that (2.41) holds for $n = K$. So we have

$$p_n = \left(\frac{\lambda}{\mu}\right)^n p_0 \qquad (n \leqslant K),$$

or equivalently,

$$p_n = \rho^n p_0 \qquad (n \leqslant K),$$

where, once again, $\rho = \lambda/\mu$. From the boundary condition $\sum_{n=0}^{K}p_n = 1$, we have

$$1 = \sum_{n=0}^{K} \rho^n p_0,$$

or

$$p_0 = \frac{1}{\sum_{n=0}^{K} \rho^n}.$$

The equation above has a finite geometric series, whose sum is clearly given by

$$\sum_{n=0}^{K} \rho^n = \begin{cases} \dfrac{1-\rho^{K+1}}{1-\rho} & (\rho \neq 1) \\ K+1 & (\rho = 1) \end{cases} ;$$

hence

$$p_0 = \begin{cases} \dfrac{1-\rho}{1-\rho^{K+1}} & (\rho \neq 1) \\ \dfrac{1}{K+1} & (\rho = 1) \end{cases} .$$

Thus

$$p_n = \begin{cases} \dfrac{(1-\rho)\rho^n}{1-\rho^{K+1}} & (\rho \neq 1) \\ \dfrac{1}{K+1} & (\rho = 1) \end{cases} . \tag{2.42}$$

Note that the steady-state solution exists even for $\rho \geqslant 1$. Intuitively, this makes sense since the truncation prevents the process from "blowing up." We also observe that letting $K \to \infty$, that is, removing the truncation, gives

$$p_n = (1-\rho)\rho^n \qquad (n < \infty)$$

which is consistent with the previously obtained $M/M/1/\infty$ results. We must keep in mind, however, that now the finite geometric progression used in determining p_0 becomes infinite and the $\rho < 1$ restriction must be invoked for steady-state existence.

Attention is next given to development of the measures of effectiveness for this model.

We are able to obtain measures for the $M/M/1/K$ problem similar to those of $M/M/1/\infty$. We first consider the expected system size L as follows. For $\rho = 1$, this expectation is clearly

$$L = \frac{\sum_{n=0}^{K} n}{K+1} = \frac{K}{2}.$$

For $\rho \neq 1$,

$$L = \sum_{n=0}^{K} n p_n$$

$$= p_0 \rho \sum_{n=0}^{K} n \rho^{n-1}$$

$$= p_0 \rho \sum_{n=0}^{K} \frac{d}{d\rho} (\rho^n)$$

$$= p_0 \rho \frac{d}{d\rho} \left(\sum_{n=0}^{K} \rho^n \right)$$

$$= p_0 \rho \frac{d}{d\rho} \left(\frac{1 - \rho^{K+1}}{1 - \rho} \right)$$

$$= p_0 \rho \frac{1 - (K+1)\rho^K + K\rho^{K+1}}{(1-\rho)^2},$$

or finally,

$$\boxed{L = \frac{\rho [1 - (K+1)\rho^K + K\rho^{K+1}]}{(1 - \rho^{K+1})(1 - \rho)}}$$

and

$$\boxed{\begin{aligned} L_q &= L - (1 - p_0) \\ &= L - \frac{\rho(1 - \rho^K)}{1 - \rho^{K+1}} \end{aligned}}.$$

The mean waiting time in the system can be obtained from Little's formula, that is, $W = L/\lambda'$, where λ' is the mean rate of customers actually entering the system and is equal to $\lambda(1 - p_K)$.[10] This effective arrival rate, λ', may also be found from Equation (2.38), namely, $\lambda' = \mu(L - L_q)$ (see Problem 2.14). The average waiting time in queue, W_q, can be obtained by either $W_q = W - 1/\mu$ or $W_q = L_q/\lambda'$. The derivations of the waiting time CDFs, $W(t)$ and $W_q(t)$, are not as neat as those for $M/M/1/\infty$, since the series used in developing the equivalent of Equation (2.31) is not infinite. However, these distributions can be obtained numerically by evaluating a finite sum (see Problem 2.15). These sums turn out to be Poisson sums, which can be found in tables. The derivation initially proceeds along the

---

[10] All relationships of Table 2.1 involving λ hold for this model as well, provided λ' replaces λ.

lines of that for $M/M/1/\infty$ in Section 2.3 as follows:

$$W_q(t) \equiv \Pr\{T_q \le t\} = \sum_{n=1}^{K-1} [\Pr\{n \text{ completions in} \le t| \text{ arrival found } n \text{ in system}\} \cdot q_n] + W_q(0),$$

where q_n is the probability of an arriving customer finding n in the system. Unlike $M/M/1/\infty$, here $q_n \ne p_n$ as an arriving customer cannot find the system in state K, since he would then be turned away. Essentially, to find q_n we must renormalize the probabilities $\{p_n\}$, omitting p_K. So

$$q_n = kp_n,$$

where k is a normalizing constant such that

$$\sum_{n=0}^{K-1} kp_n = 1.$$

Thus

$$k = \frac{1}{\sum_{n=0}^{K-1} p_n} = \frac{1}{1-p_K}$$

and

$$q_n = \frac{p_n}{1-p_K} \qquad (n \le K-1).$$

The fact that q_n is indeed proportional to p_n and all that is required is the normalizing procedure above can be proven by Bayes' theorem as follows:

$\Pr\{n \text{ in system}|\text{arrival is about to occur}\}$

$$= \frac{\Pr\{n \text{ in system}\}\Pr\{\text{arrival is about to occur}|n \text{ in system}\}}{\sum_{n=0}^{K} [\Pr\{n \text{ in system}\}\Pr\{\text{arrival is about to occur}|n \text{ in system}\}]}$$

$$= \lim_{\Delta t \to 0} \frac{p_n[\lambda\Delta t + o(\Delta t)]}{\sum_{n=0}^{K-1} p_n[\lambda\Delta t + o(\Delta t)] + p_K(0)}$$

$$= \lim_{\Delta t \to 0} \frac{p_n[\lambda + o(\Delta t)/\Delta t]}{\sum_{n=0}^{K-1} p_n[\lambda + o(\Delta t)/\Delta t]}$$

$$= \frac{\lambda p_n}{\lambda \sum_{n=0}^{K-1} p_n} = \frac{p_n}{1-p_K}.$$

Continuing on with the development of $W_q(t)$ in a fashion similar to $M/M/1/\infty$, we have

$$W_q(t) = \sum_{n=1}^{K-1} q_n \int_0^t \frac{\mu(\mu x)^{n-1}}{(n-1)!} e^{-\mu x} \, dx + W_q(0)$$

$$= \sum_{n=0}^{K-2} q_{n+1} \int_0^t \frac{\mu(\mu x)^n}{n!} e^{-\mu x} \, dx + q_0$$

$$= \sum_{n=0}^{K-2} q_{n+1} \left[1 - \int_t^\infty \frac{\mu(\mu x)^n}{n!} e^{-\mu x} \, dx \right] + q_0$$

$$= \sum_{n=0}^{K-2} q_{n+1} - \sum_{n=0}^{K-2} q_{n+1} \int_t^\infty \frac{\mu(\mu x)^n}{n!} e^{-\mu x} \, dx + q_0.$$

From the analyses similar to those performed on Equation (1.20) which lead to the Poisson CDF, we can see that

$$\int_t^\infty \frac{\mu(\mu x)^n}{n!} e^{-\mu x} \, dx = \sum_{i=0}^n \frac{(\mu t)^i e^{-\mu t}}{i!}.$$

Thus

$$W_q(t) = \sum_{n=0}^{K-2} q_{n+1} - \sum_{n=0}^{K-2} q_{n+1} \sum_{i=0}^n \frac{(\mu t)^i e^{-\mu t}}{i!} + q_0$$

$$= 1 - \sum_{n=0}^{K-2} q_{n+1} \sum_{i=0}^n \frac{(\mu t)^i e^{-\mu t}}{i!}.$$

Knowing ρ, we can get the p_n, hence the q_n. Furthermore, the series

$$\sum_{i=0}^n \frac{(\mu t)^i e^{-\mu t}}{i!}$$

is merely the CDF of a Poisson distribution with parameter μt and therefore for any t, $\Pr\{T_q \leqslant t\} \equiv W_q(t)$ can be evaluated using tables of the Poisson CDF. In fact, cumulative Poisson sums have even been graphed [see Thorndyke (1926)].

The CDF of the waiting time in system can similarly be found and also contains cumulative Poisson sums. Thus from a computational aspect, calculations concerning the probabilities for waiting time in queue or system are readily obtainable.

Let us consider once again H. R. Cutt's situation of Example 2.1. Suppose now that his customers will not wait if they find no seats available; that is, they will go to another barber. What now are the expected system size, expected queue length, expected waiting time for service, and expected time spent in the shop for those customers who enter the shop? Using the results of this section, we find that

$$L = \frac{[5/6]\left[1 - 6(5/6)^5 + 5(5/6)^6\right]}{[1/6]\left[1 - (5/6)^6\right]}$$

$$\doteq 1.97 \qquad \text{(as compared to 5.0 in Example 2.1)},$$

and

$$L_q \doteq 1.97 - \frac{[5/6]\left[1 - (5/6)^5\right]}{1 - (5/6)^6}$$

$$\doteq 1.22 \qquad \text{(as compared to 4.17 in Example 2.1)}.$$

To calculate W we need $\lambda' = \lambda(1 - p_5)$. So

$$p_5 = \frac{[5/6]^5\left[1 - (5/6)\right]}{1 - (5/6)^6}$$

$$\doteq 0.10,$$

and therefore

$$W = \frac{L}{\lambda(1 - p_5)}$$

$$\doteq \frac{1.97}{(5)(0.9)} \doteq 0.438 \text{ hr}$$

$$\doteq 26.3 \text{ min} \qquad \text{(as compared to 60 min)}$$

with

$$W_q = W - \frac{1}{\mu} \doteq 0.438 - 0.167$$

$$\doteq 0.271 \text{ hr} \doteq 16.3 \text{ min} \qquad (\text{as compared to 50 min}).$$

Cutt would also like to know how many customers he loses on the average. To find this he must determine the probability of an arriving customer finding four people waiting (five in the shop). Multiplying this by the average arrival rate λ will give the expected number of customers who arrive to a full shop and leave without entering. Therefore we have

$$\lambda p_5 \doteq 5(0.10) \doteq 0.5 \text{ customers/hr.}$$

So although the system congestion is greatly reduced (smaller L, L_q, W, and W_q), Cutt is losing, on the average, one customer every 2 hr.

2.6 TRANSIENT BEHAVIOR

In this section we consider the transient behavior of two specific queueing systems, namely, $M/M/1/1$ (no one allowed to wait) and $M/M/1/\infty$. This discussion is restricted to these two models, since the mathematics becomes extremely complicated with the slightest relaxation of Poisson-exponential assumptions, and it is our feeling that the exhibition of some fairly simple results is sufficient for our purposes. Even these two transient derivations vary greatly in difficulty. The $M/M/1/1$ solution can be found fairly easily, but the problem becomes much more complicated when the restriction on waiting room is relaxed.

2.6.1 Transient Behavior of $M/M/1/1$

The derivation of the transient probabilities, $\{p_n(t)\}$, that at an arbitrary time t there are n customers in a single-channel system with Poisson input, exponential service, and no waiting room is a straightforward procedure, since $p_n(t) = 0$ for all $n > 1$. It begins in the usual fashion:

$$\Pr\{n = 1 \text{ at } t + \Delta t\}$$

$$= \Pr\{n = 1 \text{ at } t\} \cdot \Pr\{0 \text{ departures and } 0 \text{ arrivals in } (t, t + \Delta t) | n = 1 \text{ at } t\}$$

$$+ \Pr\{n = 0 \text{ at } t\} \cdot \Pr\{1 \text{ arrival in } (t, t + \Delta t) | n = 0 \text{ at } t\}$$

$$+ o(\Delta t),$$

or

$$p_1(t+\Delta t) = p_1(t)(1-\mu\Delta t) + p_0(t)\lambda\Delta t + o(\Delta t), \tag{2.43}$$

and

$$\Pr\{n=0 \text{ at } t+\Delta t\} = \Pr\{n=0 \text{ at } t\}\cdot\Pr\{0 \text{ arrivals in } (t,t+\Delta t)|n=0 \text{ at } t\}$$

$$+\Pr\{n=1 \text{ at } t\}\cdot\Pr\{1 \text{ departure and 0 arrivals in } (t,t+\Delta t)|n=1 \text{ at } t\}$$

$$+o(\Delta t)$$

or

$$p_0(t+\Delta t) = p_0(t)(1-\lambda\Delta t) + p_1(t)\mu\Delta t + o(\Delta t). \tag{2.44}$$

Equations (2.43) and (2.44) may be rewritten as

$$p_1(t+\Delta t) - p_1(t) = -\mu\Delta t p_1(t) + \lambda\Delta t p_0(t) + o(\Delta t)$$

and

$$p_0(t+\Delta t) - p_0(t) = -\lambda\Delta t p_0(t) + \mu\Delta t p_1(t) + o(\Delta t).$$

If then both sides of each of the aforementioned equations are divided by Δt and the limit taken as Δt goes to 0, it can be seen that

$$\frac{dp_1(t)}{dt} = -\mu p_1(t) + \lambda p_0(t) \tag{2.45}$$

and

$$\frac{dp_0(t)}{dt} = -\lambda p_0(t) + \mu p_1(t). \tag{2.46}$$

These differential-difference equations can be solved easily in view of the fact that it is always true that

$$p_0(t) + p_1(t) = 1.$$

Hence Equation (2.45) is equivalent to

$$p_1'(t) = -\mu p_1(t) + \lambda[1 - p_1(t)],\cdot$$

where $p_1'(t) = dp_1(t)/dt$. So

$$p_1'(t) + (\lambda+\mu)p_1(t) = \lambda.$$

This is just an ordinary first order linear differential equation with constant coefficients, whose solution can clearly be seen from the discussion of

Section 1.8 (or Appendix 2, Section A2.1.4) to be

$$p_1(t) = Ce^{-(\lambda+\mu)t} + \frac{\lambda}{\lambda+\mu}.$$

To determine C, we use the boundary value of $p_1(t)$ at $t=0$, that is, $p_1(0)$. Thus

$$C = p_1(0) - \frac{\lambda}{\lambda+\mu},$$

and consequently

$$p_1(t) = \frac{\lambda}{\lambda+\mu}[1 - e^{-(\lambda+\mu)t}] + p_1(0)e^{-(\lambda+\mu)t}. \qquad (2.47)$$

Also,

$$p_0(t) = 1 - p_1(t),$$

and with a little algebra,

$$p_0(t) = \frac{\mu}{\lambda+\mu}[1 - e^{-(\lambda+\mu)t}] + p_0(0)e^{-(\lambda+\mu)t},$$

since $p_0(0) = 1 - p_1(0)$.

The steady-state solution can be found directly from (2.45) and (2.46) in the usual way by letting the derivatives equal zero and then with the use of the fact that $p_0 + p_1 = 1$, solving for p_0 and p_1 (see Section 2.5 and let $K=1$). Alternatively, the equilibrium solution can be found as the limit of the transient as t goes to ∞. That is,

$$p_1 = \lim_{t\to\infty} p_1(t)$$

and

$$p_0 = 1 - p_1.$$

So, from (2.47),

$$p_1 = \frac{\rho}{(\rho+1)}$$

since

$$\lim_{t\to\infty} e^{-(\lambda+\mu)t} = 0;$$

hence

$$p_0 = \frac{1}{\rho+1}.$$

Existence of the equilibrium solution is always assured, independent of the value of $\rho = \lambda/\mu$.

To get an even better feel for the behavior of this queueing system for small values of time, let us graph Equation (2.47). First rewrite (2.47) in the form

$$p_1(t) = a + be^{-ct},$$

where, of course,

$$\begin{cases} a = \dfrac{\lambda}{\lambda + \mu} = \dfrac{\rho}{\rho + 1} \\[3mm] b = \dfrac{-\lambda}{\lambda + \mu} + p_1(0) \\[3mm] c = \lambda + \mu \end{cases}.$$

Therefore we would clearly have Figure 2.2 in the event, for example, that b were greater than 0. We see that $p_1(t)$ is asymptotic to $a = \rho/(\rho + 1) = p_1$.

In addition, it can be interestingly observed that if the initial probability $p_1(0)$ is assumed to be the stationary probability p_1, then b becomes 0 and $p_1(t)$ becomes equal to this equilibrium value of p_1. In other words, the queueing process can be translated into the steady state at any time by making the assumption that the process is already in equilibrium. This property is, in fact, true for any ergodic queueing system, independent of any assumptions about its parameters.

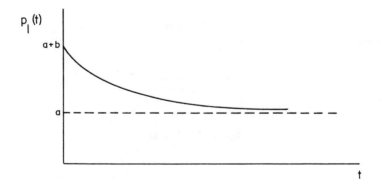

Fig. 2.2 Illustration of transient solution.

2.6.2 Transient Behavior of $M/M/1/\infty$

The transient derivation for $M/M/1/\infty$ is quite a complicated procedure. The solution of this problem post-dated that of the basic Erlang work by nearly half a century, with the first published solution due to Lederman and Reuter (1954), in which they used spectral analysis for the general birth–death process. In the same year, an additional paper appeared on the solution of this problem by Bailey (1954), and later one by Champernowne (1956). Bailey's approach to the time-dependent problem was via generating functions for the partial differential equation and Champernowne's was via advanced combinatorial methods. It is Bailey's approach that has been the most popular over the years and this is basically the one we take.

To begin, let it be assumed that the initial system size at time 0 is i. That is, if $N(t)$ denotes the number in the system at time t, then $N(0)=i$. The differential-difference equations governing the system size, given by (2.5), are

$$
\begin{cases}
p_n'(t) = -(\lambda+\mu)p_n(t) + \lambda p_{n-1}(t) + \mu p_{n+1}(t) & (n>0) \\
p_0'(t) = -\lambda p_0(t) + \mu p_1(t)
\end{cases}
$$

As indicated, we solve these time-dependent equations with probability generating functions. Therefore define

$$
P(z,t) = \sum_{n=0}^{\infty} p_n(t)z^n \qquad (z \text{ complex}),
$$

such that the summation is convergent in and on the unit circle, that is, for $|z| \leqslant 1$. When (2.5) is multiplied through by z^n for *all* $n \geqslant 0$, and then summed on n, from $n=0$ to ∞, it is found that

$$
\sum_{n=1}^{\infty} p_n'(t)z^n + p_0'(t)z^0 = -\lambda \sum_{n=1}^{\infty} p_n(t)z^n - \lambda p_0(t)z^0 - \mu \sum_{n=1}^{\infty} p_n(t)z^n
$$

$$
+ \lambda \sum_{n=1}^{\infty} p_{n-1}(t)z^n + \mu \sum_{n=1}^{\infty} p_{n+1}(t)z^n + \mu p_1(t)z^0,
$$

or equivalently,

$$
\sum_{n=0}^{\infty} p_n'(t)z^n = -\lambda \sum_{n=0}^{\infty} p_n(t)z^n - \mu \sum_{n=1}^{\infty} p_n(t)z^n
$$

$$
+ \lambda \sum_{n=1}^{\infty} p_{n-1}(t)z^n + \mu \sum_{n=0}^{\infty} p_{n+1}(t)z^n.
$$

This can be rewritten as

$$\sum_{n=0}^{\infty} p_n'(t)z^n = -\lambda \sum_{n=0}^{\infty} p_n(t)z^n - \left[\mu \sum_{n=0}^{\infty} p_n(t)z^n - \mu p_0(t) \right]$$

$$+ \lambda z \sum_{n=1}^{\infty} p_{n-1}(t)z^{n-1} + \frac{\mu}{z} \sum_{n=0}^{\infty} p_{n+1}(t)z^{n+1}.$$

Realizing that

$$\sum_{n=0}^{\infty} p_n(t)z^n = \sum_{n=1}^{\infty} p_{n-1}(t)z^{n-1} = \sum_{n=-1}^{\infty} p_{n+1}(t)z^{n+1} = P(z,t),$$

$$\sum_{n=0}^{\infty} p_{n+1}(t)z^{n+1} = \sum_{n=-1}^{\infty} p_{n+1}(t)z^{n+1} - p_0(t) = P(z,t) - p_0(t),$$

and

$$\sum_{n=0}^{\infty} p_n'(t)z^n = \frac{\partial P(z,t)}{\partial t},$$

we have

$$\frac{\partial P(z,t)}{\partial t} = -\lambda P(z,t) - \mu[P(z,t) - p_0(t)]$$

$$+ \lambda z P(z,t) + \frac{\mu}{z}[P(z,t) - p_0(t)]$$

$$= \frac{1-z}{z}[(\mu - \lambda z)P(z,t) - \mu p_0(t)], \tag{2.48}$$

subject to the condition that

$$P(z,0) = z^i.$$

In order to solve this partial differential equation (PDE), let us define the Laplace transforms (see Appendix 3, Section A3.1) with respect to time of $P(z,t)$ and $p_i(t)$ as

$$\mathcal{L}\{P(z,t)\} \equiv \bar{P}(z,s) = \int_0^{\infty} e^{-st} P(z,t) \, dt$$

and

$$\mathcal{L}\{p_i(t)\} \equiv \bar{p}_i(s) = \int_0^{\infty} e^{-st} p_i(t) \, dt \qquad [\text{Re}(s) > 0].$$

The transform of $\partial P(z,t)/\partial t$ can be found simply by integration by parts as

$$\int_0^\infty e^{-st}\frac{\partial P(z,t)}{\partial t}\,dt = \left[e^{-st}P(z,t)\right]_0^\infty + s\int_0^\infty e^{-st}P(z,t)\,dt$$

$$= -z^i + s\bar{P}(z,s).$$

Therefore if the Laplace transform (LT) is taken of both sides of (2.48), it is found that

$$sz\bar{P}(z,s) - z^{i+1} = \left[\lambda z^2 - (\lambda+\mu)z + \mu\right]\bar{P}(z,s)$$

$$-\mu(1-z)\bar{p}_0(s),$$

or

$$\bar{P}(z,s) = \frac{z^{i+1} - \mu(1-z)\bar{p}_0(s)}{(\lambda+\mu+s)z - \mu - \lambda z^2}. \tag{2.49}$$

Since the LT $\bar{P}(z,s)$ converges in the region $|z| \leqslant 1$, $\mathrm{Re}(s) > 0$, wherever the denominator of the right-hand side of (2.49) has zeros in that region, so must the numerator. This fact is henceforth used to evaluate $\bar{p}_0(s)$. The denominator has two zeros since it is a quadratic in z and they are (as functions of s)

$$\left\{ \begin{array}{l} z_1 = \dfrac{\lambda+\mu+s - \sqrt{(\lambda+\mu+s)^2 - 4\lambda\mu}}{2\lambda} \\[4mm] \\ z_2 = \dfrac{\lambda+\mu+s + \sqrt{(\lambda+\mu+s)^2 - 4\lambda\mu}}{2\lambda} \end{array} \right. , \tag{2.50}$$

where the square root is taken so that its real part is positive. It is clear that

$$\left\{ \begin{array}{l} |z_1| < |z_2| \\[3mm] z_1 + z_2 = \dfrac{\lambda+\mu+s}{\lambda} \\[4mm] z_1 z_2 = \dfrac{\mu}{\lambda} \end{array} \right. . \tag{2.51}$$

In order to complete the derivation, the following important and well-known theorem of complex analysis due to Rouché is employed:

THEOREM *If $f(z)$ and $g(z)$ are functions analytic inside and on a closed contour C and if $|g(z)| < |f(z)|$ on C, then $f(z)$ and $f(z) + g(z)$ have the same number of zeros inside C.*

A proof of this theorem may be found in any complex-variable book [for example, see Churchill (1960)].

For $|z| = 1$ and $\text{Re}(s) > 0$ we see that

$$|f(z)| \equiv |(\lambda + \mu + s)z| = |\lambda + \mu + s| > \lambda + \mu \geqslant |\mu + \lambda z^2| \equiv |g(z)|;$$

hence, from Rouché's theorem, $(\lambda + \mu + s)z - \mu - \lambda z^2$ has only *one* zero in the unit circle. This zero is obviously z_1, since $|z_1| < |z_2|$. Thus equating the numerator of the right-hand side of (2.49) to zero for $z = z_1$ gives

$$\bar{p}_0(s) = \frac{z_1^{i+1}}{\mu(1 - z_1)};$$

hence we have finally from (2.49) and the last two lines of (2.51)

$$\bar{P}(z,s) = \frac{z^{i+1} - (1-z)z_1^{i+1}/(1-z_1)}{\lambda(z - z_1)(z_2 - z)} \tag{2.52}$$

$$= \frac{1}{\lambda z_2(1 - z/z_2)} \left[\frac{z^{i+1}(1-z_1) - (1-z)z_1^{i+1}}{(z-z_1)(1-z_1)} \right]$$

$$= \frac{1}{\lambda z_2(1 - z/z_2)} \left[\frac{(z^{i+1} - z_1^{i+1})(1-z_1) + (z-z_1)z_1^{i+1}}{(z-z_1)(1-z_1)} \right]$$

$$= \frac{1}{\lambda z_2(1 - z/z_2)} \left[z^i \left(\frac{1 - (z_1/z)^{i+1}}{1 - z_1/z} \right) + \frac{z_1^{i+1}}{1 - z_1} \right]$$

$$= \frac{1}{\lambda z_2(1 - z/z_2)} \left[z^i \{1 + z_1/z + \cdots + (z_1/z)^i\} + \frac{z_1^{i+1}}{1 - z_1} \right].$$

Writing now $(1 - z/z_2)^{-1}$ as $\sum_{k=0}^{\infty} (z/z_2)^k$ and separating terms yields

$$\bar{P}(z,s) = \frac{1}{\lambda z_2} \left(z^i + z_1 z^{i-1} + \cdots + z_1^i \right) \sum_{k=0}^{\infty} \left(\frac{z}{z_2} \right)^k$$

$$+ \frac{z_1^{i+1}}{\lambda z_2(1 - z_1)} \sum_{k=0}^{\infty} \left(\frac{z}{z_2} \right)^k \qquad (|z/z_2| < 1). \qquad (2.53)$$

Now $\bar{p}_n(s)$ is the coefficient of z^n in the LT of the generating function $P(z,t)$, since by property 1 of Table A3.1, the LT is a linear operator; hence

$$\mathcal{L}\{ P(z,t) \} = \mathcal{L}\left\{ \sum p_n(t) z^n \right\} = \sum \mathcal{L}\{ p_n(t) z^n \} = \sum z^n \mathcal{L}\{ p_n(t) \} = \sum \bar{p}_n(s) z^n.$$

The contribution made to this coefficient by the second term of (2.53) is

$$\frac{z_1^{i+1}}{\lambda z_2^{n+1}(1 - z_1)} = \frac{z_1^{i+1}}{\lambda z_2^{n+1}} (1 + z_1 + z_1^2 + \cdots) = \frac{1}{\lambda} \left(\frac{\lambda}{\mu} \right)^{n+1} \sum_{l=n+i+2}^{\infty} \left(\frac{\mu}{\lambda} \right)^l \frac{1}{z_2^l}$$

$$(2.54)$$

since $|z_1| < 1$ and $z_1 z_2 = \mu/\lambda$ from (2.51).

The first term on the right-hand side of (2.53) yields the remaining coefficients of z^n. But that contribution is different for $n \geqslant i$ and $n < i$. When $n \geqslant i$, all the terms of

$$\frac{1}{\lambda z_2} \left(z^i + z_1 z^{i-1} + \cdots + z_1^i \right)$$

contribute to the coefficient of z^n, specifically, an amount equal to

$$\frac{1}{\lambda z_2} \left(\frac{1}{z_2^{n-i}} + \frac{z_1}{z_2^{n-i+1}} + \frac{z_1^2}{z_2^{n-i+2}} + \cdots + \frac{z_1^i}{z_2^n} \right),$$

which again noting that $z_1 z_2 = \mu/\lambda$ can be rewritten as

$$\frac{1}{\lambda} \left[\frac{1}{z_2^{n-i+1}} + \frac{\mu/\lambda}{z_2^{n-i+3}} + \frac{(\mu/\lambda)^2}{z_2^{n-i+5}} + \cdots + \frac{(\mu/\lambda)^i}{z_2^{n+i+1}} \right].$$

Putting this together with (2.54) gives

$$\bar{p}_n(s) = \frac{1}{\lambda}\left[\frac{1}{z_2^{n-i+1}} + \frac{\mu/\lambda}{z_2^{n-i+3}} + \frac{(\mu/\lambda)^2}{z_2^{n-i+5}} + \cdots + \frac{(\mu/\lambda)^i}{z_2^{n+i+1}} \right.$$

$$\left. + \left(\frac{\lambda}{\mu}\right)^{n+1} \sum_{l=n+i+2}^{\infty} \left(\frac{\mu}{\lambda z_2}\right)^l \right] \quad (2.55)$$

for $n \geqslant i$.

Thus to find $p_n(t)$ for $n \geqslant i$ we must find the inverse of the LT given in (2.55) which we can do term by term since the LT is a linear operator. At this point, the reader should keep in mind the form of z_2 given in (2.50). We must invert terms of the type

$$\frac{K}{z_2^m} = K\left[\frac{\lambda+\mu+s+\sqrt{(\lambda+\mu+s)^2-4\lambda\mu}}{2\lambda} \right]^{-m}.$$

It can be shown that (see Appendix 3, Table A3.2, Number 38)

$$\left(\frac{s+\sqrt{s^2-4\lambda\mu}}{2\lambda} \right)^{-n} = \mathcal{L}\left\{ n\left(\frac{\lambda}{\mu}\right)^{n/2} t^{-1}I_n\left(2\sqrt{\lambda\mu}\,t\right) \right\}, \quad (2.56)$$

where

$$I_n(y) = \sum_{k=0}^{\infty} \frac{(y/2)^{n+2k}}{k!(n+k)!} \quad (n > -1)$$

is the modified Bessel function of the first kind. We present here some properties of Bessel functions which are useful in filling in the details of the derivations to follow.

(i) $I_n(y) = k^{-n}J_n(ky)$, where $J_n(y)$ is the regular Bessel function.

(ii) $(2n/y)I_n(y) = I_{n-1}(y) - I_{n+1}(y)$.

(iii) $I_{-n}(y) = I_n(y)$, n an integer.

(iv) $I_n(y) = y^n/2^n n! + o(y^n)$.

(v) $I_n(y) = e^y/\sqrt{2\pi y} + o(1/y)$.

Using LT properties 1 and 5 from Table A3.1 and (2.56) on (2.55) [see Lederman and Reuter (1956) for all the numerical details], we arrive at the final result for $n \geqslant i$, which is

$$p_n(t) = e^{-(\lambda + \mu)t}\left[\left(\frac{\mu}{\lambda}\right)^{(i-n)/2} I_{n-i}\left(2\sqrt{\lambda\mu}\ t\right)\right.$$

$$+ \left(\frac{\mu}{\lambda}\right)^{(i-n+1)/2} I_{n+i+1}\left(2\sqrt{\lambda\mu}\ t\right)$$

$$\left. + \left(1 - \frac{\lambda}{\mu}\right)\left(\frac{\lambda}{\mu}\right)^n \sum_{l=n+i+2}^{\infty}\left(\frac{\mu}{\lambda}\right)^{l/2} I_l\left(2\sqrt{\lambda\mu}\ t\right)\right].$$

It also turns out that the foregoing is the solution for $n < i$. To indicate that this is the case (precise details are left for Problem 2.23), let us go back and examine Equation (2.53). When the multiplication of $(z^i + z_1 z^{i-1} + \cdots + z_1^i)$ and $\sum_{k=0}^{\infty}(z/z_2)^k$ in the first term of (2.53) is carried out, only those terms of $(z^i + z_1 z^{i-1} + \cdots + z_1^i)$ which give rise to exponents of z that are less than i are counted. But from that point on the derivation is exactly the same as the previous one and ends up leading to the same component in the transform due to a large number of cancellations.

The second term in (2.53) contributes the same as for the $n \geqslant i$ case, namely [see (2.54)],

$$\frac{1}{\lambda}\left(\frac{\lambda}{\mu}\right)^{n+1} \sum_{l=n+i+2}^{\infty}\left(\frac{\mu}{\lambda}\right)^l \frac{1}{z_2^l},$$

which is then appropriately inverted. Thus, finally, we have

$$p_n(t) = e^{-(\lambda + \mu)t}\left[\left(\frac{\mu}{\lambda}\right)^{(i-n)/2} I_{n-i}\left(2\sqrt{\lambda\mu}\ t\right)\right.$$

$$+ \left(\frac{\mu}{\lambda}\right)^{(i-n+1)/2} I_{n+i+1}\left(2\sqrt{\lambda\mu}\ t\right)$$

$$\left. + \left(1 - \frac{\lambda}{\mu}\right)\left(\frac{\lambda}{\mu}\right)^n \sum_{l=n+i+2}^{\infty}\left(\frac{\mu}{\lambda}\right)^{l/2} I_l\left(2\sqrt{\lambda\mu}\ t\right)\right] \qquad (n \geqslant 1). \quad (2.57)$$

It is clear from the foregoing analysis which involved first obtaining a PDE for the probability generating function $P(z,t)$ as given by (2.48), then finding the LT of $P(z,t)$ as given by (2.52), expanding it in a power series whose coefficients are the LT of $p_n(t)$ as given by (2.55), and finally the complex inversion process to obtain $p_n(t)$, that obtaining solutions for more complex models is, in general, extremely arduous and often not even possible.

As an interesting illustration emphasizing the properties of the Bessel functions, we show that (2.57) tends to the steady-state result given by (2.11) as $t \to \infty$ when $\lambda < \mu$. First we employ one of the asymptotic properties given earlier [property (v)], namely,

$$I_n(y) = \frac{e^y}{\sqrt{2\pi y}} + o\left(\frac{1}{y}\right),$$

which is equivalent to the asymptotic approximation

$$I_n(y) \sim \frac{e^y}{\sqrt{2\pi y}}.$$

When $y = 2\sqrt{\lambda\mu}\, t$, this becomes

$$I_n\left(2\sqrt{\lambda\mu}\, t\right) \sim \frac{e^{2\sqrt{\lambda\mu}\, t}}{\sqrt{4\pi\sqrt{\lambda\mu}\, t}}.$$

When this expression is substituted into the first two terms of (2.57) and the definition of I_n given previously as an infinite series is substituted into the third term, it is found that

$$p_n(t) \sim e^{-(\lambda+\mu)t} \left\{ \frac{(\mu/\lambda)^{(i-n)/2} e^{2\sqrt{\lambda\mu}\, t}}{\sqrt{4\pi t}\,(\lambda\mu)^{\frac{1}{4}}} + \frac{(\mu/\lambda)^{(i-n+1)/2} e^{2\sqrt{\lambda\mu}\, t}}{\sqrt{4\pi t}\,(\lambda\mu)^{\frac{1}{4}}} \right.$$

$$\left. + \left(1 - \frac{\lambda}{\mu}\right)\left(\frac{\lambda}{\mu}\right)^n \sum_{l=n+i+2}^{\infty}\left[\left(\frac{\mu}{\lambda}\right)^{l/2} \sum_{k=0}^{\infty} \frac{\left(\sqrt{\lambda\mu}\, t\right)^{l+2k}}{k!(l+k)!} \right] \right\}$$

$$= \frac{e^{-(\sqrt{\lambda}-\sqrt{\mu})^2 t}}{\sqrt{4\pi t}}\, \frac{\mu^{(i-n-1/2)/2}}{\lambda^{(i-n+1/2)/2}} + \frac{e^{-(\sqrt{\lambda}-\sqrt{\mu})^2 t}}{\sqrt{4\pi t}}\, \frac{\mu^{(i-n+1/2)/2}}{\lambda^{(i-n+3/2)/2}}$$

$$+ e^{-(\lambda+\mu)t}\left(1-\frac{\lambda}{\mu}\right)\left(\frac{\lambda}{\mu}\right)^n \sum_{l=n+i+2}^{\infty}\left[\left(\frac{\mu}{\lambda}\right)^{l/2} \sum_{k=0}^{\infty} \frac{\left(\sqrt{\lambda\mu}\, t\right)^{l+2k}}{k!(l+k)!} \right].$$

The first two terms of the aforementioned equation clearly go to 0 as t approaches ∞ since $e^{-(\sqrt{\lambda}-\sqrt{\mu})^2 t}/\sqrt{t}$ goes to 0 independent of the values of λ and μ. The third term behaves somewhat differently due to the presence of infinite series and can be rewritten as

$$\left(1-\frac{\lambda}{\mu}\right)\left(\frac{\lambda}{\mu}\right)^n e^{-(\lambda+\mu)t}\sum_{k=0}^{\infty}\left[\frac{(\lambda t)^k}{k!}\sum_{l=n+i+2}^{\infty}\left(\frac{\mu}{\lambda}\right)^{l/2}\frac{\lambda^{l/2}\mu^{l/2+k}t^{l+k}}{(l+k)!}\right]$$

$$=\left(1-\frac{\lambda}{\mu}\right)\left(\frac{\lambda}{\mu}\right)^n\sum_{k=0}^{\infty}\left[\frac{(\lambda t)^k e^{-\lambda t}}{k!}e^{-\mu t}\sum_{l=n+i+2}^{\infty}\frac{(\mu t)^{l+k}}{(l+k)!}\right]$$

$$=\left(1-\frac{\lambda}{\mu}\right)\left(\frac{\lambda}{\mu}\right)^n\sum_{k=0}^{\infty}\left[\frac{(\lambda t)^k e^{-\lambda t}}{k!}e^{-\mu t}\sum_{m=n+i+2+k}^{\infty}\frac{(\mu t)^m}{m!}\right]$$

$$=\left(1-\frac{\lambda}{\mu}\right)\left(\frac{\lambda}{\mu}\right)^n\sum_{k=0}^{\infty}\left[\frac{(\lambda t)^k e^{-\lambda t}}{k!}e^{-\mu t}\left(e^{\mu t}-\sum_{m=0}^{n+i+1+k}\frac{(\mu t)^m}{m!}\right)\right]$$

$$=\left(1-\frac{\lambda}{\mu}\right)\left(\frac{\lambda}{\mu}\right)^n\sum_{k=0}^{\infty}\left[\frac{(\lambda t)^k e^{-\lambda t}}{k!}\left(1-e^{-\mu t}\sum_{m=0}^{n+i+1+k}\frac{(\mu t)^m}{m!}\right)\right].$$

Realizing that $\sum_{k=0}^{\infty}[(\lambda t)^k e^{-\lambda t}/k!]$ is the total sum of Poisson probabilities which is one, the expression above reduces to

$$\left(1-\frac{\lambda}{\mu}\right)\left(\frac{\lambda}{\mu}\right)^n\left[1-\sum_{k=0}^{\infty}\left(\frac{(\lambda t)^k e^{-\lambda t}}{k!}e^{-\mu t}\sum_{m=0}^{n+i+1+k}\frac{(\mu t)^m}{m!}\right)\right].$$

In taking the limit as $t\to\infty$, we see there are terms of the form $e^{-(\lambda+\mu)t}t^m$ which go to 0 (this can readily be seen by using L'Hôpital's rule m times). Hence the result when $t\to\infty$ is merely

$$\left(1-\frac{\lambda}{\mu}\right)\left(\frac{\lambda}{\mu}\right)^n.$$

When $\lambda/\mu=1$, $p_n=0$ and when $\lambda/\mu>1$, $p_n<0$ for all n, so that only when $\lambda/\mu<1$ do we get a valid steady-state probability distribution. This then agrees with our previous steady-state results of (2.11).

2.7 BUSY PERIOD ANALYSIS

A busy period is defined to begin with the arrival of a customer to an idle channel and to end when the channel next becomes idle. A busy cycle is the sum of a busy period and an adjacent idle period. Since the arrivals are here to be assumed to follow a Poisson process, the distribution of the idle period is exponential with mean $1/\lambda$; hence the CDF of the busy cycle for the $M/M/1$ is the convolution of this negative exponential with the CDF of the busy period itself. Therefore the CDF of the busy period is sufficient to describe the busy cycle also and is found as follows.

The CDF of the busy period is determined by considering the original $M/M/1$ differential-difference equations given in (2.5) with an absorbing barrier imposed at zero system size and an initial size of 1. Then, with a little thought, it should be clear that $p_0(t)$ will, in fact, be the required busy period CDF and $p_0'(t)$ the density. The necessary equations are

$$\begin{cases} p_0'(t) = \mu p_1(t) & \text{[because of the absorbing barrier]} \\ p_1'(t) = -(\lambda+\mu)p_1(t)+\mu p_2(t) & \text{[because of the absorbing barrier]}. \\ p_n'(t) = -(\lambda+\mu)p_n(t)+\lambda p_{n-1}(t)+\mu p_{n+1}(t) & \text{[same as (2.5)]} \end{cases}$$

In a fashion identical to our earlier $M/M/1$ transient derivation, it can be shown (see Problem 2.25) that the LT of the generating function is

$$\bar{P}(z,s) = \frac{z^2-(1-z)(\mu-\lambda z)(z_1/s)}{\lambda(z-z_1)(z_2-z)}, \tag{2.58}$$

where z_1 and z_2 have the same values as before. Now $\bar{p}_0(s)=\mathcal{L}\{p_0(t)\}$ is the very first coefficient of the power series expansion of (2.58) which is $\bar{P}(0,s)$. Thus

$$\bar{p}_0(s) = \bar{P}(0,s)$$

$$= \frac{-\mu z_1/s}{-\lambda z_1 z_2}$$

$$= \frac{\mu}{\lambda s z_2}$$

$$= \frac{2\mu}{s\left(\lambda+\mu+s+\sqrt{(\lambda+\mu+s)^2-4\lambda\mu}\,\right)}.$$

From LT property 2a of Table A3.1, we have

$$\mathcal{L}\{p_0'(t)\} = s\bar{p}_0(s) - p_0(0) = s\bar{p}_0(s).$$

Using this along with (2.56) and property 5 of Table A3.1 we have

$$p_0'(t) = (2\mu)\mathcal{L}^{-1}\left\{\left(\lambda + \mu + s + \sqrt{(\lambda + \mu + s)^2 - 4\lambda\mu}\right)^{-1}\right\}$$

$$= \frac{\mu}{\lambda}\left(\frac{\lambda}{\mu}\right)^{\frac{1}{2}} t^{-1} I_1\left(2\sqrt{\lambda\mu}\, t\right) e^{-(\lambda + \mu)t}$$

$$= \frac{\sqrt{\mu/\lambda}\, e^{-(\lambda + \mu)t} I_1\left(2\sqrt{\lambda\mu}\, t\right)}{t}.$$

To get the average length of the busy period, we simply find the value of the negative of the derivative of the transform of $p_0'(t) = s\bar{p}_0(s)$ evaluated at $s = 0$, since

$$\frac{d}{ds}s\bar{p}_0(s)\bigg|_{s=0} = \frac{d}{ds}\bar{p}_0(s)\bigg|_{s=0} = \frac{d}{ds}\int_0^\infty e^{-st}p_0'(t)\,dt\bigg|_{s=0} = -\int_0^\infty t p_0'(t)\,dt.$$

The derivative is found as follows:

$$\frac{d}{ds}\bar{p}_0(s) = 2\mu\frac{d\left[\left(\lambda + \mu + s + \sqrt{(\lambda + \mu + s)^2 - 4\lambda\mu}\right)^{-1}\right]}{ds}$$

$$= -2\mu\frac{1 + \left[(\lambda + \mu + s)^2 - 4\lambda\mu\right]^{-1/2}[\lambda + \mu + s]}{\left[\lambda + \mu + s + \sqrt{(\lambda + \mu + s)^2 - 4\lambda\mu}\right]^2}.$$

Evaluation of this at zero gives

$$E[T_{\text{busy}}] = \frac{1}{\mu - \lambda}.$$

PROBLEMS

2.1. Derive Equations (2.6) by using the fact that in the steady state the expected rate at which transitions occur out of any state n must equal the rate at which transitions come into state n from elsewhere (this is said to be a derivation by detailed or stochastic balance).

2.2. Supply the steps that yield Equations (2.7).

2.3. For the following generating functions (not necessarily *probability* generating functions) write the sequence which they generate:

(a) $G(z) = 1/(1-z)$.
(b) $G(z) = z/(1-z)$.
(c) $G(z) = e^z$.

2.4. Show that the moment generating function of the sum of independent random variables is equal to the product of their respective moment generating functions.

2.5. Use the result of Problem 2.4 to show that:

(a) The sum of two independent Poisson random variables is a Poisson random variable.

(b) The sum of two independent and identical exponential random variables has a gamma distribution.

(c) The sum of two independent normal random variables is a normal random variable.

2.6. Use the method of operators to solve the difference equation

$$p_{n+4} - 10p_{n+3} + 35p_{n+2} - 50p_{n+1} + 24p_n = 0,$$

subject to $p_0 = 0$, $p_1 = 1$, $p_2 = 0$, and $p_3 = 1$.

2.7. Consider the following questions concerning the composite distribution of line waiting time $w_q(t)$ for an $M/M/1$ model.

(a) Show that $w_q(t)$ obeys the summability-to-one criterion of a valid probability distribution.

(b) Find $E[T_q | T_q > 0]$, that is, the expected time one must wait in the queue, given one has to wait at all.

2.8. Derive $w(t)$ and W (the total-waiting-time density and its expected value) as given by Equations (2.33) and (2.34).

2.9. Equation (1.1) is valid for any queueing system and has been used by Ross (1970) to prove Little's formula for the $GI/G/1$ queue. His approach is to plot the cumulative count of arrivals on the same graph as the cumulative count of departures. Then it can be seen that the area between these two step functions from the beginning of a busy period to the beginning of the next (a busy cycle) is the accumulated total of the waiting times of all the customers who have entered into the system during this busy cycle. Use this argument to derive an empirical version of Little's formula over a busy cycle.

2.10. What effect does doubling λ and μ have on L, L_q, and W in an $M/M/1$ model?

2.11. A graduate research assistant "moonlights" at the short-order counter in the student union snack bar in the evenings. He is the only one on duty at the counter during the hours he works. Arrivals to the counter seem to follow the Poisson distribution with mean of 10/hr. Each customer is served one at a time and the service time follows an exponential distribution with a mean of 4 min. Answer the following questions.

(a) What is the probability of having a queue?
(b) What is the average queue length?
(c) What is the average time a customer spends in the system?
(d) What is the probability of a customer spending more than 5 min in the queue before being waited on?
(e) The graduate assistant would like to spend his idle time grading papers. If he can grade, on an average, twenty-two papers in an hour, how many papers per hour can he average while working his shift?

2.12. A rent-a-car maintenance facility has capabilities for routine maintenance (oil change, lubrication, minor tune-up, wash, etc.) for only one car at a time. Cars arrive for this routine maintenance according to a Poisson process at a mean rate of three per day, and service time to perform this maintenance has an exponential distribution with a mean of $(7/24)$ days. It costs the company a fixed $75 a day to operate the facility. The company estimates loss in profit on a car of $5/day for every day the car is tied up in the shop. The company, by changing certain procedures and hiring faster mechanics, can decrease the mean service time to $1/4$ day. This also increases their operating costs. Up to what value can the operating cost increase before it is no longer economically attractive to make the change?

2.13. Before parts are assembled into a final product, they must be painted. The parts arrive at the painting center, which consists of an automatic painting machine, in a random fashion as a Poisson process at an average rate of λ/hr. It is also observed that the painting process appears to be exponential with an average time of service equal to $1/\mu$ hr. It is estimated that it costs the company about $$C_1$/unit/hr spent tied up in the painting center. The cost of owning and operating the painting machine is strictly a function of the speed of its operation, and, in particular, one that works at an average rate of μ is charged at the rate of $\$\mu C_2/$hr, whether or not it is always in operation. How large should μ be to allow this firm to minimize the cost of the painting operation?

2.14. Show that the two expressions for the effective mean arrival rate, λ', in the $M/M/1/K$ model are equal; that is, $\lambda(1 - p_k) = \mu(L - L_q)$.

2.15. Find the probability that a customer's wait in queue exceeds 20 min for an $M/M/1/3$ model with $\lambda = 4/$hr and $1/\mu = 15$ min.

2.16. A small drive-it-through-yourself car wash, in which the next car cannot go through the washing procedure until the car in front is completely finished, has a capacity to hold on its grounds a maximum of ten cars (including the one in wash). The company has found its arrivals to be Poisson with mean rate of 20 cars/hr, and its service times to be exponential with a mean of 12 min. What is the average number of cars lost to the car wash firm every 10-hour day as a result of its capacity limitation?

2.17. Under the assumption that customers will not wait if no seats are available, Barber Cutt can rent, on Saturday, the conference room of a small computer software firm adjacent to his shop for $4.00 (cost of cleanup crew on a Saturday). His shop is open on Saturdays from 8:00 a.m. to 2:00 p.m. and his marginal profit is $2.25 per haircut. This office can seat an additional four people. Should Cutt rent?

2.18. Find the Laplace transforms of the following functions:

(a) e^{kt}.
(b) $t^{k-1}/\Gamma(k)$.
(c) $\sin kt$.
(d) $\int_0^a [(1 - \cos tx)x^2]dx$.

2.19. Use the properties of Laplace transforms and the tables in Appendix 3 to find the functions whose Laplace transforms are the following:

(a) $(s + 1)/(s^2 + 2s + 2)$.
(b) $1/(s^2 - 3s + 2)$.
(c) $1/[s^2(s^2 + 1)]$.
(d) $e^{-s}/(s + 1)$.

2.20. Prove property 5 of Table A3.1, namely, that if $\bar{f}(s)$ is the Laplace transform of $f(t)$, then $\bar{f}(s + a)$ is the Laplace transform of $e^{-at}f(t)$.

2.21. Show Property 1 of Table A3.1, namely, that $\mathcal{L}\{\Sigma_i a_i f_i(t)\} = \Sigma_i a_i \bar{f}_i(s)$.

2.22. Derive the Poisson frequency function from Equation (1.15) with the use of the generating function $P(z,t) \equiv \sum_{n=0}^{\infty} p_n(t) z^n$.

2.23. Finish the $M/M/1/\infty$ transient derivation for $n < i$.

2.24. Show that the probability that the number in an $M/M/1/\infty$ system at an arbitrary point in time is greater than or equal to j is given by

$$\sum_{n=j}^{\infty} p_n(t) = e^{-(\lambda+\mu)t} \left[\sum_{n=j-i}^{\infty} \left(\frac{\lambda}{\mu}\right)^{n/2} I_n(2\sqrt{\lambda\mu}\, t) + \sum_{n=j+i+1}^{\infty} \left(\frac{\lambda}{\mu}\right)^{j-n/2} I_n(2\sqrt{\lambda\mu}\, t) \right].$$

2.25. Derive Equation (2.58).

2.26. Using expected-value arguments directly, show that in a $M/M/1$ queue, the expected lengths of the idle and busy periods are $1/\lambda$ and $1/(\mu-\lambda)$, respectively.

Chapter 3

SIMPLE
MARKOVIAN BIRTH–DEATH
QUEUEING MODELS

The main intent of this chapter is to broaden the class of queueing models that are amenable to analytic methods. The particular models to be considered here are the special types of Markovian models known as birth–death processes. Markov models are essentially those in which future behavior depends only on the present and not the past, and birth–death models can be characterized further by the additional property that the only possible changes of system state at any point in time are ± 1 (the models of Chapter 2 fall into this category). The approach used for each model is basically via a Chapman-Kolmogorov (C-K) equation. All Markov processes[1] satisfy a C-K equation (as well as certain non-Markov processes) and it is usually a fairly simple matter to derive one. This will lead to a straightforward steady-state derivation, though, more often than not, the transient will be difficult to obtain. The methodology used in Chapter 2 for $M/M/1$ models serves as a good example. It is essentially this methodology which we now apply to more general models.

A Markov chain is a Markov process with discrete state space, and the C-K equation for a Markov chain takes the following form for any times $s > t > u \geqslant 0$ and states j and k:

$$p_{jk}(u,s) = \sum_{\substack{\text{states} \\ i}} p_{ji}(u,t)p_{ik}(t,s), \qquad (3.1)$$

where $p_{jk}(u,s)$ is the probability of moving from state j to k in time beginning at u and ending at s, and the summation is over all states of the chain. This result should be fairly intuitive and says that the chain can reach state k at time s by starting from state j at time u and stopping off at time t at any other possible state i.

---

[1] A Markov process is a stochastic process with the Markovian memoryless property; that is, the $\Pr\{a < X(t) \leqslant b | X(t_1) = x_1, X(t_2) = x_2, \ldots, X(t_n) = x_n\} = \Pr\{a < X(t) \leqslant b | X(t_n) = x_n\}$ where $t_1 < t_2 < \cdots < t_n < t$ (see Appendix 4).

There is an additional theory which takes one from the C-K equation to two differential equations, which are called the forward and backward equations. If the transition probability functions $p_{jk}(u,s)$ of the chain have the additional properties that there exist continuous functions $q_j(t)$ and $q_{jk}(t)$ such that

$$
\left\{
\begin{array}{l}
(i) \quad \Pr\{\text{a change of state in } (t,t+\Delta t)\} = 1 - p_{jj}(t,t+\Delta t) \\[2mm]
\qquad\qquad\qquad\qquad\qquad = q_j(t)\Delta t + o(\Delta t) \\[4mm]
\quad (\text{that is, the probability that the system leaves state } j \text{ in a} \\
\quad \text{small interval of width } \Delta t \text{ is basically linearly proportional} \\
\quad \text{to } \Delta t, \text{ with proportionality constant a function of } j \text{ and } t) \\[4mm]
(ii) \quad p_{jk}(t,t+\Delta t) = q_{jk}(t)\Delta t + o(\Delta t)
\end{array}
\right. \qquad (3.2)
$$

then Equation (3.1) leads to (see Problem 3.1)

$$
\frac{\partial}{\partial t} p_{jk}(u,t) = -q_k(t)p_{jk}(u,t) + \sum_{i \neq k} p_{ji}(u,t)q_{ik}(t) \qquad (3.3a)
$$

and

$$
\frac{\partial}{\partial u} p_{jk}(u,t) = q_j(u)p_{jk}(u,t) - \sum_{i \neq j} q_{ji}(u)p_{ik}(u,t). \qquad (3.3b)
$$

These two differential equations are known, respectively, as Kolmogorov's forward and backward equations. Both (3.3a) and (3.3b) are quite helpful in the development of Markovian models and they are referred to later in the text.

Since the state space of any queue is the nonnegative integers (or a proper subset thereof), the Markovian nature of the system may be more specifically categorized as a continuous-parameter (time) Markov chain (see Appendix 4) with states $0,1,2,\ldots$. A good percentage of the relevant models have the additional birth–death property that net changes across infinitesimal time intervals can never be other than $-1, 0,$ or $+1$, such that

$$
\left\{
\begin{array}{ll}
\Pr\{\text{increase from } E_n \text{ to } E_{n+1} \text{ in } (t,t+\Delta t)\} = \lambda_n \Delta t + o(\Delta t) & (n \geqslant 0) \\[2mm]
\Pr\{\text{decrease from } E_n \text{ to } E_{n-1} \text{ in } (t,t+\Delta t)\} = \mu_n \Delta t + o(\Delta t) & (n \geqslant 1)
\end{array}
\right. .
$$

Hence

$$
\Pr\{\text{no change in } (t,t+\Delta t)\} = 1 - (\lambda_n + \mu_n)\Delta t + o(\Delta t);
$$

therefore

$$\left\{ \begin{array}{l} q_{n,n+1}(t) = \lambda_n \\[2mm] q_{n,n-1}(t) = \mu_n \qquad (\mu_n \neq 0), \\[2mm] q_n(t) = \lambda_n + \mu_n \quad (q_0(t) = \lambda_0) \end{array} \right.$$

so that it is possible to get either a forward or backward Kolmogorov equation. When $\lambda_n = \lambda$, $\mu_n = \mu$, we have the $M/M/1$ models of Chapter 2 and, indeed, the equations of (2.5) are forward Kolmogorov equations.[2]

These birth–death processes are studied in some depth in the following paragraphs in a general format. Then specific applications of these processes to queueing are the subject of Sections 3.2 through 3.8. Additional and more advanced Markovian models will be the subject of Chapter 4.

3.1 BIRTH–DEATH PROCESSES

The analysis here is very similar to that of Section 2.1. We begin by looking at the state probabilities, $p_n(t) = \Pr\{\text{state of process is } E_n \text{ at time } t\}$, for an arbitrary birth–death process.[3] We remember that such a process is Markovian and that instantaneous changes in the system state can only amount to an increase (birth) or decrease (death) of one. The probability of a birth occurring in a small interval of length Δt which began with the system in state E_n is assumed to be $\lambda_n \Delta t + o(\Delta t)$, while that of a death is assumed to be $\mu_n \Delta t + o(\Delta t)$ $(n \geq 1)$, independent of λ_n and t.

To write a difference equation for $p_n(t)$, let us first consider how the system may get to state E_n at time $t + \Delta t$. To do so, the system might have been in E_n at t and had no net change during Δt (i.e., had 0 births and 0 deaths and/or 1 birth and 1 death). Or, the system might have found itself in E_{n-1} and had a birth, or in E_{n+1} and had a death. Hence, for $n \geq 1$,

$$\begin{aligned} p_n(t + \Delta t) = \; & p_n(t)[1 - \lambda_n \Delta t][1 - \mu_n \Delta t] \\ & + p_n(t)[\lambda_n \Delta t][\mu_n \Delta t] \\ & + p_{n+1}(t)[\mu_{n+1}\Delta t][1 - \lambda_{n+1}\Delta t] \\ & + p_{n-1}(t)[\lambda_{n-1}\Delta t][1 - \mu_{n-1}\Delta t] \\ & + o(\Delta t) \qquad (n \geq 1). \end{aligned}$$

[2] In their most general form, λ_n and μ_n may be functions of time, and the resultant process is referred to as nonhomogeneous. However, in this text only time independent λ_n and μ_n (homogeneous birth–death process) are considered.

[3] Recall that for basic queueing models E_n is synonymous to the system size n.

For $n = 0$,

$$p_0(t + \Delta t) = p_0(t)[1 - \lambda_0 \Delta t] + p_1(t)[\mu_1 \Delta t][1 - \lambda_1 \Delta t] + o(\Delta t).$$

The corresponding differential-difference equations are found by dividing through by Δt and taking the limit as $\Delta t \to 0$. They are

$$\begin{cases} \dfrac{dp_n(t)}{dt} = -(\lambda_n + \mu_n)p_n(t) + \mu_{n+1}p_{n+1}(t) + \lambda_{n-1}p_{n-1}(t) & (n \geq 1) \\[4mm] \dfrac{dp_0(t)}{dt} = -\lambda_0 p_0(t) + \mu_1 p_1(t) \end{cases} \tag{3.4}$$

and are forms of the forward Kolmogorov equation. Letting $\lambda_n = \lambda$ and $\mu_n = \mu$ yields (2.5).

The stationary or steady-state solution is found in the same manner as for $M/M/1$, which was presented in Section 2.1. Since $p_n(t)$ is to be independent of time, $dp_n(t)/dt$ is zero and (3.4) becomes

$$\begin{cases} 0 = -(\lambda_n + \mu_n)p_n + \mu_{n+1}p_{n+1} + \lambda_{n-1}p_{n-1} & (n \geq 1) \\[4mm] 0 = -\lambda_0 p_0 + \mu_1 p_1 \end{cases}$$

or

$$\begin{cases} p_{n+1} = \dfrac{\lambda_n + \mu_n}{\mu_{n+1}} p_n - \dfrac{\lambda_{n-1}}{\mu_{n+1}} p_{n-1} & (n \geq 1) \\[4mm] p_1 = \dfrac{\lambda_0}{\mu_1} p_0 \end{cases} \tag{3.5}$$

We proceed to the solution by iteration on (3.5), namely,

$$p_2 = \frac{\lambda_1 + \mu_1}{\mu_2} p_1 - \frac{\lambda_0}{\mu_2} p_0$$

$$= \frac{\lambda_1 + \mu_1}{\mu_2} \frac{\lambda_0}{\mu_1} p_0 - \frac{\lambda_0}{\mu_2} p_0$$

$$= \frac{\lambda_1 \lambda_0}{\mu_2 \mu_1} p_0,$$

$$p_3 = \frac{\lambda_2 + \mu_2}{\mu_3} p_2 - \frac{\lambda_1}{\mu_3} p_1$$

$$= \frac{\lambda_2 + \mu_2}{\mu_3} \frac{\lambda_1 \lambda_0}{\mu_2 \mu_1} p_0 - \frac{\lambda_1}{\mu_3} \frac{\lambda_0}{\mu_1} p_0$$

$$= \frac{\lambda_2 \lambda_1 \lambda_0}{\mu_3 \mu_2 \mu_1} p_0, \dots .$$

The pattern which appears to be emerging is that

$$p_n = \frac{\lambda_{n-1} \lambda_{n-2} \cdots \lambda_0}{\mu_n \mu_{n-1} \cdots \mu_1} p_0 \qquad (n \geqslant 1)$$

$$= p_0 \prod_{i=1}^{n} \frac{\lambda_{i-1}}{\mu_i} \tag{3.6}$$

and we guess that this is the correct formula. To verify that (3.6) is, in fact, proper, mathematical induction will be used on Equation (3.5).

First, the formula is clearly correct for $n = 0$. We have also shown it to be true for $n = 1, 2,$ and 3. If it is assumed to be true for $n = k (\geqslant 0)$, then it is seen that

$$p_{k+1} = \frac{\lambda_k + \mu_k}{\mu_{k+1}} p_0 \prod_{i=1}^{k} \frac{\lambda_{i-1}}{\mu_i} - \frac{\lambda_{k-1}}{\mu_{k+1}} p_0 \prod_{i=1}^{k-1} \frac{\lambda_{i-1}}{\mu_i}$$

$$= \frac{p_0 \lambda_k}{\mu_{k+1}} \prod_{i=1}^{k} \frac{\lambda_{i-1}}{\mu_i} + p_0 \frac{\mu_k}{\mu_{k+1}} \prod_{i=1}^{k} \frac{\lambda_{i-1}}{\mu_i} - p_0 \left[\prod_{i=1}^{k} \frac{\lambda_{i-1}}{\mu_i} \right] \frac{\mu_k}{\mu_{k+1}}$$

$$= p_0 \prod_{i=1}^{k+1} \frac{\lambda_{i-1}}{\mu_i},$$

and the proof by induction is complete. Since $\sum_{n=0}^{\infty} p_n$ must be 1, that is,

$$1 = p_0 \left[1 + \sum_{n=1}^{\infty} \prod_{i=1}^{n} \frac{\lambda_{i-1}}{\mu_i} \right], \tag{3.7}$$

a necessary and sufficient condition for the existence of a steady-state solution is the convergence of the infinite series

$$\sum_{n=1}^{\infty} \prod_{i=1}^{n} \frac{\lambda_{i-1}}{\mu_i}.$$

For $M/M/1$, $\lambda_i = \lambda$ and $\mu_i = \mu$, and the series simplifies to the recognizable $\sum_{n=1}^{\infty} (\lambda/\mu)^n$ so that the condition is again $\lambda/\mu = \rho < 1$.

As a further illustration of these concepts, let it be assumed that $\lambda_n = \lambda$ and $\mu_n = n\mu$. Then, from (3.6),

$$p_n = p_0 \prod_{i=1}^{n} \frac{\lambda}{i\mu}$$

$$= p_0 \frac{\lambda^n}{n! \mu^n}.$$

The series $\sum_{n=1}^{\infty} \prod_{i=1}^{n} (\lambda_{i-1}/\mu_i)$ becomes $\sum_{n=1}^{\infty} [(\lambda/\mu)^n / n!]$ which is obviously $e^{\lambda/\mu} - 1$ for all finite values of λ/μ; thus $p_0 = e^{-\lambda/\mu}$. So, finally $p_n = e^{-\lambda/\mu}(\lambda/\mu)^n / n!$, which is a Poisson distribution with parameter λ/μ, and holds for *all* finite values of λ/μ. This particular model is discussed further in Sections 3.5.1 and 3.7.

The transient solutions to birth–death processes are, for the most part, extremely difficult to obtain. There are a few isolated cases for which tractable solutions do exist (i.e., no worse than that for $M/M/1$), and we discuss them in the context of the particular model involved, rather than at this juncture.

3.2 QUEUES WITH PARALLEL CHANNELS ($M/M/c$)

We now turn our attention to the multiserver model in which each server has an independently and identically distributed exponential service-time distribution, with the arrival process again assumed to be Poisson. We consider the case in which there are c servers and make use of the theory developed in Section 3.1, specifically Equation (3.6), to develop this $M/M/c$ model. Clearly, since the input is Poisson and the service exponential, we have a birth–death process. Hence $\lambda_n = \lambda$ for all n, and it remains to determine μ_n prior to being able to use Equation (3.6). Thus consider the mean service rate of this system. If there are more than c customers in the system, all c servers are busy and each is putting out at a mean rate μ, and the mean system output rate is thus $c\mu$. When there are fewer than c customers in the system, say $n < c$, only n of the c servers are busy and the system is putting out at a mean rate $n\mu$. Hence μ_n may be written as

$$\mu_n = \begin{cases} n\mu & (1 \leqslant n < c) \\ c\mu & (n \geqslant c) \end{cases} . \tag{3.8}$$

Utilizing (3.8) in (3.6) and the fact that $\lambda_n = \lambda$ for all n, we obtain

$$p_n = \begin{cases} \dfrac{\lambda^n}{n!\mu^n} p_0 & (1 \leqslant n \leqslant c) \\ \dfrac{\lambda^n}{c^{n-c}c!\mu^n} p_0 & (n \geqslant c) \end{cases} . \tag{3.9}$$

In order to find p_0, we must again use the boundary condition

$$\sum_{n=0}^{\infty} p_n = 1,$$

which gives

$$p_0 \left[\sum_{n=0}^{c-1} \frac{\lambda^n}{n!\mu^n} + \sum_{n=c}^{\infty} \frac{\lambda^n}{c^{n-c}c!\mu^n} \right] = 1.$$

Define $r = \lambda/\mu$ and $\rho = r/c = \lambda/c\mu$. Then we have

$$p_0 \left[\sum_{n=0}^{c-1} \frac{r^n}{n!} + \sum_{n=c}^{\infty} \frac{r^n}{c^{n-c}c!} \right] = 1.$$

Consider now the series

$$\sum_{n=c}^{\infty} \frac{r^n}{c^{n-c}c!} = \frac{r^c}{c!} \sum_{n=c}^{\infty} \left(\frac{r}{c} \right)^{n-c}$$

$$= \frac{r^c}{c!} \sum_{m=0}^{\infty} \left(\frac{r}{c} \right)^m$$

$$= \frac{r^c}{c!} \frac{1}{(1 - r/c)} \qquad (r/c = \rho < 1). \tag{3.10}$$

Therefore we can write

$$
p_0 = \left[\sum_{n=0}^{c-1} \frac{r^n}{n!} + \frac{cr^c}{c!(c-r)} \right]^{-1} \quad (r/c = \rho < 1)
$$

$$
= \left[\sum_{n=0}^{c-1} \frac{1}{n!} \left(\frac{\lambda}{\mu}\right)^n + \frac{1}{c!} \left(\frac{\lambda}{\mu}\right)^c \left(\frac{c\mu}{c\mu - \lambda}\right) \right]^{-1}
$$

(3.11)

Note that, here, the condition for existence of a steady-state solution is $\lambda/(c\mu) < 1$; that is, the mean arrival rate must be less than the mean maximum potential service rate of the system, which is intuitively what we would expect. Also note that when $c = 1$, (3.11) reduces to the $M/M/1/\infty$ equation.

We can now derive measures of effectiveness for the $M/M/c/\infty$ model utilizing the steady-state probabilities given by Equations (3.9) and (3.11) in a manner similar to that used for the $M/M/1/\infty$ model in Section 2.2. We first consider the expected queue size L_q as it is computationally easier to determine than L, since we have only to deal with p_n's for $n \geqslant c$. Thus, by definition,

$$
L_q = \sum_{n=c}^{\infty} (n-c)p_n
$$

$$
= \sum_{n=c}^{\infty} \frac{n}{c^{n-c}c!} r^n p_0 - \sum_{n=c}^{\infty} \frac{c}{c^{n-c}c!} r^n p_0 \quad .
$$

(3.12)

Next, consider the first series on the right-hand side of Equation (3.12):

$$
\frac{p_0}{c!} \sum_{n=c}^{\infty} \frac{nr^n}{c^{n-c}} = \frac{p_0}{c!} \left[\frac{r^{c+1}}{c} \right] \left[\sum_{n=c}^{\infty} (n-c)\left(\frac{r}{c}\right)^{n-c-1} + \sum_{n=c}^{\infty} c\left(\frac{r}{c}\right)^{n-c-1} \right]
$$

$$
= \frac{p_0}{c!} \left[\frac{r^{c+1}}{c} \right] \left[\frac{1}{(1-r/c)^2} + \frac{c^2/r}{1-r/c} \right],
$$

(3.13)

since the derivative of the geometric series $\sum_{n=0}^{\infty} x^n$ is $\sum_{n=0}^{\infty} nx^{n-1}$ and has the sum $1/(1-x)^2$.

The second series on the right-hand side of Equation (3.12) can be found from Equation (3.10) as

$$\frac{p_0}{c!} \sum_{n=c}^{\infty} \frac{cr^n}{c^{n-c}} = \frac{p_0 cr^c}{c!(1-r/c)}. \tag{3.14}$$

Finally, substituting Equations (3.13) and (3.14) into Equation (3.12) we get

$$L_q = p_0 \frac{r^{c+1}/c}{c!} \left[\frac{1}{(1-r/c)^2} + \frac{c^2/r}{1-r/c} - \frac{c^2/r}{1-r/c} \right]$$

or simply

$$\boxed{\begin{aligned} L_q &= \left[\frac{r^{c+1}/c}{c!(1-r/c)^2} \right] p_0 \\ &= \left[\frac{(\lambda/\mu)^c \lambda\mu}{(c-1)!(c\mu-\lambda)^2} \right] p_0 \end{aligned}} \quad . \tag{3.15}$$

We could next go on to find L in a similar fashion by using

$$L = \sum_{n=0}^{\infty} n p_n.$$

However, it is easier to employ Little's formula to find W_q, then use W_q to find W ($W = W_q + 1/\mu$), and finally to utilize Little's formula again to find L. Thus using Little's formula and Equation (3.15) we get[4]

$$\boxed{W_q = \frac{L_q}{\lambda} = \left[\frac{(\lambda/\mu)^c \mu}{(c-1)!(c\mu-\lambda)^2} \right] p_0} \quad . \tag{3.16}$$

[4]We again assume a FIFO discipline, but for multichannel queues, FIFO must be interpreted as first in, first out of queue and into service (the usual first come, first served). Because of the multichannel facilities, it is not necessarily true that a customer first into service will be first out of the system.

Now using the relationship between W and W_q gives

$$W = \frac{1}{\mu} + \left[\frac{(\lambda/\mu)^c \mu}{(c-1)!(c\mu-\lambda)^2} \right] p_0 \quad , \qquad (3.17)$$

and using Little's formula on Equation (3.17) yields finally[5]

$$L = \lambda W = \frac{\lambda}{\mu} + \left[\frac{(\lambda/\mu)^c \lambda\mu}{(c-1)!(c\mu-\lambda)^2} \right] p_0$$

$$= r + \left[\frac{r^{c+1}/c}{c!(1-r/c)^2} \right] p_0 \qquad . \qquad (3.18)$$

Although the probability distributions of the waiting times, $W(t)$ and $W_q(t)$, are not needed to get W and W_q, it is desirable to obtain them whenever possible so that questions concerning probabilities of waits greater than specified amounts can be answered. Therefore we next turn our attention to the derivation of these distributions, and proceed in a manner similar to that of Section 2.3. Let T_q represent the random variable "time spent waiting in the queue" and $W_q(t)$ denote its CDF. Therefore we have

$$W_q(0) = \Pr\{ T_q = 0 \}$$

$$= \Pr\{ c - 1 \text{ or less in the system} \}$$

$$= \sum_{n=0}^{c-1} p_n$$

$$= p_0 \sum_{n=0}^{c-1} \frac{r^n}{n!} .$$

[5]This result could have been obtained by using $L = L_q + \lambda/\mu$ which holds whenever Little's formula is valid as shown in Section 2.4 and mentioned in Table 2.1.

It is now necessary to evaluate $\sum_{n=0}^{c-1} r^n/n!$. This can be done by using Equation (3.11) and we find

$$\sum_{n=0}^{c-1} \frac{r^n}{n!} = \frac{1}{p_0} - \frac{cr^c}{c!(c-r)} \qquad (3.19)$$

so that

$$W_q(0) = p_0 \left[\frac{1}{p_0} - \frac{cr^c}{c!(c-r)} \right]$$

$$= 1 - \frac{cr^c p_0}{c!(c-r)}$$

$$= 1 - \frac{c(\lambda/\mu)^c}{c!(c-\lambda/\mu)} p_0.$$

For $T_q > 0$, then,

$$W_q(t) = \Pr\{T_q \leqslant t\}$$

$$= \sum_{n=c}^{\infty} [\Pr\{n-c+1 \text{ completions in } \leqslant t | \text{arrival}$$

$$\text{found } n \text{ in system}\} \cdot p_n] + W_q(0) \qquad (t>0).$$

Now when $n \geqslant c$, the system average output rate is Poisson with a mean $c\mu$, so that the time between successive completions is exponential with mean $1/(c\mu)$, and the distribution of the time for $n-c+1$ completions is Erlang type $(n-c+1)$. Thus we can write

$$W_q(t) = p_0 \sum_{n=c}^{\infty} \frac{r^n}{c^{n-c}c!} \int_0^t \frac{\mu c(\mu cx)^{n-c}}{(n-c)!} e^{-\mu cx} dx + W_q(0) \qquad (t>0)$$

$$= p_0 \frac{r^c}{(c-1)!} \int_0^t \mu e^{-\mu cx} \sum_{n=c}^{\infty} \frac{(\mu rx)^{n-c}}{(n-c)!} dx + W_q(0)$$

$$= p_0 \frac{r^c}{(c-1)!} \int_0^t \mu e^{-\mu cx} e^{\mu rx} dx + W_q(0)$$

$$W_q(t) = p_0 \frac{r^c}{(c-1)!} \int_0^t \mu e^{-\mu x(c-r)} dx + W_q(0)$$

$$= \frac{r^c(1 - e^{-\mu(c-r)t})}{(c-1)!(c-r)} p_0 + W_q(0) \quad (t > 0)$$

$$= \frac{(\lambda/\mu)^c(1 - e^{-(\mu c - \lambda)t})}{(c-1)!(c - \lambda/\mu)} p_0 + W_q(0) \quad (t > 0).$$

Summarizing,

$$W_q(t) = \begin{cases} 1 - \dfrac{c(\lambda/\mu)^c}{c!(c - \lambda/\mu)} p_0 & (t=0) \\ \dfrac{(\lambda/\mu)^c(1 - e^{-(\mu c - \lambda)t})}{(c-1)!(c - \lambda/\mu)} p_0 + W_q(0) & (t>0) \end{cases} \quad . \quad (3.20)$$

The reader should note that letting $c = 1$ reduces (3.20) to the equation for $W_q(t)$ of the $M/M/1$ model given by (2.31). This, of course, is true for all $M/M/c$ equations.

We leave as an exercise (see Problem 3.5) to show that

$$W_q = E[T_q] = \int_0^\infty t\, dW_q(t) = \left[\frac{(\lambda/\mu)^c \mu}{(c-1)!(c\mu - \lambda)^2} \right] p_0, \quad (3.21)$$

as given by (3.16). Once again, as in the $M/M/1$ case, $dW_q(t)/dt$ for $t > 0$ does not integrate to one over the range $(0, \infty)$ and is not a true density function by itself. However, as before, the addition of the point $t = 0$ does yield a valid composite probability distribution, namely,

$$w_q(t) = \begin{cases} 1 - \dfrac{c(\lambda/\mu)^c}{c!(c - \lambda/\mu)} p_0 & (t=0) \\ \dfrac{(\lambda/\mu)^c \mu e^{-(\mu c - \lambda)t}}{(c-1)!} p_0 & (t>0) \end{cases} \quad . \quad (3.22)$$

In a similar manner, the total waiting-time-in-system density function can be derived (see Problem 3.7) and is given as

$$
w(t) = \frac{\mu e^{-\mu t}[\lambda - c\mu + \mu W_q(0)] - [1 - W_q(0)][\lambda - c\mu]\mu e^{-(c\mu - \lambda)t}}{\lambda - (c-1)\mu}
$$

$$(t > 0)$$

.(3.23)

We now illustrate the foregoing developments with an example.

Example 3.1

County Hospital's ophthalmology clinic offers free glaucoma tests every Tuesday evening. There are three ophthalmologists on duty. A glaucoma test takes, on the average, 20 min, and the actual time is found to be approximately exponentially distributed around this average. Clients arrive according to a Poisson process with a mean of 6/hr, and patients are taken on a first-come, first-served basis. The hospital planners are interested in knowing: (1) how many people, on the average, are waiting; (2) the average amount of time a patient spends at the clinic; and (3) the average percentage idle time of each of the doctors. Thus we wish to calculate L_q, W, and the percentage idle time of a server. We first proceed by calculating p_0, since this factor appears in all the formulas derived for the measures of effectiveness. We have that $c = 3$, $\lambda = 6/hr$, and $\mu = 1/20$ min $= 3/hr$. Thus $r = \lambda/\mu = 2$, and, from Equation (3.11),

$$
p_0 = \left[1 + 2 + \frac{(2)^2}{2!} + \frac{3(2)^3}{3!(3-2)} \right]^{-1} = \frac{1}{9}.
$$

Now we find from Equation (3.15) that

$$
L_q = \left[\frac{(2)^4/3}{3!(1-2/3)^2} \right]\left(\frac{1}{9}\right) = \frac{8}{9} \doteq 0.889,
$$

and from Equation (3.17) that

$$
W = \frac{1}{3} + \left[\frac{(2)^3 \cdot 3}{2!(3 \cdot 3 - 6)^2} \right]\left(\frac{1}{9}\right) = \frac{13}{27} \text{ hrs}
$$

$$\doteq 28.9 \text{ min.}$$

To find the average percentage idle time of a doctor, we realize that when the system is empty all three doctors are idle, when one person only is in the clinic, two of the three doctors are idle, when two people are in the clinic only one doctor is idle, and, of course, when three or more are in the clinic all doctors are busy. Assuming random selection of doctors when more than one is idle, we get, using Equation (3.9),

$$\Pr\{\text{any server idle}\} = 1p_0 + \left(\frac{2}{3}\right)p_1 + \left(\frac{1}{3}\right)p_2$$

$$= \frac{1}{9} + \left(\frac{2}{3}\right)\left(\frac{2}{1!}\right)\left(\frac{1}{9}\right) + \left(\frac{1}{3}\right)\left(\frac{2^2}{2!}\right)\left(\frac{1}{9}\right)$$

$$= \frac{1}{3}.$$

Thus one third of his time a doctor has no patients to see. Note also that this turns out to be equal to $(1 - \lambda/c\mu)$ and further that this is not a coincidence (see Problem 3.8). In fact, the probability of any server being idle in any $M/M/c$ system is always $1 - \lambda/c\mu$, or equivalently, the probability of any server being busy (average fraction of a server's time spent serving) is $\lambda/c\mu$. The factor $\lambda/c\mu$ is often referred to as the system utilization factor or traffic intensity. One additional observation should be made, namely, that the fraction of time there is at least one idle server is not $1 - \lambda/c\mu$, but $\sum_{n=0}^{c-1} p_n$. In this example the fraction of time at least one doctor is idle is

$$\Pr\{\text{at least one server idle}\} = p_0 + p_1 + p_2$$

$$= \frac{5}{9}.$$

Thus over half the time, at least one of the doctors is idle, while any particular doctor finds himself idle 33.3% of the time.

3.2.1 Busy Period

It is not too difficult to extend conceptually the notion of the busy period (see Section 2.7) to the multichannel case. Recall that for one channel a busy period is defined to begin with the arrival of a customer to an idle channel and to end when the channel next becomes idle. In an analogous fashion, let us define an i-channel busy period for $M/M/c$ ($0 \leqslant i \leqslant c$) to begin with an arrival to the system at an instant when there are $i - 1$ in the system to the very next point in time when the system size dips to $i-1$. Let us say that the case where $i=1$ (an arrival to an empty system) defines the system busy period. In a fashion similar to that of Section 2.7, let us use $T_{b,i}$ to denote the random variable "length of the i-channel busy period." Then the CDF of $T_{b,i}$ is determined by considering the original $M/M/c$ differential-difference equations of (3.4) with an absorbing barrier imposed at a system size of $i-1$ and an initial size of i. Then it should be clear that $p_{i-1}(t)$ will, in fact, be the required CDF and $p'_{i-1}(t) = dp_{i-1}(t)/dt$ its density. The necessary equations are

$$
\begin{cases}
p'_{i-1}(t) = i\mu p_i(t) & \text{[because of absorbing barrier]} \\[2mm]
p'_i(t) = -(\lambda+i\mu)p_i(t) + (i+1)\mu p_{i+1}(t) & \text{[because of absorbing barrier]} \\[2mm]
p'_n(t) = -(\lambda+n\mu)p_n(t) + \lambda p_{n-1}(t) + (n+1)\mu p_{n+1}(t) & (i \leq n < c) \\[2mm]
p'_n(t) = -(\lambda+c\mu)p_n(t) + \lambda p_{n-1}(t) + c\mu p_{n+1}(t) & (n \geq c)
\end{cases}
$$

Proceeding further along the lines of Section 2.7 at this stage would get us bogged down in great algebraic detail. Suffice it to say that any resultant CDF would be in terms of modified Bessel functions (as it was in Section 2.7), but with enough time and patience, $p'_i(t)$, $\bar{p}_i(s)$, and $E[T_{b,i}]$ can be gotten.

3.3 QUEUES WITH PARALLEL CHANNELS AND TRUNCATION $(M/M/c/K)$

We now take up a parallel-server model in which a limit is placed on the number allowed in the system, namely, the $M/M/c/K$ model. Utilizing Equation (3.6) with

$$
\lambda_n = \begin{cases} \lambda & (0 \leq n < K) \\ 0 & (n \geq K) \end{cases}
$$

and

$$
\mu_n = \begin{cases} n\mu & (0 \leq n < c) \\ c\mu & (c \leq n \leq K) \end{cases} ,
$$

we get

$$
p_n = \begin{cases}
\dfrac{\lambda^n}{n\mu(n-1)\mu\cdots(1)\mu}p_0 & (0 \leq n < c) \\[4mm]
\dfrac{\lambda^n}{\underbrace{c\mu c\mu\cdots c\mu}_{n-c\ \text{terms}} c\mu(c-1)\mu(c-2)\mu\cdots(1)\mu}p_0 & (c \leq n \leq K)
\end{cases} ,
$$

or equivalently,

$$p_n = \begin{cases} \dfrac{1}{n!}\left(\dfrac{\lambda}{\mu}\right)^n p_0 & (0 \leqslant n < c) \\[3mm] \dfrac{1}{c^{n-c}c!}\left(\dfrac{\lambda}{\mu}\right)^n p_0 & (c \leqslant n \leqslant K) \end{cases}. \tag{3.24}$$

The boundary condition $\sum_{n=0}^{K} p_n = 1$ will yield p_0; that is,

$$\sum_{n=0}^{c-1} \frac{1}{n!}\left(\frac{\lambda}{\mu}\right)^n p_0 + \sum_{n=c}^{K} \frac{1}{c^{n-c}c!}\left(\frac{\lambda}{\mu}\right)^n p_0 = 1;$$

hence

$$p_0 = \left[\sum_{n=0}^{c-1} \frac{1}{n!}\left(\frac{\lambda}{\mu}\right)^n + \sum_{n=c}^{K} \frac{1}{c^{n-c}c!}\left(\frac{\lambda}{\mu}\right)^n\right]^{-1}.$$

To simplify, consider $\sum_{n=c}^{K}(1/c^{n-c}c!)r^n$, where $r = \lambda/\mu$:

$$\sum_{n=c}^{K} \frac{1}{c^{n-c}c!}r^n = \frac{r^c}{c!}\sum_{n=c}^{K}\left(\frac{r}{c}\right)^{n-c}$$

$$= \begin{cases} \dfrac{r^c}{c!}\dfrac{1-\rho^{K-c+1}}{1-\rho} & (\rho = r/c = \lambda/c\mu \neq 1) \\[3mm] \dfrac{r^c}{c!}(K-c+1) & (\rho = 1) \end{cases}.$$

Thus

$$p_0 = \begin{cases} \left[\displaystyle\sum_{n=0}^{c-1}\frac{1}{n!}\left(\frac{\lambda}{\mu}\right)^n + \frac{(\lambda/\mu)^c}{c!}\frac{1-(\lambda/c\mu)^{K-c+1}}{1-\lambda/c\mu}\right]^{-1} & (\lambda/c\mu \neq 1) \\[5mm] \left[\displaystyle\sum_{n=0}^{c-1}\frac{1}{n!}\left(\frac{\lambda}{\mu}\right)^n + \frac{(\lambda/\mu)^c}{c!}(K-c+1)\right]^{-1} & (\lambda/c\mu = 1) \end{cases}.$$

$$\tag{3.25}$$

We leave as an exercise (see Problem 3.15) to show that taking the limit as

$K\to\infty$ in Equations (3.24) and (3.25) and restricting $\lambda/c\mu<1$ yield the results obtained for the $M/M/c/\infty$ model given by Equations (3.9) and (3.11). Also it is noted that letting $c=1$ in (3.24) and (3.25) yields the results for the $M/M/1/K$ model given by Equation (2.42) of Section 2.5.

We next proceed to find the expected queue length and system size as follows:

$$L_q = \sum_{n=c}^{K} (n-c)p_n$$

$$= \frac{p_0}{c!} \sum_{n=c}^{K} \frac{(n-c)r^n}{c^{n-c}} \qquad [\text{from}(3.24)]$$

$$= \frac{p_0(c\rho)^c \rho}{c!} \sum_{n=c}^{K} (n-c)\rho^{n-c-1}$$

$$= \frac{p_0(c\rho)^c \rho}{c!} \sum_{i=0}^{K-c} i\rho^{i-1}$$

$$= \frac{p_0(c\rho)^c \rho}{c!} \frac{d}{d\rho}\left[\frac{1-\rho^{K-c+1}}{1-\rho}\right],$$

or

$$\boxed{L_q = \frac{p_0(c\rho)^c \rho}{c!(1-\rho)^2}\left[1-\rho^{K-c+1}-(1-\rho)(K-c+1)\rho^{K-c}\right]} \quad . \quad (3.26)$$

To obtain the expected system size recall that

$$L_q = \sum_{n=c}^{K} (n-c)p_n = \sum_{n=c}^{K} np_n - c\sum_{n=c}^{K} p_n$$

$$= \sum_{n=0}^{K} np_n - \sum_{n=0}^{c-1} np_n - c\sum_{n=c}^{K} p_n$$

$$= L - \sum_{n=0}^{c-1} np_n - c\left(1-\sum_{n=0}^{c-1} p_n\right)$$

$$= L - \sum_{n=0}^{c-1} (n-c)p_n - c.$$

Hence

$$L = L_q + c - \sum_{n=0}^{c-1} (c-n)p_n \, ,$$

or

$$L = L_q + c - p_0 \sum_{n=0}^{c-1} \frac{(c-n)(\rho c)^n}{n!} \, . \tag{3.27}$$

Once again, as in the $M/M/1/K$ model, the waiting-time CDFs, $W(t)$ and $W_q(t)$, are somewhat complicated, since the series are finite, but they can be expressed, as was the case for $M/M/1/K$, in terms of cumulative Poisson sums. However, their expected values can be readily obtained by use of Little's formula; that is,

$$W = \frac{L}{\lambda'}, \qquad \lambda' = \lambda(1-p_K) \tag{3.28}$$

and

$$W_q = W - \frac{1}{\mu} \, , \tag{3.29}$$

or

$$W_q = \frac{L_q}{\lambda'} \, . \tag{3.30}$$

Example 3.2

Consider a state automobile inspection station with three inspection stalls, each with room for only one car. It is reasonable to assume that cars wait in such a way that when a stall becomes vacant, the car at the head of the line pulls up to it. The station can accommodate at most four cars waiting (seven in the station) at one time (see Figure 3.1). The arrival pattern is Poisson with a mean of one car every minute during the peak periods. The service time is exponential with mean 6 min. I. M. Fussy, the chief inspector, wishes to know the average number in the system during peak periods, the average wait (including service), and the expected number per hour that cannot enter the station because of full capacity.

Using minutes as the basic time unit, $\lambda = 1$ and $\mu = 1/6$. Thus we have $r = 6$ and $\rho = 2$. We next calculate p_0 from Equation (3.25) and find that

$$p_0 = \left[\sum_{n=0}^{2} \left(\frac{6^n}{n!} \right) + \frac{6^3}{3!} \frac{1-2^5}{1-2} \right]^{-1}$$

$$= 1/1141 \doteq 0.00088.$$

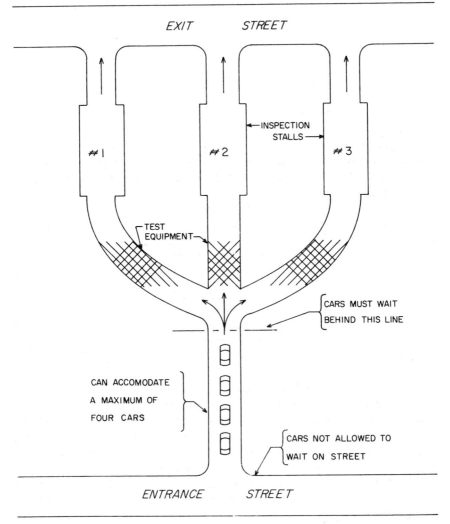

Fig. 3.1 Inspection center layout.

From Equation (3.26) we get

$$L_q = \frac{p_0(6^3)(2)}{3!}\left[1 - 2^5 + 5(2^4)\right]$$

$$= 3528p_0 \doteq 3.09 \text{ cars,}$$

and (3.27) yields

$$L \doteq 3.09 + 3 - p_0 \sum_{n=0}^{2} \left[\frac{(3-n)6^n}{n!} \right]$$

$$\doteq 6.06.$$

To find the average wait during peak periods, Equation (3.28) is used and it is found that

$$W = \frac{L}{\lambda(1 - p_7)}$$

$$= \frac{L}{1 - p_0 6^7 / (3^4 3!)}$$

$$\doteq 12.3 \text{ min.}$$

The expected number of cars per hour that cannot enter the station is given by

$$60 \lambda p_K = 60 p_7$$

$$= \frac{60 p_0 6^7}{(3^4 3!)}$$

$$\doteq 30.4 \text{ cars/hr.}$$

Might this suggest an alternative setup for the inspection station?

3.4 ERLANG'S FORMULA $(M/M/c/c)$

The special case of the truncated queue $M/M/c/K$ for which $K = c$, that is, where no line is allowed to form, gives rise to a stationary distribution which is known as Erlang's first formula and can be readily obtained from Equations (3.24) and (3.25) with $K = c$ as

$$p_n = \frac{(\lambda/\mu)^n / n!}{\sum_{i=0}^{c} (\lambda/\mu)^i / i!} \qquad (0 \leqslant n \leqslant c) \qquad . \qquad (3.31)$$

The resultant formula for p_c is itself called *Erlang's loss formula* and corresponds to the probability of a full system at any time in the steady state. Since the input is Poisson, the probability that an arrival is lost is equal to the probability that all channels are busy. The original physical situation which motivated Erlang in 1917 to devise this model was the simple telephone network. Incoming calls arrive as a Poisson stream, service times are mutually independent, exponential random variables, and all calls that arrive and find every trunk line busy, that is, get a busy signal, are turned away. The model has, in fact, always been of great value in the design of telephone exchanges.

But the great importance of this formula lies in the very surprising fact that (3.31) is valid for *any* $M/G/c/c$ *independent* of the form of the service-time distribution. That is to say, the steady-state system probabilities are only a function of the mean service time, but not of the underlying CDF! Erlang was also able to deduce the formula for the case when service times are constant. While this result was later shown to be correct, his proof was not quite valid. Later works by Vaulot (1927), Pollaczek (1932), Palm (1938), Kosten (1948-9), and others corrected the defect in Erlang's 1917 proof and supplied proofs for the complete arbitrariness of the service-time distribution. The newest in this sequence of papers is a fairly recent work by Takacs (1969), which supplies some additional results for the problem. We shall prove the validity of Erlang's loss formula for general service in Chapter 5, Section 5.2.2.

Doeh (1960) generated a set of curves on probability paper for the loss formula, which is from (3.31)

$$p_c = \frac{(c\rho)^c / c!}{\sum_{i=0}^{c} (c\rho)^i / i!} \qquad (\rho = \lambda/c\mu),$$

as a function of c for values of $(c\rho)$ ranging from 0.01 to 20.0. His graph is presented with permission here as Figure 3.2.[6]

Example 3.3

We illustrate the use of these graphs by the following example. In 3 years of data collection, incoming telephone calls at peak periods to Sidetrack Railways are found to have identical exponential interarrival distributions

---

[6]Note that multiplying the numerator and denominator of the right-hand side of p_c by $e^{-c\rho}$ results in the ratio of a Poisson probability and a cumulative Poisson probability, so that one could make use of Poisson tables in calculating p_c.

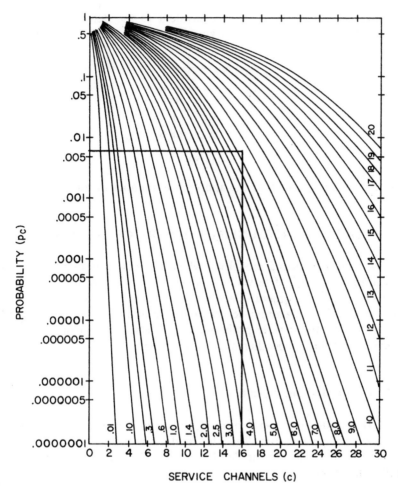

Fig. 3.2   Probability of loss (p_c) versus service channels (c) for various usage ($c\rho$).

with mean frequency of 84.3 calls/hr. An analysis of telephone conversations revealed that the average call duration was 0.103 hr. Personnel familiar with and responsible for meeting customer service are confident that a busy signal on the average of once every 2 hr can be tolerated. What is the appropriate number of telephone lines?

To solve the problem, we must first satisfy ourselves that we do, in fact, have an $M/G/c/c$ problem. A little bit of thought should convince us that we do, since when all telephone lines are in use, a busy signal results. The value of λ is 84.3/hr, the expected service time is 0.103 hr, and therefore $c\rho \doteq (84.3)(.103) = 8.7$. Each telephone line corresponds to a service channel; hence the question asked is equivalent to the determination of the number of channels which would give rise to a p_c satisfying the equation

$$\lambda p_c = \frac{1}{2hr},$$

or

$$p_c = \frac{1}{2\lambda} \doteq \frac{1}{168.6}$$

$$\doteq 0.0059.$$

From Figure 3.2 we see that the appropriate number of lines is 16.

3.5   QUEUES WITH UNLIMITED SERVICE ($M/M/\infty$)

We now treat a queueing model for which there is unlimited service, that is, an infinite number of servers available. This model is often referred to as the ample-server problem. A self-service type situation is one example of such a model. For this ample-server model, in addition to deriving steady-state results, transient state probabilities will also be developed since for this case, the transient mathematics is tractable.

3.5.1   Steady-State Results

We make use of the general birth–death results given by Equation (3.6) with $\lambda_n = \lambda$ and $\mu_n = n\mu$, for all n. This model was actually developed at the end of Section 3.1 as an example of general birth–death processes. Repeating this analysis here,

$$p_n = \frac{\lambda^n}{n\mu(n-1)\mu\cdots(1)\mu}p_0$$

$$= \frac{\lambda^n}{n!\mu^n}p_0.$$

We find p_0 as usual by using the boundary condition that $\sum_{n=0}^{\infty}p_n = 1$, which yields

$$p_0 = \left[\sum_{n=0}^{\infty}\frac{\lambda^n}{n!\mu^n}\right]^{-1}.$$

But

$$\sum_{n=0}^{\infty} \frac{1}{n!}\left(\frac{\lambda}{\mu}\right)^n = \sum_{n=0}^{\infty} \frac{r^n}{n!} \quad (r=\lambda/\mu)$$

$$= e^r;$$

thus

$$p_n = \frac{r^n e^{-r}}{n!}$$

or

$$\boxed{p_n = \frac{(\lambda/\mu)^n e^{-\lambda/\mu}}{n!} \quad (n \geqslant 0)} \quad . \tag{3.32}$$

Hence the steady-state probability distribution of n in the system is given by the Poisson with parameter λ/μ. Note that λ/μ is not in any way restricted for the existence of a steady-state solution. It also turns out (we show this in Section 5.2.3) that Equation (3.32) is valid for *any* $M/G/\infty$ model, that is, p_n depends only on the mean service time and not on the shape of the distribution. It is not surprising that this is true here in light of a similar result we mentioned previously for $M/M/c/c$, since p_n of (3.32) could have been obtained from (3.31) by taking the limit as $c \to \infty$.

The expected system size, L, is the mean of the Poisson distribution and is clearly given as

$$\boxed{L = \frac{\lambda}{\mu}} \quad . \tag{3.33}$$

Since we have as many servers as customers in the system, $L_q = 0$ and, of course, $W_q = 0$. The average waiting time in the system obviously becomes merely the average service time; hence

$$\boxed{W = \frac{1}{\mu}} \quad . \tag{3.34}$$

Also, the waiting-time distribution, $W(t)$, is identical to the service-time distribution, namely, exponential with mean $1/\mu$.

Example 3.4

Television station WDAC in a large metropolitan area wishes to know the average number of viewers they can expect on a Saturday evening prime-time program. They have found from past surveys that people turning on their television sets on Saturday evening during prime time can be described rather well by a Poisson distribution with a mean of 100,000/hr. There are five major T.V. stations in the area and it is believed that a given person chooses among these nearly at random. Surveys have also shown that the average person tunes in 90 min and that viewing times are approximately exponentially distributed.

To find L we need merely find λ/μ. The mean arrival rate (people tuning in WDAC during prime time on Saturday evening) is $100,000/5 = 20,000$/hr. The mean service time, $1/\mu$, is 90 min or 1.5 hr; thus μ is $(2/3)$/hr. The average number of viewers during prime time is then

$$L = \frac{20,000}{2/3}$$

$$= 30,000 \text{ people.}$$

3.5.2 Transient Analysis

It turns out for this model that the development of the transient solution is not too difficult a task. We must first have the differential-difference equations, which can be readily obtained from Equations (3.4) with $\lambda_n = \lambda$ and $\mu_n = n\mu$, and are

$$\begin{cases} p_n'(t) = -(\lambda + n\mu)p_n(t) + (n+1)\mu p_{n+1}(t) + \lambda p_{n-1}(t) & (n \geqslant 1) \\ p_0'(t) = -\lambda p_0(t) + \mu p_1(t), \end{cases} \tag{3.35}$$

where $p_n'(t) = dp_n(t)/dt$. As per usual, let us denote the generating function of the transient probabilities by

$$P(z,t) = \sum_{n=0}^{\infty} p_n(t)z^n.$$

If the first equation of (3.35) is multiplied by z^n, both sides of the equation summed from 1 to ∞, the second equation of (3.35) added to the result, and terms grouped (as was done in Section 2.6.2), then we would have

$$\sum_{n=0}^{\infty} p_n'(t)z^n = -\sum_{n=0}^{\infty} (\lambda + n\mu)p_n(t)z^n + \mu \sum_{n=0}^{\infty} (n+1)p_{n+1}(t)z^n$$

$$+ \lambda \sum_{n=1}^{\infty} p_{n-1}(t)z^n.$$

This equation may then be rewritten as

$$\sum_{n=0}^{\infty} p_n'(t)z^n = -\lambda \sum_{n=0}^{\infty} p_n(t)z^n - \mu z \sum_{n=1}^{\infty} p_n(t)nz^{n-1}$$

$$+ \mu \sum_{n=0}^{\infty} p_{n+1}(t)(n+1)z^n + \lambda z \sum_{n=1}^{\infty} p_{n-1}(t)z^{n-1}.$$

Since

$$\sum_{n=0}^{\infty} p_n(t)z^n = \sum_{n=1}^{\infty} p_{n-1}(t)z^{n-1} = P(z,t),$$

$$\sum_{n=1}^{\infty} p_n(t)nz^{n-1} = \sum_{n=0}^{\infty} p_{n+1}(t)(n+1)z^n = \frac{\partial P(z,t)}{\partial z},$$

and

$$\sum_{n=0}^{\infty} p_n'(t)z^n = \frac{\partial P(z,t)}{\partial t},$$

we have

$$\frac{\partial P(z,t)}{\partial t} = \mu(1-z)\frac{\partial P(z,t)}{\partial z} - \lambda(1-z)P(z,t). \tag{3.36}$$

Any equation of the form

$$p\frac{\partial P}{\partial t} + q\frac{\partial P}{\partial z} = r, \tag{3.37}$$

where p, q, and r are functions of t, z, and P, is called a *planar differential equation* [see Betz et al. (1954), for example]. Any integral of (3.37), $P = f(t,z)$, represents a surface and is called an integral surface of (3.37). This differential equation expresses the geometrical fact that the normal to the surface at a point (t_0, z_0, P_0) is perpendicular to the line through this point whose direction numbers are the values of p, q, and r evaluated at the point. This line has the equations

$$\frac{t-t_0}{p} = \frac{z-z_0}{q} = \frac{P-P_0}{r}.$$

Hence, at each point of the surface, there is defined a direction whose direction numbers, dt, dz, and dP, satisfy the equations

$$\frac{dt}{p} = \frac{dz}{q} = \frac{dP}{r}.$$

For the case at hand, $p = 1$, $q = -(1-z)\mu$, and $r = -\lambda(1-z)P(z,t)$. Hence we are to solve

$$\frac{dt}{1} = \frac{dz}{-(1-z)\mu} = \frac{dP(z,t)}{-\lambda(1-z)P(z,t)}.$$

This easily yields by direct integration the two solutions

$$(1-z)e^{-\mu t} = C_1,$$

and

$$P(z,t)e^{-z\lambda/\mu} = C_2,$$

where C_1 and C_2 are constants. Since any two constants can be related functionally (e.g., another constant K can always be found such that $C_2 = KC_1$), we have

$$C_2 = g(C_1),$$

which implies that

$$P(z,t)e^{-z\lambda/\mu} = g[(1-z)e^{-\mu t}]$$

or

$$P(z,t) = e^{z\lambda/\mu}g[(1-z)e^{-\mu t}]. \tag{3.38}$$

Now when $t = 0$, let it be assumed that the system is empty, that is, that $p_0(0) = 1$ and $p_n(0) = 0$ for $n > 0$. Thus we have that

$$P(z,0) = \sum_{n=0}^{\infty} p_n(0)z^n = 1,$$

and, from (3.38), that

$$1 = e^{z\lambda/\mu}g[(1-z)]$$

or

$$g[(1-z)] = e^{-z\lambda/\mu}.$$

Letting $y = (1-z)$ yields

$$g(y) = e^{-(1-y)\lambda/\mu}.$$

Hence

$$g[(1-z)e^{-\mu t}] = \exp\left\{[(1-z)e^{-\mu t}-1]\frac{\lambda}{\mu}\right\},$$

and, from (3.38), we finally get

$$P(z,t) = \exp\left\{\frac{z\lambda}{\mu}+[(1-z)e^{-\mu t}-1]\frac{\lambda}{\mu}\right\}$$

$$= \exp\left\{(z-1)(1-e^{-\mu t})\left(\frac{\lambda}{\mu}\right)\right\}. \tag{3.39}$$

To obtain the state probabilities $\{p_n(t)\}$ it is necessary to expand (3.39) in a power series, $a_0 + a_1 z + a_2 z^2 + \ldots$, where the a_n coefficients then are the $p_n(t)$ we desire since we are expanding a probability generating function. To do this we can use a Taylor series expansion about zero (Maclaurin series); that is,

$$a_n = \frac{1}{n!}\left.\frac{\partial^n P(z,t)}{\partial z^n}\right|_{z=0}.$$

Noting from Equation (3.39) that $P(z,t)$ is of the form $e^{C(z-1)}$, we then have $\partial^n P(z,t)/\partial z^n = C^n e^{C(z-1)}$, and[7]

$$p_n(t) = \frac{1}{n!}\left[(1-e^{-\mu t})\frac{\lambda}{\mu}\right]^n \exp\left[(z-1)(1-e^{-\mu t})\frac{\lambda}{\mu}\right]\Bigg|_{z=0}$$

$$= \frac{1}{n!}\left[(1-e^{-\mu t})\frac{\lambda}{\mu}\right]^n \exp\left[-(1-e^{-\mu t})\frac{\lambda}{\mu}\right] \quad (n \geqslant 0). \tag{3.40}$$

It is easily seen that letting $t \to \infty$ in Equation (3.40) yields the steady-state solution given by Equation (3.32), namely,

$$p_n = \frac{(\lambda/\mu)^n e^{-\lambda/\mu}}{n!}.$$

[7]Note that for this transient analysis, we were able to solve the PDE directly for the probability generating function, expand it in a power series, and "pick off" the coefficients which are the $\{p_n(t)\}$. For $M/M/1/\infty$, we could not solve the PDE directly but had to resort to getting its LT, expanding it in a power series whose coefficients were the LTs of the $\{p_n(t)\}$, which necessitated the laborious inversion procedure to finally obtain the $\{p_n(t)\}$.

3.6 FINITE SOURCE QUEUES

In previous models we have assumed that the population from which arrivals come (the calling population) is infinite, since the number of arrivals in any time interval is a Poisson random variable with a denumerably infinite sample space. We now treat a problem where the calling population is finite, say of size M, and the occurrence probabilities are functions of past behavior. A typical application of this model is that of machine servicing, where the calling population is the machines and an arrival corresponds to a machine breakdown. The repairmen (or repair crews), of course, are the servers. We assume c servers are available, that the service times are identical exponential random variables with mean $1/\mu$, and that the arrival process is described as follows. If a calling unit is not in the system at time t, the probability it will have entered by time $t+\Delta t$ is $\lambda \Delta t + o(\Delta t)$; that is, the time a calling unit spends outside the system is exponential with mean $1/\lambda$.

We can use the birth–death theory developed previously in Section 3.1, namely, the equations of (3.6), as follows:

$$\lambda_n = \begin{cases} (M-n)\lambda & (0 \leqslant n < M) \\ 0 & (n \geqslant M) \end{cases}$$

and

$$\mu_n = \begin{cases} n\mu & (0 \leqslant n < c) \\ c\mu & (n \geqslant c) \end{cases} .$$

Using (3.6) yields

$$p_n = \begin{cases} \dfrac{M!/(M-n)!}{n!} \left(\dfrac{\lambda}{\mu} \right)^n p_0 & (0 \leqslant n < c) \\ \dfrac{M!/(M-n)!}{c^{n-c}c!} \left(\dfrac{\lambda}{\mu} \right)^n p_0 & (c \leqslant n \leqslant M) \end{cases} ,$$

or equivalently,

$$p_n = \begin{cases} \dbinom{M}{n} \left(\dfrac{\lambda}{\mu} \right)^n p_0 & (0 \leqslant n < c) \\ \dbinom{M}{n} \dfrac{n!}{c^{n-c}c!} \left(\dfrac{\lambda}{\mu} \right)^n p_0 & (c \leqslant n \leqslant M) \end{cases} \tag{3.41}$$

where p_0 is found in the usual way from $\sum_{n=0}^{M} p_n = 1$, as

$$p_0 = \left[\sum_{n=0}^{c-1} \binom{M}{n} \left(\frac{\lambda}{\mu} \right)^n + \sum_{n=c}^{M} \binom{M}{n} \frac{n!}{c^{n-c} c!} \left(\frac{\lambda}{\mu} \right)^n \right]^{-1} . \quad (3.42)$$

To find the average number of customers in the system (if we are dealing with the machine breakdown problem we are interested in the average number of machines "down" for repair), we use, as before, the definition of expected value and get

$$L = \sum_{n=0}^{M} n p_n$$

$$= \sum_{n=0}^{c-1} n \binom{M}{n} \left(\frac{\lambda}{\mu} \right)^n p_0 + \sum_{n=c}^{M} n \binom{M}{n} \frac{n!}{c^{n-c} c!} \left(\frac{\lambda}{\mu} \right)^n p_0 ,$$

or finally,

$$L = p_0 \left[\sum_{n=0}^{c-1} n \binom{M}{n} \left(\frac{\lambda}{\mu} \right)^n + \frac{1}{c!} \sum_{n=c}^{M} n \binom{M}{n} \frac{n!}{c^{n-c}} \left(\frac{\lambda}{\mu} \right)^n \right] . \quad (3.43)$$

There is no neater expression for L; p_0 must first be calculated and then multiplied by the sum of two series, one of c terms and the other of $M - c + 1$ terms. An expression for L_q as a function of L can be obtained as follows:

$$L_q = \sum_{n=c}^{M} (n-c) p_n$$

$$= \sum_{n=c}^{M} n p_n - c \sum_{n=c}^{M} p_n$$

$$= L - \sum_{n=0}^{c-1} n p_n - c \left[1 - \sum_{n=0}^{c-1} p_n \right]$$

$$= L - c + \sum_{n=0}^{c-1} (c-n) p_n .$$

Hence

$$L_q = L - c + p_0 \sum_{n=0}^{c-1} (c-n) \binom{M}{n} \left(\frac{\lambda}{\mu}\right)^n . \qquad (3.44)$$

To obtain the expected waiting-time measures, W and W_q, we must find the mean rate of arrivals into the system. To do so, we note that if the system has n units in it, there are $M - n$ possible arrival units outside the system, each with a mean arrival rate of λ, so that the mean arrival rate to the system is $(M - n)\lambda$. Summing up over all possible n after weighting each term by its appropriate probability p_n we find that

$$\lambda' = \sum_{n=0}^{M} (M-n)\lambda p_n$$

$$= \lambda M \sum_{n=0}^{M} p_n - \lambda \sum_{n=0}^{M} n p_n .$$

Thus

$$\lambda' = \lambda(M-L). \qquad (3.45)$$

Equation (3.45) is certainly intuitive, since, on the average, L are in the system and hence, on the average, $M - L$ are outside and each has a mean arrival rate of λ.

Using Equation (3.45) in Little's formula we then have

$$W = \frac{L}{\lambda(M-L)} \qquad (3.46)$$

and

$$W_q = \frac{L_q}{\lambda(M-L)} . \qquad (3.47)$$

We can also, of course, obtain W from W_q or W_q from W by the relation

$$W = W_q + \frac{1}{\mu}.$$

We saw from Section 2.4, Equation (2.38), and also from Problem 2.14 that the above relation, together with Little's formula, provides another way of obtaining λ', namely,

$$\lambda' = \mu(L - L_q) \qquad (3.48)$$

Equation (3.48) is true for all queueing models for which Little's formula holds (see Problem 3.27).

Example 3.5

The W. E. Finish Machine Shop Company has five turret lathes. These machines break down periodically and the company has two repairmen to service the lathes when they break down. When a lathe is fixed, the time until the next breakdown is exponentially distributed with a mean of 30 hr. The shop always has enough of a work backlog to ensure that all lathes in operating condition will be working. The repair time for each repairman is exponentially distributed with a mean of 3 hr. The shop manager wishes to know the average number of lathes operational at any given time, the expected "down-time" of a lathe that requires repair, and the expected percent idle time of each repairman.

To answer any of these questions, we must first calculate p_0 from Equation (3.42). In this example, $M=5$, $c=2$, $\lambda=1/30$, and $\mu=1/3$, and thus $\lambda/\mu=1/10$. From (3.42) we have

$$
p_0 = \left[\sum_{n=0}^{1} \binom{5}{n}\left(\frac{1}{10}\right)^n + \sum_{n=2}^{5} \binom{5}{n} \frac{n!}{2^{n-2}2!}\left(\frac{1}{10}\right)^n \right]^{-1}
$$

$$
\doteq [1 + 0.5 + 0.10 + 0.015 + 0.0015 + 0.000075]^{-1} \doteq 0.62.
$$

The average number of operational lathes is merely $M - L$; hence, from Equation (3.43) we have

$$
L = p_0 \left[\sum_{n=0}^{1} n\binom{5}{n}\left(\frac{1}{10}\right)^n + \frac{1}{2!}\sum_{n=2}^{5} n\binom{5}{n}\frac{n!}{2^{n-2}}\left(\frac{1}{10}\right)^n \right]
$$

$$
\doteq 0.62[1(0.5) + 2(0.10) + 3(0.015) + 4(0.0015) + 5(0.000075)]
$$

$$
\doteq 0.47.
$$

Thus $5 - 0.47$ or 4.53 machines, on the average, are in operating condition. The expected down-time can be found from (3.46) and is

$$
W \doteq \frac{0.47}{(1/30)(4.53)} \doteq 3.11 \text{ hr.}
$$

The average percent idle time of each server is

$$p_0 + \tfrac{1}{2}p_1 \doteq 0.62 + \frac{1}{2}\binom{5}{1}\left(\frac{1}{10}\right)(0.62) \doteq 0.78,$$

or each repairman is idle, on the average, 78% of the time.[8]

The manager, because of the long idle time, is interested in knowing the answer to the same questions if he reduces his repair force to one man. The results are

$$\begin{cases} p_0 \doteq 0.56 \quad \text{(average percentage idle} = 56\%) \\[2mm] L \doteq 0.63 \\[2mm] M - L \doteq 4.37 \\[2mm] W \doteq 4.32 \text{ hr} \end{cases}$$

Since under both situations over four machines are expected operational at any time and the down-time increase is about one hour when reducing to only one repairman, the manager might well decide to take one repairman off lathes and put him to work elsewhere.

This model can be generalized to include the use of spares. We assume now that we have Y spares on hand, so that when a machine fails, a spare is immediately substituted for it. If it happens that all spares are being used and a breakdown occurs, then the system becomes short. When a machine is repaired, it then becomes a spare (unless the system is short in which case the repaired machine goes immediately into service). For this model, λ_n is slightly different and is given by

$$\lambda_n = \begin{cases} M\lambda & (0 \leqslant n < Y) \\ (M - n + Y)\lambda & (Y \leqslant n < Y + M) \\ 0 & (n \geqslant Y + M) \end{cases} \cdot$$

Again considering c repairmen, we have

$$\mu_n = \begin{cases} n\mu & (0 \leqslant n < c) \\ c\mu & (n \geqslant c) \end{cases} \cdot$$

[8]Calculations by hand for finite source queueing models are quite tedious, especially if M is large, because of having to evaluate finite sums term by term. There are tables available for finite source models [see, for example, Peck and Hazelwood (1958)].

We first assume $c \leqslant Y$ and use Equation (3.6) once more to find that

$$
p_n = \begin{cases}
\dfrac{M^n \lambda^n}{n! \mu^n} p_0 & (0 \leqslant n < c) \\[2ex]
\dfrac{M^n \lambda^n}{c^{n-c} c! \mu^n} p_0 & (c \leqslant n < Y) \\[2ex]
\dfrac{[M^Y M! / (M-n+Y)!] \lambda^n}{c^{n-c} c! \mu^n} p_0 & (Y \leqslant n \leqslant Y+M)
\end{cases} \, ,
$$

or equivalently,

$$
p_n = \begin{cases}
\dfrac{M^n}{n!} \left(\dfrac{\lambda}{\mu} \right)^n p_0 & (0 \leqslant n < c) \\[2ex]
\dfrac{M^n}{c^{n-c} c!} \left(\dfrac{\lambda}{\mu} \right)^n p_0 & (c \leqslant n < Y) \\[2ex]
\dfrac{M^Y M!}{(M-n+Y)! c^{n-c} c!} \left(\dfrac{\lambda}{\mu} \right)^n p_0 & (Y \leqslant n \leqslant Y+M)
\end{cases} \qquad \cdot \ (3.49)
$$

If Y is very large, we essentially have an infinite calling population with mean arrival rate $M\lambda$. Letting Y go to infinity in (3.49) yields the $M/M/c/\infty$ results of (3.9) with $M\lambda$ for λ.

When $c > Y$, we have

$$
p_n = \begin{cases}
\dfrac{M^n}{n!} \left(\dfrac{\lambda}{\mu} \right)^n p_0 & (0 \leqslant n \leqslant Y) \\[2ex]
\dfrac{M^Y M!}{(M-n+Y)! n!} \left(\dfrac{\lambda}{\mu} \right)^n p_0 & (Y+1 \leqslant n < c) \\[2ex]
\dfrac{M^Y M!}{(M-n+Y)! c^{n-c} c!} \left(\dfrac{\lambda}{\mu} \right)^n p_0 & (c \leqslant n \leqslant Y+M)
\end{cases} \qquad \cdot \ (3.50)
$$

It should be noted that when $Y = 0$ (no spares), Equation (3.50) reduces to Equation (3.41).

The empty-system probability, p_0, can be found as previously for the no-spares model by using $\sum_{n=0}^{Y+M} p_n = 1$, so that p_0 is made up of the sums of finite series. The same is true for L and L_q, and we can obtain results analogous to Equations (3.42), (3.43), and (3.44). To obtain results for W and W_q comparable to Equations (3.46) and (3.47), we must first obtain the *effective* mean arrival rate λ'. To get this, one can use Equation (3.48). However, λ' can be directly obtained using logic similar to that used for

Equation (3.45), namely,

$$\lambda' = \sum_{n=0}^{Y-1} M\lambda p_n + \sum_{n=Y}^{Y+M} (M-n+Y)\lambda p_n$$

$$= M\lambda \sum_{n=0}^{Y-1} p_n + M\lambda \sum_{n=Y}^{Y+M} p_n - \lambda \sum_{n=Y}^{Y+M} (n-Y)p_n$$

$$= \lambda \left[M - \sum_{n=Y}^{Y+M} (n-Y)p_n \right]. \tag{3.51}$$

The calculations for the spares model proceed in a similar fashion to those of the no-spares case.

As a final point to close this section, suppose that we desired the waiting-time distribution in addition to the mean given by Equation (3.46). The standard procedure in the past has been to find the waiting time of an arriving customer conditioned on the existence of n in the system at the point of arrival, and then unconditioning with respect to the stationary distribution $\{q_n\}$, where $\{q_n\}$ are the state probabilities given an arrival occurs. As was the case for $M/M/c/K$, $q_n \neq p_n$ here and therefore, before we can obtain the CDF of system or line wait, we have to relate the general-time probability p_n to the probability that an *arrival* finds n in the system, q_n. For the general finite-source queue (machine-repair problem without spares) the two probabilities are related as

$$q_n = \frac{(M-n)p_n}{k},$$

where k is an appropriate normalizing constant determined from $\sum_{n=0}^{M-1} q_n = 1$. To prove this, we again use Bayes' theorem as in the $M/M/1/K$ situation of Section 2.5:

$\Pr\{\,n \text{ in system}|\text{arrival is about to occur}\}$

$$= \frac{\Pr\{\,n \text{ in system}\}\Pr\{\text{arrival is about to occur}|n \text{ in system}\}}{\sum_n [\Pr\{\,n \text{ in system}\}\Pr\{\text{arrival is about to occur}|n \text{ in system}\}]}$$

$$= \lim_{\Delta t \to 0} \frac{p_n[(M-n)\lambda\Delta t + o(\Delta t)]}{\sum_n p_n[(M-n)\lambda\Delta t + o(\Delta t)]}$$

$$= \lim_{\Delta t \to 0} \frac{p_n[(M-n)\lambda + o(\Delta t)/(\Delta t)]}{\sum_n p_n[(M-n)\lambda + o(\Delta t)/(\Delta t)]}$$

$$= \frac{(M-n)p_n}{\sum_n (M-n)p_n}$$

$$= \frac{(M-n)p_n}{(M-L)} .$$

The waiting-time distributions again turn out to be in terms of cumulative Poisson sums as in the $M/M/c/K$ model (see, for example, Section 2.5 for $M/M/1/K$). The analysis for $c = 1$ proceeds as follows:

$$W_q(t) \equiv \Pr\{T_q \leqslant t\}$$

$$= \sum_{n=1}^{M-1} [\Pr\{n \text{ completions in } \leqslant t | \text{arrival found } n \text{ in system}\} \cdot q_n] + W_q(0)$$

$$= \sum_{n=1}^{M-1} q_n \int_0^t \frac{\mu(\mu x)^{n-1}}{(n-1)!} e^{-\mu x} dx + q_0$$

$$= \sum_{n=0}^{M-2} q_{n+1} \int_0^t \frac{\mu(\mu x)^n}{n!} e^{-\mu x} dx + q_0$$

$$= \sum_{n=0}^{M-2} q_{n+1} \left[1 - \int_t^\infty \frac{\mu(\mu x)^n}{n!} e^{-\mu x} dx \right] + q_0 .$$

Using the same analysis on the integral as in Sections 2.5 and 1.8, Equation (1.20), we get

$$W_q(t) = 1 - \sum_{n=0}^{M-2} q_{n+1} \sum_{i=0}^{n} \frac{(\mu t)^i e^{-\mu t}}{i!} .$$

3.7   STATE-DEPENDENT SERVICE

In this section we treat queues with state-dependent service; that is, the mean service rate depends on the state of the system (number in the system). In many real situations, the server (or servers) may speed up when seeing a long line forming. On the other hand, it may happen if the server is inexperienced that he becomes flustered and his mean rate actually decreases as the system becomes more congested. It is these types of situations that are now considered.

The first model we consider is one in which the server has two mean rates, say "slow" and "fast." He works at the slow rate until there are k in the system, at which point he switches to his fast rate. (The service mechanism here might also be a machine with two speeds.) We still assume the service times are Markovian, but the mean rate μ_n now explicitly depends on the system state n. Furthermore, no limit on the number in the system is imposed. Thus μ_n is given as

$$\mu_n = \begin{cases} \mu_1 & (1 \leqslant n < k) \\ \mu & (n \geqslant k) \end{cases}. \tag{3.52}$$

Assuming the arrival process is Poisson with parameter λ and utilizing Equation (3.6), we have

$$p_n = \begin{cases} \left(\dfrac{\lambda}{\mu_1}\right)^n p_0 & (0 \leqslant n < k) \\ \dfrac{\lambda^n}{\mu_1^{k-1}\mu^{n-k+1}} p_0 & (n \geqslant k) \end{cases}. \tag{3.53}$$

From the boundary condition $\sum_{n=0}^{\infty} p_n = 1$, we find that

$$p_0 = \left[\sum_{n=0}^{k-1} \left(\frac{\lambda}{\mu_1}\right)^n + \sum_{n=k}^{\infty} \frac{\lambda^n}{\mu_1^{k-1}\mu^{n-k+1}} \right]^{-1},$$

or

$$p_0 = \left[\frac{1-\rho_1^k}{1-\rho_1} + \frac{\rho\rho_1^{k-1}}{1-\rho} \right]^{-1} \qquad (\rho_1 = \lambda/\mu_1, \ \rho = \lambda/\mu < 1). \tag{3.54}$$

The reader should note that if $\mu_1 = \mu$, Equations (3.53) and (3.54) reduce to those of the $M/M/1/\infty$ model, given by Equation (2.11) of Section 2.1.

To find the expected system size, we proceed as in Section 2.2. That is,

$$L = \sum_{n=0}^{\infty} n p_n$$

$$= p_0 \left[\sum_{n=0}^{k-1} n\rho_1^n + \sum_{n=k}^{\infty} n\rho_1^{k-1} \rho^{n-k+1} \right]$$

$$= p_0 \left[\rho_1 \sum_{n=0}^{k-1} n\rho_1^{n-1} + \rho_1 \left(\frac{\rho_1}{\rho} \right)^{k-2} \sum_{n=k}^{\infty} n\rho^{n-1} \right]$$

$$= p_0 \left[\rho_1 \frac{d}{d\rho_1} \sum_{n=0}^{k-1} \rho_1^n + \rho_1 \left(\frac{\rho_1}{\rho} \right)^{k-2} \frac{d}{d\rho} \sum_{n=k}^{\infty} \rho^n \right]$$

$$= p_0 \left[\rho_1 \frac{d}{d\rho_1} \left(\frac{1-\rho_1^k}{1-\rho_1} \right) + \rho_1 \left(\frac{\rho_1}{\rho} \right)^{k-2} \frac{d}{d\rho} \left(\frac{1}{1-\rho} - \frac{1-\rho^k}{1-\rho} \right) \right] \quad (\rho < 1)$$

or finally,

$$\boxed{ L = p_0 \left\{ \frac{\rho_1 [1 + (k-1)\rho_1^k - k\rho_1^{k-1}]}{(1-\rho_1)^2} + \frac{\rho\rho_1^{k-1}[k - (k-1)\rho]}{(1-\rho)^2} \right\} }. \quad (3.55)$$

We can proceed in a manner similar to that of Section 2.2 to find L_q by

$$\boxed{ L_q = L - (1 - p_0) }, \quad (3.56)$$

and can find W and W_q from Little's formula as

$$\boxed{ W = \frac{L}{\lambda} }$$

and

$$\boxed{ W_q = \frac{L_q}{\lambda} }.$$

Note that here the relation $W = W_q + 1/\mu$ cannot be used, since μ is not constant but depends on the system-state switch point, k. However, from

Equations (3.56) and Little's formula, one can see that

$$W = W_q + \frac{1-p_0}{\lambda},$$

which implies that the expected service time is $(1-p_0)/\lambda$.

Example 3.6

Sonny Schine and Harvey Goode have invented and applied for a patent on a machine which polishes automobiles. They have formed a partnership called the Goode-Schine Garage, and rent an old building in which they have set up their machine. Since this is a parttime job for both men, the garage is open on Saturdays only. Customers are taken on a first-come, first-served basis, and since their garage is in a low density population and traffic area, there is virtually no limit on the number of customers who can wait. The car-polishing machine can run at two speeds. At the low speed, it takes 40 min, on the average, to polish a car. On the high speed, it takes only 20 min on the average. Once a switch is made, the actual times can be assumed to follow an exponential distribution.

It is estimated that customers will arrive according to a Poisson process with a mean interarrival time of 30 min. Sonny Schine has had a course in queueing theory and decides to calculate the effect of two policies: switching to high speed if there are any customers waiting (that is, two or more in the system) versus switching to high speed only when more than one customer is waiting (three or more in the system). The machine speeds can be switched at any time, even while the machine is in operation. It is desired to know the average waiting time under the two policies.

It is necessary, therefore, to calculate W for the case when $k=2$ and then for $k=3$. We must first calculate p_0 from Equation (3.54), then L from Equation (3.55), and finally W from Little's formula. The computations follow. We first calculate ρ and ρ_1 to be

$$\rho_1 = \frac{\lambda}{\mu_1} = \frac{1/30}{1/40} = \frac{4}{3}$$

$$\rho = \frac{\lambda}{\mu} = \frac{1/30}{1/20} = \frac{2}{3}.$$

For Case 1, $k=2$ and

$$
\left\{
\begin{aligned}
p_0 &= \left[\frac{1-(4/3)^2}{1-(4/3)} + \frac{(2/3)(4/3)}{1-(2/3)} \right]^{-1} = \frac{1}{5} = 0.20 \\
\\
L &= \frac{1}{5}\left[\frac{(4/3)\left[1+(4/3)^2-2(4/3)\right]}{(-1/3)^2} + \frac{(2/3)(4/3)[2-(2/3)]}{(1/3)^2} \right]. \\
\\
&= \frac{12}{5} = 2.40 \text{ cars} \\
\\
W &= \frac{L}{\lambda} = \frac{2.40}{1/30} = 72 \text{ min} = 1 \text{ hr and } 12 \text{ min}
\end{aligned}
\right.
$$

For Case 2, $k=3$ and

$$
\left\{
\begin{aligned}
p_0 &= \left[\frac{1-(4/3)^3}{1-(4/3)} + \frac{(2/3)(4/3)^2}{1-(2/3)} \right]^{-1} = \frac{3}{23} \doteq 0.13 \\
\\
L &= \frac{3}{23}\left[\frac{(4/3)\left[1+2(4/3)^3-3(4/3)^2\right]}{(-1/3)^2} + \frac{(2/3)(4/3)^2[3-2(2/3)]}{(1/3)^2} \right]. \\
\\
&= \frac{204}{69} \doteq 2.96 \text{ cars} \\
\\
W &= \frac{L}{1/30} \doteq 89 \text{ min} = 1 \text{ hr and } 29 \text{ min}
\end{aligned}
\right.
$$

Schine feels that the average wait of 15 more minutes for switching speeds at three rather than two would not have any adverse effect on his clientele. However, it costs more to run the machine at the higher speed. In fact, it is estimated that it costs \$5 per operating hour to operate the machine at low speed and \$8 per operating hour to operate at high speed. Thus the expected cost of operation when switching at k is given by

$$
C(k) = 5 \sum_{n=1}^{k-1} p_n + 8 \sum_{n=k}^{\infty} p_n \ ,
$$

or

$$C(k) = 5 \sum_{n=1}^{k-1} \rho_1^n p_0 + 8\left(1 - \sum_{n=0}^{k-1} \rho_1^n p_0\right).$$

For Case 1 we have

$$C(2) = 5\left(\frac{4}{3}\right)\left(\frac{1}{5}\right) + 8\left[1 - \frac{1}{5} - \left(\frac{4}{3}\right)\left(\frac{1}{5}\right)\right] \doteq \$5.60/\text{hr},$$

while for Case 2 the average operating cost per hour is

$$C(3) = 5\left(\frac{3}{23}\right)\left[\frac{4}{3} + \left(\frac{4}{3}\right)^2\right] + 8\left[1 - \frac{3}{23} - \left(\frac{4}{3}\right)\left(\frac{3}{23}\right) - \left(\frac{4}{3}\right)^2\left(\frac{3}{23}\right)\right]$$

$$\doteq \$5.74/\text{hr}.$$

Thus it is cheaper to switch at $k=2$ even though the hourly cost per operating hour is higher since switching at $k=2$ rather than $k=3$ yields a higher p_0; hence the system is idle more often which more than makes up for the higher high-speed operating cost. If, however, the high-speed operating cost were even higher, it might turn out that switching at $k=3$ could be more economical (see Problem 3.32). It should be noted that the operating costs of \$5.60/hr and \$5.74/hr, respectively, are not costs per operating hour, but rather average costs per hour, including hours or portions of hours when the machine is idle. It should further be noted that to obtain a true optimal operating policy, the cost of customer wait should also be included. This topic will be taken up in greater detail in Chapter 7, Section 7.2.1.

The model above can be generalized to a c-server system with a rate switch at $k>c$ (see Problem 3.34). One can also generalize the model by having two rate switches (or even more if desired) at k_1 and k_2 ($k_2 < k_1$), say (see Problem 3.35), and could also derive results for a multiserver, multirate, state-dependent model of this type.

We next present a model with state-dependent service where the mean service rate changes whenever the system size changes. We again assume a single-server, Markovian state-dependent-service model, this time with μ_n given by

$$\mu_n = n^\alpha \mu. \tag{3.57}$$

Once again utilizing Equation (3.6) we have

$$p_n = \frac{p_0 \lambda^n}{n^\alpha (n-1)^\alpha (n-2)^\alpha \cdots (1)^\alpha \mu^n}$$

$$= \frac{p_0 \lambda^n}{(n!)^\alpha \mu^n}$$

$$= \frac{p_0 r^n}{(n!)^\alpha} \qquad (r = \lambda/\mu), \tag{3.58}$$

where

$$p_0 = \left[\sum_{n=0}^{\infty} \frac{r^n}{(n!)^\alpha} \right]^{-1}. \tag{3.59}$$

The infinite series of (3.59) converges for any r as long as $\alpha > 0$, but it is not obtainable in closed form unless $\alpha = 1$.[9] Thus to evaluate p_0, numerical methods must be used. One procedure to evaluate p_0 would be to calculate

$$\hat{p}_0(N) = \left[\sum_{n=0}^{N} \frac{r^n}{(n!)^\alpha} \right]^{-1}$$

for successively increasing values of N and terminate the procedure [call $\hat{p}_0(N') = p_0$] when two successive $\hat{p}_0$ values differ by less than some pre-specified arbitrarily small value, that is, when $\hat{p}_0(N') - \hat{p}_0(N'+1) < \epsilon$. How-ever, this does not provide us with any bound on the error produced by discarding the tail of the series $\sum_{N'+1}^{\infty} [r^n/(n!)^\alpha]$. Nevertheless, intuitively we feel this a reasonable procedure.

However, when $\alpha \geqslant 1$, we can, in fact, obtain a rigorous bound on the error caused by discarding $\sum_{N'+1}^{\infty} [r^n/(n!)^\alpha]$ in estimating p_0. To do this we note for $\alpha \geqslant 1$ and any $N \geqslant 0$ that

$$\sum_{n=N}^{\infty} \frac{r^n}{(n!)^\alpha} \leqslant \sum_{n=N}^{\infty} \frac{r^n}{n!}.$$

---

[9] When $\alpha = 1$, $\sum_{n=0}^{\infty} r^n/n! = e^r$. This reduces to the ample-server model presented in Section 3.5.

Now $\sum_{n=N}^{\infty} r^n/n!$ is the tail of the series e^r; therefore, given any arbitrary $\epsilon > 0$, there exists an M such that

$$\sum_{n=M}^{\infty} \frac{r^n}{n!} = e^r - \sum_{n=0}^{M-1} \frac{r^n}{n!} < \epsilon. \tag{3.60}$$

Thus

$$\sum_{n=M}^{\infty} \frac{r^n}{(n!)^{\alpha}} < \epsilon$$

and

$$\left[\sum_{n=0}^{M-1} \frac{r^n}{(n!)^{\alpha}} + \epsilon \right]^{-1} < \left[\sum_{n=0}^{\infty} \frac{r^n}{(n!)^{\alpha}} \right]^{-1} = p_0,$$

or

$$\sum_{n=0}^{M-1} \frac{r^n}{(n!)^{\alpha}} + \epsilon > \frac{1}{p_0},$$

which gives

$$\sum_{n=0}^{M-1} \frac{r^n}{(n!)^{\alpha}} > \frac{1}{p_0} - \epsilon.$$

Letting $N' = M - 1$ we have

$$\hat{p}_0(N') = \left[\sum_{n=0}^{N'} \frac{r^n}{(n!)^{\alpha}} \right]^{-1} < \frac{1}{(1/p_0) - \epsilon} = \frac{p_0}{1 - \epsilon p_0} < \frac{p_0}{1 - \epsilon}. \tag{3.61}$$

Also

$$\hat{p}_0(N') = \left[\sum_{n=0}^{N'} \frac{r^n}{(n!)^{\alpha}} \right]^{-1} > \left[\sum_{n=0}^{\infty} \frac{r^n}{(n!)^{\alpha}} \right]^{-1} = p_0. \tag{3.62}$$

Hence, from (3.61) and (3.62),

$$p_0 < \hat{p}_0(N') < \frac{p_0}{1-\epsilon} \doteq p_0(1+\epsilon), \tag{3.63}$$

since, for ϵ very small, $1/(1-\epsilon) \doteq 1+\epsilon$. Thus for any prespecified ϵ, one can find an M from (3.60), namely, the smallest M which satisfies (3.60). If we let N' equal $M-1$ and calculate the estimate of p_0 by $\hat{p}_0(N')$, we are guaranteed an error within the bounds given by (3.63).

To illustrate this model, suppose the Goode-Schine boys of Example 3.6 can modify their machine for an infinite number of speed switches, with mean rate given by

$$\mu_n = n^2\mu,$$

where μ equals μ_1 of Example 3.6, namely, $(1/40)$. However, the cost of operating the machine per operating hour is estimated to be \$10, regardless of the particular speed at which it is operating. Is it economical to modify the machine? The average hourly cost of operation under the modification would be

$$10(1-p_0). \tag{3.64}$$

Prior to calculating p_0, we must first specify ϵ. Let us set ϵ at 0.001. Thus, from (3.63), we have

$$p_0 < \hat{p}_0(N') \leqslant 1.001p_0,$$

or equivalently,

$$0.999\hat{p}_0(N') \leqslant p_0 < \hat{p}_0(N'). \tag{3.65}$$

Using (3.60) we must find M such that

$$e^{\frac{4}{3}} - \sum_{n=0}^{M-1} \frac{(4/3)^n}{n!} < 0.001$$

or

$$\sum_{n=0}^{M-1} \frac{(4/3)^n}{n!} > 3.7937 - 0.001 \doteq 3.7927.$$

The smallest M satisfying the inequality is $M=8$ since $\sum_{n=0}^{7}(4/3)^n/n! = 3.7934$ while $\sum_{n=0}^{6}(4/3)^n/n! = 3.7919$.

Now p_0 is estimated by

$$\hat{p}_0 = \left[\sum_{n=0}^{7} \frac{(4/3)^n}{(n!)^2} \right]^{-1} \doteq 0.3510,$$

and the average hourly operating cost is estimated at[10]

$$10(1 - \hat{p}_0) \doteq 10(1 - 0.3510) \doteq \$6.49.$$

Hence it does not pay them to modify the machine.

3.8 QUEUES WITH IMPATIENCE

The intent of this section is to discuss the effects of customer impatience upon the development of waiting lines of the $M/M/c$ type. These concepts may be easily extended to other Markovian models in a reasonably straightforward fashion and will not be explicitly pursued. However, some examples of impatience are discussed later for the $M/G/1$ queue.

A customer is said to be impatient if he tends to join the queue only when a short wait is expected and tends to remain in line if his wait has been sufficiently small. The impatience that results from an excessive wait is just as important in the total queueing process as the arrivals and departures. When this impatience becomes sufficiently strong and customers leave before being serviced, the manager of the enterprise involved must take action to reduce the congestion to levels that customers can tolerate. The models subsequently developed find practical application in this attempt of management to provide adequate service for its customers with tolerable waiting.

Impatience generally takes three forms. The first is balking, the reluctance of a customer to join a queue upon arrival, the second reneging, the reluctance to remain in line after joining and waiting, and the third jockeying between lines when each of a number of parallel lines has its own queue.

3.8.1 $M/M/1$ Balking

In real practice, it is often likely that arrivals become discouraged when the queue is long and may not wish to wait. One such model could be $M/M/c/K$; that is, if people see K ahead of them in the system, they do not join. Generally, unless K is the result of a physical restriction such as no more places to park or room to wait, people will not act quite like that voluntarily. Rarely do all customers have exactly the same discouragement limit all the time.

---

[10]Using (3.65) we know the average operating cost will be between \$6.490 and
$10[1 - (0.3510)(0.999)] = \$6.494$.

Haight (1957) considers another model of balking for $M/M/1$ in which there is a length K_q which is the greatest queue length at which an arrival would not balk. This length K_q is a random variable whose distribution is the same for all customers. There is a third well-known approach to $M/M/1$ balking in which there is no fixed length K_q associated with the problem but rather a series of monotonically decreasing functions of length which multiply the arrival rate λ. There can be balking for all queue lengths, while in the previous case there is balking only for queue lengths greater than K_q. It turns out that there is a mathematical equivalence between the stationary distributions associated with these latter two cases, and we begin by looking at Haight's model

As defined in the foregoing, K_q is the greatest length at which an arrival does not balk and has the same distribution for all arriving customers, say $F(n) = \Pr\{K_q \leqslant n\}$. So $F(n-1)$ is the balking distribution in the sense that it is the probability that the arrival refuses to join when n are in the system. An arriving customer joins the queue if $K_q \geqslant n$; that is,

$$\Pr\{\text{arrival joins the queue}\} = \Pr\{K_q \geqslant n\}$$

$$= 1 - \Pr\{K_q < n\}$$

$$= 1 - F(n-1) \qquad (n \geqslant 0; \ F(-1) \equiv 0).$$

Let us define $1 - F(n)$ as $G(n)$, so that $G(n-1)$ is now the probability that an arrival joins the queue when there are n in the system.

This process is still birth–death, but the arrival rate must now be adjusted to $\lambda_n = \lambda G(n-1)$. Hence we know from (3.6) that

$$p_n = p_0 \prod_{i=1}^{n} \frac{\lambda_{i-1}}{\mu_i}$$

$$= p_0 \prod_{i=1}^{n} \frac{\lambda G(i-2)}{\mu}.$$

Since $G(-1) = 1 - F(-1) = 1$, we have

$$p_n = p_0 \left(\frac{\lambda}{\mu}\right)^n \prod_{i=1}^{n-1} G(i-1) \qquad (n > 1), \tag{3.66}$$

and

$$p_1 = p_0 \frac{\lambda}{\mu}.$$

We get p_0 in the usual way as

$$p_0 = \left[1 + \frac{\lambda}{\mu} + \sum_{n=2}^{\infty} \left(\frac{\lambda}{\mu} \right)^n \prod_{i=1}^{n-1} G(i-1) \right]^{-1},$$

and convergence of the infinite series above is a necessary and sufficient condition for ergodicity, with $\lambda/\mu < 1$ sufficient only.

It should be clear from (3.66) that

$$p_{n+1} = \rho G(n-1) p_n \qquad (\rho = \lambda/\mu). \tag{3.67}$$

Hence if $p_n \neq 0$ and $p_{n+1} = 0$, then $G(n-1) = 0$ and $p_m = 0$ for all $m > n$; that is, there is complete balking for system sizes $\geqslant n$.[11]

Example 3.7

It is estimated from observation of an $M/M/1$ balking situation that

$$G(n) = \begin{cases} \dfrac{1}{n+1} & (0 \leqslant n < k) \\ 0 & (n \geqslant k) \end{cases}.$$

The average arrival rate is one per hour and the mean service time is equal to 30 min. We desire to find the stationary probabilities associated with this problem, and then to find L. Equation (3.66) gives

$$p_n = \begin{cases} p_0 \rho & (n = 1) \\ p_0 \rho^n \prod_{i=1}^{n-1} G(i-1) & (n > 1) \end{cases}.$$

Here

$$\rho = \left(\frac{1}{\text{hr}} \right) \left(\frac{1}{2} \text{hr} \right)$$

$$= \frac{1}{2},$$

---

[11]Note that the $M/M/1/K$ model is a special case of this model with

$$G(n) = \begin{cases} 1 & (0 \leqslant n < K-1) \\ 0 & (n \geqslant K-1) \end{cases}$$

and hence

$$p_n = \begin{cases} \dfrac{p_0}{2} & (n=1) \\[2ex] \dfrac{p_0}{2^n} \displaystyle\prod_{i=1}^{n-1} \dfrac{1}{i} & (2 \leqslant n \leqslant k+1) \\[2ex] 0 & (n>k+1) \end{cases},$$

with

$$p_0 = \left[1 + \frac{1}{2} + \sum_{n=2}^{k+1} \left(\frac{1}{2}\right)^n \prod_{i=1}^{n-1} \frac{1}{i} \right]^{-1}.$$

That is,

$$p_n = \begin{cases} \dfrac{p_0}{2^n(n-1)!} & (1 \leqslant n \leqslant k) \\[2ex] 0 & (n>k) \end{cases}$$

and

$$p_0 = \left\{ 1 + \sum_{n=1}^{k+1} [2^n(n-1)!]^{-1} \right\}^{-1}.$$

Therefore

$$L = p_0 \sum_{n=1}^{k+1} \frac{n}{2^n(n-1)!}.$$

Example 3.8

Suppose it were ascertained from data that balking is present and that the system sizes are binomially distributed, namely,

$$p_n = \binom{k}{n} p^n (1-p)^{k-n} \qquad (0 \leqslant n \leqslant k; 0<p<1).$$

We wish now to find the impatience function, $G(n)$, which gives rise to this output. The utilization factor ρ is found as

$$\rho = \frac{p_1}{p_0} = \frac{kp}{1-p}.$$

From (3.67) we see that

$$G(n-1) = \frac{p_{n+1}}{\rho p_n};$$

hence

$$G(n-1) = \frac{\binom{k}{n+1} p^{n+1}(1-p)^{k-n-1}}{[kp/(1-p)]\binom{k}{n} p^n (1-p)^{k-n}}$$

$$= \begin{cases} \dfrac{k-n}{nk+k} & (0 < n \le k) \\ 0 & (k < n) \end{cases}.$$

It is interesting to note that $G(n)$ is independent of p, which is a parameter of p_n. Regardless of the value of p associated with the distribution of system size, $G(n)$ depends only on the fixed value of k and the variable n.

For the third approach to balking models we have a series of monotonically decreasing functions of the length multiplying the average rate λ. Let b_n be this function such that $\lambda_n = b_n \lambda$ and

$$0 \le b_{n+1} \le b_n \le 1 \qquad (n > 0; \ b_0 \equiv 1).$$

Hence, from Equation 3.6, we may write

$$p_n = p_0 \prod_{i=1}^{n} \frac{\lambda_{i-1}}{\mu_i}$$

$$= p_0 \left(\frac{\lambda}{\mu}\right)^n \prod_{i=1}^{n} b_{i-1}. \qquad (3.68)$$

Possible examples that may be useful for the discouragement function b_n are $1/(n+1)$, $1/(n^2+1)$, and $e^{-\alpha n}$. The discouragement function b_n corresponds to $G(n-1)$ of the previous model, whose probabilities are given in Equation (3.66), and we observe that the two formulas are virtually identical. The only difference is that (3.68) may be nonzero for all values of n, while (3.66) becomes zero for all $n \ge K_q + 1$. People are not always discouraged because of queue size but may attempt to estimate how long they would have to wait. If the queue is moving quickly, then the person may join a long one. On the other hand, if the queue is slow-moving, he will become discouraged even if the line is relatively short. Now if n people are in the system, an estimate for the average waiting time might be n/μ, if the customer had an idea of μ. We usually do, so a very possible balking function might be $b_n = e^{-\alpha n/\mu}$.

3.8.2 $M/M/1$ Reneging

Customers who tend to be impatient may not always be discouraged by excessive queue size, but may instead join the queue to see how long their wait may become, all the time retaining the prerogative to renege if their estimate of their total wait is intolerable. Haight (op. cit.) has proposed a single-channel birth–death model where both reneging and the simple balking of the previous section exist, which gives rise to a reneging function $r(n)$ defined by

$$r(n) = \lim_{\Delta t \to 0} [\Pr\{\text{unit reneges during } \Delta t \text{ when there are } n \text{ customers}$$

$$\text{in the system}\}/\Delta t]$$

$$[r(0) \equiv 0 \equiv r(1)].$$

This new process is still birth–death, but the death rate must now be adjusted to $\mu_n = \mu + r(n)$. Thus it follows from (3.6) that

$$p_n = p_0 \prod_{i=1}^{n} \frac{\lambda_{i-1}}{\mu_i}$$

$$= p_0 \prod_{i=1}^{n} \frac{\lambda G(i-2)}{\mu + r(i)}$$

$$= p_0 \lambda^n \prod_{i=1}^{n} \frac{G(i-2)}{\mu + r(i)} \qquad (n \geqslant 1),$$

where

$$p_0 = \left[1 + \sum_{n=1}^{\infty} \lambda^n \prod_{i=1}^{n} \frac{G(i-2)}{\mu + r(i)} \right]^{-1}.$$

A good possibility for the reneging function $r(n)$ could be $e^{\alpha n/\mu}$, $n \geq 2$. A waiting customer would probably estimate his average system waiting time as n/μ if $n-1$ customers were in front of him, assuming an estimate for μ were available. Again, the probability of a renege would be estimated by a function of the form $e^{\alpha n/\mu}$. Further work on Markovian impatience beyond the level of this book has been published by Barrer (1957a, b), and the reader is referred to these, if interested.

As was mentioned in the introduction to this section, there is yet an additional form of impatience called jockeying, that is, moving back and forth among the several subqueues before each of several multiple channels. Though this phenomenon is quite interesting and clearly applicable to numerous real-life situations, jockeying is analytically difficult to pursue very far, especially when we have more than two channels, since the probability functions become too complicated and the general concepts become hazy. Even though partial results can be obtained for the two-channel case, no particular insight to the multichannel cases is gained from it. If, however, the reader is specifically interested in this subject, he is referred to Koenigsberg (1966).

PROBLEMS

3.1 Verify Equations (3.3a) and (3.3b) by using (3.2) in (3.1). (*Hint*: To obtain (3.3a), let $s = t + \Delta t$. What is required to obtain (3.3b)?)

3.2 Find the probability that k or more are in an $M/M/c/\infty$ system for any $k \geq c$.

3.3 For the $M/M/c/\infty$ model, give an expression for p_n in terms of p_c, rather than p_0, and then derive L_q and $W_q(t)$ in terms of ρ and p_c.

3.4 Show for $M/M/c/\infty$ that $L = L_q + \lambda/\mu$ by an expected value argument. (*Hint*: Let N represent the random variable (RV) "number in the system" and N_q the RV "number in queue." We then have $N_q = N - N$ when $N < c$ and $N_q = N - c$ when N is c or greater. Now utilize Equations (3.9) and (3.11) to prove the desired result.)

3.5 Show that calculating W_q by $\int_0^\infty t \, dW_q(t)$ for the $M/M/c/\infty$ model yields the same result as given by Equation (3.16).

3.6 Derive for $M/M/c/\infty$ the conditional density function $w_q(t|t > 0)$, that is, the density of the waiting time in queue for those who wait. Also find the expected wait in queue for those who wait.

3.7 For $M/M/c/\infty$, derive $w(t)$, the density function of the total time spent in the system, and use it to calculate W. Check your result with

Equation (3.17). (*Hint*: If the approach of deriving the CDF $W(t)$ is used, two cases must be considered, one when there are less than c in the system and an arrival can go into service immediately and the other when c or more are in the system. For the latter, a convolution of an Erlang and an exponential is required. This is quite difficult. It is easier here to get $w(t)$ directly by $w(t)\,dt \doteq \Pr\{t \leqslant T \leqslant t + dt\}$ and realizing that the probability of a service in dt is $\mu\,dt$.)

3.8 Show that the probability of any server being busy is $\lambda/c\mu$ for an $M/M/c$ model.

3.9 The Deputy Inspector General for Inspection and Safety administers the Air Force Accident and Incident Investigation and Reporting Program. It has established 25 investigation teams to analyze and evaluate each accident/incident to make sure it is properly reported to accident investigation boards. Each of these teams is dispatched to the locale of the accident or incident as each requirement for such support occurs. Support is only rendered those commands who have neither the facilities nor qualified personnel to conduct such services. Each accident/incident will require a team being dispatched for a random amount of time, apparently exponential with mean of 3 weeks. Requirements for such support are received by the Deputy Inspector General's office as a Poisson process with mean rate of 347/yr. At any given time, two teams are not available due to personnel leaves, sickness, and so on. Find the expected time spent by an accident/incident in and waiting for evaluation.

3.10 An organization is presently involved in the establishment of a telecommunication center so that they may experience a more rapid outgoing message capability. Overall, the center is responsible for the transmission of outgoing messages, and receives and distributes incoming messages. The center manager at this time is primarily concerned with determining the number of transmitting personnel required at the new center. Outgoing message transmitters are responsible for making minor corrections to messages, assigning numbers when absent from original message forms, maintaining an index of codes, a 30-day file of outgoing messages, and the actual transmission of the message. It has been predetermined that this process is exponential and requires a mean processing time of 28 min/ message. Transmission personnel will operate at the center 7 hr/day, 5 days/week. All outgoing messages will be processed in the order they are received and follow a Poisson process with a mean rate of 21/7hr-day. Processing on messages requiring transmission must be started within an average of 2 hr from the time they arrive at the center. Determine the minimum number of transmitting personnel to accomplish this service criterion. If the service criterion were to have the probability of any message waiting for the start of transmission for more than 3 hr to be less than 0.05, how many transmitting personnel are required?

3.11   A small branch bank has two tellers, one for receipts and one for withdrawals. Customers arrive to each teller's cage according to a Poisson distribution with a mean of 20/hr. (Total mean arrival rate to the bank is 40/hr.) The service time of each teller is exponential with a mean of 2 min. The bank manager is considering changing the setup to allow each teller to handle both withdrawals and deposits to avoid the situations which arise from time to time when the queue is sizable in front of one teller while the other is idle. However, since the tellers would have to handle both receipts and withdrawals, their efficiency would decrease to a mean service time of 2.4 min. Compare the present system to the proposed system with respect to the total expected number of people in the bank, the expected time a customer would have to spend in the bank, the probability of a customer having to wait more than 5 min, and the average idle time of the tellers.

3.12   The Hott Too Trott Heating and Air Conditioning Company must choose between operating one of two types of service shops for maintaining its trucks. It estimates that trucks will arrive at the maintenance facility according to a Poisson distribution with mean rate of one every 40 min and feels that this rate is independent of which facility is chosen. In the first type of shop, there are dual facilities operating in parallel; each facility can service a truck in 30 min on the average (the service time follows an exponential distribution). In the second type there is a single facility, but it can service a truck in 15 min on the average (service times are also exponential in this case). To help management to decide, they ask their operations research analyst, Mr. C. Raf Tee, to answer the following:

(a) How many trucks, on the average, will be in each of the two types of facilities?

(b) How long, on the average, will a truck spend in each of the two types of facilities?

(c) Management calculates that each minute a truck must spend in the shop reduces contribution to profit by two dollars. They also know from previous experience in running dual facility shops that the cost of operating such a facility is one dollar per minute (including labor, overhead, etc.). What would the operating cost per minute have to be for operating the second type (single facility) shop in order for there to be no difference between the two types of shops?

3.13   The ComPewter Company, which leases electronic data processing (EDP) equipment, considers it necessary to overhaul its equipment once a year. Alternative 1 is to provide two separate maintenance stations where all work is done by hand (one machine at a time) for a total annual cost of

$150,000. The maintenance time for a machine has an exponential distribution with a mean of 6 hr. Alternative 2 is to provide one maintenance station with mostly automatic equipment involving an annual cost of $200,000. In this case, the maintenance time for a machine has an exponential distribution with a mean of 3 hr. For both alternatives, the machines arrive according to a Poisson input with a mean arrival rate of one every 8 hr (since the company leases such a large number of machines we can consider the machine population as infinite). The cost of idle time per machine is $30/hr. Which alternative should the company choose? Assume that the maintenance facilities are always open and that they work $(24)(365) = 8760$ hr/yr.

3.14 Show that $M/M/1/\infty$ is always better with respect to L than an $M/M/2/\infty$ with the same ρ.

3.15 Show for the $M/M/c/K$ model that taking the limit for p_n and p_0 as $K \to \infty$ and restricting $\lambda/c\mu < 1$ in Equations (3.24) and (3.25) yield the results obtained for the $M/M/c/\infty$ model as given in Equations (3.9) and (3.11).

3.16 Show by using Equations (3.25), (3.27), and (3.28) that Equations (3.29) and (3.30) give the same results, that is, that $L - L_q = \lambda'/\mu$.

3.17 The Fowler-Heir Oil Company operates a crude oil unloading port at their major refinery. The port has six unloading berths and four unloading crews. When all berths are full, arriving ships are diverted to an overflow facility 20 miles down river. Tankers arrive according to a Poisson process with a mean of one every 2 hr. It takes an unloading crew, on the average, ten hours to unload a tanker, the unloading time following an exponential distribution. Tankers waiting for unloading crews are served on a first-come, first-served basis. Company mangement wishes to know the following.

(a) On the average, how many tankers are at the port?
(b) On the average, how long does a tanker spend at the port?
(c) What is the average arrival rate at the overflow facility?
(d) The company is considering building another berth at the main port. It is estimated that construction and maintenance costs would amount to X dollars per year. The company estimates that to divert a tanker to the overflow port when the main port is full costs Y dollars. What is the relation between X and Y for which it would pay for the company to build an extra berth at the main port?

3.18 Fly-Bynite Airlines has a telephone exchange with three lines, each manned by a clerk during their busy periods. During their peak 3 hours per 24 hr period, many callers are unable to get into the exchange (there is no provision for callers to hold if all servers are busy). The company estimates, because of severe competition, that 60% of the callers not getting through use another airline. If the number of calls during these peak periods is roughly Poisson with a mean of 20 calls/hr and each clerk spends on an average 6 min with a caller, his service time being approximately exponentially distributed, and the average customer spends $70 per trip, what is the average daily loss due to the limited service facilities? (We may assume that the number of people not getting through during off-peak hours is negligible.) If a clerk's pay and fringe benefits cost the company $8/hr and a clerk must work an 8-hr shift, what is the optimum number of clerks to employ? The three peak hours occur during the 8-hr day shift. At all other times, one clerk can handle all the traffic, and since the company never closes the exchange, exactly one clerk is used on the off-shifts. Assume that the cost of adding lines to the exchange is negligible.

3.19   Show that the results obtained for the ample-server model $(M/M/\infty)$ can also be developed by taking the limit as $c \to \infty$ in the results for the $M/M/c/\infty$ model.

3.20   Derive the steady-state $M/M/\infty$ solution directly from the transient.

3.21   Find the mean number in an $M/M/\infty$ system at any point in time from Equation (3.39).

3.22   Find the partial differential equation corresponding to (3.36) for the time-dependent birth–death process where λ_n is replaced by $\lambda(t)$ and $\mu_n = n\mu$. Then solve for the generating function, assuming that the initial state is chosen according to a Poisson distribution.

3.23   The Good Writers Correspondence Academy offers a go-at-your-own-pace correspondence course in good writing. New applications are accepted at any time and the applicant can enroll immediately. Past records indicate applications follow a Poisson distribution with a mean of 8/month. An applicant's mean completion time is found to be 10 weeks, with the distribution of completion times being exponential. On the average, how many "pupils" are enrolled in the school at any given time?

3.24   An application of an $M/M/\infty$ model to the field of *inventory control* is as follows. A manufacturer of a very expensive, rather infrequently demanded item uses the following inventory control procedure. He keeps a safety stock of S units on hand. The customer demand for units can be described by a Poisson process with mean λ. Every time a request for a unit is made (a customer demand) an order is placed on the factory to manufacture another (this is called a one-for-one ordering policy). The amount of time required to manufacture a unit is exponential with mean $1/\mu$. There is a carrying cost for inventory on shelf of h dollars per unit per unit time held on shelf (representing capital tied up in inventory which could be invested and earning interest, insurance costs, spoilage, etc.) and a shortage cost of p dollars per unit (a shortage occurs when a customer requests a unit and there is none on shelf, that is, safety stock is depleted to zero). It is assumed that customers who request an item but find that there is none immediately available will wait until stock is replenished by orders due in (this is called backordering or backlogging); thus one can look at the charge p as a discount given to the customer because he must wait for his request to be satisfied.

The problem, then, becomes one of finding the optimal value of S which minimizes total expected costs per unit time; that is, find the S which minimizes

$$E[C] = h \sum_{z=1}^{S} zp(z) + p\lambda \sum_{z=-\infty}^{0} p(z) \qquad (\$/\text{unit time}),$$

where z is the steady-state on-hand inventory level ($+$ means items on shelf, $-$ means items in backorder) and $p(z)$ its probability frequency function. Note that $\sum_{z=1}^{S} zp(z)$ is the average value of the safety stock and $\lambda\sum_{z=-\infty}^{0} p(z)$ is the expected number of backorders per unit time since $\sum_{z=-\infty}^{0} p(z)$ is the percentage of time there is no on-shelf safety stock and λ is the average request rate. If $p(z)$ could be determined, one could optimize $E[C]$ with respect to S.

(a) Show the relationship between Z and N, where N denotes the number of orders outstanding, that is, the number of orders currently being processed at the factory. Hence relate $p(z)$ to p_n.

(b) Show that $\{p_n\}$ are the steady-state probabilities of an $M/M/\infty$ queue if one considers the order processing procedure as the queueing system. State explicitly what the input and service mechanisms are.

(c) Find the optimum S for $\lambda = 8/\text{month}$, $1/\mu = 3$ days, $h = \$50/\text{unit}/\text{month}$ held and $p = \$500/\text{unit}$ backordered.

3.25 For the $M/M/\infty$ model, find the transient solution, $p_n(t)$, when the initial state is one, that is, $p_1(0) = 1$ and $p_n(0) = 0$, $n \neq 1$.

3.26 For Problem 3.25, plot $p_0(t)$ and $p_3(t)$ as functions of time (assuming the initial state is zero). Indicate the steady-state values p_0 and p_3 on the graph.

3.27 Show that Equation (3.48) yields the effective mean arrival rates, λ', for the following models:

(a) $\lambda' = \lambda$, $M/M/1/\infty$, $M/M/c/\infty$.
(b) $\lambda' = \lambda(1 - p_K)$, $M/M/1/K$, $M/M/c/K$.
(c) $\lambda' = \lambda(M - L)$, M-machine, c-repairmen model.

3.28 A coin operated dry-cleaning store has five machines. The operating characteristics of the machines are such that any machine breaks down according to a Poisson process with mean breakdown rate of one per day. A repairman can fix a machine according to an exponential distribution with a mean repair time of one-half a day. Currently three repairmen are on duty. The manager, Lew Cendirt, has the option of replacing these three repairmen with a super-repairman whose salary is equal to the total of the three regulars, but who can fix a machine in one-third the time, that is, in one-sixth of a day. Should he be hired?

3.29 Suppose that each of five machines in a given shop breaks down according to a Poisson law at an average rate of one every 10 hr, and the failures are repaired one at a time by two maintenance men operating as two channels, such that each machine has an exponentially distributed servicing requirement of mean 5 hr.

(a) What is the probability that exactly one machine will be "up" at any one time?

(b) If performance of the workmen is measured by the ratio of average waiting time to average service time, what is this measure for the current situation?

(c) What is the answer to (a) if an identical spare machine is put on reserve?

3.30 Find the steady-state probabilities for a machine repair problem with M machines, Y spares, c repairmen ($c \leqslant Y$) but with the following discipline: if no spares are on hand and a machine fails ($n = Y + 1$), the remaining $M - 1$ machines running are stopped until a machine is repaired; that is, if the machines are to run, there must be M running simultaneously.

3.31 Very often in real-life modeling, even when the calling population is finite, an infinite source model is used as an approximation. To compare the two models, calculate L for Example 3.5 assuming the calling population (number of machines) is infinite. Also calculate L for an exact model when the number of machines equals 10 and 5, respectively, for $M\lambda = 1/3$ in both cases, and compare to the calculations from an approximate infinite source model. How do you think ρ affects the approximation? (*Hint*: When using an infinite source model as an approximation to a finite source model, λ must be set to $M\lambda$.)

3.32 Find the average operating costs per hour of Example 3.6 when

(a) $C_1 =$ low-speed cost $= \$5/$operating hour; $C_2 =$ high-speed cost $= \$10/$operating hour.

(b) $C_1 = \$5/$operating hour; $C_2 = \$12/$operating hour.

(c) Evaluate (b) for $k = 4$, and what now is the best policy?

3.33   Assume we have a two-state, state-dependent service model as in Section 3.7 with $\rho_1 = 4/3$ and $\rho = 2/3$. Suppose that the customers are lawn-treating machines owned by the Green Thumb Lawn Service Company and these machines require, at random times, greasing on the company's two-speed greasing machine. Furthermore, suppose that the cost per operating hour of the greaser at the lower speed, C_1, is \$5 and at the high speed, C_2 is \$22. Also the company estimates the cost of downtime of a lawn treater to be \$1/hr. What is the optimal switch point, k? (*Hint*: Try several values of k starting at $k = 1$ and compute the total expected cost).

3.34   Derive the steady-state system-size probabilities for a c-server Poisson input, exponential state-dependent service model where the mean service rate switches from μ_1 to μ when $k > c$ are in the system.

3.35   Derive the steady-state system-size probabilities for a single-server Poisson input, exponential state-dependent-service model with mean rates μ_1 $(1 \leqslant n < k_1)$, μ_2 $(k_1 \leqslant n < k_2)$ and μ $(n \geqslant k_2)$.

3.36   It is known for an $M/M/1$ balking situation that the stationary distribution is given by the negative binomial

$$p_n = \binom{N+n-1}{N-1} x^n (1+x)^{-N-n} \qquad (n \geqslant 0, \, x > 0, \, N > 1).$$

Find L, L_q, W, W_q, and $G(n)$.

3.37   For an $M/M/1$ balking model, it is known that $b_n = e^{-\alpha n / \mu}$. Find p_n (for all n).

3.38   Suppose that the $M/M/1$ reneging model of Section 3.8.2 had the balking function $G(n)$ of Example 3.7 and a reneging function $r(n) = n/\mu$. Find the stationary system-size distribution.

3.39   Invent a clever company name and send it to the authors. But do not send one solely to harass them and do not be gross in your language.

Chapter 4
ADVANCED MARKOVIAN
MODELS

This chapter continues the development of models that are amenable to analytic methods and is concerned especially with Markovian problems of the non-birth–death type. That is, we allow changes of more than one over infinitesimal time intervals, but insist on retaining the memoryless Markovian property. The Chapman-Kolmogorov, and backward and forward equations of Chapter 3 are, of course, still valid, and are the essence of the approach to solution for these non-birth–death Markovian problems.

4.1. BULK INPUT $(M^{[X]}/M/1)$

In continuation of our relaxation of the simple assumptions underlying $M/M/1$, let it now be assumed, in addition to the fact that the arrival stream forms a Poisson process, that the actual number of customers in any arriving module is a random variable X, which may take on any positive integral value less than ∞ with probability c_x. It should be clear that this new queueing problem, let it be called $M^{[X]}/M/1$, is still Markovian in the sense that future behavior is a function only of the present and not the past.

We now recall the discussions of Section 1.9, particularly those of the common generalizations of the Poisson process. If λ_x is the arrival rate of the Poisson process of batches of size X, then clearly $c_x = \lambda_x/\lambda$, where λ is the composite arrival rate of all batches and is equal to $\sum_{i=1}^{\infty}\lambda_i$. This total process, which arises from the overlap of the set of Poisson processes with rates $\{\lambda_x, x = 1, 2, \dots\}$, as previously mentioned in Section 1.9 is a *multiple* or *compound Poisson process*.

A set of Chapman-Kolmogorov equations can be derived for this problem in the usual manner with little difficulty, and they give rise to the differential-difference equations

$$\begin{cases} \dfrac{dp_n(t)}{dt} = -(\lambda+\mu)p_n(t) + \mu p_{n+1}(t) + \lambda \sum_{k=1}^{n} p_{n-k}(t)c_k \qquad (n \geqslant 1) \\[2ex] \dfrac{dp_0(t)}{dt} = -\lambda p_0(t) + \mu p_1(t) \end{cases}$$

The last term in the equation for $dp_n(t)/dt$ comes from the fact that a total of n in the system at time $t + \Delta t$ can arise from $n - k$ present at time t with a batch of size k arriving in the subsequent interval of length Δt. If the steady state is then assumed, we obtain the $M^{[X]}/M/1$ stationary equations[1]

$$\begin{cases} 0 = -(\lambda + \mu)p_n + \mu p_{n+1} + \lambda \sum_{k=1}^{n} p_{n-k} c_k & (n \geqslant 1) \\[2mm] 0 = -\lambda p_0 + \mu p_1 \end{cases} \qquad (4.1)$$

To solve the system of equations given by (4.1), let us use a generating-function approach, where

$$P(z) = \sum_{n=0}^{\infty} p_n z^n \qquad (|z| \leqslant 1)$$

and

$$C(z) = \sum_{n=0}^{\infty} c_n z^n \qquad (|z| \leqslant 1)$$

$$= \sum_{n=1}^{\infty} c_n z^n$$

will be the generating functions of the steady-state probabilities $\{p_n\}$ and the batch-size distribution $\{c_n\}$, respectively. If each equation of (4.1) is then multiplied by the appropriate z^n, and if the resultant equations are summed, it is found that

$$0 = -\lambda \sum_{n=0}^{\infty} p_n z^n - \mu \sum_{n=1}^{\infty} p_n z^n + \frac{\mu}{z} \sum_{n=1}^{\infty} p_n z^n + \lambda \sum_{n=1}^{\infty} \sum_{k=1}^{n} p_{n-k} c_k z^n. \quad (4.2)$$

We observe that $\sum_{k=1}^{n} p_{n-k} c_k$ is the probability function for the sum of the steady-state system size and batch size, since this is merely a convolution formula for discrete random variables. It can easily be shown that the generating function of this sum is the product of the respective generating functions (a basic property of all generating functions), namely,

$$\sum_{n=1}^{\infty} \sum_{k=1}^{n} p_{n-k} c_k z^n = \sum_{k=1}^{\infty} c_k z^k \sum_{n=k}^{\infty} p_{n-k} z^{n-k}$$

$$= C(z)P(z).$$

[1]It is not a very difficult matter to extend the results of this section to the $M^{[X]}/M/c$ model. This would be done in the same manner as $M/M/1$ is extended to $M/M/c$.

Hence (4.2) may be rewritten as

$$0 = -\lambda P(z) - \mu[P(z) - p_0] + \frac{\mu}{z}[P(z) - p_0] + \lambda C(z)P(z),$$

and thus

$$\boxed{P(z) = \frac{\mu p_0(1-z)}{\mu(1-z) - \lambda z[1 - C(z)]} \qquad (|z| \leq 1)} \qquad (4.3)$$

We can obtain p_0 with the use of the condition that $P(1) = 1$. That is,

$$1 = \lim_{z \to 1} P(z) = \lim_{z \to 1} \left\{ \frac{\mu p_0(1-z)}{\mu(1-z) - \lambda z[1 - C(z)]} \right\}.$$

But since the numerator and denominator are both 0, we use L'Hôpital's rule and find that

$$\lim_{z \to 1} P(z) = \lim_{z \to 1} \left[\frac{-\mu p_0}{-\mu - \lambda + \lambda C(z) + \lambda z \, dC(z)/dz} \right]$$

$$= \frac{-\mu p_0}{-\mu + \lambda E[X = \text{batch size}]};$$

hence

$$p_0 = 1 - \frac{\lambda E[X]}{\mu}$$

$$= 1 - \rho \qquad (\rho \equiv \lambda E[X]/\mu), \qquad (4.4)$$

with $\rho < 1$ sufficient for stationarity. Given $C(z)$ then, the stationary distribution $\{p_n\}$ would be obtained by using any method to recover the coefficients in the series expansion of $P(z)$ which would come from (4.3).

For example, let us illustrate the foregoing results by assuming that the batch sizes are distributed geometrically, that is, that

$$c_x = (1-\alpha)\alpha^{x-1} \qquad (0 < \alpha < 1).$$

Then

$$C(z) = (1-\alpha) \sum_{n=1}^{\infty} \alpha^{n-1} z^n \qquad (|z| \leq 1)$$

$$= \frac{z(1-\alpha)}{1 - \alpha z} \qquad (|\alpha z| < 1),$$

and thus from Equation (4.3), we find that

$$P(z) = \frac{\mu p_0(1-z)}{\mu(1-z) - \lambda z[1 - z(1-\alpha)/(1-\alpha z)]}$$

where

$$p_0 = 1 - \frac{\lambda E[X]}{\mu}$$

$$= 1 - \frac{\lambda}{\mu(1-\alpha)}$$

$$= 1 - \rho.$$

Algebraic simplification gives

$$P(z) = \frac{\mu p_0(1-z)}{\mu(1-z) - \lambda z(1-z)/(1-\alpha z)}$$

$$= \frac{\mu(1-\rho)(1-\alpha z)}{\mu(1-\alpha z) - \lambda z}$$

$$= \frac{(1-\rho)(1-\alpha z)}{1 - [\alpha + (1-\alpha)\rho]z}$$

$$= (1-\rho)\left\{ \frac{1}{1 - [\alpha + (1-\alpha)\rho]z} - \frac{\alpha z}{1 - [\alpha + (1-\alpha)\rho]z} \right\}.$$

Therefore, utilizing the formula for the sum of a geometric series,

$$P(z) = (1-\rho)\left(\sum_{n=0}^{\infty} \{ [\alpha + (1-\alpha)\rho]z \}^n - \sum_{n=0}^{\infty} \alpha[\alpha + (1-\alpha)\rho]^n z^{n+1} \right),$$

and

$$p_n = (1-\rho)\left\{ [\alpha + (1-\alpha)\rho]^n - \alpha[\alpha + (1-\alpha)\rho]^{n-1} \right\}$$

$$= (1-\rho)[\alpha + (1-\alpha)\rho]^{n-1}[(1-\alpha)\rho] \qquad (n>0).$$

Example 4.1

Let us consider a multistage, machine-line process which produces an assembly in quantity. After the first stage many items are found to have one or more defects which must be repaired before they enter the second stage. It is the job of one workman to make the necessary adjustments to put the assembly back into the stream of the process. The number of defectives per item is registered automatically, and it exceeds two an extremely small number of times. The interarrival times for both units with one and two defective parts are found to be governed closely by exponential distributions, with parameters $\lambda_1 = 1/\mathrm{hr}$ and $\lambda_2 = 2/\mathrm{hr}$, respectively. There are so many different types of parts which have been found defective that an exponential distribution does actually provide a good fit for the workman's service-time distribution, with mean $1/\mu = 10$ min.

But it is subsequently noted that the rates of defectives have increased, although not continuously. It is therefore decided to put another workman on the job who would concentrate on repairing those units with two defectives, while the original workman works only on singles. When to add the additional man will be decided on the basis of a cost analysis.

Now there are a number of alternative cost structures available, and it is decided by management that the expected cost of the system to the company will be based upon the average delay time of assemblies in for repair, which is directly proportional to the average number of units in the system, L. To find L under the assumption that there are only two possible batch sizes, we substitute $C(z) = (\lambda_1 z + \lambda_2 z^2)/\lambda$ into (4.3) and then evaluate the derivative of $P(z)$ at $z = 1$. So

$$P(z) = \frac{\mu(1-\rho)(1-z)}{\mu(1-z) - z(\lambda - \lambda_1 z - \lambda_2 z^2)}$$

and

$$P'(z) = \mu(1-\rho)$$

$$\times \left\{ \frac{-\mu(1-z) + z(\lambda - \lambda_1 z - \lambda_2 z^2) - (1-z)(-\mu - \lambda + 2\lambda_1 z + 3\lambda_2 z^2)}{\left[\mu(1-z) - z(\lambda - \lambda_1 z - \lambda_2 z^2)\right]^2} \right\}.$$

Thus using L'Hôpital's rule twice, we get

$$L = P'(1) = \lim_{z \to 1} P'(z)$$

$$= \lim_{z \to 1} \mu(1-\rho) \left\{ \frac{2\lambda_1 z + 6\lambda_2 z^2 - 2\lambda_1 - 6\lambda_2 z}{2\left[\mu(1-z) - z(\lambda - \lambda_1 z - \lambda_2 z^2)\right]\left[-\mu - \lambda + 2\lambda_1 z + 3\lambda_2 z^2\right]} \right\}$$

$$= \lim_{z \to 1} \frac{\mu(1-\rho)}{2}$$

$$\times \left\{ \frac{2\lambda_1 + 12\lambda_2 z - 6\lambda_2}{\left(-\mu - \lambda + 2\lambda_1 z + 3\lambda_2 z^2\right)^2 + \left(\mu - \mu z - \lambda z + \lambda_1 z^2 + \lambda_2 z^3\right)\left(2\lambda_1 + 6\lambda_2 z\right)} \right\}$$

$$= \frac{\mu(1-\rho)(\lambda_1 + 3\lambda_2)}{(\mu - \lambda_1 - 2\lambda_2)^2}, \tag{4.5}$$

with

$$\rho = \frac{\lambda_1 + 2\lambda_2}{\mu}.$$

If we define λ_1/μ as ρ_1 and λ_2/μ as ρ_2, then the expected batch size, $E[X]$, is $(\lambda_1 + 2\lambda_2)/\lambda$, and ρ may be written as

$$\rho = \rho_1 + 2\rho_2.$$

Thus dividing the numerator and denominator of (4.5) by $\mu^2(1-\rho)$,

$$L = \frac{\rho_1 + 3\rho_2}{1 - \rho_1 - 2\rho_2}. \tag{4.6}$$

For the example then, $\rho = 5/6$, $\rho_1 = 1/6$, and $\rho_2 = 2/6$, which yields $L = 7$.

If C_1 is the cost per unit time per waiting repair and C_2 the cost of a workman per unit time, then the expected cost C of a single-server system is

$$C = C_1 L + C_2$$

$$= C_1 \frac{\rho_1 + 3\rho_2}{1 - \rho_1 - 2\rho_2} + C_2.$$

Then if a second repairman sets up a separate service channel, the additional cost of his time is incurred, over and above the cost of the items in the queue. In this case, one would now have two queues. The singlet line would be a standard $M/M/1$, but the doublets would not. However, a single Poisson stream of doublets is merely a special case of the multiple Poisson bulk-input model with $\lambda_1 = 0$. The expected number of required repairs in the system is then the sum of the expected values of the two streams. Since the first stream is a standard $M/M/1$, its expected length is

$$L_1 = \frac{\rho_1}{1 - \rho_1}.$$

To get the expected length of the second line we let $\rho_1 = 0$ in Equation (4.6). So

$$L_2 = \frac{3\rho_2}{1 - 2\rho_2},$$

and thus

$$L = L_1 + L_2$$

$$= \frac{\rho_1}{1 - \rho_1} + \frac{3\rho_2}{1 - 2\rho_2}.$$

Therefore the new expected cost is

$$C^* = C_1 \left(\frac{\rho_1}{1 - \rho_1} + \frac{3\rho_2}{1 - 2\rho_2} \right) + 2C_2.$$

Hence any decision is based upon the comparative magnitude of C and C^*, and an additional channel is invoked whenever

$$C^* < C,$$

or

$$C_1 \left(\frac{\rho_1}{1 - \rho_1} + \frac{3\rho_2}{1 - 2\rho_2} \right) + C_2 < C_1 \left(\frac{\rho_1 + 3\rho_2}{1 - \rho_1 - 2\rho_2} \right),$$

that is,

$$C_2 < C_1 \left(\frac{\rho_1 + 3\rho_2}{1 - \rho_1 - 2\rho_2} - \frac{\rho_1}{1 - \rho_1} - \frac{3\rho_2}{1 - 2\rho_2} \right),$$

and removed when the inequality is reversed.

4.2. BULK SERVICE $(M/M^{[Y]}/1)$

For the main problem of the section, let it be assumed that arrivals occur to a single-channel facility as an ordinary Poisson process, that they are served FIFO, that there is no waiting capacity constraint, and that these customers are served K at a time, except when less than K are in the system and ready for service, at which time all units are served. The amount of time required for the service of any batch is an exponentially distributed random variable, whether or not the batch is of full size K. This model will henceforth be known by the notation $M/M^{[K]}/1$. A slight variation on this theme which we also consider is a model in which the batch size service must be exactly K and if the number present when the server becomes idle is less than K, he waits until K accumulates.

The basic model is, of course, a non-birth–death Markovian problem. The Kolmogorov equations are[2]

$$
\begin{cases}
\dfrac{dp_n(t)}{dt} = -(\lambda+\mu)p_n(t) + \lambda p_{n-1}(t) + \mu p_{n+K}(t) & (n \geqslant 1) \\[2mm]
\dfrac{dp_0(t)}{dt} = -\lambda p_0(t) + \mu \displaystyle\sum_{i=1}^{K} p_i(t)
\end{cases}
$$

When the steady state is assumed and the derivatives set to zero, it is found that

$$
\begin{cases}
0 = \mu p_{n+K} - (\lambda+\mu)p_n + \lambda p_{n-1} & (n \geqslant 1) \\[2mm]
0 = \mu p_K + \mu p_{K-1} + \ldots + \mu p_1 - \lambda p_0
\end{cases}
\tag{4.7}
$$

The first equation of (4.7) may be rewritten in operator notation as (see Section 2.1.4 p. 49)

$$
[\mu D^{K+1} - (\lambda+\mu)D + \lambda]p_n = 0 \qquad (n \geqslant 0);
\tag{4.8}
$$

hence if $(r_1, \ldots, r_{K+1})$ are the roots of the operator or characteristic equation, then

$$
p_n = \sum_{i=1}^{K+1} C_i r_i^n \qquad (n \geqslant 0).
$$

Since $\sum_{n=0}^{\infty} p_n = 1$, each r_i must be less than one or $C_i = 0$ for all r_i not less than one. So let us now determine the number of roots less than one. For this an appeal is made to Rouché's theorem (see Section 2.6.2), and it is found that there is, in fact, one, and only one, root (say r_0) in $(0, 1)$ [see Problem 4.1]. So

$$
p_n = C r_0^n \qquad (n \geqslant 0, \ 0 < r_0 < 1).
\tag{4.9}
$$

Using the boundary condition that $\sum p_n$ must total one, we find that

$$
C = p_0 = 1 - r_0;
$$

---

[2]These results can also be extended to the c-channel case in the usual way; however, the computations become more difficult.

hence

$$p_n = (1 - r_0)r_0^n \qquad (n \geq 0, 0 < r_0 < 1)$$.

Measures of effectiveness for this model can be obtained in the usual way. Since the stationary solution is so similar to that of $M/M/1$, we can immediately write that

$$L = \frac{r_0}{1 - r_0}$$

and that

$$W = \frac{r_0}{\lambda(1 - r_0)}.$$

Let us now assume that the batch size must be exactly K and if not, the server waits until such time to start. Then Equations (4.7) must be slightly rewritten to read

$$\begin{cases} 0 = \mu p_{n+K} - (\lambda + \mu)p_n + \lambda p_{n-1} & (n \geq K) \\ 0 = \mu p_{n+K} - \lambda p_n + \lambda p_{n-1} & (1 \leq n < K) \\ 0 = -\lambda p_0 + \mu p_K \end{cases} \qquad (4.10)$$

The first equation of this second approach to bulk service is identical to that of the first approach; hence

$$p_n = Cr_0^n \qquad (n \geq K - 1, 0 < r_0 < 1).$$

The obtaining of C (and p_0) is a more complicated procedure here since the foregoing formula for p_n is valid only for $n \geq K - 1$.

From the steady-state equations,

$$p_K = \frac{\lambda}{\mu} p_0 = Cr_0^K,$$

and therefore

$$C = \frac{p_0 \lambda}{\mu r_0^K}.$$

To get p_0 now, we must use the $K - 1$ stationary equations given in (4.10) as

$$0 = \mu p_{n+K} - \lambda p_n + \lambda p_{n-1}.$$

Substituting the aforementioned formula for p_n, $n \geqslant K$, into these $K-1$ equations, it can be seen that

$$p_0 r_0^n = p_n - p_{n-1} \qquad (1 \leqslant n < K).$$

This is a nonhomogeneous linear difference equation (see Appendix 2) whose solution is

$$p_n = C_1 + C_2 r_0^n,$$

where

$$C_1 = p_0 - C_2$$

and

$$C_2 = -\frac{p_0 r_0}{1-r_0}.$$

Thus

$$p_n = \begin{cases} \dfrac{p_0(1-r_0^{n+1})}{1-r_0} & (1 \leqslant n < K) \\[2mm] \dfrac{p_0 \lambda}{\mu} r_0^{n-K} & (n \geqslant K) \end{cases} \tag{4.11}$$

To get p_0 we use the usual boundary condition that $\sum_{n=0}^{\infty} p_n = 1$. Hence from Equation (4.11),

$$1 = p_0 + p_0 \sum_{n=1}^{K-1} \frac{1-r_0^{n+1}}{1-r_0} + \frac{p_0 \lambda}{\mu} \sum_{n=K}^{\infty} r_0^{n-K}$$

$$= p_0 \left[1 + \frac{K-1}{1-r_0} - \frac{r_0^2(1-r_0^{K-1})}{(1-r_0)^2} + \frac{\lambda}{\mu(1-r_0)} \right]$$

$$= \frac{p_0}{\mu(1-r_0)^2} \left[-\mu r_0 + \mu K - \mu K r_0 + \mu r_0^{K+1} + \lambda(1-r_0) \right]$$

$$= \frac{p_0}{\mu(1-r_0)^2} \left(\mu r_0^{K+1} - \mu r_0 - \lambda r_0 + \lambda + \mu K - \mu K r_0 \right).$$

But, from the characteristic equation in (4.8), we know that

$$\mu r_0^{K+1} - \mu r_0 - \lambda r_0 + \lambda = 0.$$

Thus

$$p_0^{-1} = \frac{1}{\mu(1-r_0)^2} [\mu K(1-r_0)]$$

$$= \frac{K}{1-r_0}. \tag{4.12}$$

Results for p_n could have also been obtained via the probability generating function. The generating function can be shown to be (see Problem 4.3)

$$P(z) = \frac{(1-z^K)\sum_{n=0}^{K-1} p_n z^n}{\rho z^{K+1} - (\rho+1)z^K + 1}, \tag{4.13}$$

where $\rho = \lambda/\mu$. It is necessary to eliminate the $p_n, n=0,1,\ldots,K-1$, from the right-hand side of (4.13) to make use of $P(z)$. To do this we again appeal to Rouché's theorem (see Section 2.6.2). The generating function $P(z)$ has the property that it must converge inside the unit circle.[3] We notice that the denominator of $P(z)$ has $K+1$ zeros. Applying Rouché's theorem to the denominator (see Problem 4.4) tells us that K of these lie on or within the unit circle. One zero of the denominator is $z=1$; thus $K-1$ lie within and must coincide with those of $\sum_{n=0}^{K-1} p_n z^n$ for $P(z)$ to converge, so that when a zero appears in the denominator it is cancelled by one in the numerator. Hence this leaves one zero of the denominator (since there are a total of $K+1$) which lies outside the unit circle. We denote this by z_0.

Now dividing the denominator by $(z-1)(z-z_0)$ results in a polynomial with $K-1$ roots inside the unit circle. These must therefore match the roots of $\sum_{n=0}^{K-1} p_n z^n$, so that the two polynomials must differ by at most a multiplicative constant, and we may therefore write that

$$\sum_{n=0}^{K-1} p_n z^n = A \frac{\rho z^{K+1} - (\rho+1)z^K + 1}{(z-1)(z-z_0)}.$$

---

[3]The generating function must be analytic; that is, all its derivatives must exist and the series of the derivatives, $\sum_{n=0}^{\infty} n p_n z^{n-1}$, must converge.

Substituting the right-hand side into (4.13) yields

$$P(z) = \frac{A(1-z^K)}{(z-1)(z-z_0)}$$

$$= \frac{A}{z_0-z} \sum_{n=0}^{K-1} z^n.$$

Utilizing the property that $P(1)=1$ we have

$$1 = \frac{AK}{z_0-1},$$

or

$$A = \frac{z_0-1}{K};$$

thus

$$P(z) = \frac{(z_0-1)\sum\limits_{n=0}^{K-1} z^n}{(z_0-z)K}. \tag{4.14}$$

In order to obtain p_n, it is necessary to expand (4.13) in a Maclaurin series such that

$$p_n = \frac{1}{n!} \frac{d^n}{dz^n} \left[\frac{(z_0-1)\sum\limits_{n=0}^{K-1} z^n}{(z_0-z)K} \right]\Bigg|_{z=0} \qquad (1\leqslant n) \tag{4.15}$$

and

$$p_0 = P(0) = \frac{z_0-1}{Kz_0} = \frac{1-(1/z_0)}{K},$$

which can be shown to be (see Problem 4.5)

$$p_n = \begin{cases} \dfrac{1-(1/z_0)^{n+1}}{K} & (0\leqslant n<K) \\[3mm] \dfrac{(1/z_0)^{n+1}(z_0^K-1)}{K} & (n\geqslant K) \end{cases} \tag{4.16}$$

Comparing the p_0 obtained from (4.15) with that of (4.12) we see the relation between the root z_0 and the root r_0 as

$$r_0 = \frac{1}{z_0}.$$

Using this relation one can see that (4.16) agrees with (4.11), realizing that $(\lambda/\mu)p_0 = p_K$ from (4.10) and that in (4.16) for $n \geqslant K$, p_n can be written as

$$p_n = \frac{(1/z_0)^{K+1}(z_0^K - 1)}{K}\left(\frac{1}{z_0}\right)^{n-K} = p_K\left(\frac{1}{z_0}\right)^{n-K}.$$

Example 4.2

The Drive-It-Through-Yourself Car Wash decides to change its operating procedure. It will install new machinery which will permit the washing of two cars at once (and one if no other cars wait) and will move to a new location which will effectively have no waiting capacity limitation. The company expects arrivals to be Poisson with mean 20/hr, and its service times to be exponential (a function of car size) with a mean of 5 min. What average line length should they anticipate?

The model for this bulk-service problem is the first of the two presented. The given parameters are $\lambda = 20/\text{hr}$, $\mu = 1/(5 \text{ min}) = 12/\text{hr}$, and $K = 2$. The operator equation is therefore

$$12r^3 - 32r + 20 = 3r^3 - 8r + 5 = 0.$$

One root is, of course, $r = 1$, and division by the factor $r - 1$ leaves

$$3r^2 + 3r - 5 = 0,$$

which has the roots

$$r = \frac{-3 \pm \sqrt{9 + 60}}{6}.$$

We select that positive root with absolute value less than 1, namely,

$$r_0 = \frac{-3 + \sqrt{69}}{6} \doteq 0.884.$$

Therefore

$$L \doteq \frac{0.884}{0.116} \doteq 7.6 \text{ cars},$$

and[4]

$$L_q = L - \frac{\lambda}{\mu} \doteq 7.6 - \frac{20}{12} \doteq 5.9 \text{ cars.}$$

4.3. ERLANGIAN MODELS ($M/E_K/1$ AND $E_K/M/1$)

Up to now, all probabilistic queueing models studied have assumed Poisson input (exponential interarrival times) and exponential service times. In many practical situations, however, the exponential assumptions may be rather limiting, especially the assumption concerning service times being distributed exponentially. In this section, we allow for a more general probability distribution for describing the input process or the service mechanism.

To begin, consider a random variable X which has the gamma probability density

$$f(x) = \frac{1}{\Gamma(\alpha)\beta^\alpha} x^{\alpha-1} e^{-x/\beta} \qquad (\alpha, \beta > 0; \ 0 \leqslant x < \infty),$$

where $\Gamma(\alpha)$ is the usual gamma function,[5] and α and β are the parameters of the distribution. The mean, $E[X]$, and variance, $Var[X]$, are given as

$$\begin{cases} E[X] = \alpha\beta \\ Var[X] = \alpha\beta^2 \end{cases}.$$

If we further consider a special class of these gamma probability distributions where α and β are related by

$$\begin{cases} \alpha = k \\ \beta = \dfrac{1}{k\mu} \end{cases},$$

where k is any arbitrary positive integer and μ any arbitrary positive constant, we obtain the Erlang family of probability distributions, namely,

$$f(x) = \frac{(\mu k)^k}{(k-1)!} x^{k-1} e^{-k\mu x} \qquad (0 < x < \infty).$$

---

[4] Recall this relation is true for all queues which satisfy Little's formulas $L = \lambda W$ and $L_q = \lambda W_q$.

[5] $\Gamma(\alpha) = \displaystyle\int_0^\infty x^{\alpha-1} e^{-x} dx.$

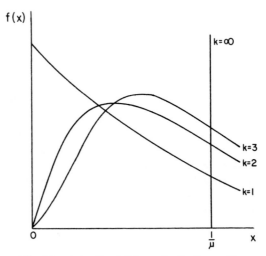

Fig. 4.1   A family of Erlang distributions with mean $1/\mu$

The parameters of the Erlang are, of course, k and μ, and the mean and variance are given by

$$\left\{ \begin{array}{l} E[X] = \dfrac{1}{\mu} \\[2em] Var[X] = \dfrac{1}{k\mu^2} \end{array} \right. .$$

For any particular value of k, the resulting Erlang is referred to as an Erlang type k distribution. Figure 4.1 illustrates the effect of k on the Erlang family of distributions.

The Erlang family of probability distributions provides much more modeling flexibility than does the exponential. In fact, the exponential is a special Erlang, namely, type 1. As k increases, the Erlang becomes more symmetrical, and as it approaches ∞, the Erlang becomes deterministic with value $1/\mu$. Thus in practical situations where observed data might not bear out the exponential distribution assumption, the Erlang can provide greater flexibility by being better able to represent the real world.

Another reason why the Erlang is useful in queueing analyses is its relation to the exponential distribution with the latter's Markovian property. The Erlang distribution itself is of course non-Markovian (Section 1.9 showed that the exponential distribution was unique in the Markovian property). However, as was shown in Problem 2.5b, one can readily see

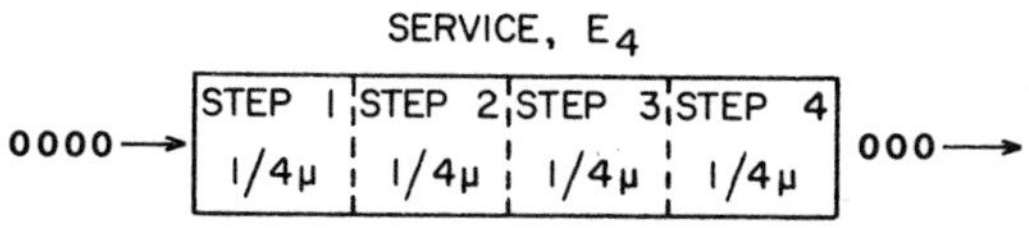

SERVICE, E_4

Fig. 4.2 Use of the Erlang for phased service

that the sum of k independently and identically distributed exponential random variables with mean $1/k\mu$ yields an Erlang type k distribution. It is this relation, as we shall see a little later, that allows the analysis of queueing models with Erlangian input or service to be performed.

The relation of the Erlang to exponential distributions also allows us to describe queueing models where the service (or arrivals) may be a series of identical phases. For example, suppose in performing a laboratory test, the lab technician must perform four steps, each taking the same mean time (say $1/4\mu$), with the times distributed exponentially. This is represented pictorially in Figure 4.2. The overall service function is Erlang type 4 with mean $1/\mu$. If the input process were Poisson, we would have an $M/E_4/1$ model.

It should be noted explicitly by the reader what this model implies. First, all steps (or phases) of the service are independent and *identical*. Second, only one customer at a time is in the service mechanism; that is, a customer enters phase 1 of the service, then progresses through the remaining phases and must complete the last phase before the next customer enters the first phase. This rules out assembly-line-type models where as soon as a customer finishes a phase of service another can enter it. Finally, while the phase representation for the Erlang may serve as a realistic model for some situations, the Erlang is not restricted to modeling only situations where there are actually phases of service. The Erlang has a more general use in that it has greater "flexibility" than the exponential; hence it is better able to "fit" observed data in many situations in which the exponential cannot.

4.3.1. Erlang Service Model ($M/E_k/1$)

We now consider a model in which the service time has an Erlang type k distribution. Even though the service may not actually consist of k phases, it is convenient in analyzing this model to consider the Erlang as being made up of k exponential phases, each with mean $1/k\mu$. Let $p_{n,i}(t)$ represent the probability of n in the system and the customer in service being in phase i ($i = 1, 2, \ldots, k$), where we now number the phases backward; that is, k is the first phase of service and 1 is the last (a customer leaving phase 1 actually leaves the system). We can write the following difference equations, where all $o(\Delta t)$ terms are ignored:

$$
\begin{cases}
p_{n,i}(t+\Delta t) = p_{n,i}(t)(1-\lambda\Delta t - k\mu\Delta t) + p_{n,i+1}(t)k\mu\Delta t \\[2mm]
\qquad + p_{n-1,i}(t)\lambda\Delta t \qquad (n\geqslant 2;\ 1\leqslant i\leqslant k-1) \\[3mm]
p_{n,k}(t+\Delta t) = p_{n,k}(t)(1-\lambda\Delta t - k\mu\Delta t) + p_{n+1,1}(t)k\mu\Delta t \\[2mm]
\qquad + p_{n-1,k}(t)\lambda\Delta t \qquad (n\geqslant 2) \\[3mm]
p_{1,i}(t+\Delta t) = p_{1,i}(t)(1-\lambda\Delta t - k\mu\Delta t) \\[2mm]
\qquad + p_{1,i+1}(t)k\mu\Delta t \qquad (n=1;\ 1\leqslant i\leqslant k-1) \\[3mm]
p_{1,k}(t+\Delta t) = p_{1,k}(t)(1-\lambda\Delta t - k\mu\Delta t) + p_{2,1}(t)k\mu\Delta t \\[2mm]
\qquad + p_0(t)\lambda\Delta t \qquad (n=1) \\[3mm]
p_0(t+\Delta t) = p_0(t)(1-\lambda\Delta t) + p_{1,1}(t)k\mu\Delta t \qquad (n=0)
\end{cases}
\qquad (4.17)
$$

Following the usual procedure of obtaining the differential-difference equations and then the steady-state difference equations, we have

$$
\begin{cases}
0 = -(\lambda+k\mu)p_{n,i} + k\mu p_{n,i+1} + \lambda p_{n-1,i} \qquad (n\geqslant 2;\ 1\leqslant i\leqslant k-1) \\[3mm]
0 = -(\lambda+k\mu)p_{n,k} + k\mu p_{n+1,1} + \lambda p_{n-1,k} \qquad (n\geqslant 2) \\[3mm]
0 = -(\lambda+k\mu)p_{1,i} + k\mu p_{1,i+1} \qquad (1\leqslant i\leqslant k-1) \\[3mm]
0 = -(\lambda+k\mu)p_{1,k} + k\mu p_{2,1} + \lambda p_0 \\[3mm]
0 = -\lambda p_0 + k\mu p_{1,1}
\end{cases}
\qquad (4.18)
$$

These equations are not particularly easy to handle. Furthermore, after obtaining the $p_{n,i}$, to get the steady-state probabilities of n customers in the system, it is necessary to calculate

$$p_n = \sum_{i=1}^{k} p_{n,i} \quad . \tag{4.19}$$

It is not too difficult, however, to obtain the expected measures of effectiveness (L, L_q, W, W_q) from (4.18) and we next proceed to accomplish this.

Consider the following generating function:

$$G(z) = \sum_{n=1}^{\infty} \sum_{i=1}^{k} z^{k(n-1)+i} p_{n,i} + p_0 \quad . \tag{4.20}$$

From $G(z)$ we can obtain W_q, the expected time spent in the queue waiting for service, as follows. Letting T_q represent the time spent in the queue (a random variable with $E[T_q] = W_q$) we have, because of the Markovian property of the phases of service, that

$$W_q = E[T_q] = E[N_q]\frac{1}{\mu} + E[I]\frac{1}{k\mu},$$

where N_q is the number in the queue and I the phase of service the customer being served is in (N_q, I are random variables). Thus

$$W_q = \sum_{n=1}^{\infty} \sum_{i=1}^{k} \left[\frac{k(n-1)+i}{k\mu} \right] p_{n,i} + 0 p_0$$

$$= \frac{1}{k\mu} G'(z)\big|_{z=1} \quad . \tag{4.21}$$

We must now find $G(z)$ in closed form. To accomplish this, we multiply the first equation of (4.18) by $z^{k(n-1)+i}$, the second by z^{kn}, the third by z^i, the fourth by z^k, and the fifth by z^0, sum up all terms and obtain (see Problem 4.7)

$$0 = \frac{1}{z}[G(z) - p_0] - (1+r)G(z) + p_0 + rz^k G(z) \quad (r = \lambda/k\mu) \quad . \tag{4.22}$$

Thus we have

$$G(z) = \frac{p_0(1-z)}{1 - z(1+r) + rz^{k+1}}.$$

To find p_0 we use the boundary condition that $G(1) = 1$, and employing

L'Hôpital's rule we have

$$G(1) = \lim_{z \to 1} G(z)$$

$$= \lim_{z \to 1} \left[\frac{-p_0}{-(1+r) + (k+1)rz^k} \right]$$

$$= \frac{p_0}{1 - kr}.$$

Hence

$$\frac{p_0}{i - kr} = 1$$

and

$$p_0 = 1 - kr = 1 - \rho \qquad (\rho \equiv \lambda/\mu). \tag{4.23}$$

Now to find W_q we need

$$G'(z)\big|_{z=1}$$

$$= \frac{-[1 - z(1+r) + rz^{k+1}] - (1-z)[-(1+r) + (k+1)rz^k]}{[1 - z(1+r) + rz^{k+1}]^2} \left(1 - \frac{\lambda}{\mu} \right)\bigg|_{z=1}.$$

To find $G'(z)\big|_{z=1}$ we use L'Hôpital's rule twice to get

$$G'(1) = \frac{(k+1)\rho}{2(1-\rho)}.$$

Hence from (4.21), we have

$$W_q = \frac{1}{k\mu} G'(1)$$

$$= \frac{(k+1)\rho}{2k\mu(1-\rho)}$$

or

$$\boxed{W_q = \frac{k+1}{2k} \frac{\lambda}{\mu(\mu - \lambda)}} \qquad . \tag{4.24}$$

From (4.24) we can get W by

$$W = W_q + \frac{1}{\mu},$$

and using Little's formulas we also obtain

$$\boxed{\begin{array}{l} L_q = \lambda W_q = \dfrac{k+1}{2k} \dfrac{\lambda^2}{\mu(\mu-\lambda)} \\[2ex] L = \lambda W \end{array}} \tag{4.25}$$

We now consider the more complicated procedure for finding the steady-state probability distribution. To obtain the steady-state probabilities, one can consider the following generating functions:

$$H(z,y) = \sum_{n=1}^{\infty} \sum_{i=1}^{k} z^n y^i p_{n,i}$$

and

$$P(x) = \sum_{n=0}^{\infty} x^n p_n$$

$$= p_0 + \sum_{n=1}^{\infty} x^n \sum_{i=1}^{k} p_{n,i}.$$

We note that

$$P(x) = p_0 + H(x,1)$$

and further that $P(x)$ is the probability generating function for p_n while $H(z,y)$ is the probability generating function for $p_{n,i}$. It can be shown that using (4.18) and the appropriate multiplications and then summing eventually yields (see Problem 4.8)

$$H(z,y) = \frac{rzy(1-z)p_0}{1-y(1+r-rz)} \frac{1 - y^k(1+r-rz)^k}{1 - z(1+r-rz)^k}, \tag{4.26}$$

where p_0, from (4.23), is $1-\rho$. Hence we have

$$P(x) = p_0 + \frac{rx(1-x)p_0}{r(x-1)} \frac{1 - (1+r-rx)^k}{1 - x(1+r-rx)^k}$$

$$= \frac{(1-x)(1-\rho)}{1 - x(1+r-rx)^k}. \tag{4.27}$$

Thus if (4.27) is expanded in a Maclaurin series, the p_n can be obtained (see Problem 4.13).

While the procedures above are straightforward in that they are directly analogous to our previous approaches in analyzing exponential service problems via generating function methodology, it is useful, at this point, to present an alternative means of deriving results for the $M/E_k/1$ model. Let us reconsider Equations (4.18) and make the transformation (n,i) $=(n-1)k+i$. This can be interpreted as the number of phases of service in the system since $n-1$ customers are waiting, each requiring k phases of service and the customer in service requires i phases of service. Making this transformation in (4.18) yields

$$\begin{cases} 0 = -(\lambda+k\mu)p_{(n-1)k+i} + k\mu p_{(n-1)k+i+1} + \lambda p_{(n-2)k+i} & (n \geq 1;\ 1 \leq i \leq k) \\ 0 = -\lambda p_0 + k\mu p_1 \end{cases} \tag{4.28}$$

where any p with a negative subscript is assumed to be zero. It can be seen quite readily that by writing out the top equation in (4.28) sequentially starting at $n=1, i=1$, (4.28) can be simplified as

$$\begin{cases} 0 = -(\lambda+k\mu)p_n + k\mu p_{n+1} + \lambda p_{n-k} & (n \geq 1) \\ 0 = -\lambda p_0 + k\mu p_1 \end{cases}, \tag{4.29}$$

where n is the number of phases in the system and again a negative subscript indicates the term is zero.

We note that (4.29) is equivalent to (4.1) for the bulk input model, $M^{[X]}/M/1$ of Section 4.1, where the bulk size X is a constant equal to k ($c_k = 1, c_x = 0, x \neq k$), and μ is replaced by $k\mu$. Thus we can use the results of Section 4.1 to yield the desired quantities for the Erlang-service model. Denoting by $\mu^{(B)}$ the mean service rate for the bulk model so that $\mu^{(B)} = k\mu$, we can see immediately from (4.4) that

$$p_0 = 1 - \frac{\lambda E[X]}{\mu^{(B)}}$$

$$= 1 - \frac{\lambda k}{\mu^{(B)}}$$

$$= 1 - \frac{\lambda}{\mu},$$

which yields (4.23). Also, letting $p_n^{(B)}$ represent the probability of n in the bulk input system, we have that p_n, the probability of n in the Erlang-service system, is given by

$$p_n = \sum_{j=(n-1)k+1}^{nk} p_j^{(B)} \qquad (\mu^{(B)} = k\mu; \; n \geqslant 1), \qquad (4.30)$$

with $p_0 = p_0^{(B)}$.

To find L_q for the Erlang model we first relate it to $L_q^{(P)}$, the expected number of phases of service in the queue, as follows. If there are, on the average, L_q customers in the queue, there must be, on the average, kL_q phases of service in the queue since each customer in reality desires k phases of service. Hence

$$L_q = \frac{L_q^{(P)}}{k}. \qquad (4.31)$$

Also, we can relate $L_q^{(P)}$ to $L^{(P)}$, the average number of phases of service in the system, by

$$L^{(P)} = L_q^{(P)} + E[\text{number of phases in service}], \qquad (4.32)$$

since the number in the system must be equal to the number in service and the number waiting for service, and the expected value operator is additive. The expected number of phases in service is given by

$$E[\text{number of phases in service}] = \frac{\lambda(k+1)}{2\mu}, \qquad (4.33)$$

since when the system is empty (probability $p_0 = 1 - \lambda/\mu$) there are zero phases of service, but when the system is not empty (probability λ/μ) a customer is in service. It is equally likely that this customer is in any one of the k possible phases, thus the expected number of phases, given the system is not empty, is

$E[\text{number of phases of service remaining}|\text{system not empty}]$

$$= k\frac{1}{k} + (k-1)\frac{1}{k} + \ldots + 1\frac{1}{k}$$

$$= \frac{1}{k}(1 + 2 + \ldots + k) = \frac{k(k+1)}{2k}$$

$$= \frac{(k+1)}{2}. \qquad (4.34)$$

Hence multiplying (4.34) by λ/μ gives (4.33), and therefore from (4.32) and (4.33), we have

$$L^{(P)} = L_q^{(P)} + \frac{(k+1)\lambda}{2\mu}.$$

Now from (4.31) and the above, we have

$$L_q = \frac{L^{(P)} - (k+1)\lambda/2\mu}{k}. \tag{4.35}$$

We can get $L^{(P)}$ from the $L^{(B)}$ of the bulk model as follows. We use the same methodology for Equation (4.29) as was done in Section 4.1 for Equation (4.1). The probability generating function $P(z)$ is obtained for the bulk model probabilities and then $L^{(B)}$ is found from $P'(1)$. We can use (4.3) directly to obtain $P(z)$, realizing that $C(z)$ is merely equal to z^k, since the batch size is fixed at k. Hence from (4.3),

$$P(z) = \frac{\mu^{(B)}p_0(1-z)}{\mu^{(B)}(1-z) - \lambda z(1-z^k)},$$

$$P'(z) = \frac{-[\mu^{(B)}(1-z) - \lambda z(1-z^k)]\mu^{(B)}p_0 \quad -\mu^{(B)}p_0(1-z)[-\mu^{(B)} - \lambda + \lambda(k+1)z^k]}{[\mu^{(B)}(1-z) - \lambda z(1-z^k)]^2}.$$

Using L'Hôpital's rule twice and the relation $\mu^{(B)} = k\mu$ gives

$$L^{(P)} = P'(1; \mu^{(B)} = k\mu) = \frac{(k+1)\lambda}{2(\mu-\lambda)}.$$

Substituting the above into (4.35) we have

$$L_q = \left[\frac{(k+1)\lambda}{2(\mu-\lambda)} - \frac{(k+1)\lambda}{2\mu} \right]\frac{1}{k}$$

$$= \frac{k+1}{2k}\frac{\lambda^2}{\mu(\mu-\lambda)},$$

which is the same result as given by (4.25). The other expected value measures of effectiveness, L, W, and W_q, can be obtained from L_q by using Little's formulas and the relation between W and W_q as done previously.

The purpose of deriving these results again using the "method of phases" was to show that the prior results obtained for the bulk input queueing model of Section 4.1 can be utilized in obtaining results for the Erlang service models. While we have shown here the relation between $M^{[X]}/M/1$ and $M/E_k/1$, a similar relation holds between $M/M^{[Y]}/1$ and $E_k/M/1$ so that the previous bulk results of Section 4.2 can be useful in deriving results about the Erlang arrival model to be treated in the following section. Prior to considering an Erlang arrival model, we illustrate Erlang service models by the following three examples.

Example 4.3

The Grabeur-Munee Savings and Loan has a drive-up window. During the busy periods for drive-up service, customers arrive according to a Poisson distribution with a mean of 4/hr. From observations on the teller's performance, the mean service time is estimated to be 10 min, with a variance of 25 min^2. It is felt that the Erlang would be a reasonable assumption for the distribution of the teller's service time. Also, since the savings and loan building (and drive-up window) is located in a large shopping center, there is virtually no limit on the number of vehicles that can wait. The company officials wish to know, on the average, how long a customer must wait until he reaches the window for service, and how many vehicles are waiting for service.

The appropriate model, of course, is an $M/E_k/1$ model. To determine k, we first note that $\mu = (1/10)/\text{min}$ and that

$$\sigma^2 = \frac{1}{k\mu^2} = 25,$$

which yields

$$k = \frac{1}{25\mu^2} = \frac{100}{25} = 4.$$

Thus we have an $M/E_4/1$ model and from (4.24) we have

$$W_q = \frac{5}{8} \frac{4}{6(6-4)} \doteq 0.208 \text{ hr}$$

$$\doteq 12.5 \text{ min.}$$

The expected number waiting for service is

$$L_q = \lambda W_q \doteq 0.832.$$

Example 4.4

A small heating oil distributor has only one truck. The capacity of the truck is such that after delivering to a customer, it must return to be refilled. Customers call in for deliveries on the average of once every 50 min during the height of the heating season. The distribution of time between calls has been found to be approximately exponential. It has also been observed that it takes, on an average, 20 min for the truck to get to a customer and an average of 20 min for the truck to return. Both the time for driving out and the time for returning have approximately exponential distributions. The time it takes to unload and load the truck has been included in the driving times. W. A. R. Mup, the manager, is considering the possibility of purchasing an additional truck and would like to know, under the current situation, how long on an average a customer has to wait from the time he places a call until the truck arrives. All customers are served on a first-come, first-served basis.

The service time in this problem corresponds to the time a truck is tied up and is made up of two identical exponential stages, one consisting of loading and traveling out to the customer, and one consisting of unloading and traveling back to the terminal. Hence we have an $M/E_2/1$ model, where $\lambda = (1/50)/\text{min}$ or $(6/5)/\text{hr}$ and $\mu = (1/40)/\text{min}$ or $(3/2)/\text{hr}$. The average time a customer must wait from the time he places his call until his service starts (the truck begins loading for his delivery) is W_q and is given by (4.24) as

$$W_q = \frac{3}{4} \frac{6/5}{(3/2)[(3/2)-(6/5)]}$$

$$= 2 \text{ hr.}$$

The average time it takes the truck to arrive once it is dispatched (loading for a given customer commences) is one half an average service time, that is, 20 min. Thus the average wait from the time a customer calls in until the truck arrives and begins unloading is 2 hr and 20 min.

Example 4.5

A manufacturer of a special electronic guidance system component has a quality control check point at the end of the production line to assure that the component is properly calibrated. If it fails the test, the component is sent to a special repair center where it is readjusted. There are two men at this center and each can adjust a component in an average of 5 min, their repair time being exponentially distributed. The average number of rejects from the quality control point per hour is 18 and follows a Poisson distribution. The company can lease one machine which can adjust the component to the same degree of accuracy as the men in *exactly* $2\frac{2}{3}$ min, that is, no variation in repair time. The machine leasing costs are roughly equivalent to the salary and fringe benefit costs for the men. (If the men are replaced, they can be used elsewhere in the company so there will be no morale or labor problems.) Should the company lease the machine?

We wish to compare the expected waiting time, W, and system size, L, under each alternative. Alternative one, keeping the two men, is an $M/M/2$ model, while alternative two, leasing the machine, is an $M/D/1$ model. The calculations for alternative one are as follows.

$M/M/2$: $\lambda = 18/\text{hr}$

$$\mu = (1/5)/\text{min} = 12/\text{hr}$$

$$W_q = \frac{\mu(\lambda/\mu)^2 p_0}{(2-1)!(2\mu-\lambda)^2} \qquad [\text{from } (3.16)]$$

$$= \frac{3}{4}p_0$$

$$p_0 = \left[\sum_{n=0}^{1} \left(\frac{1}{n!}\right)\left(\frac{\lambda}{\mu}\right)^n + \frac{1}{2!}\left(\frac{\lambda}{\mu}\right)^2 \frac{2\mu}{2\mu-\lambda} \right]^{-1} \qquad [\text{from } (3.11)]$$

$$= \frac{1}{7}$$

$$W_q = \frac{3}{28}\text{ hr} \doteq 6.4 \text{ min}$$

$$W \doteq 6.4 + 5 = 11.4 \text{ min}$$

$$L = \lambda W \doteq \frac{18(11.4)}{60} \doteq 3.42.$$

$M/D/1$: to obtain the results for an $M/D/1$ model we can use

$\lim_{k \to \infty} M/E_k/1$. Hence

$$\mu = [(8/3)/\min]^{-1} = (45/2)/\text{hr}$$

$$W_q = \lim_{k \to \infty} \left(\frac{k+1}{2k} \frac{\lambda}{\mu(\mu-\lambda)} \right) \qquad [\text{from } (4.24)]$$

$$= \frac{1}{2} \frac{\lambda}{\mu(\mu-\lambda)} = \frac{4}{45} \text{ hr} = 5\frac{1}{3} \text{ min}$$

$$W = 5\frac{1}{3} + 2\frac{2}{3} = 8 \text{ min}$$

$$L = \lambda W = \frac{18(8)}{60} = 2.40.$$

Thus alternative two (the machine) shows to be preferable.

As a final observation, it should be noted that it is an easy matter to derive steady-state difference equations like (4.29) in the event that there are c channels (that is, an $M/E_k/c$ model). All the measures of effectiveness can then be obtained in terms of p_0, as was done for $M/M/c$, and it would not be too difficult then to solve problems like Example 4.3 when more than one channel is involved. A reference for some of the computational aspects of $M/E_k/c$ is Mayhugh and McCormick (1968).

4.3.2 Erlang Arrival Model ($E_k/M/1$)

As mentioned in the previous section, we utilize the results of the second bulk-service model of Section 4.2 to develop results for this Erlang input model. We assume that the interarrival times are Erlang type k distributed, with a mean of $1/\lambda$. We can look therefore at an arrival having passed through k phases, each with a mean time of $1/k\lambda$, prior to actually entering the system. Here we number the phases frontwards from 0 to $k-1$. Again, we remind the reader that this is a device convenient for analysis that does not have to correspond to the actual arrival mechanism —the only assumption on interarrival times is that they follow an Erlang type k distribution with mean $1/\lambda$.

We define the state variable as the number of arrival phases in the system so that we desire to find the probability of n arrival phases in the system in the steady state, which we denote by $p_n^{(P)}$. It is an easy matter once we have this to obtain the probability of n customers in the system by utilizing a relation similar to (4.30), that is,

$$p_n = \sum_{j=nk}^{nk+k-1} p_j^{(P)}. \tag{4.36}$$

We can get $p_n^{(P)}$ from $p_n^{(B)}$, where $p_n^{(B)}$ is the steady-state probability of n in the bulk-service model given by (4.11) or (4.16) but with λ replaced by $k\lambda$ (why?).

We can, for this model, also derive the waiting-time distribution (recall this is a composite distribution because of the nonzero probability of no wait at all for service to commence—see Section 2.3) as follows. Denoting by q_n the probability that an arrival into the system finds n customers already there, q_n can be found as

$$q_n = \Pr\{n \text{ in system} | \text{arrival about to occur}\}$$

$$= \Pr\{n \text{ in system} | \text{arrival in phase } k-1\}$$

$$= \frac{\Pr\{n \text{ in system and arrival in phase } k-1\}}{\Pr\{\text{arrival in phase } k-1\}}$$

$$= \frac{p_{nk+k-1}^{(P)}}{1/k}$$

since it is equally likely that an arrival is in any one of the k phases and hence

$$\Pr\{\text{arrival in phase } k-1\} = \frac{1}{k}.$$

Thus we have

$$q_n = kp_{nk+k-1}^{(P)}. \tag{4.37}$$

Now if there are n customers in the system upon arrival, the conditional waiting time is the time it takes to serve these n people, which is the convolution of n exponentials, each with mean $1/\mu$. This yields an Erlang type n distribution and the unconditional waiting time "density" can be written as

$$w_q(t) = \sum_{n=1}^{\infty} q_n \frac{\mu^n}{(n-1)!} t^{n-1} e^{-\mu t} \qquad (t > 0), \tag{4.38}$$

where q_n is given by (4.37). The probability of no wait for service upon arrival is given by

$$w_q(0) \equiv \Pr\{\text{no wait for service}\} = q_0 = kp_{k-1}^{(P)}.$$

Obtaining concise expressions for L, L_q, W, and W_q is not possible; however, for any specific problem these can be calculated by first obtaining $p_n^{(P)}$ from the bulk-service model using (4.11) or (4.16) and then using (4.36) to calculate p_n for the Erlang arrival model. The mean system size L can then be obtained by

$$L = \sum_{n=1}^{\infty} np_n$$

or by the use of Little's formula, first obtaining W_q from

$$W_q = \int_0^{\infty} tw_q(t)\,dt$$

after calculating q_n from (4.37) and then using the usual relation

$$W = W_q + \frac{1}{\mu}.$$

Of course, L_q can be obtained by Little's formula.

Example 4.6

Arrivals coming to a single-server queueing system are found to have an Erlang type 2 distribution with mean interarrival time of 30 min. The mean service time is 25 min, and service times are exponentially distributed. Find the steady-state system-size probabilities, and the expected-value measures of effectiveness for this system.

The inputs required to answer these questions are $\lambda = 1/(30 \text{ min}) = 2/\text{hr}$, $\mu = 1/(25 \text{ min}) = (12/5)/\text{hr}$, and $k = 2$. We must first find the root r_0 needed for (4.11) in order to find $p_n^{(P)}$. This root can be found using the operator equation found in (4.8). Note that this operator equation is valid for both of the bulk-service models, and it is the latter model which we require for an $E_k/M/1$ model. Repeating (4.8) with λ replaced by $k\lambda$ we have

$$\mu r^{k+1} - (k\lambda + \mu)r + k\lambda = 0$$

or

$$\frac{12}{5}r^3 - \left(4 + \frac{12}{5}\right)r + 4 = 0.$$

Upon simplifying we have

$$12r^3 - 32r + 20 = 0$$

or

$$3r^3 - 8r + 5 = 0.$$

This happens to be the same operator equation as found in Example 4.2 and the positive root with absolute value less than 1 is

$$r_0 \doteq 0.884.$$

Now utilizing (4.11) with λ replaced by $k\lambda = 4$, $\mu = 12/5$, and $k = 2$, we get

$$p_n^{(P)} \doteq \begin{cases} 1.88 p_0^{(P)} & (n = 1) \\ 1.67(0.884)^{n-2} p_0^{(P)} & (n \geqslant 2) \end{cases},$$

and from (4.12) we have

$$p_0^{(P)} \doteq 0.06.$$

The above are the phase probabilities. To obtain p_n, the system size probabilities for the Erlang arrival model, we compute

$$p_n = \sum_{j=nk}^{nk+k-1} p_j^{(P)},$$

or

$$\begin{cases} p_0 = \sum_{j=0}^{1} p_j^{(P)} = p_0^{(P)} + p_1^{(P)} \doteq 2.88(0.06) \doteq 0.17 \\ \\ p_n = \sum_{j=2n}^{2n+1} 1.67(0.06)(0.884)^{j-2} \\ \\ \quad = 0.10 \sum_{j=2n}^{2n+1} (0.884)^{j-2} \qquad (n \geqslant 1) \end{cases}$$

Thus

$$L = p_1 + 2p_2 + 3p_3 + \ldots$$

$$= 0.10 \sum_{j=2}^{3} (0.884)^{j-2} + 0.2 \sum_{j=4}^{5} (0.884)^{j-2} + 0.3 \sum_{j=6}^{7} (0.884)^{j-2} + \ldots$$

$$\doteq 3.78,$$

and

$$W = \frac{L}{\lambda} \doteq \frac{3.78}{2} = 1.89 \text{ hr.}$$

Some results can be obtained for $E_k/M/c$ and $E_k/E_l/c$, but they involve greater numerical computation. The $E_k/M/c$ model can be handled by generalizing the bulk service equations of (4.7) to allow for multiple servers. However, $E_k/E_l/c$ requires greater numerical manipulations. The reader interested in multichannel Erlang models is referred to Hillier and Lo (1972).

4.4 PRIORITY QUEUE DISCIPLINES

Up to this point, all the models considered have had the property that units proceed to service on a first-come, first-served basis. This is obviously not the only manner of service, and there are many alternatives, such as last-come, first-served, selection in random order, and selection by priority. In priority schemes customers with the highest priorities are selected for service ahead of those, with lower priorities, independent of their time of arrival into the system. There are two further refinements possible in priority situations, namely, preemption and nonpreemption. In preemptive cases, the customer with the highest priority is allowed to enter service immediately even if another with lower priority is already present in service when the higher customer arrives to the system. In addition, a decision has to be made whether to continue the preempted customer's service from the point of preemption when resumed or to start anew. On the other hand, a priority discipline is said to be nonpreemptive if there is no interruption and the highest priority customer just goes to the head of the queue to wait his turn.

Obviously, a very considerable portion of real-life queueing situations contains priority considerations. Priority queues are generally more difficult to model than nonpriority situations, but nevertheless the priority models should not be oversimplified merely to permit solution. Full consideration of priorities is absolutely essential when considering costs of queueing systems and optimal design. These latter subjects will themselves be taken up in greater detail in Chapter 7, Section 7.2.

4.4.1 Single Exponential Channel with Priorities

Let us begin by assuming that customers arrive as a Poisson process to a single exponential channel and that upon arrival to the system each unit will be designated to be a member of one of two priority classes [see Morse (1958)]. The usual convention is to number the priority classes so that the smaller the number, the higher the priority. Let it further be assumed that the (Poisson) arrivals of the first or higher priority have mean rate λ_1 and that the second or lower priority units have mean rate λ_2, such that $\lambda \equiv \lambda_1 + \lambda_2$. We also suppose that the first-priority items have the right to be served ahead of the others, but that there is no preemption.

In view of the foregoing assumptions, a system of differential-difference equations may be established for $p_{mnr}(t) \equiv \Pr\{$at time t, m units of priority 1 and n units of priority 2 are in the system, and a unit of priority $r = 1$ or 2 is in service$\}$. These then lead to the following stationary difference equations (see Problem 4.19) in the event that $\rho = \lambda/\mu < 1$:

$$
\begin{cases}
0 = -\lambda p_0 + \mu(p_{101} + p_{012}) \\[2mm]
0 = -(\lambda + \mu)p_{101} + \lambda_1 p_0 + \mu(p_{201} + p_{112}) \\[2mm]
0 = -(\lambda + \mu)p_{012} + \lambda_2 p_0 + \mu(p_{111} + p_{022}) \\[2mm]
0 = -(\lambda + \mu)p_{m01} + \lambda_1 p_{m-1,0,1} + \mu(p_{m+1,0,1} + p_{m12}) \\[2mm]
0 = -(\lambda + \mu)p_{0n2} + \lambda_2 p_{0,n-1,2} + \mu(p_{1n1} + p_{0,n+1,2}) \\[2mm]
0 = -(\lambda + \mu)p_{1n1} + \lambda_2 p_{1,n-1,1} + \mu(p_{2n1} + p_{1,n+1,2}) \\[2mm]
0 = -(\lambda + \mu)p_{m12} + \lambda_1 p_{m-1,1,2} \\[2mm]
0 = -(\lambda + \mu)p_{mn1} + \lambda_1 p_{m-1,n,1} + \lambda_2 p_{m,n-1,1} \\[1mm]
\qquad + \mu(p_{m+1,n,1} + p_{m,n+1,2}) \qquad (m > 1, n > 0) \\[2mm]
0 = -(\lambda + \mu)p_{mn2} + \lambda_1 p_{m-1,n,2} + \lambda_2 p_{m,n-1,2} \qquad (m > 0, n > 1)
\end{cases}
\tag{4.39}
$$

It should be clear that p_0 is still $1 - \rho$, since the ordering of service in no way affects the probability of idleness and that

$$p_n = \sum_{m=0}^{n-1} (p_{n-m,m,1} + p_{m,n-m,2}) = (1-\rho)\rho^n \qquad (n>0).$$

Also, since the percentage of time the system is busy is ρ, the percentage of time it is busy with a type r customer would be $\rho\lambda_r/\lambda$, so that

$$\sum_{m=1}^{\infty} \sum_{n=0}^{\infty} p_{mn1} = \frac{\lambda_1}{\mu},$$

and

$$\sum_{m=0}^{\infty} \sum_{n=1}^{\infty} p_{mn2} = \frac{\lambda_2}{\mu}.$$

It turns out, however, that obtaining a reasonable solution to these stationary equations is a very difficult matter, as we might have anticipated in view of the triple subscripts. The most that we can do is obtain expected values via two-dimensional generating functions. So define

$$P_{m1}(z) = \sum_{n=0}^{\infty} z^n p_{mn1},$$

$$P_{m2}(z) = \sum_{n=1}^{\infty} z^n p_{mn2},$$

$$H_1(y,z) = \sum_{m=1}^{\infty} y^m P_{m1}(z) \qquad \text{[with } H_1(1,1) = \lambda_1/\mu\text{]},$$

$$H_2(y,z) = \sum_{m=0}^{\infty} y^m P_{m2}(z) \qquad \text{[with } H_2(1,1) = \lambda_2/\mu\text{]},$$

and

$$H(y,z) = H_1(y,z) + H_2(y,z) + p_0$$

$$= \sum_{m=1}^{\infty} \sum_{n=0}^{\infty} y^m z^n p_{mn1} + \sum_{m=0}^{\infty} \sum_{n=1}^{\infty} y^m z^n p_{mn2} + p_0$$

$$= \sum_{m=1}^{\infty} \sum_{n=1}^{\infty} y^m z^n (p_{mn1} + p_{mn2}) + \sum_{m=1}^{\infty} y^m p_{m01} + \sum_{n=1}^{\infty} z^n p_{0n2} + p_0,$$

where $H(y,z)$ then is merely the joint generating function for the two priorities, regardless of which type is in service. Note that $H(y,y) = p_0/(1-\rho y)$ [with $H(1,1)=1$] since $H(y,z)$ collapses to the generating function of $M/M/1$ when z is set equal to y and thus no priority distinction is made [see (2.14)]. Hence if L_1 and L_2 are used to denote the mean number of customers present in system for each of the two priority classes, then

$$\frac{\partial H(y,z)}{\partial y}\Bigg|_{y=z=1} = L_1$$

$$= L_{q_1} + \frac{\lambda_1}{\mu}$$

$$= \lambda_1 W_1,$$

and

$$\frac{\partial H(y,z)}{\partial z}\Bigg|_{y=z=1} = L_2$$

$$= L_{q_2} + \frac{\lambda_2}{\mu}$$

$$= \lambda_2 W_2,$$

where L_{q_1} and L_{q_2} are the respective mean queue lengths.[6] If we then multiply Equations (4.39) by the appropriate powers of y and z, and sum accordingly, it is found that

$$\left[1 + \rho - \frac{\lambda_1 y}{\mu} - \frac{\lambda_2 z}{\mu} - \frac{1}{y}\right] H_1(y,z) = \frac{H_2(y,z)}{z} + \frac{\lambda_1 y p_0}{\mu} - P_{11}(z) - \frac{P_{02}(z)}{z}$$

$$(4.40)$$

and

$$\left[1 + \rho - \frac{\lambda_1 y}{\mu} - \frac{\lambda_2 z}{\mu}\right] H_2(y,z) = P_{11}(z) + \frac{P_{02}(z)}{z} - \left[\rho - \frac{\lambda_2 z}{\mu}\right] p_0. \quad (4.41)$$

To determine fully the generating functions H_1 and H_2, we need to know the values of $P_{11}(z)$, $P_{02}(z)$, and p_0. One such equation relating p_{11}, p_{02}, and p_0 may be found by summing z^n $(n = 2, 3, \ldots)$ times the equation of (4.39) which involves p_{0n2}, and then using the first and third equations of (4.39)

---

[6]The use of Little's formula in these Poisson priority situations can be defended since the expected number of customers of equal priority to accumulate during any individual's waiting time is that priority's arrival rate divided into its mean waiting time.

thus resulting in

$$P_{11}(z) = \left(1 + \rho - \frac{\lambda_2 z}{\mu} - \frac{1}{z}\right)P_{02}(z) + \left(\rho - \frac{\lambda_2 z}{\mu}\right)p_0.$$

Substitution of this equation into (4.40) and (4.41) then gives H_1 and H_2 as a function of p_0 and P_{02}, and thus also $H(y,z)$ as

$$H(y,z) = H_1(y,z) + H_2(y,z) + p_0$$

$$= \frac{(1-y)p_0}{1 - y - \rho y(1 - z - \lambda_1 y/\lambda + \lambda_1 z/\lambda)}$$

$$+ \frac{(1 + \rho - \rho z + \lambda_1 z/\mu)(z-y)P_{02}(z)}{z[1 + \rho - \lambda_1 y/\mu - \lambda_2 z/\mu][1 - y - \rho y(1 - z - \lambda_1 y/\lambda + \lambda_1 z/\lambda)]}.$$

By employing the condition that $H(1,1) = 1$, it is found that

$$P_{02}(1) = \frac{\lambda_2 p_0/\mu}{(1 + \lambda_1/\mu)(1 - \rho)}.$$

We next take the partial derivatives of H with respect to both y and z and then evaluate at $(1,1)$ to find the means L_1 and L_2. In so doing, the exact functional relationship for $P_{02}(z)$ turns out not to be needed, and $P_{02}(1)$ alone suffices. Without suffering through the details of the differentiation, the final desired results are

$$L_1 = \frac{(\lambda_1/\mu)(1 + \rho - \lambda_1/\mu)}{1 - \lambda_1/\mu}$$

$$L_{q_1} = \frac{\rho\lambda_1/\mu}{1 - \lambda_1/\mu}$$

$$W_{q_1} = \frac{\lambda}{\mu(\mu - \lambda_1)} \tag{4.42}$$

$$L_2 = \frac{(\lambda_2/\mu)(1 - \lambda_1/\mu + \rho\lambda_1/\mu)}{(1 - \rho)(1 - \lambda_1/\mu)}$$

$$L_{q_2} = \frac{\rho\lambda_2/\mu}{(1 - \rho)(1 - \lambda_1/\mu)}$$

$$W_{q_2} = \frac{\lambda}{(\mu - \lambda)(\mu - \lambda_1)}$$

Some interesting and important observations may be made about these formulas. First, it should be noted that, as expected, units of priority two wait in queue longer than the higher priority units, such that

$$W_{q_2} = \frac{\mu}{\mu - \lambda} W_{q_1}.$$

We can also see that as $\rho \to 1$, the expectations L_2, L_{q_2}, and $W_{q_2} \to \infty$, while the corresponding means for the first priority approach finite limits. These first priority expectations go to ∞ only when $\lambda_1/\mu \to 1$. That is, the first-priority units need not accumulate, even though a steady-state does not exist. In addition, it should be noted that the expected wait in line for first-priority units in the absence of second-priority customers is smaller than their wait in the presence of the lower-priority units by a factor of λ_1/λ. In fact, the relative performance of the higher priority compared to the lower increases as $\rho \to 1$ and $\lambda_1 \to 0$. However, had the units of higher priority had the power of preemption, then the two expected waits above would be identical. Also, the average number in queue is $L_q = L_{q_1} + L_{q_2} = \rho^2/(1-\rho)$ (the same as for the nonpriority case), which implies that the unconditional average wait, $W_q = (\lambda_1/\lambda)W_{q_1} + (\lambda_2/\lambda)W_{q_2}$, is the same as that for the nonpriority case.

Similar equations can be obtained for the model which slightly generalizes the foregoing by serving the priority one customers at the rate μ_1 and the priority two customers at μ_2. These results are [details may be found in Morse (op. cit)]:

$$L_{q_1} = \frac{\lambda_1}{\mu_1} \hat{\rho} \frac{\lambda_1/\lambda + (\lambda_2/\lambda)(\mu_1^2/\mu_2^2)}{1 - \lambda_1/\mu_1}$$

$$L_{q_2} = \frac{(\lambda_2/\mu_1)\hat{\rho}}{1 - \lambda_1/\mu_1} \frac{\lambda_1/\lambda + (\lambda_2/\lambda)(\mu_1^2/\mu_2^2)}{1 - \lambda_1/\mu_1 - \lambda_2/\mu_2}$$

$$L_q = L_{q_1} + L_{q_2}$$

$$= \frac{(\lambda_1/\mu_1)\hat{\rho} + (\lambda_2\lambda/\mu_2^2)}{1 - \lambda_1/\mu_1 - \lambda_2/\mu_2} \frac{1 - (\lambda_1/\mu_1)[\lambda_1/\lambda + (\lambda_2/\lambda)(\mu_1/\mu_2)]}{1 - \lambda_1/\mu_1} \quad (\hat{\rho} = \lambda/\mu_1)$$

$$(4.43)$$

The foregoing may be presented in slightly different form by introducing the actual system utilization factor, say ρ^*, where

$$\rho^* = \% \text{ system busy time}$$

$$= \frac{\lambda_1}{\mu_1} + \frac{\lambda_2}{\mu_2}.$$

Observe that $\rho^* < \hat{\rho}$ if

$$\frac{\lambda_1}{\mu_1} + \frac{\lambda_2}{\mu_2} < \frac{\lambda}{\mu_1} = \frac{\lambda_1 + \lambda_2}{\mu_1},$$

or

$$\mu_2 > \mu_1,$$

and, similarly, $\rho^* > \hat{\rho}$ when $\mu_2 < \mu_1$. Then, for example, the third equation of (4.43) could be rewritten as

$$L_q = \frac{(\rho^*)^2}{1 - \rho^*} \frac{\lambda_1/\lambda + (\lambda_2/\lambda)(\mu_1^2/\mu_2^2)}{[\lambda_1/\lambda + (\lambda_2/\lambda)(\mu_1/\mu_2)]^2} \frac{1 - \lambda_1\rho^*/\lambda}{1 - \lambda_1\hat{\rho}/\lambda}. \tag{4.44}$$

Further insight can be gained into this latest model by comparing it, on the one hand, to the first model of this section, which assumed constant service rates, with results given in Equations (4.42), and, on the other hand, to a model with no priorities but unequal service rates for customers of two major types, say, for example, adults and children. The analysis of the latter model is not a terribly difficult matter (see Problem 4.20), with the expected line lengths given as

$$
\begin{aligned}
L_{q_1} &= \frac{\lambda_1}{\mu_1}\hat{\rho} \frac{1 - (1 - \mu_1/\mu_2)(\lambda_2/\mu_2)}{1 - \lambda_1/\mu_1 - \lambda_2/\mu_2} \\[2mm]
L_{q_2} &= \frac{\lambda_2}{\mu_1}\hat{\rho} \frac{\mu_1^2/\mu_2^2 + (1 - \mu_1/\mu_2)(\lambda_1/\mu_2)}{1 - \lambda_1/\mu_1 - \lambda_2/\mu_2} \\[2mm]
L_q &= \frac{(\lambda_1/\mu_1)\hat{\rho} + (\lambda_2/\mu_2)\hat{\rho}(\mu_1/\mu_2)}{1 - \lambda_1/\mu_1 - \lambda_2/\mu_2} \\[2mm]
&= \frac{(\rho^*)^2}{1 - \rho^*} \frac{\lambda_1/\lambda + (\lambda_2/\lambda)(\mu_1^2/\mu_2^2)}{[\lambda_1/\lambda + (\lambda_2/\lambda)(\mu_1/\mu_2)]^2}
\end{aligned}
\tag{4.45}
$$

As a "sidelight," it is interesting to note that the L_q of (4.45) is always greater than that of the standard $M/M/1$ with mean service time equal to the weighted average of the respective means; that is, $1/\mu = (\lambda_1/\lambda)/\mu_1 + (\lambda_2/\lambda)/\mu_2$ (see Problem 4.21).

Now to compare the L_q's of (4.43) [or (4.44)], the two-priority, two-rate case, with the L_q's of (4.42), the two-priority, one-rate case, we assume that the single rate chosen for use in (4.42) lies somewhere between the μ_1 and

μ_2 of the other model. One possibility is to choose $1/\mu$ equal to the weighted average, but the choice could be otherwise. When the weighted average is used, the ρ^* of (4.44) becomes equivalent to the ρ of (4.42). If the value of μ is chosen to equal the larger of μ_1 and μ_2, then it can easily be shown that all three measures, L_{q_1}, L_{q_2}, and L_q, are less in (4.42) than they are in (4.43), while this comparison is completely reversed when $\mu = \min(\mu_1, \mu_2)$ (see Problem 4.22). Therefore, as we would intuitively expect, any comparison between L_{q_1}, L_{q_2}, or L_q of (4.42) with those of (4.43) in the event that μ lies strictly between μ_1 and μ_2 is going to be a function of the exact values of the parameters involved, namely, μ_1, μ_2, μ, λ_1, and λ_2.

In comparing the two-priority, two-rate case of (4.43) now to the no-priority, two-rate case of (4.45), it can be shown (after much algebraic manipulation) that the imposition of priorities increases the mean number of priority two items waiting, while shortening the mean queue for priority one units (see Problem 4.23). Thus if a queueing system is to be designed in such a way as to reduce expected waits for one particular group of customers, then this group should be given priority. This is especially good if the nonpriority units (essentially priority two) have a higher mean service time.

A comparison now of Equations (4.44) and (4.45) would show that the L_q's (thus the W_q's also) differ by the factor

$$\frac{1 - \lambda_1 \rho^* / \lambda}{1 - \lambda_1 \hat{\rho} / \lambda}.$$

When this ratio is larger than one, there is more waiting if priorities are imposed than otherwise, while when the ratio is less than one, the imposition of priorities would reduce the mean values. The ratio will exceed one whenever $\rho^* < \hat{\rho}$. But from our previous discussion, this is equivalent to requiring

$$\frac{1}{\mu_2} < \frac{1}{\mu_1}.$$

Hence the numerator is greater than the denominator as long as the expected service time for the priority units is greater than that for the nonpriority customers, and vice versa when $1/\mu_2 > 1/\mu_1$.

The foregoing comparative results have very important implications for the design of queueing systems and specifically give rise to an optimal design rule called "priority assignment by shortest processing time" or simply "the shortest processing time (SPT) rule" [see Schrage and Miller (1966)]. That is, if the design criterion of a queue is the reduction of the total number waiting, or equivalently the overall mean delay, then if it is at all possible, it pays to give priority to that group of customers which seems to have the faster service rate. The degree of improvement is then a function of the ratio of the two mean arrival rates, namely, λ_1/λ_2.

To summarize the aforementioned comparisons between the three main models of this section, we present Table 4.1.

Example 4.7

To further our analysis let us consider the following problem: our friend the barber H. R. Cutt has decided to explore the possibility of giving priority to customers who wish only a trim. Cutt estimates that the time required to trim a customer (needed one-third of the time) is still exponential, but with a mean of 5 min, and that the removal of these customers from the total population increases the mean of the nonpriority class to 12.5 min leaving the distribution exponential. If Cutt measures his performance by the value of the mean waiting time (remember that the mean throughput will not change as long as λ remains constant, but lower waits may increase λ in the future), would this change in policy reduce average line waits in any way?

So we have $\lambda_1 = \lambda/3 = (5/3)/\text{hr}$ and $\lambda_2 = 2\lambda/3 = (10/3)/\text{hr}$, while $\mu_1 = 12/\text{hr}$ and $\mu_2 = (24/5)/\text{hr}$. These give $\hat{\rho} = \lambda/\mu_1 = (\lambda_1 + \lambda_2)/\mu_1 = 2/5$. Then the substitution of the appropriate values of the constants into Equations (4.43) gives

$$L_{q_1} = \frac{75}{248} \doteq 0.302,$$

$$L_{q_2} = \frac{225}{62} \doteq 3.63,$$

and

$$L_q \doteq 3.93.$$

Hence

$$W_q = \frac{L_q}{\lambda} \doteq 47 \text{ min.}$$

The correct model to compare this with is the no-priority two-service-rate model of (4.45) rather than the original $M/M/1$ example of Section 2.3 which gave a wait in queue of 50 min. The calculations following (4.45)

TABLE 4.1 COMPARISON OF PRIORITY MODELS

Model	Results in Equations
(1) 2 priority, 1 service rate	(4.42)
(2) 2 priority, 2 service rates	(4.43)
(3) no priority, 2 service rates	(4.45)

ˈersus	$M/M/1$	(2)	(3)
(1)	$\lambda=\lambda_1+\lambda_2$ $L_{q_1}<L_{q_1}\ (M/M/1)$ $L_{q_2}>L_{q_2}\ (M/M/1)$ $L_q=L_q\ (M/M/1)$	L_{q_1} $\left.\begin{array}{l}\\ L_{q_2}\\ \\ L_q\end{array}\right\}$ all comparisons are a function of the values of the parameters involved, $\mu_1,\mu_2,\mu,\lambda_1,$ and λ_2^{a}	Not applicable
(2)	$\lambda=\lambda_1+\lambda_2$ $\dfrac{1}{\mu}=\dfrac{\lambda_1/\lambda}{\mu_1}+\dfrac{\lambda_2/\lambda}{\mu_2}$ $L_q>L_q\ (M/M/1),\quad \mu_1<\mu_2$ $L_q<L_q\ (M/M/1),\quad \mu_1>\mu_2^{b}$	—	$L_{q_1}:\ (2)<(3)$ $L_{q_2}:\ (2)>(3)$ $L_q:\ (2)<(3),\ \mu_1>\mu_2$ $\qquad\ (2)>(3),\ \mu_1<\mu_2$
(3)	$\lambda=\lambda_1+\lambda_2$ $\dfrac{1}{\mu}=\dfrac{\lambda_1/\lambda}{\mu_1}+\dfrac{\lambda_2/\lambda}{\mu_2}$ $L_q>L_q\ (M/M/1)$	—	—

[a] μ used in (4.42) chosen to lie between μ_1 and μ_2.
[b] See Problem (4.20).

187

give

$$
\left\{
\begin{array}{l}
\rho^* = \dfrac{5}{6} \\[2ex]
L_q \doteq 4.7 \\[2ex]
W_q \doteq \dfrac{4.7}{5} \doteq 0.94 \text{ hr} \doteq 56 \text{ min}
\end{array}
\right. \qquad .
$$

Thus introducing a priority system for the customers requiring less service (faster service rate) not only reduces their average waiting, but also the average wait of *all* customers, thus illustrating the effect of the SPT rule.

Note that using the two-service-rate model of (4.45) gives a slightly larger W_q than would be obtained ignoring the fact that customers are of two types (which would be the case if we considered the results of Section 2.5, that is, the straight $M/M/1$ model which gave a $W_q = 50$). This agrees with our previous statement to that effect immediately following the development of Equation (4.45) and also with the results as given in Table 4.1.

4.4.2. Nonpreemptive Markovian Systems with Many Priorities

As observed in the previous section, the determination of stationary probabilities in a nonpreemptive Markovian system is an exceedingly difficult matter, well near impossible when the number of priorities exceeds two. In light of this and the difficulty of handling multiindexed generating functions when there are more than two priority classes, an alternative approach to obtaining the mean-value measures L and W is used, namely, a direct expected-value procedure.

Suppose that items of the kth priority (the smaller the number, the higher the priority) arrive before a single channel according to a Poisson distribution with parameter λ_k $(k = 1, 2, \ldots, r)$ and that these customers wait on a first-come, first-served basis within their respective priorities. Let the service distribution for the kth priority be exponential with mean $1/\mu_k$. Whatever the priority of a unit in service, it completes its service before another item is admitted.

We begin by defining

$$
\rho_k = \frac{\lambda_k}{\mu_k} \qquad (1 \leqslant k \leqslant r)
$$

and

$$\sigma_k = \sum_{i=1}^{k} \rho_i \qquad (\sigma_0 \equiv 0, \ \sigma_r \equiv \rho).$$

The system is stationary for $\sigma_r = \rho < 1$. Then suppose that a customer of priority i arrives at the system at time t_0 and enters service at time t_1. Its line wait is thus $T_q = t_1 - t_0$. At t_0 assume that there are n_1 customers of priority one in the line ahead of this new arrival, n_2 of priority two, n_3 of priority three, and so on. Let S_0 be the time required to finish the item already in service and S_k be the total time required to serve n_k. During the new customer's waiting time T_q, (say) n_k' items of priority $k < i$ will arrive and go to service ahead of this current arrival. If S_k' is the total service time of all the n_k', then it can be seen that

$$T_q = \sum_{k=1}^{i-1} S_k' + \sum_{k=1}^{i} S_k + S_0.$$

If expected values are taken on both sides of the foregoing, then we find that

$$W_q^{(i)} \equiv E[T_q] = \sum_{k=1}^{i-1} E[S_k'] + \sum_{k=1}^{i} E[S_k] + E[S_0].$$

Since $\sigma_{i-1} < \sigma_i$ for all i, $\rho < 1$ implies that $\sigma_{i-1} < 1$ for all i.

To find $E[S_0]$, we observe that the combined service distribution is the mixed exponential, which is formed from the law of total probability as

$$B(t) = \sum_{k=1}^{r} \frac{\lambda_k}{\lambda} (1 - e^{-\mu_k t}),$$

where

$$\lambda = \sum_{k=1}^{r} \lambda_k.$$

The random variable "remaining time of service," S_0, has the value 0 if the system is idle; hence

$$E[S_0] = \Pr\{\text{system is busy}\} \ E[S_0 | \text{busy system}].$$

But the probability that the system is busy is

$$\lambda(\text{expected service time}) = \lambda \sum_{k=1}^{r} \frac{\lambda_k}{\lambda} \frac{1}{\mu_k}$$

$$= \rho,$$

and

$E[S_0|\text{system busy}] =$

$\sum_{k=1}^{r} E[S_0|\text{system busy with } k \text{ type customer}]\cdot \text{Pr}\{\text{customer has priority } k\}$

$$= \sum_{k=1}^{r} \frac{1}{\mu_k} \frac{\rho_k}{\rho}.$$

Therefore

$$E[S_0] = \rho \sum_{k=1}^{r} \frac{1}{\mu_k} \frac{\rho_k}{\rho}$$

$$= \sum_{k=1}^{r} \frac{\rho_k}{\mu_k}. \qquad (4.46)$$

Since n_k and the service times of individual customers, $S_k^{(n)}$, are independent,

$$E[S_k] = E[n_k S_k^{(n)}]$$

$$= E[n_k]E[S_k^{(n)}]$$

$$= \frac{E[n_k]}{\mu_k}.$$

Utilizing Little's formula then gives

$$E[S_k] = \frac{\lambda_k W_q^{(k)}}{\mu_k}$$

$$= \rho_k W_q^{(k)}.$$

Similarly,

$$E[S_k'] = \frac{E[n_k']}{\mu_k},$$

and then utilizing the uniform property of the Poisson we have

$$E[S_k'] = \frac{\lambda_k W_q^{(i)}}{\mu_k}.$$

Therefore

$$W_q^{(i)} = W_q^{(i)} \sum_{k=1}^{i-1} \rho_k + \sum_{k=1}^{i} \rho_k W_q^{(k)} + E[S_0],$$

or

$$W_q^{(i)} = \frac{\sum_{k=1}^{i} \rho_k W_q^{(k)} + E[S_0]}{1 - \sigma_{i-1}}. \tag{4.47}$$

The solution to Equation (4.47) was found by Cobham (1954), after whom much of this analysis follows, by induction on i, after a general pattern emerged upon iteration (see Problem 4.24). That solution is

$$W_q^{(i)} = \frac{E[S_0]}{(1 - \sigma_{i-1})(1 - \sigma_i)}. \tag{4.48}$$

Using Equation (4.46) finally gives

$$W_q^{(i)} = \frac{\sum_{k=1}^{r} (\rho_k / \mu_k)}{(1 - \sigma_{i-1})(1 - \sigma_i)}. \tag{4.49}$$

Note that (4.49) holds as long as $\sigma_r = \sum_{k=1}^{r} \rho_k < 1$.
We also know, therefore, from Little's formula, that[7]

$$L_q^{(i)} = \lambda_i W_q^{(i)}$$

$$= \frac{\lambda_i \sum_{k=1}^{r} (\rho_k / \mu_k)}{(1 - \sigma_{i-1})(1 - \sigma_i)}$$

and that the total expected system size is

$$L_q = \sum_{i=1}^{r} L_q^{(i)}$$

$$= \sum_{i=1}^{r} \frac{\lambda_i \sum_{k=1}^{r} (\rho_k / \mu_k)}{(1 - \sigma_{i-1})(1 - \sigma_i)}.$$

[7]The case $r = 2$ does in fact check with the earlier result given by Equation (4.43).

Results for higher moments were obtained by Kesten and Runnenburg (1957) and the interested reader is referred to that reference for the appropriate derivations and formulas.

It now would seem quite logical to ask how the value of L_q obtained for the $M/M/1$ priority model compares to that of the ordinary $M/M/1$ case with service time equal to the average over all priorities (as was done earlier in the two-priority case), namely,

$$\frac{1}{\mu} = \sum_{i=1}^{r} \frac{\lambda_i}{\lambda} \frac{1}{\mu_i}.$$

It should seem intuitive that the average wait of an item in a priority situation should be different from its wait if the discipline were nonpriority. But what about the weighted average of the waits taken over all priorities,

$$\overline{W}_q = \sum_{i=1}^{r} \frac{\lambda_i W_q^{(i)}}{\lambda}?$$

For after all, if $\overline{W}_q$ were the same as the nonpriority W_q, then the two L_q's would have to be equal.

But some further thought and analogy with the two-priority case should convince us that the expected-value measures are the same if, and only if, the μ_i's are identical. That is to say,

$$\overline{W}_q = \frac{\lambda}{\mu(\mu - \lambda)}$$

if, and only if, $\mu_i \equiv \mu$ for all i (see Problem 4.25). In fact, it turns out that if higher priority units have faster service rates, then the average wait over all units (also the average system size measures) is less than a nonpriority system with a constant service time of

$$\frac{1}{\mu} = \sum_{i=1}^{r} \frac{(\lambda_i/\lambda)}{\mu_i},$$

again illustrating the SPT rule mentioned earlier. If the opposite is true, that is, lower priorities have faster service, then the "equivalent" nonpriority model gives the lower average wait and system size (see Problem 4.26). These differences increase as saturation is approached. If priorities and service rate rankings are mixed, then the result depends upon the pairings of priorities and service rates, and the actual numerical values of the average service times.

This argument is essentially the same as the one given in the two-priority case to justify the SPT rule. That is, if the overriding requirement in the design of a queueing system is the reduction of the delay for one specific set of items, then this class should be given priority. This becomes especially profitable if the urgent set of items takes less time to serve on the average. If, however, the criterion for design is simply to reduce the average wait in queue of all units or to reduce the total number waiting, then it helps to give priority to that class of units which tends to have the faster service rate if such is discernible. For a further discussion of the effect of priorities on delay, the reader is referred to Morse (1958) and Jaiswal (1968).

Some additional comments on where there are *no* differences between the state probabilities and measures of effectiveness for $M/M/1/\infty/$FIFO and $M/M/1/\infty/$PRI (as well as $M/M/1/\infty/$GD): the same state probabilities and measures hold for arbitrary queue disciplines provided (1) all arrivals stay in the queue until served; (2) the mean service time of all units is the same; (3) the server completes service before it starts on the next item; and (4) the service channel always admits a waiting customer immediately upon the completion of another.

The analysis for the multiple-channel case is very similar to that of the preceding model except that it must now be assumed that service is governed by identical exponential distributions for each priority at each of c channels. Unfortunately for multichannels we must assume no service-time distinction between priorities or else the mathematics becomes quite intractable.

Let us define

$$\rho_k = \frac{\lambda_k}{c\mu} \qquad (1 \leqslant k \leqslant r)$$

and

$$\sigma_k = \sum_{i=1}^{k} \rho_i \qquad (\sigma_r \equiv \rho = \lambda/c\mu).$$

Again the system is completely stationary for $\rho < 1$, and

$$W_q^{(i)} = \sum_{k=1}^{i-1} E[S_k'] + \sum_{k=1}^{i} E[S_k] + E[S_0],$$

where, as before, S_k is the time required to serve n_k items of the kth priority in the line ahead of the item, S_k' is the service time of the n_k' items of priority k which arrive during $W_q^{(i)}$, and S_0 is the amount of time remaining until the next server becomes available. The first two terms of the right-hand side of the $W_q^{(i)}$ equation are exactly the same as the single-channel case, except that the system service rate $c\mu$ is used in place of the single-service rate μ_k throughout the argument.

To derive $E[S_0]$, consider

$$E[S_0] = \Pr\{\text{all channels busy}\} \cdot E[S_0 | \text{all channels busy}].$$

The probability that all channels are busy is [from Equation (3.9)]

$$\sum_{n=c}^{\infty} p_n = p_0 \sum_{n=c}^{\infty} \frac{(c\rho)^n}{c^{n-c}c!}$$

$$= \frac{p_0(c\rho)^c}{c!(1-\rho)},$$

and

$$E[S_0 | \text{all channels busy}] = \frac{1}{c\mu}$$

from the memorylessness of the exponential. Thus from Equation (3.11),

$$E[S_0] = \frac{(c\rho)^c}{c!(1-\rho)(c\mu)} \left[\sum_{n=0}^{c-1} \frac{(c\rho)^n}{n!} + \frac{(c\rho)^c}{c!(1-\rho)} \right]^{-1}.$$

Therefore from (4.49),

$$W_q^{(i)} = \frac{E[S_0]}{(1-\sigma_{i-1})(1-\sigma_i)}$$

$$= \frac{\left[c!(1-\rho)(c\mu) \sum_{n=0}^{c-1} (c\rho)^{n-c}/n! + c\mu \right]^{-1}}{(1-\sigma_{i-1})(1-\sigma_i)},$$

and the expected line wait taken over all priorities is thus

$$W_q = \sum_{i=1}^{r} \frac{\lambda_i}{\lambda} W_q^{(i)}.$$

4.4.3 Preemptive Priorities

Let us now extend the Markovian model of Section 4.4.1 to permit units of the higher priority to preempt. We then have to decide whether or not the ejected items lose all service performed before ejection. But since it is to be assumed that service is exponential, such a question is irrelevant in view of memorylessness. In addition, since the customer served will always be of priority one when at least one unit of that priority is present, we may drop the use of the third subscript of p_{mnr} as a service-customer indicator and instead use it to introduce a new priority, henceforth called 3 or lowest. So p_{mnr} is now the steady-state probability that there are m units of priority one in the system with arrival rate λ_1 and service rate μ_1, n units of priority two in the system with arrival rate λ_2 and service rate μ_2, and r units of priority three, with arrival rate λ_3 and service rate μ_3.

Under the aforementioned assumptions, a system of difference equations may be derived for the stationary probabilities ($\lambda = \lambda_1 + \lambda_2 + \lambda_3$ and $\rho = \lambda_1/\mu_1 + \lambda_2/\mu_2 + \lambda_3/\mu_3 < 1$), namely,

$$
\begin{cases}
0 = -\lambda p_{000} + \mu_1 p_{100} + \mu_2 p_{010} + \mu_3 p_{001} \\[2mm]
0 = -(\lambda + \mu_1)p_{m00} + \lambda_1 p_{m-1,0,0} + \mu_1 p_{m+1,0,0} \\[2mm]
0 = -(\lambda + \mu_2)p_{0n0} + \mu_1 p_{1n0} + \lambda_2 p_{0,n-1,0} + \mu_2 p_{0,n+1,0} \\[2mm]
0 = -(\lambda + \mu_3)p_{00r} + \mu_1 p_{10r} + \mu_2 p_{01r} + \lambda_3 p_{0,0,r-1} + \mu_3 p_{0,0,r+1} \\[2mm]
0 = -(\lambda + \mu_1)p_{mn0} + \lambda_1 p_{m-1,n,0} + \lambda_2 p_{m,n-1,0} + \mu_1 p_{m+1,n,0} \\[2mm]
0 = -(\lambda + \mu_1)p_{m0r} + \lambda_1 p_{m-1,0,r} + \lambda_3 p_{m,0,r-1} + \mu_1 p_{m+1,0,r} \\[2mm]
0 = -(\lambda + \mu_2)p_{0nr} + \mu_1 p_{1nr} + \lambda_2 p_{0,n-1,r} + \mu_2 p_{0,n+1,r} + \lambda_3 p_{0,n,r-1} \\[2mm]
0 = -(\lambda + \mu_1)p_{mnr} + \lambda_1 p_{m-1,n,r} + \lambda_2 p_{m,n-1,r} + \lambda_3 p_{m,n,r-1} + \mu_1 p_{m+1,n,r}
\end{cases} \qquad (4.50)
$$

(There will be a total of 2^r equations when the number of preemptive priorities is r.)

The solution to this system of stationary equations is quite complicated, but suffice it to say that the steady-state generating function can be obtained (see Problem 4.27), and, from it, the mean number of priority i

waiting in the system is found for $\rho_i \equiv \lambda_i/\mu$ to be

$$E[N_q^{(i)}] = \frac{\rho_i \sum\limits_{n=1}^{i} \rho_n}{\left(1 - \sum\limits_{n=1}^{i-1} \rho_n\right)\left(1 - \sum\limits_{n=1}^{i} \rho_n\right)}$$

when $\sum_{i=1}^{3} \rho_i < 1$. In addition, the variance can be found to be

$$\mathrm{Var}[N_q^{(i)}] = 2\rho_i^3 \sum_{n=1}^{i-1} \frac{\rho_n}{\left(1 - \sum\limits_{n=1}^{i-1} \rho_n\right)\left(1 - \sum\limits_{n=1}^{i} \rho_n\right)}$$

$$+ \frac{\rho_i^2 + \rho_i\left(1 - \sum\limits_{n=1}^{i-1} \rho_n\right)\left(1 - \sum\limits_{n=1}^{i} \rho_n\right)}{\left(1 - \sum\limits_{n=1}^{i-1} \rho_n\right)^2 \left(1 - \sum\limits_{n=1}^{i} \rho_n\right)}.$$

Both of these results generalize to $r > 3$ priorities.

There are many variations on these themes, and if the reader is interested in this problem area, he is referred to Jaiswal (op. cit). But to close this section one final model will be presented which is an interesting extension by Phipps (1956) of Cobham's models of Section 4.4.2. Phipps allowed the number of priorities to be continuous such that the priority of a given unit is assigned according to some measure of the length of time needed to service this unit.[8] In particular, let this unit arrive as a Poisson process with mean rate λ_t ($\int_0^\infty \lambda_t dt = \lambda$) and be serviced according to an exponential distribution with mean time $1/\mu_t$. Then the total service cumulative distribution is

$$B(t) = \int_0^\infty \frac{\lambda_t}{\lambda}(1 - e^{-\mu_t t})\,dt$$

$$= 1 - \frac{1}{\lambda}\int_0^\infty \lambda_t e^{-\mu_t t}dt.$$

The expected waiting time $W_q^{(t)}$ of a customer with type t priority is by

[8]This might, indeed, be the case in a situation such as the loading of programs onto a computer.

analogy with the discrete case, Equation (4.48), given as

$$W_q^{(t)} = \frac{E[S_0]}{\left[1 - \lambda \int_0^t y \, dB(y)\right]^2}.$$

It can be shown in general (this will be taken up in Chapter 5—see Problem 5.5) that

$$E[S_0] = \frac{\lambda}{2} E[S^2]$$

for any single-channel queue with Poisson input. Thus

$$W_q^{(t)} = \frac{(1/2) \int_0^\infty \lambda_t t^2 \mu_t e^{-\mu_t t} dt}{\left(1 - \int_0^t \lambda_y y \mu_y e^{-\mu_y y} dy\right)^2}.$$

The expected number of customers in the line is thus

$$L_q = \lambda \int_0^\infty W_q^{(t)} dt$$

$$= \left[\frac{\lambda}{2} \int_0^\infty \lambda_t t^2 \mu_t e^{-\mu_t t} dt\right] \left[\int_0^\infty \left(1 - \int_0^t \lambda_y y \mu_y e^{-\mu_y y} dy\right)^{-2} dt\right].$$

If μ_t is assumed to be the constant μ and λ_t to be λ, then the foregoing simplifies to

$$W_q^{(t)} = \frac{(\lambda/\mu^2)}{\left(1 - \lambda \int_0^t y \mu e^{-\mu y} dy\right)^2}$$

$$= \frac{(\lambda/\mu^2)}{\left\{1 - (\lambda/\mu)[1 - e^{-\mu t}(1 + \mu t)]\right\}^2}$$

and

$$L_q = \left(\frac{\lambda}{\mu}\right)^2 \int_0^\infty \left\{1 - \frac{\lambda}{\mu}[1 - e^{-\mu t}(1 + \mu t)]\right\}^{-2} dt.$$

4.5   SERIES QUEUES

In this section we consider models in which there are a series of service stations through which each calling unit must progress prior to leaving the system. Some examples of such series queueing situations (sometimes referred to as tandem queues) are manufacturing or assembly line processes in which units must proceed through a series of work stations, each station preforming a given task, a registration process (such as university registration) where the registrant must visit a series of "desks" (his advisor, his department chairman, the cashier, etc.), or a clinic physical examination procedure where the patient goes through a series of stages (lab tests, electrocardiogram, chest x-ray, etc.). In the following subsections several types of series queueing models are analyzed.

4.5.1   Queue Output

The first series model to be considered is a sequence of queues with no restriction on the waiting room's capacity between stations. Such a situation is pictured in Figure 4.3. We further assume that the calling units arrive according to a Poisson process, mean λ, and the service time of each server at station i ($i = 1, 2, \ldots, n$) is exponential with mean $1/\mu_i$. One can readily see that since there is no restriction on waiting between stations, each station can be analyzed separately as a single stage (nonseries) queueing model.

The first station is an $M/M/c_1/\infty$ model. It is necessary to find the output distribution (distribution of times between successive departures) in order to find the input distribution (times between successive arrivals) to the next station. It turns out rather surprisingly that the departure time

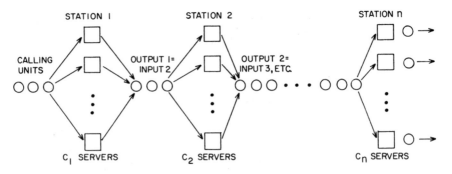

Fig. 4.3   Series queue, infinite waiting room

distribution from an $M/M/c/\infty$ queue is identical to the interarrival time distribution, namely, exponential with mean $1/\lambda$; hence all stations are $M/M/c_i/\infty$ models. Thus the results of Section 3.2 can be used on each station individually and a complete analysis of this type of series situation is possible. It remains then to show that the departure times are exponential with parameter λ.

Consider an $M/M/c/\infty$ queue in steady state. Let $N(t)$ now represent the number of customers in the system at a time t after the last departure. Since we are considering steady state, we have

$$\Pr\{N(t)=n\}=p_n. \qquad (4.51)$$

Furthermore, let T represent the random variable "time between successive departures" (interdeparture time), and

$$F_n(t)=\Pr\{N(t)=n \text{ and } T>t\}. \qquad (4.52)$$

So $F_n(t)$ is the joint probability that there are n in the system at a time t after the last departure and that t is less than the interdeparture time; that is, another departure has not as yet occurred. The cumulative distribution of the random variable T which shall be denoted as $C(t)$ is given by

$$C(t)\equiv\Pr\{T\leqslant t\}=1-\sum_{n=0}^{\infty}F_n(t) \qquad (4.53)$$

since

$$\sum_{n=0}^{\infty}F_n(t)=\Pr\{T>t\}$$

is the marginal complementary cumulative distribution of T. To find $C(t)$, it is necessary to first find $F_n(t)$.

We can write the following difference equations concerning $F_n(t)$:

$$\begin{cases} F_n(t+\Delta t)=(1-\lambda\Delta t)(1-c\mu\Delta t)F_n(t) \\ \qquad\qquad +\lambda\Delta t(1-c\mu\Delta t)F_{n-1}(t)+o(\Delta t) \qquad (c\leqslant n) \\ F_n(t+\Delta t)=(1-\lambda\Delta t)(1-n\mu\Delta t)F_n(t) \\ \qquad\qquad +\lambda\Delta t(1-n\mu\Delta t)F_{n-1}(t)+o(\Delta t) \qquad (1\leqslant n\leqslant c) \\ F_0(t+\Delta t)=(1-\lambda\Delta t)F_0(t)+o(\Delta t) \end{cases}$$

Proceeding in the usual fashion we can obtain the differential-difference equations as

$$\begin{cases} \dfrac{dF_n(t)}{dt} = -(\lambda + c\mu)F_n(t) + \lambda F_{n-1}(t) & (c \leqslant n) \\[3mm] \dfrac{dF_n(t)}{dt} = -(\lambda + n\mu)F_n(t) + \lambda F_{n-1}(t) & (1 \leqslant n \leqslant c) \quad \cdot (4.54) \\[3mm] \dfrac{dF_0(t)}{dt} = -\lambda F_0(t) \end{cases}$$

Using the boundary condition

$$F_n(0) \equiv \Pr\{N(0) = n \text{ and } T > 0\}$$

$$= \Pr\{N(0) = n\} = p_n \qquad [\text{from } (4.51)]$$

and methodology similar to that of Section 1.8 (see Problem 4.28) where the Poisson process is derived, we find that the solution to (4.54) is

$$F_n(t) = p_n e^{-\lambda t}. \qquad (4.55)$$

The reader can easily verify that (4.55) is a solution to (4.54) by substitution, recalling that for $M/M/c/\infty$ models,

$$p_{n+1} = \begin{cases} \dfrac{\lambda}{(n+1)\mu} p_n & (1 \leqslant n \leqslant c) \\[3mm] \dfrac{\lambda}{c\mu} p_n & (c \leqslant n) \end{cases} \qquad [\text{see } (3.9)].$$

To obtain $C(t)$, the cumulative distribution of the interdeparture times, we use (4.55) in (4.53) to get

$$C(t) = 1 - \sum_{n=0}^{\infty} p_n e^{-\lambda t}$$

$$= 1 - e^{-\lambda t} \sum_{n=0}^{\infty} p_n$$

$$= 1 - e^{-\lambda t}$$

and

$$c(t) = \frac{dC(t)}{dt},$$

so that

$$c(t) = \lambda e^{-\lambda t}$$ (4.56)

It is also true that the random variables $N(t)$ and T are independent and furthermore that successive interdeparture times are independent of each other [see Problem (4.29)]. So we see that the output distribution is identical to the input distribution and not at all affected by the exponential service mechanism. Intuitively one would expect the means to be identical since we are in steady state so that the input rate must equal the output rate, but it is not quite so intuitive that the variances and indeed the distributions be identical. Nevertheless it is true and proves extremely useful in analyzing series queues where the initial input rate is Poisson, the service at all stations is exponential, and there is no restriction on queue size between stations.

We now illustrate a series queueing situation of the type above with an example.

Example 4.8

Cary Meback, the president of a large Virginia supermarket chain, is experimenting with a new store design and has remodeled one of his stores as follows. Instead of the usual checkout counter design, the store has been remodeled to include a checkout "lounge." As customers complete their shopping, they enter the lounge with their carts and, if all checkers are busy, they receive a number. They then park their carts and take a seat. When a checker is free, the next number is called and the customer with that number enters the available checkout counter. The store has been enlarged so that for practical purposes, there is no limit on either the number of shoppers that can be in the food aisles or the number that can wait in the lounge, even during peak periods.

The management estimates that during peak hours customers arrive according to a Poisson process at a mean rate of 40/hr and it takes a customer, on the average, $\frac{3}{4}$ hr to fill his shopping cart, the filling times being approximately exponentially distributed. Furthermore, the checkout times are also approximately exponentially distributed with a mean of 4 min, regardless of the particular checkout counter (during peak periods each counter has a cashier and bagger, hence the relatively low mean checkout time).

Meback wishes to know the following:

(i) What is the minimum number of checkout counters required in operation during peak periods?

(ii) If it is decided to add one more than the minimum number of counters required in operation, what is the average waiting time in the lounge? How many people, on the average, will be in the lounge? How many people, on the average, will be in the entire supermarket?

This situation can be modeled by a two-station series queue. The first is the food portion of the supermarket. Since it is self-service and arrivals are Poissonian, we have an $M/M/\infty$ model with $\lambda = 40$ and $\mu = 4/3$. The second station is an $M/M/c$ model, since the output of an $M/M/\infty$ queue is identical to its input. Hence the input to the checkout lounge is Poissonian with a mean of 40/hr also. Since for steady-state convergence, $c\mu > \lambda$, the minimum number of checkout counters, c_m, is given by

$$c_m > \frac{\lambda}{\mu} = \frac{40}{15} \doteq 2.67;$$

hence c_m must be 3.

If it is decided to have four counters in operation, we have at the checkout stations an $M/M/4/\infty$ model with $\lambda = 40$ and $\mu = 15$. Meback desires to know W_q and L_q for the $M/M/4/\infty$ model and the average total number in the supermarket which is the sum of the L's for each model. Using Equations (3.11) and (3.16) for an $M/M/c/\infty$ model, we get

$$p_0 = \left[\sum_{n=0}^{3} \frac{1}{n!} \left(\frac{8}{3} \right)^n + \frac{1}{4!} \left(\frac{8}{3} \right)^4 \left(\frac{4}{4 - 8/3} \right) \right]^{-1}$$

$$\doteq 0.06$$

and

$$W_q \doteq \left[\frac{(8/3)^4 15}{3!(60 - 40)^2} \right] (0.06)$$

$$\doteq 0.019 \text{ hr}$$

$$\doteq 1.14 \text{ min.}$$

To get L_q we use Little's formula and find

$$L_q = \lambda W_q \doteq 40(0.019) = 0.76,$$

or, on the average, less than one person will be waiting in the lounge for a checker to become free.

The total number of people in the system, on the average, is the L for this $M/M/4/\infty$ model plus the L for the $M/M/\infty$ model. For the checkout station we get

$$L = \lambda W = \lambda \left(W_q + \frac{1}{\mu} \right)$$

$$\doteq 40 \left(0.019 + \frac{4}{60} \right)$$

$$\doteq 3.44.$$

For the supermarket proper we have from equation (3.33) of the $M/M/\infty$ model that

$$L = \frac{\lambda}{\mu} = \frac{40}{(4/3)} = 30.$$

Hence the average number of customers in the store during peak hours is 33.44, if Meback decides on four checkout counters in operation. He might do well to perform similar calculations for three checkout counters operating to see how much congestion increases (see Problem 4.30).

For series queues, therefore, as long as there are no capacity limitations between stations and the input is Poisson, results can be rather easily obtained. Furthermore, it can be shown (see Problem 4.31) that the joint probability that there are n_1 at station one, n_2 at station two,..., n_j at station j is merely the product, $p_{n_1} p_{n_2} \cdots p_{n_j}$. It is not even necessary to have the customer flow strictly in "line"; that is, it is possible to find situations where customers can enter the system at any station, and proceed from that station to one of several others according to some probability. For example, suppose that for station i $(i = 1, 2, \ldots, N)$ customers arrive from outside the system according to a Poisson process with parameter λ_i and that q_{ik} represents the probability of a customer having been serviced at station i going next to station k. The $M/M/c/\infty$ analysis goes through for each station where the input to station i has parameter

$$\lambda_i' = \lambda_i + \sum_{j=1}^{N} q_{ji} \lambda_j'.$$

Of course, the Poisson assumption is crucial here, since the reason for being able to analyze the process in this manner lies in the ability to divide or mix Poisson streams without destroying the Poissonian character of the resulting stream. The fact that the distribution of the number of customers in any station is independent of the others is also critical.

The analysis for series queues when there are limits on the capacity at a station (except for the case where the only limit is at the last station in a pure series flow situation and arriving customers who exceed the capacity are shunted out of the system—see Problem 4.32) are much more complex. This results from the blocking effect; that is, a station downstream comes up to capacity and thereby prevents any further processing at upstream stations which feed it. We treat some of these types of models in the next section.

4.5.2 Series Queues with Blocking

We consider first a simple sequential two-station single-server-at-each-station model where no queue is allowed to form at either station. If a customer is in station two, and service is completed at station one, the station one customers must wait there until the station two customer is completed; that is, the system is *blocked*. Arrivals at station one when the system is blocked are turned away. Also if a customer is in process at station one, even if station two is empty, arriving customers are turned away, since the system is a sequential one; that is, all customers require service at one and then service at two.

We wish to find the steady-state probability, p_{n_1,n_2}, of n_1 in the first station and n_2 in the second station. For this model, the possible states are given below in Table 4.2.

Assuming arrivals to the system (station one) are Poisson with parameter λ and service is exponential with parameters μ_1 and μ_2, respectively, we can

TABLE 4.2   POSSIBLE SYSTEM STATES

n_1,n_2	Description
0, 0	System empty
1, 0	Customer in process at 1 only
0, 1	Customer in process at 2 only
1, 1	Customers in process at 1 and 2
b, 1	Customer in process at 2 and a customer finished at 1 but waiting for 2 to become available, that is, system blocked

write the following difference equations (omitting $o(\Delta t)$ terms) as

$$
\begin{cases}
p_{0,0}(t+\Delta t) = p_{0,0}(t)(1-\lambda\Delta t) + p_{0,1}(t)(1-\lambda\Delta t)\mu_2\Delta t \\[2mm]
p_{1,0}(t+\Delta t) = p_{1,0}(t)(1-\mu_1\Delta t) + p_{1,1}(t)(1-\mu_1\Delta t)\mu_2\Delta t + p_{0,0}(t)\lambda\Delta t \\[2mm]
p_{0,1}(t+\Delta t) = p_{0,1}(t)(1-\lambda\Delta t)(1-\mu_2\Delta t) + p_{1,0}(t)\mu_1\Delta t + p_{b,1}(t)\mu_2\Delta t \\[2mm]
p_{1,1}(t+\Delta t) = p_{1,1}(t)(1-\mu_1\Delta t)(1-\mu_2\Delta t) + p_{0,1}(t)\lambda\Delta t(1-\mu_2\Delta t) \\[2mm]
p_{b,1}(t+\Delta t) = p_{b,1}(t)(1-\mu_2\Delta t) + p_{1,1}(t)\,\mu_1\Delta t(1-\mu_2\Delta t)
\end{cases}
$$

The usual procedure leads to the steady-state equations

$$
\begin{cases}
0 = -\lambda p_{0,0} + \mu_2 p_{0,1} \\[2mm]
0 = -\mu_1 p_{1,0} + \mu_2 p_{1,1} + \lambda p_{0,0} \\[2mm]
0 = -(\lambda+\mu_2)p_{0,1} + \mu_1 p_{1,0} + \mu_2 p_{b,1} \\[2mm]
0 = -(\mu_1+\mu_2)p_{1,1} + \lambda p_{0,1} \\[2mm]
0 = -\mu_2 p_{b,1} + \mu_1 p_{1,1}
\end{cases}
\tag{4.57}
$$

Using the boundary equation, $\sum\sum p_{n_1,n_2}=1$, we have six equations in five unknowns [there is some redundancy in (4.57); hence we can solve for the five steady-state probabilities]. Equation (4.57) can be used to get all probabilities in terms of $p_{0,0}$ and the boundary condition can be used to find $p_{0,0}$. If we let $\mu_1=\mu_2$, the results are (see Problem 4.34)

$$
\begin{cases}
p_{1,0} = \dfrac{\lambda(\lambda+2\mu)}{2\mu^2}p_{0,0} \\[4mm]
p_{0,1} = \dfrac{\lambda}{\mu}p_{0,0} \\[4mm]
p_{1,1} = \dfrac{\lambda^2}{2\mu^2}p_{0,0} \\[4mm]
p_{b,1} = \dfrac{\lambda^2}{2\mu^2}p_{0,0} \\[4mm]
p_{0,0} = \dfrac{2\mu^2}{3\lambda^2+4\mu\lambda+2\mu^2}
\end{cases}
\tag{4.58}
$$

It is easy to see how the problem expands if one allows limits other than zero on queue length or considers more stations. For example, if one customer is allowed to wait between stations, this results in seven state probabilities for which to solve, utilizing seven equations and a boundary condition (see Problem 4.35). The complexity results from having to write a difference equation for each possible system state. Conceptually, however, these types of series queueing situations can be attacked via the methodology presented above.

We treat one last set of series situations before moving on.[9] We first consider a two-station sequential series queue where no waiting is allowed between the stations but where a queue with no limit is permitted in front of the first station. The arrival process to station one is Poisson, mean λ, while service at each station is exponential with parameters μ_1 and μ_2, respectively. It is necessary to use three subscripts on p, the first for the number at the first station (including the customer in service), the last for the number in the second station (1 or 0), and the middle to indicate a blocked system (0 or b). For example, $p_{4,0,1}$ would indicate three people in the queue at station one, the station one customer being served, and a customer in service at station two, while $p_{4,b,1}$ indicates four customers in the station one queue and a customer finished service at station one waiting for the customer in station two to finish service (note that when blocking occurs, the first subscript applies to number in queue only). The difference equations are then

$$p_{0,0,0}(t+\Delta t)=p_{0,0,0}(t)(1-\lambda\Delta t)+p_{0,0,1}(t)(1-\lambda\Delta t)\mu_2\Delta t,$$

$$p_{0,0,1}(t+\Delta t)=p_{0,0,1}(t)(1-\lambda\Delta t)(1-\mu_2\Delta t)+p_{0,b,1}(t)(1-\lambda\Delta t)\mu_2\Delta t$$
$$+p_{1,0,0}(t)(1-\lambda\Delta t)\mu_1\Delta t,$$

$$p_{0,b,1}(t+\Delta t)=p_{0,b,1}(t)(1-\lambda\Delta t)(1-\mu_2\Delta t)+p_{1,0,1}(1-\lambda\Delta t)\mu_1\Delta t,$$

and, for $n>1$,

$$p_{n,0,0}(t+\Delta t)=p_{n,0,0}(t)(1-\lambda\Delta t)(1-\mu_1\Delta t)+p_{n-1,0,0}(t)\lambda\Delta t(1-\mu_1\Delta t)$$
$$+p_{n,0,1}(t)(1-\lambda\Delta t)\mu_2\Delta t,$$

$$p_{n,0,1}(t+\Delta t)=p_{n,0,1}(t)(1-\lambda\Delta t)(1-\mu_1\Delta t)(1-\mu_2\Delta t)$$
$$+p_{n-1,0,1}(t)\lambda\Delta t(1-\mu_1\Delta t)(1-\mu_2\Delta t)$$
$$+p_{n,b,1}(t)(1-\lambda\Delta t)(1-\mu_1\Delta t)\mu_2\Delta t+p_{n+1,0,0}(t)(1-\lambda\Delta t)\mu_1\Delta t,$$

[9]These models can be found in Hunt (1956).

$$p_{n,b,1}(t+\Delta t) = p_{n,b,1}(t)(1-\lambda\Delta t)(1-\mu_2\Delta t) + p_{n-1,b,1}(t)\lambda\Delta t(1-\mu_2\Delta t)$$

$$+ p_{n+1,0,1}(t)(1-\lambda\Delta t)\mu_1\Delta t(1-\mu_2\Delta t).$$

These lead, in the usual manner to the following steady-state difference equations:

$$\begin{cases} 0 = -(\lambda+\mu_1)p_{n,0,0} + \lambda p_{n-1,0,0} + \mu_2 p_{n,0,1} \\[2mm] 0 = -(\lambda+\mu_1+\mu_2)p_{n,0,1} + \lambda p_{n-1,0,1} + \mu_2 p_{n,b,1} + \mu_1 p_{n+1,0,0} \\[2mm] 0 = -(\lambda+\mu_2)p_{n,b,1} + \lambda p_{n-1,b,1} + \mu_1 p_{n+1,0,1} \\[2mm] 0 = -\lambda p_{0,0,0} + \mu_2 p_{0,0,1} \\[2mm] 0 = -(\lambda+\mu_2)p_{0,0,1} + \mu_2 p_{1,b,1} + \mu_1 p_{1,0,0} \\[2mm] 0 = -(\lambda+\mu_2)p_{0,b,1} + \mu_1 p_{1,0,1} \end{cases} \left.\begin{matrix} \\ \\ (n>1) \\ \\ \\ \\ \\ \end{matrix}\right\} \quad . \quad (4.59)$$

Consider the case where $\mu_1 = \mu_2$. We also make a simplifying change in notation as follows, since the last two subscripts take on only one of three possible sets of values. Let

$$p_{n,1} = p_{n,0,0} ,$$

$$p_{n,2} = p_{n,0,1} ,$$

and

$$p_{n,3} = p_{n,b,1} .$$

Equations (4.59) then become

$$\begin{cases} 0 = -(\lambda+\mu)p_{n,1} + \lambda p_{n-1,1} + \mu p_{n,2} \\[2mm] 0 = -(\lambda+2\mu)p_{n,2} + \lambda p_{n-1,2} + \mu p_{n,3} + \mu p_{n+1,1} \quad (n>1) \\[2mm] 0 = -(\lambda+\mu)p_{n,3} + \lambda p_{n-1,3} + \mu p_{n+1,2} \\[2mm] 0 = -\lambda p_{0,1} + \mu p_{0,2} \\[2mm] 0 = -(\lambda+\mu)p_{0,2} + \mu p_{1,3} + \mu p_{1,1} \\[2mm] 0 = -(\lambda+\mu)p_{0,3} + \mu p_{1,2} \end{cases} \quad . \quad (4.60)$$

Using the method of linear operators and "upping" the first subscript by one, we have on the first three equations of (4.60)

$$
\begin{cases}
0 = [-(\lambda+\mu)D+\lambda]p_{n,1}+\mu Dp_{n,2} \\
0 = [-(\lambda+2\mu)D+\lambda]p_{n,2}+\mu Dp_{n,3}+\mu D^2p_{n,1}, \quad (n>0) \\
0 = [-(\lambda+\mu)D+\lambda]p_{n,3}+\mu D^2p_{n,2}
\end{cases}
$$

where D is the linear operator, $Dp_{n,i}=p_{n+1,i}$. These equations can be written in matrix form (after dividing by μ) as

$$
\begin{pmatrix}
-(\rho+1)D+\rho & D & 0 \\
D^2 & -(\rho+2)D+\rho & D \\
0 & D^2 & -(\rho+1)D+\rho
\end{pmatrix}
\begin{pmatrix}
p_{n,1} \\
p_{n,2} \\
p_{n,3}
\end{pmatrix} = 0,
$$

where $\rho=\lambda/\mu$. For a solution to exist, the determinant of coefficients must vanish; that is,

$$
\begin{vmatrix}
-(\rho+1)D+\rho & D & 0 \\
D^2 & -(\rho+2)D+\rho & D \\
0 & D^2 & -(\rho+1)D+\rho
\end{vmatrix} = 0.
$$

Upon evaluating the determinant we get the quartic equation

$$
2(\rho+1)D^4-\left[(\rho+1)^2(\rho+2)+2\rho\right]D^3+\left[\rho(\rho+1)^2+2\rho(\rho+1)(\rho+2)\right]D^2
$$

$$
-\rho^2(3\rho+4)D+\rho^3=0. \tag{4.61}
$$

Although in general it is not easy to find the roots of a quartic equation, in this case we are fortunate in that $D=1$ is a root [the reader can easily verify this by substitution in (4.61)]. One is generally a root to an operator equation arising from a queueing analysis. From the theory of equations, all rational roots are of the form a/b where a must be a divisor of the coefficient of D^0 and b must be a divisor of the coefficient of D^4. This leads us to the possible root $D=\rho/(\rho+1)$, which can in fact be verified as a root by substitution. Knowing two roots it is possible, by synthetic division, to reduce (4.61) to a quadratic equation where the two remaining roots can be found from the quadratic formula as

$$
D=(\rho/4)\left[(\rho+3)\pm\sqrt{\rho^2+6\rho+1}\,\right].
$$

Thus the four roots are

$$
\begin{cases}
r_1 = 1 \\[2mm]
r_2 = \dfrac{\rho}{\rho+1} \\[2mm]
r_3 = \dfrac{\rho}{4}\left(\rho+3+\sqrt{\rho^2+6\rho+1}\ \right) \\[2mm]
r_4 = \dfrac{\rho}{4}\left(\rho+3-\sqrt{\rho^2+6\rho+1}\ \right)
\end{cases}
$$

[The reader is referred to Borofsky (1950) for further details of solutions to quartic equations.] The solutions for $p_{n,i}$ ($i=1,2,3$) are given by linear combinations of the nth powers of the roots: that is,

$$
p_{n,i} = \sum_{j=1}^{4} d_{ij} r_j^n \qquad (n>0), \tag{4.62}
$$

where the constants $\{d_{ij}\}$ must be determined from boundary conditions. Because all $p_{n,i}$ must sum to one, any root $\geqslant 1$ must be deleted from (4.62) to have convergence. Thus r_1 is omitted from consideration and also ρ must be such that $r_2, r_3, r_4 < 1$. In order for r_2, r_3, and r_4 to be <1, ρ must be $<2/3$ (r_3 is limiting). Thus

$$
p_{n,i} = d_{i2}\left(\frac{\rho}{\rho+1}\right)^n + d_{i3}\left[\frac{\rho}{4}\left(\rho+3+\sqrt{\rho^2+6\rho+1}\ \right)\right]^n
$$

$$
+ d_{i4}\left[\frac{\rho}{4}\left(\rho+3-\sqrt{\rho^2+6\rho+1}\ \right)\right]^n \quad (1\leqslant i\leqslant 3;\ n>0;\ \rho<2/3).
$$

$$
\tag{4.63}
$$

In (4.63) there are nine unknown $\{d_{ij}\}$ to be determined. This can be done by using the three boundary equations given by the last three equations of (4.60), the extra boundary condition $\Sigma_i\Sigma_n p_{n,i}=1$, and an appropriate number of the first three equations of (4.60) as follows. We refer to the individual equations of (4.60) sequentially; that is, (1) refers to the first, (2) to the second,...,(6) to the last. From (4) of (4.60) we can relate $p_{0,2}$ to $p_{0,1}$ (note $p_{0,1}$ is, in the old notation, $p_{0,0,0}$, the probability of a completely empty system). Now substituting (4.63) into (5) of (4.60), we have an equation in the unknowns of d_{32}, d_{33}, d_{34}, d_{12}, d_{13}, d_{14}, and the probability $p_{0,1}$ (using the previous relation on $p_{0,2}$). From (6) of (4.60), we have an equation in d_{32}, d_{33}, d_{34}, d_{22}, d_{23}, d_{24}. Thus, so far, we have two equations in the nine unknowns and $p_{0,1}$. We now utilize (1), (2), and (3) of (4.60) seven times—let $n = 2, 3, 4$ in (1) and $n = 2, 3$ in (2) and (3), and each time substitute in for $p_{n,i}$ the right side of equation (4.63) to make nine equations in the nine d_{ij} unknowns (and $p_{0,1}$). Thus the $\{d_{ij}\}$ can be determined as functions of $p_{0,1}$ (hence all $p_{n,i}$ can be determined as functions of $p_{0,1}$). Finally, the condition $\Sigma_i \Sigma_n p_{n,i} = 1$ can be used to find $p_{0,1}$. This procedure is quite complicated to do in terms of ρ, but for any given ρ ($\rho < 2/3$) it can be done, albeit with some difficulty, since one must solve nine equations in nine unknowns and carry $p_{0,1}$.

The expected system size L is given by

$$L = \sum_{n=0}^{\infty} n p_{n,1} + \sum_{n=0}^{\infty} (n+1) p_{n,2} + \sum_{n=0}^{\infty} (n+2) p_{n,3}. \tag{4.64}$$

One can readily see the increasing complexities in extending this model to include more stations and/or allowing greater capacities between the stations. However, it is possible by a simpler analysis to determine the maximum allowable ρ for steady-state existence for these more complicated models, and it is interesting to see how this value (call it $\rho_{\max}$) behaves as the model becomes more complex. We illustrate the approach on the model above showing $\rho_{\max}$ to be two-thirds as we determined previously and present some results for the more complex models.

The idea behind the analysis is to determine the percent of time the first station will be blocked. If we denote this by p_B, and if the mean service rate of the first station is μ_1, effectively the maximum mean service rate of the first station, because of the blocking that occurs, can be only $(1 - p_B)\mu_1$; hence λ must be $< (1 - p_B)\mu_1$ for ergodicity. Thus

$$\rho_{\max} = \frac{\lambda}{\mu_1} = 1 - p_B. \tag{4.65}$$

Now we know p_B can be found by

$$p_B = \sum_{n=0}^{\infty} p_{n,3};$$

however, the trick is to find p_B without having to obtain $p_{n,3}$. We do this by defining three new probabilities as follows

$$q_i = \sum_{n=0}^{\infty} p_{n,i} \qquad (i = 1, 2, 3);$$

that is, q_i is the probability that the system is in state i ($i = 1$ indicates no one waiting or in second-station service, $i = 2$ indicates a person in second-station service, and $i = 3$ indicates a potentially blocked situation).

We can form difference equations on the q_i as follows:

$$\begin{cases} q_1(t+\Delta t) = [q_1(t) - p_{0,1}(t)][1 - \mu_1 \Delta t] + p_{0,1}(t) + q_2(t)\mu_2 \Delta t \\ \\ q_2(t+\Delta t) = [q_2(t) - p_{0,2}][1 - \mu_1 \Delta t][1 - \mu_2 \Delta t] + p_{0,2}(t)[1 - \mu_2 \Delta t] \\ \qquad\qquad + [q_1(t) - p_{0,1}(t)]\mu_1 \Delta t + q_3(t)\mu_2 \Delta t \\ \\ q_3(t+\Delta t) = q_3(t)[1 - \mu_2 \Delta t] + [q_2(t) - p_{0,2}(t)]\mu_1 \Delta t \end{cases} \cdot (4.66)$$

Proceeding in the usual manner, (4.66) leads to the steady-state difference equations

$$\begin{cases} 0 = -\mu_1 q_1 + \mu_1 p_{0,1} + \mu_2 q_2 \\ 0 = -(\mu_1 + \mu_2)q_2 + \mu_1 p_{0,2} + \mu_1 q_1 - \mu_1 p_{0,1} + \mu_2 q_3 \\ 0 = -\mu_2 q_3 + \mu_1 q_2 - \mu_1 p_{0,2} \end{cases} \cdot (4.67)$$

Now letting $\mu_1 = \mu_2 = \mu$, we have

$$\begin{cases} 0 = -q_1 + p_{0,1} + q_2 \\ 0 = -2q_2 + p_{0,2} + q_1 - p_{0,1} + q_3 \\ 0 = -q_3 + q_2 - p_{0,2} \end{cases} \cdot$$

As the system approaches saturation, $p_{0,1}$ and $p_{0,2}$ become smaller and smaller and eventually can be neglected giving

$$\begin{cases} q_1 = q_2 \\ 2q_2 = q_1 + q_3 \\ q_3 = q_2 \end{cases},$$

or

$$q_1 = q_2 = q_3 = \frac{1}{3}.$$

The fraction of time that the first station is free to serve customers is then

$$1 - p_B = \frac{q_1 + q_2}{q_1 + q_2 + q_3} \tag{4.68}$$

or

$$1 - p_B = \frac{2}{3};$$

hence from (4.65),

$$\rho_{max} = \frac{2}{3},$$

which is the same answer obtained from consideration of the roots of the operator equation determined previously.

Since it is the first station in this type of model which has blocking more frequently than subsequent stations, the limit on ρ for the first station is the limit for the system. For the case where $\mu_1 \neq \mu_2$, we find, using (4.65), (4.68), and (4.66), that

$$\rho_{max} = 1 - p_B = \frac{(\mu_2/\mu_1)q_2 + q_2}{(\mu_2/\mu_1)q_1 + q_2 + (\mu_2/\mu_1)q_2}$$

or

$$\rho_{max} = \frac{\mu_2(\mu_1 + \mu_2)}{\mu_1^2 + \mu_1 \mu_2 + \mu_2^2},$$

and thus the limit on the mean arrival rate is

$$\lambda_{max} = \mu_1 \rho_{max}.$$

This procedure can be followed for a three-station series queue of this type (albeit with more complicated algebra, since now there are eight q_i; see Problem 4.36) and it can be shown [see Hunt (op. cit.)] that

$$\rho_{max} = \frac{N}{D},$$

where

$$N = \mu_2 \mu_3 (\mu_2 + \mu_3)(\mu_1^4 + 2\mu_1^3 \mu_2 + 3\mu_1^3 \mu_3 + \mu_1^2 \mu_2^2 + 4\mu_1^2 \mu_2 \mu_3$$

$$+ 3\mu_1^2 \mu_3^2 + \mu_1 \mu_2^2 \mu_3 + 4\mu_1 \mu_2 \mu_3^2 + \mu_1 \mu_3^3 + \mu_2^2 \mu_3^2 + \mu_3^3 \mu_2)$$

and

$$D = \mu_1^5(\mu_2^2 + \mu_2 \mu_3 + \mu_3^2) + \mu_1^4(2\mu_2^3 + 5\mu_2^2 \mu_3 + 5\mu_2 \mu_3^2 + 3\mu_3^3)$$
$$+ \mu_1^3(\mu_2^4 + 5\mu_2^3 \mu_3 + 8\mu_2^2 \mu_3^2 + 7\mu_2 \mu_3^3 + 3\mu_3^4)$$
$$+ \mu_1^2(\mu_2^4 \mu_3 + 5\mu_2^3 \mu_3^2 + 8\mu_2^2 \mu_3^3 + 5\mu_2 \mu_3^4 + \mu_3^5)$$
$$+ \mu_1(\mu_2^4 \mu_3^2 + 5\mu_2^3 \mu_3^3 + 5\mu_2^2 \mu_3^4 + \mu_2 \mu_3^5)$$
$$+ (\mu_2^4 \mu_3^3 + 2\mu_2^3 \mu_3^4 + \mu_2^2 \mu_3^5).$$

For equal service rates at all stations, ρ_{max} reduces to

$$\rho_{max} = \frac{22}{39} \doteq 0.5641.$$

Also for four stations (now there are 21 q_i) one finds

$$\rho_{max} \doteq 0.5115.$$

As we would expect, ρ_{max} decreases as the number of stations in the series system increases, since the more stations, the greater the probability of a blockage.

Further results on ρ_{max} are given in Hunt (op. cit) for a two-station system, with an arbitrary fixed capacity of K for the second station (queue capacity of $K-1$) as

$$\rho_{max} = \frac{\mu_2(\mu_1^{K+1} - \mu_2^{K+1})}{\mu_1^{K+2} - \mu_2^{K+2}},$$

and for $\mu_1 = \mu_2$ this reduces to

$$\rho_{max} = \frac{K+1}{K+2}.$$

Note that if $K=1$ (no waiting capacity), ρ_{max} becomes the "familiar" two-thirds. Results are also given for a three-station system with $K-2$ (queue capacity of one allowed) and equal mean service rates as

$$\rho_{max} \doteq 0.6705.$$

Finally Hunt obtains some results for a series model in which an infinite queue is allowed at the first station (as in the above models) but no queue at subsequent stations and no vacant facilities are allowed; that is, the line

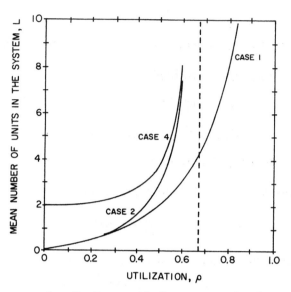

Fig. 4.4   The mean number of units present in the system as a function of utilization. All curves are for two stages. The dotted line, $\rho=2/3$, shows the maximum possible utilization for cases 2 and 4. The maximum possible utilization for case 1 is unity.

moves all at once as a unit. For two, three, and four stations, respectively, $\{\rho_{max}\}$ of $2/3$, $6/11$, and $12/25$ are obtained. If the reader is interested in more detail he is referred to Hunt. We end this section by presenting a graph contained in Hunt's paper (see Figure 4.4)[10] comparing expected system size, L, for three different two-station series models. Case 1 represents infinite queues allowed in front of each station (the model treated in Section 4.5.1), case 2 represents no queue allowed in front of station two [L given by (4.64)], and case 4 represents the last model referred to (no queue allowed at station two and the line moves all at once as a unit).

4.6   CYCLIC QUEUES

The cyclic queue problem is one in which there is a set of queues in tandem serving a fixed population of size (say) M, where a unit leaving the last phase of service returns and waits in an ordered fashion for service in the queue of the first phase, going through the queue and the service channel of the phases in sequence, and so on. There are numerous possible variations on this theme. We may, for example, remove the requirement that units proceed in order and instead allow them to move between the

[10]Reprinted with permission of *Operations Research*.

tandem queues according to some preassigned probability distribution (this is called feedback), or even combine the series models of Section 4.5 with the simple cyclic queue to come up with a situation in which units proceed through a sequence of tandem queues, but some customers require additional service and must cycle back upon completion of all normal tandem service. A perfect example of this latter case might be programs submitted for run on a computer, since a program would normally be required to run through a sequence of operations, which, however, might not all be successful. We only explore the basic cyclic model here, and any reader with further interest in the subject is referred to the literature, for example, Saaty (1961).

To begin, let n_j be the number of customers waiting and in service at the jth phase (of a total of k), such that $\sum_{j=1}^{k} n_j = M$. Each service center will be assumed to consist of one exponential channel with mean rate μ_j for the jth server. Hence if $p(n_1, n_2, \ldots, n_k)$ is used to denote the probability that n_j units are in the jth queue, $j = 1, \ldots, k$, then it is easy to show (see Problem 4.37) that there is one stationary equation and that it is

$$0 = -p(n_1, n_2, \ldots, n_k) \sum_{j=1}^{k} \mu_j$$

$$+ \sum_{j=1}^{k} \mu_j p(n_1, \ldots, n_j + 1, n_{j+1} - 1, \ldots, n_k),$$

where the kth phase is understood to be linked to the first. Also if $n_j = 0$ for some j, then the jth term does not appear, since there then cannot be a service completion, and if $n_{i+1} - 1 < 0$, the ith term is again zero.

By iteration (see Problem 4.38), for example, the solution to the aforementioned steady-state equation is found to be

$$p(n_1, \ldots, n_k) = \frac{\mu_1^{M - n_1}}{\mu_2^{n_2} \cdots \mu_k^{n_k}} p(M, 0, \ldots, 0) \quad , \qquad (4.69)$$

where $p(M, 0, \ldots, 0)$ is to be found by summing $p(n_1, \ldots, n_k)$ over all possible vectors or partitions $(n_1, \ldots, n_k)$ whose elements sum to M. Observe that the fraction of time during which the jth phase is idle, say $1 - \rho_j$, is obtained by putting n_j equal to zero in $p(n_1, \ldots, n_k)$ and then summing over all partitions of M in which $n_j = 0$. The average number of customers in the jth phase is given by

$$L^{(j)} = \sum_{\mathcal{P}_j} n_j p(n_1, \ldots, n_j, \ldots, n_k),$$

where $\mathcal{P}_j$ is the set of partitions in which $n_j \neq 0$, while the expected number of customers in the jth line is given by

$$L_q^{(j)} = L^{(j)} - \rho_j.$$

For example, let us assume that all the μ_j are equal, with value μ. Then

$$p(n_1, \ldots, n_k) = \frac{\mu^{M-n_1}}{\mu^{M-n_1}} p(M, 0, \ldots, 0)$$

$$= p(M, 0, \ldots, 0),$$

and hence all the probabilities are equal and have the value

$$p(n_1, \ldots, n_k) = (\text{total number of possible } k - \text{partitions of } M)^{-1}$$

$$= \left[\binom{M+k-1}{M} \right]^{-1} = \frac{(k-1)! M!}{(M+k-1)!},$$

which comes from the well-known combinatorial result for combinations with replacement. The jth utilization factor, found by summing the probabilities over all partitions with $n_j = 0$, is given by

$$\rho_j = 1 - \sum_{\mathcal{P}_j} p(n_1, \ldots, n_k)$$

$$= 1 - \binom{M+k-2}{M} \frac{(k-1)! M!}{(M+k-1)!}$$

$$= 1 - \frac{(M+k-2)!}{M!(k-2)!} \frac{(k-1)! M!}{(M+k-1)!}$$

$$= 1 - \frac{k-1}{M+k-1}$$

$$= \frac{M}{M+k-1} \qquad (\text{for all } j).$$

In addition, it should be obvious that

$$L^{(j)} = \frac{M}{k} \qquad (\text{for all } j),$$

and therefore

$$L_q^{(j)} = \frac{M}{k} - \frac{M}{M+k-1}$$

$$= \frac{M(M-1)}{k(M+k-1)}.$$

The average line waiting time at the jth phase is found to be

$$W_q^{(j)} = \left(\frac{L^{(j)}}{\mu} \right) \cdot \Pr\{\text{channel is busy}\}$$

$$= \frac{M}{k\mu} \frac{M}{M+k-1};$$

hence the expected system wait at j is

$$W^{(j)} = \frac{M}{k\mu} \frac{M}{M+k-1} + \frac{1}{\mu}.$$

Therefore the average time required for a customer to complete a full cycle is

$$\sum_{j=1}^{k} W^{(j)} = \frac{M^2}{\mu(M+k-1)} + \frac{k}{\mu}.$$

PROBLEMS

4.1 Show that

$$\mu r^{K+1} - (\lambda + \mu)r + \lambda = 0$$

has exactly one root in the interval $(0, 1)$ using Rouché's theorem. [Hint: Refer to Section 2.6.2 and set $g = r^{K+1}$ and $f = -(\lambda/\mu + 1)r + \lambda/\mu$.]

4.2 The moonlighting graduate assistant of Problem 2.11 decides that a more correct model for his short-order counter is an $M^{[X]}/M/1$ where the batch sizes are 1, 2, or 3 with equal probability, such that five batches arrive per hour on the average (this maintains the previous total arrival rate of 10 customers per hour). The mean service time remains at 4 min. Compare the average queue length of this model with that of the previous $M/M/1$ model.

4.3 Derive the probability generating function for the bulk-service model as given in Equation (4.13).

4.4 Apply Rouché's theorem to the denominator of Equation (4.13) to show that K zeros lie on or within the unit circle. [Hint: Show that K zeros lie on or within $|z| = 1 + \delta$ by defining $f(z)$ and $g(z)$ such that $f(z) + g(z)$ equals the denominator.]

4.5 Show that Equation (4.15) yields Equation (4.16) for $K = 2$.

4.6 A ferry loads cars for delivery across a river and must have a full ferry load of 10 cars. Loading is instantaneous and round-trip time is an exponential random variable with mean of 15 min. The cars arrive at the shore as a Poisson process with a mean of 30/hr. On the average, how many autos are waiting on the shore at any instant for a ferry?

4.7 Derive Equation (4.22). [*Hint*: Consider first the sum of the first terms of all the equations of (4.18) multiplied by the appropriate factor, then the second terms of all equations, and last the third terms of all equations.]

4.8 Derive Equation (4.26). [*Hint*: Multiply the equations of (4.18) whose first terms are $p_{n,i}$ by $z^n y^i$, respectively, and sum over all n and i.]

4.9 Derive the steady-state probability that a customer is in phase i for an $M/E_k/1/1$ model, that is, an Erlang service model where no queue is allowed to form.

4.10 Consider the $M/E_k/c/c$ model.

(a) Derive the steady-state difference equation for this model. [*Hint*: Let $p_{n;s_1,s_2,\ldots,s_k}$ represent the probability of n in the system with s_1 channels in phase 1, s_2 in phase 2, etc.]

(b) Show that $p_n = p_0 \rho^n/n!$, $\rho = \lambda/\mu$, is a solution to the problem. [*Hint*: First show that $Ar^n(1/s_1!s_2!\ldots s_k!)$ is a solution to the equation of (a). Then show that

$$p_n = \sum_{s_1+s_2+\ldots+s_k=n} p_{n;s_1,s_2,\ldots,s_k} = A\rho^n/n!$$

by utilizing the multinomial expansion $(x_1+x_2+\ldots+x_k)^n$, then setting $x_1=x_2=\ldots=x_k=1$.]

(c) Compare this result with the $M/M/c/c$ results of Section 3.4, Equation (3.31), and comment.

4.11 Explain intuitively why the $M^{[k]}/M/1/\infty$ bulk-arrival model can be used to represent the $M/E_k/1/\infty$ model when customers are considered to be phases of service to be completed.

4.12 Consider a single-server model with Poisson input and an average customer arrival rate of 30/hr. Currently service is provided by a mechanism that takes *exactly* 1.5 min. Suppose the system can be served instead by a mechanism that has an exponential service-time distribution. What must be the mean service time for this mechanism to ensure (a) the same average time a customer spends in the system, or (b) the same average number of customers in the system?

4.13 For an $M/E_3/1$ model with $\lambda=6$ and $\mu=8$, find the probability of more than two in the system. [*Hint*: Expand (4.27) in a Maclaurin's series to get p_1 and p_2.]

4.14 A large producer of television sets has a policy of checking all sets prior to shipping them from the factory to a warehouse. Each line (color, portable, etc.) has its own expert inspector. At times, the highest volume line (black and white portables) has experienced a "bottleneck" condition (at least in the management's opinion) and a detailed study was made of the inspection performance. Sets were found to arrive at the inspector's station according to a Poisson distribution with a mean of 5/hr. In carrying out the inspection, the inspector performs 10 separate tests, each taking, on an average, 1 min. The times for each test were found to be approximately exponentially distributed. Find the average waiting time a set experiences, the average number of sets in the system, and the average idle time of the inspector.

4.15 The Rhodehawgg Trucking Company has a choice of hiring one of two individuals to operate their single-channel truck-washing facility. In studying their performances it was found that one man's times for completely washing a truck were approximately exponentially distributed with a mean rate of 6/day, while the other man's times were distributed according to an Erlang type 2 with a mean rate of 5/day. Which man should be hired?

4.16 Isle-Air Airline offers air shuttle service between San Juan and Charlotte Amalie every 2 hr. The procedure calls for no advanced reservations but for passengers to come directly to the gate from which the shuttle leaves to purchase their tickets. It is found that passengers arrive according to a Poisson distribution with a mean of 18/hr. There is one agent at the gate check-in counter and a time study provided the following fifty observations on processing time in minutes:

> 4.00, 1.44, 4.44, 1.74, 1.16, 4.20, 3.59, 2.14, 3.54, 2.56,
> 5.53, 2.02, 3.06, 1.66, 3.23, 4.84, 7.99, 3.07, 1.24, 3.40,
> 5.01, 2.78, 1.62, 5.19, 5.09, 3.78, 1.52, 3.94, 1.96, 6.20,
> 3.67, 3.37, 1.84, 1.60, 1.31, 5.64, 0.99, 3.06, 1.24, 3.11,
> 4.57, 0.90, 2.78, 1.64, 2.43, 5.26, 2.11, 4.27, 3.36, 4.76.

On the average, how many are in the queue waiting for tickets and what is the average wait in the queue? [*Hint*: Find the sample mean and variance of the observed service times and see what distribution might fit.]

4.17 Check the calculations for p_0, p_1, and p_2 in Example 4.6 by using Equation (4.16).

4.18   A generalization of the inventory-control procedure of Problem 3.24 is as follows. Again using S as the safety stock value, the policy is to place an order for an amount Q when the on-hand plus on-order inventory reaches a level s ($Q = S - s$). Note that the one-for-one policy of Problem 3.24 is a special case for which $Q = 1$ or equivalently $s = S - 1$. This policy is called a trigger point-order quantity policy and is sometimes also referred to as a continuous review (s, S) policy.

Generally, a manufacturing set-up cost of K dollars per order placed is also included so that an additional cost term of $K\lambda / Q$ ($/unit time) is included in $E(C)$.

For this situation, again assuming Poisson demand and exponential leadtimes, describe the queueing model appropriate to the order processing procedure. Then relate the steady-state probabilities resulting from the order-processing queueing model to $p(z)$, the probability distribution of on-hand inventory. Finally, discuss the optimization procedure for $E(C)$, now a function of two variables (either s and Q, S and Q, or S and s) and the practicality of using this type of analysis.

4.19   Derive the stationary equations, (4.39), for the single exponential channel with two priorities.

4.20   Derive Equation (4.45).

4.21   Show that L_q of (4.45) is always greater than L_q of $M/M/1/\infty$, that is, $L_q = \rho / 1 - \rho$ with $\rho = \lambda_1 / \mu_1 + \lambda_2 / \mu_2$. [*Hint*: Use differential calculus.]

4.22   Compare the L_q's of the two-priority, two-rate case with those of the two-priority, one-rate case when the μ of the latter equals min (μ_1, μ_2).

4.23   Verify the results of cell (Model 2, Model 3) in Table 4.1.

4.24   Carry out the induction that leads from Equation (4.47) to (4.48).

4.25   Show that $\overline{W}_q$ of Section 4.4.2 becomes the W_q of $M/M/1/\infty/$FIFO when $\mu_i \equiv \mu$ for all i.

4.26   Consider the $M/M/1/\infty/$PRI model of Section 4.4.1. Let the number of priorities be two, with input rates $\lambda_1 = 1 = \lambda_2$. If the service rate of the higher-priority items is $\mu_1 = 3$ and that of the lower $\mu_2 = 2$, show that the expected line wait W_q is lower than that of the ordinary $M/M/1/\infty$ with $1/\mu = 1/2\mu_1 + 1/2\mu_2$. Check your answer by using the method of Section 4.4.2.

What happens when the slower-serviced items get the higher priority?

4.27   Find the generating function for the steady-state probabilities given by Equation (4.50).

4.28   Find the solution [Equation (4.55)] to Equations (4.54) using Equation (1.17), Equation (3.9), and the boundary condition, $F_n(0) = p_n$, according to the methodology of Section 1.8.

4.29 Show that the number in the system at a time t after the last departure, $N(t)$, and the time between successive departures, T, are independent random variables. [*Hint*: Find the marginal probability distribution of $N(t)$ from (4.52) and show that the product of this and $c(t)$ gives the joint probability distribution.] Also show that successive interdeparture times are independent.

4.30 For Example 4.8 calculate the same performance measures using three checkout counters operating. If you were Meback's advisor, what would you recommend concerning the number of counters to have in operation?

4.31 For a two-station series queue (single server at each station) with Poisson input to the first with parameter λ, exponential service at each station with parameters μ_1 and μ_2, respectively, and no limit on number in system at either station, show that the steady-state probability that there are n_1 in the first-station system (queue and service) and n_2 in the second-station system is given by

$$p_{n_1,n_2} = p_{n_1} p_{n_2} = \rho_1^{n_1} \rho_2^{n_2} (1 - \rho_1)(1 - \rho_2).$$

[*Hint*: Find the differential equations for $p_{n_1,n_2}(t)$ and then from these the steady-state difference equations. Show that $\rho_1^{n_1} \rho_2^{n_2} p_{0,0}$ is a solution to these equations by substitution and then find $p_{0,0}$ by the boundary condition $\sum_{n_1=0}^{\infty} \sum_{n_2=0}^{\infty} p_{n_1,n_2} = 1$.]

4.32 Consider a three-station series queueing system (single server at each station) with Poisson input, parameter λ, and exponential service, parameters μ_1, μ_2, μ_3, respectively. There is no capacity limit on the queue in front of the first two stations, but at the third there is a limit of K allowed (including service). If K are in the third station, then any subsequent arrivals are shunted out of the system. Find the expected number in the system (all three stations) and the expected time spent in the system by a customer who completes all three stages of service.

4.33 The Won Hung Rhee Chinese carry-out restaurant serves two dishes, chow-mein and spare-ribs. There are two separate windows, one for chow-mein and one for spare-ribs. Customers arrive according to a Poisson process with a mean rate of 20/hr. Sixty percent go to the chow-mein window while 40% go to the rib window. Twenty percent of those who come from the chow-mein window with their order go next to the rib window; the other 80% leave the restaurant. Ten percent of those who purchase ribs then go to the chow-mein window while the other 90% leave. It takes on the average 4 min to fill a chow-mein order and 5 min to fill a spare-rib order, the service times being exponential. How many on the average are in the restaurant? What is the average wait at each window? If a person wants both chow-mein and ribs, how long, on the average, does he spend in the restaurant?

4.34 Derive the results of (4.58) from (4.57) and the boundary condition.

4.35 Derive the steady-state difference equations for a sequential two-station, single-server system, with Poisson input of parameter λ and exponential service with parameters μ_1, μ_2, respectively, where no queue is allowed in front of station one and at most one customer is allowed to wait between the stations. Blockage occurs when there is a customer waiting at station two and a customer completed at station one. For the case $\mu_1 = \mu_2$, solve for the steady-state probabilities.

4.36 What are the eight q_i for the three-station series queueing model with blocking described in Section 4.5.2?

4.37 Derive the basic steady-state equation for the simple cyclic queue given in Section 4.6.

4.38 Derive Equation (4.69).

4.39 Find the effective mean arrival rate to each phase of the cyclic system of Section 4.6.

Chapter 5

MODELS WITH GENERAL ARRIVAL OR SERVICE PATTERNS

For the models treated in this chapter, a Chapman-Kolmogorov analysis as in the previous chapters is not possible, since we no longer have a Markov process because of the relaxation of the exponential assumption on interarrival times and/or service times. However, for many of the models considered here, while we no longer have a Markov process, there is, nevertheless, imbedded within this non-Markov stochastic process a Markov chain (referred to as an imbedded Markov chain). For these types of models, we can employ some of the theory of Markov chains (see Appendix 4) for our analysis. We first treat the case of Poisson input and general service with a single server.

5.1 SINGLE-SERVER QUEUES WITH POISSON INPUT AND GENERAL SERVICE ($M/G/1$)

Consider the system immediately after a customer's service is completed and service is about to commence on the next customer in queue. The service times are independently and identically distributed random variables with any arbitrary probability distribution. We denote the cumulative distribution function (CDF) by $B(t)$ and the density function, if it exists, by $b(t)$. The arrival process is, as before, Poisson with parameter λ. The imbedded stochastic process $X(t_i)$, where X denotes the number in the system and $t_1, t_2, t_3, \cdots$ are the successive times of completion of service, can be shown to be Markovian as follows. Since t_i is the completion time of the ith customer, then $X(t_i)$ is the number of customers the ith customer leaves behind as he departs. Since the state space is discrete, let us use a subscript notation so that X_i represents the number of customers remaining in the system as the ith customer departs. We can then write for all $n > 0$ that

$$X_{n+1} = \begin{cases} X_n - 1 + A_{n+1} & (X_n \geqslant 1) \\ A_{n+1} & (X_n = 0) \end{cases}, \qquad (5.1)$$

where X_n is the number in the system at the nth departure point and A_{n+1} is the number of customers who arrived during the service time, $S^{(n+1)}$, of the $(n+1)$st customer.

The random variable $S^{(n+1)}$ by assumption is independent of previous service times and the length of the queue, so let it henceforth be denoted by S. Since arrivals are Poissonian, the random variable A_{n+1} depends only on S and not on the queue or on the time of service initiation, so let it henceforth be denoted by A. Then it follows that[1]

$$\Pr\{A = a\} = \int_0^\infty \Pr\{A = a | S = t\} \, dB(t) \qquad (5.2)$$

and

$$\Pr\{A = a | S = t\} = \frac{e^{-\lambda t}(\lambda t)^a}{a!}, \qquad (5.3)$$

so that

$$\Pr\{X_{n+1} = j | X_n = i\} = \Pr\{A = j - i + 1\}$$

$$= \begin{cases} \displaystyle\int_0^\infty \frac{e^{-\lambda t}(\lambda t)^{j-i+1}}{(j-i+1)!} \, dB(t) & (j \geqslant i-1, i \geqslant 1) \\ 0 & (j < i-1, i \geqslant 1) \end{cases}. \qquad (5.4)$$

If a departing customer leaves an empty system, the system state remains zero until an arrival comes. Thus the transition probabilities for the case $i = 0$ are identical to those for $i = 1$. Hence we can readily see that the imbedded process is Markovian, since only the indices (i, j) are involved in (5.4), and furthermore since the state variable is discrete, it is a Markov chain. This allows the utilization of Markov-chain theory in the analysis of the $M/G/1$ model. Prior to doing this, however, since it is possible to derive the expected-value measures of effectiveness through a direct expected-value argument, we treat this analysis next.

---

[1]For generality, we use the Stieltjes integral. In cases where the density function exists (which will be most of the cases encountered) $dB(t)$ can be replaced by $b(t) dt$, thus yielding the more familiar Riemann integral of elementary calculus.

5.1.1 Expected-Value Measures of Effectiveness: The Pollaczek-Khintchine Formula.

Using a straightforward expected-value argument the expected system size, L, will be derived. From this we can then obtain the other measures in the usual way.

Again considering the queue just as a customer departs and assuming for the time being that a steady-state solution exists, we can rewrite (5.1) as

$$X_{n+1} = X_n - U(X_n) + A, \qquad (5.5)$$

where

$$U(X_n) = \begin{cases} 1 & (X_n > 0) \\ 0 & (X_n = 0) \end{cases}.$$

Since we are assuming a steady-state solution, we can take expected values of these random variables, noting that $E(X_{n+1}) = E(X_n) = L^{(D)}$ (the superscript (D) is used on L to denote that it represents the expected steady-state system size at departure points and not at any general time as does L). Taking the expectation of both sides of (5.5) yields

$$L^{(D)} = L^{(D)} - E[U(X_n)] + E[A]$$

or

$$E[U(X_n)] = E[A].$$

But, from (5.2) and (5.3), we have

$$E[U(X_n)] = E[A] = \int_0^\infty E[A|S=t] \, dB(t)$$

$$= \int_0^\infty \lambda t \, dB(t) = \lambda E[S] = \frac{\lambda}{\mu} \equiv \rho. \qquad (5.6)$$

Squaring (5.5) gives

$$X_{n+1}^2 = X_n^2 + U^2(X_n) + A^2 - 2X_n U(X_n) - 2AU(X_n) + 2AX_n. \qquad (5.7)$$

By taking expected values of (5.7), noting that $E[X_{n+1}^2] = E[X_n^2]$, we then have that

$$0 = E[U^2(X_n)] + E[A^2] - 2E[X_n U(X_n)] - 2E[AU(X_n)] + 2E[AX_n].$$

But $U^2(X_n) = U(X_n)$, and $X_n U(X_n) = X_n$. Then using (5.6) yields

$$0 = \rho + E[A^2] - 2L^{(D)} - 2\rho^2 + 2L^{(D)}\rho$$

or

$$L^{(D)} = \frac{\rho - 2\rho^2 + E[A^2]}{2(1-\rho)}. \tag{5.8}$$

Now

$$E[A^2] = Var[A] + (E[A])^2$$

$$= Var[A] + \rho^2,$$

where[2]

$$Var[A] = E[Var[A|S]] + Var[E[A|S]].$$

Now, by the use of Equation (5.3), the equation above becomes

$$Var[A] = E[\lambda S] + Var[\lambda S]$$

$$= \rho + \lambda^2 \sigma_S^2, \tag{5.9}$$

where σ_S^2 is the variance of the service-time distribution. Thus by use of (5.9) in (5.8), we finally get

$$L^{(D)} = \rho + \frac{\rho^2 + \lambda^2 \sigma_S^2}{2(1-\rho)}. \tag{5.10}$$

It can be shown (this is discussed more fully in a later section) that although $L^{(D)}$ is the expected steady-state system size at departure points, it is equivalent to the expected steady-state system size at an arbitrary point in time, L. That is to say,

$$\boxed{L = \rho + \frac{\rho^2 + \lambda^2 \sigma_S^2}{2(1-\rho)}} \; . \tag{5.11}$$

Equation (5.11) is often referred to as the Pollaczek-Khintchine (P-K) formula.[3] From it, the expected wait W in the system can be obtained via Little's formula. Next, W_q can be obtained by the usual relation $W = W_q +$

---

[2] See Parzen (1962), p. 55.

[3] It is also common to find $W_q = (L/\lambda) - (1/\mu) = (\rho^2 + \lambda^2 \sigma_S^2)/[2\lambda(1-\rho)]$ referred to as the P-K formula, although these are essentially equivalent due to Little's formula.

$1/\mu$. Finally, L_q can be found by once again utilizing Little's formula, or the equivalent relation, $L_q = L - \lambda/\mu$.

It is only necessary, therefore, to know the mean and variance of the service-time distribution to obtain the expected-value measures of effectiveness of an $M/G/1$ queue. These are, indeed, powerful results since this information concerning the service mechanism is usually readily available or can be estimated without great difficulty. For example, letting $\sigma_S^2 = 0$ gives results for the $M/D/1$ model. The reader can easily check this by comparing (5.11) with $\sigma_S^2 = 0$ to L for the $M/E_k/1$ model of Section 4.3.1 with $k = \infty$.

Example 5.1

Consider a single-server, Poisson-input queue with mean arrival rate of 10/hr. Currently, the server works according to an exponential distribution with mean service time of 5 min. Management has available a training course which will result in an improvement (decrease) in the variance of the service time but at a slight increase in the mean. After completion of the course, it is estimated that the mean service time will increase to 5.5 min but the standard deviation will decrease from 5 min (the exponential case) to 4 min. Management would like to know whether they should have the server undergo further training.

To answer the question, we will compare L and W for each case, the first model being $M/M/1$ and the second $M/G/1$. For $M/M/1$ we can either use the results of Sections 2.2 and 2.3 or the P-K results above with $\sigma_S = 1/\mu$. For the $M/G/1$ we use the aforementioned P-K results. The comparisons are presented in Table 5.1.

Thus it is not profitable to have the server "better" trained. Note that reducing the standard deviation twice as much as the increase in mean resulted in considerably poorer performance. Performance is much more sensitive to mean service time than to its variation. In this example, the

TABLE 5.1 COMPARISON OF MODELS

	Present (M/M/1)	After Training (M/G/1)
L	5	8.625
W	30 min	51.750 min

mean time increases only 10% while the standard deviation decreases by 20% (a decrease in the variance of almost 36%). It would be interesting to calculate the reduction in variance required to make up for the increase of 0.5 in the mean. We can do this by solving for σ_S^2 in (5.11) as follows:

$$L = 5 = \rho + \frac{\rho^2 + \lambda^2 \sigma_S^2}{2(1-\rho)},$$

where

$$\rho = \frac{\lambda}{\mu} = 10\left(\frac{11}{2}\right)\left(\frac{1}{60}\right) = \frac{11}{12}.$$

This yields a $\sigma_S^2 < 0$ which, of course, is not possible. What this means is that even with $\sigma_S^2 = 0$ (deterministic service times), L is greater than 5, and, in fact, turns out to be

$$L = \rho + \frac{\rho^2}{2(1-\rho)} \doteq 6.0.$$

Problem 5.4 asks the reader to find the value of σ_S^2 required to yield the same L if the mean service time were increased to only 5.2 min after training.

5.1.2   Departure-Point Steady-State System-Size Probabilities

This section treats the development of the steady-state system-size probabilities at departure points. We let π_n represent the probability of n in the system at a departure point (a point of time just after a customer has completed service) after steady state is reached.[4] These probabilities, $\{\pi_n\}$, are not in general the same as the previous steady-state system-size probabilities, $\{p_n\}$, which were valid for any arbitrary point of time after steady state is reached. For the $M/G/1$ model, however, it turns out that the π's and the p's are identical, which is shown in a later section.

We have shown in Section 5.1 that the imbedded stochastic process at departure points is a Markov chain. We denote the transition probability matrix by

$$P = [p_{ij}],$$

---

[4]Again, for the time being the existence of a steady-state solution is assumed. This topic is treated in a later section.

where

$$p_{ij} = \Pr\{\text{system size immediately after a departure point is } j \mid$$

$$\text{system size after previous departure was } i\}$$

$$= \Pr\{X_{n+1} = j \mid X_n = i\}.$$

From (5.4) we have

$$p_{ij} = \int_0^\infty \frac{e^{-\lambda t}(\lambda t)^{j-i+1}}{(j-i+1)!} dB(t) \qquad (j \geqslant i-1, \; i \geqslant 1). \qquad (5.12)$$

Simplification results by defining

$$k_n = \Pr\{n \text{ arrivals during a service time } S = t\}$$

$$= \int_0^\infty \frac{e^{-\lambda t}(\lambda t)^n}{n!} dB(t), \qquad (5.13)$$

so that p_{ij} can be seen to equal k_{j-i+1} and

$$P = [p_{ij}] = \begin{bmatrix} k_0 & k_1 & k_2 & \cdots \\ k_0 & k_1 & k_2 & \cdots \\ 0 & k_0 & k_1 & \cdots \\ 0 & 0 & k_0 & \cdots \\ \cdot & \cdot & \cdot & \\ \cdot & \cdot & \cdot & \cdots \\ \cdot & \cdot & \cdot & \end{bmatrix}. \qquad (5.14)$$

Assuming steady state is achievable, the steady-state probability vector, $\pi = \{\pi_n\}$, can be found as the solution to the stationary equation (see Appendix 4)

$$\pi P = \pi. \qquad (5.15)$$

This yields (see Problem 5.6)

$$\pi_i = \pi_0 k_i + \sum_{j=1}^{i+1} \pi_j k_{i-j+1} \qquad (i = 0, 1, 2, \ldots) \qquad (5.16)$$

Now define the generating functions

$$\Pi(z) = \sum_{i=0}^{\infty} \pi_i z^i \qquad (|z| \leqslant 1) \tag{5.17}$$

and

$$K(z) = \sum_{i=0}^{\infty} k_i z^i. \tag{5.18}$$

Then multiplying (5.16) by z^i, summing over i, and solving for $\Pi(z)$ yields (see Problem 5.7)

$$\Pi(z) = \frac{\pi_0(1-z)K(z)}{K(z) - z}. \tag{5.19}$$

Using the fact that $\Pi(1) = 1$, along with L'Hôpital's rule, and realizing that $K(1) = 1$ and $K'(1) = \lambda(1/\mu)$, we find that

$$\pi_0 = 1 - \rho \qquad (\rho \equiv \lambda E[\text{service time}]). \tag{5.20}$$

Hence

$$\boxed{\Pi(z) = \frac{(1-\rho)(1-z)K(z)}{K(z) - z}}. \tag{5.21}$$

Since by definition $\Pi'(1)$ is the expected system size, it is possible to derive the *P-K* formula (see Problem 5.8) directly from Equation (5.21).

Equation (5.21) is as far as we can go in obtaining the $\{\pi_n\}$ (which we show, in the next section, are equivalent to the $\{p_n\}$) without making assumptions as to the specific service-time distribution. Problem 5.9 asks the reader to verify that Equation (5.21) reduces to the generating function for $M/M/1$ given by (2.16) when one assumes exponential service. We shall now present an example of an $M/G/1$ system with empirical service times.

Example 5.2

To illustrate how the results above can be utilized on an empirically determined service distribution, consider the following situation. The Bearing Straight Corporation, a single product company, makes a very specialized plastic bearing. The company is a high volume producer, and the bearing undergoes a single machining operation, performed on a specialized machine. Because of the volume of sales, the company keeps a large number of machines in operation at all times (we may assume for practical purposes that the machine population is infinite).

Machines break down according to a Poisson process with a mean of 5/hr. The company has a single expert repairman, and the machine characteristics are such that the breakdowns are due to one of two possible malfunctions. Depending on which of the malfunctions caused the break-down, it takes the expert repairman either 9 or 12 min to repair a machine. Since he is an expert and the machines are identical, any variation in these service times is miniscule and can be ignored. The type of malfunction which causes a breakdown occurs at random, but it has been observed that one-third of the malfunctions require the 12-min repair time. The company wishes to know the probability that more than three machines will be down at any given time.

The service-time mechanism can be looked at as the two-point probability distribution given in Table 5.2.

If management were interested in only the expected number of machines down, it is an easy matter to obtain this from the P-K formula, and it turns out that $L = 2.96$ (see Problem 5.10). However, to answer the question asked, it is necessary to find p_0, p_1, p_2, and p_3, since

$$\text{Pr}\{\text{more than 3 machines down}\} = \sum_{n=4}^{\infty} p_n$$

$$= 1 - \sum_{n=0}^{3} p_n.$$

The probability that no machines are down is easily obtained as

$$p_0 = \pi_0 = 1 - \rho = 1 - \frac{5}{6} = \frac{1}{6}.$$

TABLE 5.2   SERVICE-TIME
PROBABILITY DISTRIBUTION

t (min)	$b(t)$	$B(t)$
9	2/3	2/3
12	1/3	1

$$\begin{cases} E[S] = \dfrac{1}{\mu} = 10 \text{ min} \\[2mm] \sigma_S^2 = 2 \text{ min}^2 \end{cases}$$

To find the others, we can make use of (5.16) iteratively. However, an alternative procedure which does not require successively calculating lower probabilities in order to obtain a particular p_n is to first calculate the generating function given in Equation (5.21) and then to expand it in a series to obtain $p_i = \pi_i$ $(i = 1, 2, 3)$. To evaluate $\Pi(z)$, we must first find $K(z)$ from Equation (5.18), which necessitates finding k_n as given in (5.13). Since we have a two-point discrete distribution for $b(t)$, the Stieltjes integral reduces to a summation, and we have

$$k_n = \frac{1}{n!} \left\{ e^{-5(3/20)} \left[5\left(\frac{3}{20}\right) \right]^n \left(\frac{2}{3}\right) + e^{-5(1/5)} \left[5\left(\frac{1}{5}\right) \right]^n \left(\frac{1}{3}\right) \right\}$$

$$= \frac{2}{3n!} e^{-3/4} \left(\frac{3}{4}\right)^n + \frac{1}{3n!} e^{-1}.$$

Thus

$$K(z) = \frac{2}{3} e^{-3/4} \sum_{i=0}^{\infty} \frac{(3z/4)^i}{i!} + \frac{1}{3} e^{-1} \sum_{i=0}^{\infty} \frac{z^i}{i!}.$$

Although we can get $K(z)$ in closed form since the sums are the infinite series expressions for $e^{3z/4}$ and e^z, respectively, it behooves us not to do so as we ultimately desire a power series expansion for $\Pi(z)$ in order to "pick off" the probabilities. To make $K(z)$ easier to work with, define

$$c_i = \frac{2}{3} e^{-3/4} \left(\frac{3}{4}\right)^i + \frac{1}{3} e^{-1}, \tag{5.22}$$

so that $K(z)$ can be written as

$$K(z) = \sum_{i=0}^{\infty} \frac{c_i z^i}{i!}.$$

Now from (5.21), we have

$$\Pi(z) = \frac{(1-\rho)(1-z) \sum_{i=0}^{\infty} \frac{c_i z^i}{i!}}{\sum_{i=0}^{\infty} \frac{c_i z^i}{i!} - z},$$

which gives

$$\Pi(z) = \frac{(1-\rho)\left[\displaystyle\sum_{i=0}^{\infty}\frac{c_i z^i}{i!} - \sum_{i=0}^{\infty}\frac{c_i z^{i+1}}{i!}\right]}{c_0 + (c_1 - 1)z + \displaystyle\sum_{i=2}^{\infty}\frac{c_i z^i}{i!}}$$

$$= \frac{(1-\rho)\left[c_0 + \displaystyle\sum_{i=1}^{\infty}\left(\frac{c_i}{i!} - \frac{c_{i-1}}{(i-1)!}\right)z^i\right]}{c_0 + (c_1 - 1)z + \displaystyle\sum_{i=2}^{\infty}\frac{c_i z^i}{i!}}$$

$$= \frac{(1-\rho)\left[1 + \displaystyle\sum_{i=1}^{\infty}\left(\frac{c_i}{c_0 i!} - \frac{c_{i-1}}{c_0(i-1)!}\right)z^i\right]}{1 + \dfrac{(c_1 - 1)}{c_0}z + \displaystyle\sum_{i=2}^{\infty}\frac{c_i z^i}{c_0 i!}}. \tag{5.23}$$

It is necessary to have (5.23) in terms of a power series in z and not the ratio of two power series. For our example, we require coefficients up to and including z^3. These can be obtained by long division, carefully keeping track of the coefficients of terms up to and including z^3. However, it can be shown [see Abramowitz and Stegun (1964), p. 15] that the ratio of two power series is a power series, namely,

$$\frac{1 + \displaystyle\sum_{i=1}^{\infty} a_i z^i}{1 + \displaystyle\sum_{i=1}^{\infty} b_i z^i} = \sum_{i=0}^{\infty} d_i z^i,$$

where d_i can be obtained recursively from

$$d_i = \begin{cases} a_i - \displaystyle\sum_{j=1}^{i} b_j d_{i-j} & (i=1,2,\dots) \\ 1 & (i=0) \end{cases}.$$

Thus it is necessary to obtain d_1, d_2, and d_3 which, when multiplied by $(1-\rho)$, give p_1, p_2, and p_3, respectively. In terms of the $\{c_i\}$, we get

$$
\begin{cases}
p_1 = (1-\rho)\left(\dfrac{1}{c_0} - 1\right) \\[2ex]
p_2 = (1-\rho)\left(\dfrac{1-c_1}{c_0} - 1\right)\left(\dfrac{1}{c_0}\right) \\[2ex]
p_3 = (1-\rho)\left(\dfrac{1}{c_0}\right)\left[\left(\dfrac{1-c_1}{c_0}\right)\left(\dfrac{1-c_1}{c_0} - 1\right) - \dfrac{c_2}{2c_0}\right]
\end{cases}
\tag{5.24}
$$

Finally, using (5.22) to evaluate c_0, c_1, and c_2, then substituting into (5.24), with the aid of a desk calculator or digital computer, yields

$$
\begin{cases}
p_1 \doteq 0.2143 \\
p_2 \doteq 0.1773 \\
p_3 \doteq 0.1293
\end{cases}
$$

and

$$
\Pr\{\text{more than 3 machines down}\} = 1 - \sum_{n=0}^{3} p_n \doteq 0.3124.
$$

Thus more than three machines are down slightly over 31% of the time.

While many authors may leave the reader with the impression that the solution to the $M/G/1$ problem is essentially given by the generating function of Equation (5.21), this example illustrates that, even in the very simplest two-point service distribution, quite a bit of work remains to achieve final numerical results for $\{\pi_n\}$ (which are directly equivalent to $\{p_n\}$ for this model as shown in the next section).

The results above for the two-point service distribution can be rather easily generalized to a k-point service distribution, using a similar analysis [see Greenberg (1973) and Problem 5.12]. Letting the service-time distribution be given by

$$
b(t_i) = \Pr\{T = t_i\} = b_i \qquad (i = 1, 2, \ldots, k),
$$

then

$$
K(z) = \sum_{i=0}^{\infty} \frac{c_i z^i}{i!},
\tag{5.25}
$$

where

$$c_i = \sum_{j=1}^{k} b_j e^{-\lambda t_j} (\lambda t_j)^i \tag{5.26}$$

and

$$\Pi(z) = \frac{(1-\rho)\left[1 + \sum_{i=1}^{\infty} \left(\frac{c_i}{c_0 i!} - \frac{c_{i-1}}{c_0(i-1)!}\right) z^i\right]}{1 + \left(\frac{c_1}{c_0} - \frac{1}{c_0}\right) z + \sum_{i=2}^{\infty} \frac{c_i}{c_0 i!} z^i}. \tag{5.27}$$

Equation (5.27) is identical to (5.23); however, the c_i are different. Using the same relationship for the quotient of two power series as given previously, the $\{d_i\}$ and hence the $\{p_i\}$ can be obtained and are identically those given by (5.24), with c_i now given by (5.26). Of course, one could generate as many p_i as desired by continuing the procedure further, that is, by calculating more d_i's.

5.1.3 Proof that $\pi_n = p_n$

We now show that π_n, the steady-state probability of n in the system at a departure point, is equal to p_n, the steady-state probability of n in the system at an arbitrary point in time.

Consider a specific realization of the actual process over a long interval of time T. If $A_n(t)$ denotes the number of unit upward jumps (arrivals) from state n occurring in $(0,t)$ and $D_n(t)$ the number of unit downward jumps (departures) to state n in $(0,t)$, then since arrivals occur singly and customers are served singly we must have

$$|A_n(T) - D_n(T)| \leqslant 1. \tag{5.28}$$

As long as the system is unsaturated, and this is true for $\rho < 1$ as shown in the next section, the ratio of the total number of departures $D(T)$ to arrivals $A(T)$ must go to 1 as T goes to infinity, that is,

$$\lim_{T \to \infty} \frac{D(T)}{A(T)} = \lim_{T \to \infty} \frac{D(T)}{D(T)} = 1 \tag{5.29}$$

with probability one.

Now dividing (5.28) by $A(T)$ and taking the limit as T goes to infinity yields

$$\lim_{T \to \infty} \left| \frac{A_n(T)}{A(T)} - \frac{D_n(T)}{A(T)} \right| \leq \lim_{T \to \infty} \frac{1}{A(T)} = 0,$$

since $A(T)$ goes to infinity as T goes to infinity. Thus

$$\lim_{T \to \infty} \left| \frac{A_n(T)}{A(T)} - \frac{D_n(T)}{A(T)} \right| = 0,$$

which implies

$$\lim_{T \to \infty} \frac{A_n(T)}{A(T)} = \lim_{T \to \infty} \frac{D_n(T)}{A(T)}. \tag{5.30}$$

But using (5.29) we can augment the right-hand side of (5.30) so that

$$\lim_{T \to \infty} \frac{A_n(T)}{A(T)} = \lim_{T \to \infty} \frac{D_n(T)}{A(T)} \lim_{T \to \infty} \frac{A(T)}{D(T)}$$

$$= \lim_{T \to \infty} \frac{D_n(T)}{A(T)} \frac{A(T)}{D(T)}$$

$$= \lim_{T \to \infty} \frac{D_n(T)}{D(T)}. \tag{5.31}$$

Since arrivals occur at the points of a Poisson process operating independently of the state of the process, $\lim_{T \to \infty} [A_n(T)/A(T)]$ is identical to the general-time probability p_n, and by definition,

$$\pi_n = \lim_{T \to \infty} \frac{D_n(T)}{D(T)}.$$

Hence from (5.31),

$$\pi_n = p_n.$$

5.1.4 Ergodic Theory

To find conditions for the existence of the steady state for the $M/G/1$ queue (i.e., conditions under which the process is ergodic), we first obtain a sufficient condition from the theory of Markov chains in view of the Markovian nature of the departure-point process and the previously shown equality of departure-point and general-time probabilities. Then we show that this sufficient condition is also necessary by the direct use of the generating function $\Pi(z)$ given by Equation (5.19).

The Markov-chain proof relies heavily on two well-known theorems listed in Appendix 4 as Theorems A4.7 and A4.8. The former says that a discrete-time Markov chain is ergodic if it is irreducible, aperiodic, and positive recurrent, while Theorem A4.8 gives a sufficient condition for the positive recurrence of an irreducible and aperiodic chain.

The behavior of any single-channel queueing system is, at least, a function of the input parameters and the service-time distribution, say $B(t)$. Hence we would expect, as in $M/M/1$, that the existence of ergodicity depends upon the value of the utilization factor $\rho = $ (mean service time)/(mean interarrival time). The transition matrix P which characterizes the imbedded chain of the $M/G/1$ is given by Equation (5.14), remembering that

$$k_n = \Pr\{\, n \text{ arrivals during a service period}\,\}$$

$$= \frac{1}{n!} \int_0^\infty (\lambda t)^n e^{-\lambda t}\, dB(t),$$

with $\int_0^\infty t\, dB(t) = E[S]$. The problem then is to determine a fairly simple sufficient condition under which the $M/G/1$ is ergodic, and our past experiences suggest that we try $\rho = \lambda E[S] < 1$.

We show in the following that the $M/G/1$ imbedded chain is indeed irreducible, aperiodic, and positive recurrent (see Appendix 4) when $\rho < 1$, and hence, by Theorem A4.7, possesses a long-run distribution under this condition. The chain is clearly irreducible since any state can be reached from any other state. It can next be observed directly from the transition matrix that $\Pr\{\text{passing from state } k \text{ to state } k \text{ in one transition}\} = p_{kk}^{(1)} > 0$, for all k, and therefore the period of each k, defined as the GCD$\{\,n\,|\,p_{kk}^{(n)} > 0\,\}$, is one.[5] Hence the system is aperiodic. Notice that neither irreducibility nor aperiodicity depend upon ρ, but, rather, each is an inherent property of this chain. In fact, it is the positive recurrence which depends upon the value of ρ. It thus remains for us to show that the chain is positive recurrent when $\rho < 1$, for this would ensure ergodicity, and we could then apply the theorem to infer the existence of the stationary distribution.

[5]GCD stands for the greatest common divisor.

To obtain the required result, we employ Theorem A4.8 and show, using Foster's method [Foster (1953)], that the necessary inequality has the required solution when $\rho < 1$. An educated guess at this required solution is

$$x_j = \frac{j}{1-\rho} \qquad (j \geqslant 0).$$

Note that $\Sigma j p_{ij}$ is the mean "to" state from i.

Then, from the matrix P,

$$\sum_{j=0}^{\infty} p_{ij} x_j = \sum_{j=i-1}^{\infty} k_{j-i+1} \left(\frac{j}{1-\rho} \right)$$

$$= \frac{k_0(i-1)}{1-\rho} + \frac{k_1 i}{1-\rho} + \frac{k_2(i+1)}{1-\rho} + \cdots$$

$$= \frac{k_0(i-1)}{1-\rho} + \frac{k_1(i-1)}{1-\rho} + \frac{k_2(i-1)}{1-\rho} + \cdots$$

$$+ \frac{k_1}{1-\rho} + \frac{2k_2}{1-\rho} + \cdots$$

$$= (i-1) \sum_{j=0}^{\infty} \frac{k_j}{1-\rho} + \sum_{j=1}^{\infty} \frac{j k_j}{1-\rho}$$

$$= \frac{i-1}{1-\rho} + \sum_{j=1}^{\infty} \frac{j k_j}{1-\rho}.$$

But it should be noted that

$$\sum_{j=1}^{\infty} j k_j = \sum_{j=1}^{\infty} j \frac{1}{j!} \int_0^{\infty} (\lambda t)^j e^{-\lambda t} \, dB(t)$$

$$= \int_0^{\infty} e^{-\lambda t} \left[\sum_{j=1}^{\infty} \frac{1}{(j-1)!} (\lambda t)^j \right] dB(t)$$

$$= \int_0^{\infty} e^{-\lambda t} \lambda t e^{\lambda t} \, dB(t)$$

$$= \int_0^\infty \lambda t \, dB(t)$$

$$= \lambda E[S] = \rho. \tag{5.32}$$

So

$$\sum_{j=0}^\infty p_{ij} x_j = \frac{i - 1 + \rho}{1 - \rho}$$

$$= x_i - 1 \quad (x_i \geqslant 0 \quad \text{since } 1 - \rho > 0).$$

Also,

$$\sum_{j=0}^\infty p_{0j} x_j = \sum_{j=0}^\infty k_j x_j$$

$$= \sum_{j=0}^\infty \frac{k_j j}{1 - \rho}$$

$$= \frac{\rho}{1 - \rho} < \infty.$$

Hence it follows that the chain is ergodic when $\rho < 1$, and possesses identical stationary and long-run distributions.

The proof of the necessity of $\rho < 1$ for ergodicity arises directly from the existence of the generating function $\Pi(z)$ over the interval $|z| \leqslant 1$,

$$\Pi(z) = \frac{\pi_0 (1 - z) K(z)}{K(z) - z}.$$

We know that $\Pi(1)$ must be equal to one; hence

$$1 = \lim_{z \to 1} \Pi(z)$$

$$= \pi_0 \frac{-K(1)}{K'(1) - 1} \quad (\text{using L'Hôpital's rule})$$

$$= \pi_0 \frac{-1}{\rho - 1} \quad (\rho = \lambda E[S]).$$

But $\pi_0 > 0$ when the chain is ergodic and therefore $\rho - 1 < 0$; thus $\rho < 1$ is necessary and sufficient for ergodicity.

5.1.5   Special Cases ($M/E_k/1$ and $M/D/1$)

Consider first the Erlang-k service model. From (5.13) and (5.18)

$$k_n = \int_0^\infty \frac{e^{-\lambda t}(\lambda t)^n}{n!} \frac{(\mu k)^k}{(k-1)!} t^{k-1} e^{-k\mu t} \, dt$$

and

$$K(z) = \sum_{i=0}^\infty \int_0^\infty \frac{e^{-(\lambda t + k\mu t)}}{(k-1)!} \frac{(\lambda t z)^i}{i!} (\mu k)^k \, t^{k-1} \, dt$$

$$= \frac{(\mu k)^k}{(k-1)!} \int_0^\infty e^{-(\lambda + k\mu - \lambda z)t} t^{k-1} \, dt$$

$$= \frac{(\mu k)^k}{(k-1)!} \frac{(k-1)!}{(\lambda + k\mu - \lambda z)^k}$$

$$= \left[1 + \frac{\lambda}{k\mu}(1-z) \right]^{-k}$$

$$= [1 + r(1-z)]^{-k} \qquad (r = \lambda/k\mu).$$

Now, from (5.21), we have

$$\Pi(z) = \frac{(1-\rho)(1-z)[1+r(1-z)]^{-k}}{[1+r(1-z)]^{-k} - z}$$

$$= \frac{(1-\rho)(1-z)}{1 - z[1+r(1-z)]^k}. \qquad (5.33)$$

The generating function given by (5.33) is identical to that derived in Section 4.3.1 as given by Equation (4.27). Expanding this in a Maclaurin series (which can be quite cumbersome) yields the state probabilities (see Problem 4.13).

The generating function for the $M/D/1$ model can be obtained in a similar fashion. Assuming *all* service times are exactly $1/\mu$ we have

$$k_n = \frac{e^{-\rho}\rho^n}{n!} \qquad (\rho = \lambda/\mu),$$

$$K(z) = \sum_{i=0}^{\infty} \frac{e^{-\rho}\rho^i}{i!} z^i$$

$$= e^{-\rho(1-z)},$$

and[6]

$$\Pi(z) = \frac{(1-\rho)(1-z)e^{-\rho(1-z)}}{e^{-\rho(1-z)} - z}$$

$$= \frac{(1-\rho)(1-z)}{1 - ze^{\rho(1-z)}}. \tag{5.34}$$

Again, a Maclaurin expansion will yield the state probabilities, which can (with some algebraic complexity) be shown to be [see Saaty (1961), p. 155]:

$$\left\{ \begin{array}{l} p_0 = 1 - \rho \\[4pt] p_1 = (1-\rho)(e^\rho - 1) \\[4pt] \quad\vdots \\[4pt] p_n = (1-\rho)\sum_{k=1}^{n}(-1)^{n-k}e^{k\rho}\left[\dfrac{(k\rho)^{n-k}}{(n-k)!} + \dfrac{(k\rho)^{n-k-1}}{(n-k-1)!}\right] \quad (n \geqslant 2) \end{array} \right. , \tag{5.35}$$

where the second factor in p_n is ignored for $k = n$.

These results can be obtained in another manner (without requiring the differentiation needed for a Maclaurin expansion) which we illustrate as

[6]This, of course, could also have been obtained by letting $k \to \infty$ in (5.33) or by using (5.23) for the empirical service-time model letting the service time take on one value with probability one.

follows. Considering (5.34) we expand $1/(1 - ze^{\rho(1-z)})$ as a geometric series giving

$$\Pi(z) = (1-\rho)(1-z) \sum_{k=0}^{\infty} e^{k\rho(1-z)} z^k$$

$$= (1-\rho)(1-z) \sum_{k=0}^{\infty} e^{k\rho} e^{-k\rho z} z^k$$

$$= (1-\rho)(1-z) \sum_{k=0}^{\infty} e^{k\rho} \sum_{m=0}^{\infty} \frac{(-k\rho z)^m}{m!} z^k$$

$$= (1-\rho)(1-z) \sum_{k=0}^{\infty} \sum_{m=0}^{\infty} e^{k\rho} (-1)^m \frac{(k\rho)^m}{m!} z^{m+k}$$

$$= (1-\rho)(1-z) \sum_{k=0}^{\infty} \sum_{n=k}^{\infty} e^{k\rho} (-1)^{n-k} \frac{(k\rho)^{n-k}}{(n-k)!} z^n.$$

We now change the order of summation and get (graphing the region of summation on the $n-k$ plane will easily verify this)

$$\Pi(z) = (1-\rho)(1-z) \sum_{n=0}^{\infty} \sum_{k=0}^{n} e^{k\rho} (-1)^{n-k} \frac{(k\rho)^{n-k}}{(n-k)!} z^n$$

$$= (1-\rho) \left\{ \sum_{n=0}^{\infty} \sum_{k=0}^{n} e^{k\rho} (-1)^{n-k} \frac{(k\rho)^{n-k}}{(n-k)!} z^n \right.$$

$$\left. - \sum_{n=0}^{\infty} \sum_{k=0}^{n} e^{k\rho} (-1)^{n-k} \frac{(k\rho)^{n-k}}{(n-k)!} z^{n+1} \right\}$$

$$= (1-\rho) \left\{ \sum_{n=0}^{\infty} \sum_{k=0}^{n} e^{k\rho} (-1)^{n-k} \frac{(k\rho)^{n-k}}{(n-k)!} z^n \right.$$

$$\left. - \sum_{n=1}^{\infty} \sum_{k=0}^{n-1} e^{k\rho} (-1)^{n-k-1} \frac{(k\rho)^{n-k-1}}{(n-k-1)!} z^n \right\}.$$

Thus we have a power series in z and can "pick off" the p_n's. For $n=0$, we get (noting that $0^0 \equiv 1$)

$$p_0 = (1-\rho).$$

For $n=1$,

$$p_1 = (1-\rho)(e^\rho - 1).$$

For $n \geqslant 2$, the $k=0$ term of the first factor is always 0 because of $k\rho$, so that the summation on k can start at $k=1$. Hence we can write

$$p_n = (1-\rho)\left\{ \sum_{k=1}^{n} e^{k\rho}(-1)^{n-k}\frac{(k\rho)^{n-k}}{(n-k)!} + \sum_{k=1}^{n-1} e^{k\rho}(-1)^{n-k}\frac{(k\rho)^{n-k-1}}{(n-k-1)!} \right\},$$

which is equivalent to (5.35).

This same technique of series expansion can be used on (5.33) for $M/E_k/1$ to obtain closed-form expressions for the $\{p_n\}$. We leave this as an exercise (see Problem 5.14) and present the results below:

$$
\begin{aligned}
p_n = (1-\rho)\Bigg\{ & (1+r)^{nk} \\
& - \left[1 + \binom{2}{1}\binom{(n-1)k}{1}r + \binom{3}{1}\binom{(n-1)k}{2}r^2 + \cdots \right. \\
& \left. + \binom{(n-1)k+1}{1}\binom{(n-1)k}{(n-1)k}r^{(n-1)k} \right] \\
& + \left[\binom{2}{2}\binom{(n-2)k}{1}r + \binom{3}{2}\binom{(n-2)k}{2}r^2 + \cdots \right. \\
& \left. + \binom{(n-2)k+1}{2}\binom{(n-2)k}{(n-2)k}r^{(n-2)k} \right] \\
& - \left[\binom{3}{3}\binom{(n-3)k}{2}r^2 + \binom{4}{3}\binom{(n-3)k}{3}r^3 + \cdots \right.
\end{aligned}
$$

$$
+\left(\begin{array}{c}(n-3)k+1\\3\end{array}\right)\left(\begin{array}{c}(n-3)k\\(n-3)k\end{array}\right)r^{(n-3)k}\Bigg]+\cdots
$$

$$
+(-1)^{n-1}\Bigg[\left(\begin{array}{c}n-1\\n-1\end{array}\right)\left(\begin{array}{c}k\\n-2\end{array}\right)r^{n-2}+\left(\begin{array}{c}n\\n-1\end{array}\right)\left(\begin{array}{c}k\\n-1\end{array}\right)r^{n-1}
$$

$$
+\left(\begin{array}{c}n+1\\n-1\end{array}\right)\left(\begin{array}{c}k\\2\end{array}\right)r^{n}+\cdots+\left(\begin{array}{c}k+1\\n-1\end{array}\right)\left(\begin{array}{c}k\\k\end{array}\right)r^{k}\Bigg]\Bigg\}. \tag{5.36}
$$

5.1.6   Waiting Times

In this section we wish to present an assortment of important results concerning the delay times.

It has already been shown that the average system wait is related to the average system size by Little's formula $W=L/\lambda$. A natural thing to then require is either a possible relationship between higher moments or between distribution functions [or equivalently between Laplace-Stieltjes transforms (LSTs)]. It turns out that such can be done with some extra effort.

First, we go back to the early discussion of Little's formula given in Section 2.4, and note that the stationary probability for the $M/G/1$, p_n, can always be written in terms of the waiting-time CDF as

$$
p_n=\pi_n=\frac{1}{n!}\int_0^\infty (\lambda t)^n e^{-\lambda t}\,dW(t)\qquad (n\geqslant 0).
$$

If we multiply this equation by z^n, sum on n, and define the usual generating function $P(z)=\sum_{n=0}^\infty p_n z^n$, then it is found that

$$
P(z)=\sum_{n=0}^\infty p_n z^n
$$

$$
=\int_0^\infty e^{-\lambda t}\sum_{n=0}^\infty \frac{(\lambda t z)^n}{n!}\,dW(t)
$$

$$
=\int_0^\infty e^{-\lambda t(1-z)}\,dW(t)
$$

$$
=W^*[\lambda(1-z)], \tag{5.37}
$$

where $W^*(s)$ is the LST of $W(t)$. The succession of moments of system size and delay can now be easily related to each other by repeated differentiation of the aforementioned equality, $P(z) = W^*[\lambda(1-z)]$. We therefore have by the chain rule that

$$\frac{d^kP(z)}{dz^k} = \frac{(-1)^k\lambda^k d^kW^*(u)}{du^k}\bigg|_{u=\lambda(1-z)}$$

$$= (-1)^k\lambda^k(-1)^k E[T^k e^{-Tu}]\big|_{u=\lambda(1-z)} \ .$$

Hence if $L_{(k)}$ is used to denote the kth factorial moment of the system size and W_k the regular kth moment of the system waiting time, then

$$\boxed{L_{(k)} = \frac{d^kP(z)}{dz^k}\bigg|_{z=1} = \lambda^k W_k} \ . \qquad (5.38)$$

This is a very important result which provides a nice generalization of Little's formula since the higher ordinary moments can be obtained from the factorial moments.

In the $M/M/1$ queue we were able to easily obtain a simple formula for the waiting-time distribution in terms of the service-time distribution (see Section 2.3 and Problem 2.8), namely,

$$W(t) = (1-\rho) \sum_{n=0}^{\infty} \rho^n B^{(n+1)}(t), \qquad (5.39)$$

where $B(t)$ is the exponential CDF and $B^{(n+1)}(t)$ its $(n+1)$st convolution. The derivation of this result absolutely required the memorylessness of the exponential service since the arrivals catch the server in the midst of a serving period with probability equal to ρ. However, we lose memorylessness and are now going to require an alternative approach to derive a comparable result for $M/G/1$.

To do so, we begin by deriving a simple relationship between the Laplace-Stieltjes transforms of the service and waiting times, $B^*(s)$ and $W^*(s)$, respectively. From (5.37) we know that $P(z) = W^*[\lambda(1-z)]$ and from (5.21) that $P(z) = (1-\rho)(1-z)K(z)/[K(z)-z]$. But, from (5.18) and (5.13),

$$K(z) = \int_0^\infty e^{-\lambda t} \sum_{n=0}^{\infty} \frac{(\lambda t z)^n}{n!} dB(t)$$

$$= \int_0^\infty e^{-\lambda t(1-z)} dB(t)$$

$$= B^*[\lambda(1-z)].$$

Putting these three equations together, we find that

$$W^*[\lambda(1-z)] = \frac{(1-\rho)(1-z)B^*[\lambda(1-z)]}{B^*[\lambda(1-z)]-z},$$

or

$$W^*(s) = \frac{(1-\rho)sB^*(s)}{s-\lambda[1-B^*(s)]}. \qquad (5.40)$$

But, from the convolution property of transforms,

$$W^*(s) = W_q^*(s)B^*(s)$$

since

$$T = T_q + S.$$

Thus

$$\boxed{W_q^*(s) = \frac{(1-\rho)s}{s-\lambda[1-B^*(s)]}} \quad .$$

Expanding the right hand side as a geometric series[7] we have

$$W_q^*(s) = (1-\rho) \sum_{n=0}^{\infty} \left\{ \frac{\lambda}{s}[1-B^*(s)] \right\}^n$$

$$= (1-\rho) \sum_{n=0}^{\infty} \left\{ \frac{\rho\mu}{s}[1-B^*(s)] \right\}^n.$$

But it can be shown (see Appendix 3) that $\mu[1-B^*(s)]/s$ is the LST of

$$R(t) = \mu \int_0^t [1-B(x)]\,dx,$$

and therefore

$$W_q^*(s) = (1-\rho) \sum_{n=0}^{\infty} [\rho R^*(s)]^n,$$

which yields, after term-by-term inversion utilizing the convolution prop-

---

[7]This requires that $(\lambda/s)[1-B^*(s)]$ be <1, which is true for $\rho<1$ since $1-B^*(s)=1-\int_0^\infty e^{-st}\,dB(t) < 1 - \int_0^\infty(1-st)\,dB(t) = s/\mu = s\rho/\lambda < s/\lambda.$

erty, a result surprisingly similar to (5.39), namely,

$$W_q(t) = (1 - \rho) \sum_{n=0}^{\infty} \rho^n [R^{(n)}(t)] \quad . \tag{5.41}$$

Let us consider this result a little more carefully to see exactly what it is saying. First of all, what is the actual meaning of the function $R(t)$? It can, in fact, be shown using renewal theory [see Ross (1970), for example] that $R(t)$ is nothing more than the CDF of the remaining service time of the customer being served at the instant the new customer arrives. Then Equation (5.41) says that if time is re-ordered with this remaining service time as the fundamental unit, any arrival in the steady-state finds n such time units of potential service in front of him with probability $(1 - \rho)\rho^n$, giving a result remarkably like that for the $M/M/1$ queue.

Example 5.3

Let us illustrate some of these results by going back to the Bearing Straight Corporation of Example 5.2. Bearing Straight has determined that it costs them $1000/hr that a machine is down, and that an additional penalty must be incurred because of the possibility of an excessive number of machines being down. It is decided to cost this variability at $2000 × (standard deviation of customer delay). Under such a total cost structure, what is the total cost of their policy, using the parameters indicated in Example 5.2 and assuming that repair labor is a sunk cost?

Problem 5.10 asks us to show that $L \doteq 2.96$. But we are also going to need the variance of the system waits, T, where we know that

$$E[N(N-1)] \equiv L_{(2)} = \lambda^2 W_2 \qquad [\text{from (5.38)}];$$

hence

$$Var[T] = W_2 - W^2 = \frac{L_{(2)}}{\lambda^2} - \frac{L^2}{\lambda^2}.$$

To get $L_{(2)}$ the second derivative of $P(z)$ is found from (5.23) and then evaluated at $z = 1$ to be 14.50 (see Problem 5.16).

Therefore

$$Var[T] = \frac{14.50 - 8.75}{25}$$

$$= 0.23 (\text{hr})^2.$$

Thus the total cost of Bearing's policy computes as

$$C(\$/\text{hr}) = (1000)L + (2000)\sqrt{0.23}$$
$$\doteq 2960 + 960$$
$$\doteq \$3920/\text{hr}.$$

5.1.7   Busy Period Analysis

The determination of the distribution of the busy period for an $M/G/1$ queue is a somewhat more difficult matter than finding that of the $M/M/1$, particularly in view of the fact that the service-time CDF must be carried as an unknown. But it is not too much of a task to find the Laplace-Stieltjes transform (LST) of the busy-period CDF, from which we can easily obtain any number of moments.

To begin, let $G(x)$ denote the CDF of the busy period X of an $M/G/1$ with service CDF $B(t)$. Then we condition X on the length of the first service time inaugurating the busy period. Since each arrival during that service time will contribute to the busy period by having arrivals come during his service time, we can look at each arrival during the first service time of the busy period as essentially generating his own busy period. Thus we can write

$$G(x) = \int_0^x \Pr\{\text{given first service time} = t, \text{ busy}$$
$$\text{period generated by all arrivals}$$
$$\text{occurring during } t \leqslant x - t\} \, dB(t)$$

$$= \int_0^x \sum_{n=0}^{\infty} \frac{e^{-\lambda t}(\lambda t)^n}{n!} G^{(n)}(x-t) \, dB(t), \qquad (5.42)$$

where $G^{(n)}(x)$ is the n-fold convolution of $G(x)$. Next let

$$G^*(s) = \int_0^{\infty} e^{-sx} \, dG(x)$$

be the LST of $G(x)$ and $B^*(s)$ be the LST of $B(t)$.

Then, by taking the transform of both sides of (5.42), it is found that

$$G^*(s) = \int_0^{\infty} \int_0^x \sum_{n=0}^{\infty} e^{-sx} \frac{e^{-\lambda t}(\lambda t)^n}{n!} G^{(n)}(x-t) \, dB(t) \, dx.$$

Changing the order of integration gives

$$G^*(s) = \int_0^\infty \sum_{n=0}^\infty \frac{e^{-\lambda t}(\lambda t)^n}{n!} dB(t) \int_t^\infty e^{-xs}G^{(n)}(x-t)\,dx,$$

and by the convolution property,[8]

$$G^*(s) = \int_0^\infty \sum_{n=0}^\infty \frac{e^{-\lambda t}(\lambda t)^n}{n!} e^{-st}[G^*(s)]^n dB(t)$$

$$= \int_0^\infty e^{-\lambda t}e^{\lambda t G^*(s)}e^{-st}\,dB(t)$$

$$= B^*[s+\lambda-\lambda G^*(s)].$$

Hence the mean length of the busy period is found as

$$E[X] = -\frac{dG^*(s)}{ds}\bigg|_{s=0} \equiv -G^{*\prime}(0),$$

where

$$G^{*\prime}(s) = B^{*\prime}[s+\lambda-\lambda G^*(s)]\cdot[1-\lambda G^{*\prime}(s)].$$

Therefore[9]

$$E[X] = -B^{*\prime}[\lambda-\lambda G^*(0)]\cdot[1-\lambda G^{*\prime}(0)]$$

$$= -B^{*\prime}(0)\cdot\{1+\lambda E[X]\},$$

or

$$E[X] = -\frac{B^{*\prime}(0)}{1+\lambda B^{*\prime}(0)}.$$

[8]Takács (1962) has shown that this equation in transforms may, in fact, be inverted to give

$$G(x) = \sum_{n=1}^\infty \frac{1}{n}\int_0^x \frac{(\lambda t)^{n-1}}{(n-1)!}e^{-\lambda t}dB^{(n)}(t).$$

[9]Note that $B^{*\prime}[s+\lambda-\lambda G^*(s)] = dB^*(u)/du$ for $u=s+\lambda-\lambda G^*(s)$.

But $B^{*\prime}(0) = -1/\mu$; hence

$$E[X] = \frac{1/\mu}{1 - \lambda/\mu}$$

$$= \frac{1}{\mu - \lambda},$$

which, surprisingly, is exactly the same result we obtained earlier for $M/M/1$.

There is an alternative way to derive the expected value without requiring the transform, namely, using the ratio approach indicated in Problem 2.26. We have already shown that the percentage idleness over the long term is $p_0 = 1 - \rho$. Therefore

$$\frac{\rho}{1-\rho} = \frac{\lambda}{\mu - \lambda} = \frac{E[X]}{E[\text{length of idle period}]}$$

$$= \frac{E[X]}{1/\lambda},$$

from which we see that

$$E[X] = \frac{1}{\mu - \lambda}.$$

5.1.8   Some Additional Results

In this section we present some assorted additional results for $M/G/1$ queues. At first, a brief discussion of the $M/G/1/K$ queue is presented, followed in order by similarly brief discussions of the problems of impatience, priorities, output, and transience, with mention made of finite source and batching, all in the context of the $M/G/1$.

We begin now with the brief discussion of the $M/G/1/K$ queue, that is, one in which there is a finite restriction on the size of the waiting room. The analysis proceeds in a way very similar to that of the unlimited-waiting-room case. Let us thus examine each of the main results of $M/G/1/\infty$ for applicability to $M/G/1/K$.

The Pollaczek-Khintchine formula will not now hold since the expected number of (joined) arrivals during a service period must be conditioned on the system size. The best way to get the new result is directly from the steady-state probabilities since there are now only a finite number of them.

The single-step transition matrix must here be truncated at $K-1$,[10] such that

$$
P = \begin{bmatrix}
k_0 & k_1 & k_2 & \cdots & 1-\sum_{n=0}^{K-2} k_n \\[2em]
k_0 & k_1 & k_2 & \cdots & 1-\sum_{n=0}^{K-2} k_n \\[2em]
0 & k_0 & k_1 & \cdots & 1-\sum_{n=0}^{K-3} k_n \\[2em]
0 & 0 & k_0 & \cdots & 1-\sum_{n=0}^{K-4} k_n \\[1em]
 & & \vdots & & \\[1em]
0 & 0 & 0 & \cdots & 1-k_0
\end{bmatrix},
$$

which implies that the stationary equation is

$$
\pi_i = \begin{cases}
\pi_0 k_i + \displaystyle\sum_{j=1}^{i+1} \pi_j k_{i-j+1} & (i=0,1,2,\ldots,K-2) \\[1.5em]
1 - \pi_0 \displaystyle\sum_{n=0}^{K-2} k_n - \sum_{j=1}^{K-1} \pi_j \sum_{n=0}^{K-j-1} k_n & (i=K-1)
\end{cases}.
$$

These K (consistent) equations in K unknowns can then be solved for all the probabilities, and the average system size at points of departure is thus given by $L = \sum_{n=0}^{K-1} n\pi_n$.

We also notice that the first portion of the stationary equation is identical to that of the unlimited $M/G/1$, and therefore deduce that the respective stationary probabilities, say $\{\pi_i\}$ for $M/G/1/K$ and $\{\pi_i^*\}$ for $M/G/1/\infty$, must be at worst proportional for $i \leqslant K-1$; that is,

$$
\pi_i = C\pi_i^* \qquad (i=0,1,\ldots,K-1).
$$

We find C by the usual use of the summability-to-one condition and get

$$
\sum_{i=0}^{K-1} \pi_i = 1 = C \sum_{i=0}^{K-1} \pi_i^*,
$$

[10]It is not K, since we are observing just after a departure.

or
$$C = \frac{1}{\sum\limits_{i=0}^{K-1} \pi_i^*}.$$

Furthermore, we note that the probability distribution for the system size encountered by an arrival will be different from $\{\pi_n\}$ since now the state space must be enlarged to include K. Let q_n' then denote the probability that an arriving customer finds a system with n customers. [10] Then if we go back to Section 5.1.3 and the proof that $\pi_n = p_n$ for $M/G/1$, it is noted in that proof, essentially Equation (5.31), that the distribution of system sizes just prior to points of arrival (our $\{q_n\}$) is identical to the departure-point probabilities (here $\{\pi_n\}$) as long as arrivals occur singly and service is not in bulk. Such is also the case with q'_n, except that the state spaces are different. This difference is easily taken care of by first noting that Equation (5.31) is really saying

$$\pi_n = \Pr\{\text{arrival finds } n | \text{customer does in fact join}\}$$
$$= q_n \cdot \quad (0 \leqslant n \leqslant K-1)$$
$$= \frac{q_n'}{1 - q_K'}.$$

Therefore

$$q_n' = (1 - q_K')\pi_n \quad (0 \leqslant n \leqslant K-1).$$

To now get q_K', we use an approach very similar to the one used earlier for simple Markovian models when we showed [as in Section 3.6, Equation (3.48)] that the effective arrival rate must equal the effective departure rate; that is,

$$\lambda(1 - q_K') = \mu(1 - p_0).$$

Hence

$$q_K' = \frac{\rho - 1 + p_0}{\rho}$$

and

$$q_n' = \frac{(1 - p_0)\pi_n}{\rho} \quad (0 \leqslant n \leqslant K-1).$$

But since the original arrival process is Poisson, $q_n' = p_n$ for all n. Thus

---

[10] Here we are speaking about the distribution of arriving customers whether or not they join the queue as opposed to only those arrivals who join, denoted by q_n. They q'_n distribution is often of interest in its own right.

$$q_0' = p_0 = \frac{(1 - p_0)\pi_0}{\rho}$$

or

$$p_0 = \frac{\pi_0}{\pi_0 + \rho},$$

and finally

$$q_n' = \frac{\pi_n}{\pi_0 + \rho}.$$

With respect to impatience, one can easily introduce balking into $M/G/1$ by prescribing a probability b that an arrival decides to actually join the system. Then the true input process becomes a (filtered) Poisson with mean $b\lambda t$ and Equation (5.13) thus has to be rewritten as

$$k_n = \int_0^\infty \frac{e^{-b\lambda t}(b\lambda t)^n}{n!} dB(t).$$

But the rest of the analysis goes through in a parallel fashion to the regular $M/G/1$, with the probability of idleness, p_0, now equal to $1 - b\lambda/\mu$. For a more comprehensive treatment of impatience in $M/G/1$, the reader is referred to Rao (1968) who treats both balking and reneging.

Next, we look at the problem of priorities in the $M/G/1$. It turns out that the nonpreemptive, many-priority model of Section 4.4.2 can be extended very nicely to situations with nonexponential service. The definitions and derivation are exactly the same with the single exception of the value of $E[S_0]$, the expected time required to finish the item in service at the time of an arrival, the value of which must change in view of the loss of forgetfulness by the service. But we have, in fact, actually done this calculation for $M/G/1$ as Problem 5.5 and found for S equal to the service time that

$$E[S_0] = \frac{\lambda}{2} E[S^2].$$

Therefore we may write that the expected delay in line of the ith of r priorities is given by

$$W_q^{(i)} = \frac{(\lambda/2) E[S^2]}{(1 - \sigma_{i-1})(1 - \sigma_i)},$$

where we recall that

$$\sigma_i = \sum_{k=1}^i \rho_k \qquad (\rho_k = \lambda_k/\mu_k).$$

When considering series queues, as far as output is concerned, we have already shown that $M/M/1$ has Poisson output, but now would like to know whether there are any other $M/G/1$'s which also possess this property. The answer is essentially no[11] and this is shown as follows.

Define $C(t)$ as the CDF of the interdeparture times of an $M/G/1$. Then, for $B(t)$ the service CDF, it follows that

$$C(t) \equiv \Pr\{\text{interdeparture time} \leq t\}$$

$$= \Pr\{\text{system experienced no idleness during interdeparture period}\} \cdot$$
$$\Pr\{\text{interdeparture time} \leq t| \text{ no idleness}\}$$

$$+ \Pr\{\text{system experienced some idleness during interdeparture period}\} \cdot$$
$$\Pr\{\text{interdeparture time} \leq t| \text{ some idleness}\}$$

$$= \rho B(t) + (1-\rho)\int_0^t B(t-u)\lambda e^{-\lambda u}\, du$$

since the length of an interdeparture period with idleness is the sum of the idle time and service time. Now if it is assumed that $C(t) = 1 - e^{-\lambda t}$,

$$1 - e^{-\lambda t} = \rho B(t) + (1-\rho)\int_0^t B(t-u)\lambda e^{-\lambda u}\, du.$$

Then taking the Laplace-Stieltjes transform with respect to t of both sides of this equation gives

$$\frac{\lambda}{\lambda+s} = \rho B^*(s) + (1-\rho)B^*(s)\frac{\lambda}{\lambda+s}.$$

Hence

$$B^*(s) = \frac{\lambda/(\lambda+s)}{\rho + (1-\rho)\lambda/(\lambda+s)}$$

$$= \frac{\lambda}{\rho s + \lambda}$$

$$= \frac{\mu}{\mu+s},$$

---

[11]G could be degenerate with all mass at 0.

which is the LST of $1 - e^{-\mu t}$. Therefore the *only service distribution* that has exponential output is the exponential! This has very serious negative implications for the solution of series models since we have essentially shown that the output of a first stage will only be exponential, which we would like it to be, if it is $M/M/1$. However, small $M/G/1$-type series problems can be handled numerically by appropriate utilization of the formula for the CDF of the interdeparture times, $C(t)$.

Next, to get any transient results, we would have to appeal directly to the theory of Markov chains and the Chapman-Kolmogorov equation

$$p_j^{(m)} = \sum_k p_k^{(0)} p_{kj}^{(m)},$$

where $p_j^{(m)}$ is then the probability that the system state is j just after the mth customer has departed. The necessary matrix multiplications must be done with some caution since we are dealing with an $\infty \times \infty$ matrix in the unlimited-waiting-room case. But this can indeed be done by carefully truncating the transition matrix at an appropriate point where the entries have become very small [see, for example, Neuts (1973)].

To close this section, a brief mention of two additional problem types will be made, namely, finite-source and bulk queues. The finite-source $M/G/1$ is essentially the machine-repair problem with arbitrarily distributed repair times and has been solved in the literature again using an imbedded-Markov-chain approach. Any interested reader is referred to Takacs (1962) for a fairly detailed discussion of these kinds of problems. The bulk-input $M/G/1$ denoted by $M^{[X]}/G/1$, and the bulk-service $M/G/1$, denoted by $M/G^{[Y]}/1$, can also be solved with the use of Markov chains. The bulk-input model is presented in the next section but the bulk-service problem is quite a bit more messy and therefore is not treated in this book. However, the reader is referred to Prabhu (1965b) for the details of this latter model.

5.1.9 The Bulk Input Queue ($M^{[X]}/G/1$)

The $M^{[X]}/G/1$ queueing system can be described in the following manner:

(i) Customers arrive as a Poisson process with parameter λ in groups of random size C, where C has the distribution

$$\Pr\{C = n\} = c_n \qquad (n > 1)$$

and the generating function (which will be assumed to exist)

$$C(z) = E[z^C] = \sum_{n=1}^{\infty} c_n z^n \qquad (|z| \leqslant 1).$$

The probability that a total of n customers arrive in an interval of length t is thus given by

$$p_n(t) = \sum_{k=0}^{n} e^{-\lambda t} \frac{(\lambda t)^k}{k!} c_n^{(k)} \qquad (n \geqslant 0),$$

where $\{c_n^{(k)}\}$ is the k-fold convolution of $\{c_n\}$ with itself (that is, the arrivals form a compound Poisson process),

$$c_n^{(0)} \equiv \begin{cases} 1 & (n=0) \\ 0 & (n>0) \end{cases}.$$

(ii)   The customers are serviced singly by one server on a first-come, first-served basis.

(iii)   The service times of the succession of customers are independent and identically distributed random variables with CDF $B(t)$ and LST $B^*(s)$.

Let us make a slight change here from $M/G/1$ which will help us later, such that the imbedded chain we shall use will be generated by the points (therefore called regeneration points) at which either a departure occurs or an idle period is ended. This process will be called $\{X_n, n=1,2,\ldots | X_n$ = number of customers in the system immediately after the nth regeneration point$\}$, with transition matrix given by

$$\begin{bmatrix} 0 & c_1 & c_2 & \cdot & \cdot & \cdot \\ k_0 & k_1 & k_2 & \cdot & \cdot & \cdot \\ 0 & k_0 & k_1 & \cdot & \cdot & \cdot \\ 0 & 0 & k_0 & \cdot & \cdot & \cdot \\ \cdot & \cdot & \cdot & \cdot & \cdot & \cdot \\ \cdot & \cdot & \cdot & \cdot & \cdot & \cdot \\ \cdot & \cdot & \cdot & \cdot & \cdot & \cdot \end{bmatrix},$$

where

$$k_n = \Pr\{\, n \text{ arrivals during a full}$$
$$\text{service period}\,\}$$

$$= \int_0^\infty p_n(t)\, dB(t)$$

$$= \int_0^\infty \sum_{k=0}^n e^{-\lambda t} \frac{(\lambda t)^k}{k!} c_n^{(k)}\, dB(t)$$

$$= p_{i,n+i-1} \qquad (i>0).$$

The application of Foster's theorem (A48) in a fashion similar to that of Section 5.1.4 for $M/G/1$ shows that this chain is ergodic and hence possesses identical long-run and stationary distributions when

$$\rho \equiv \sum_{n=1}^\infty n k_n < 1.$$

This definition of ρ turns out to be perfectly consistent with our earlier ones since for $j>0$

$$\sum_{n=0}^\infty n k_n = \sum_{n=0}^\infty \int_0^\infty \sum_{k=0}^n e^{-\lambda t} \frac{(\lambda t)^k}{k!} n c_n^{(k)}\, dB(t).$$

Now reversing the order of summation, we have

$$\rho = \int_0^\infty \sum_{k=0}^\infty e^{-\lambda t} \frac{(\lambda t)^k}{k!} E[\, C^{(k)}\,]\, dB(t)$$

$$= E[\,C\,] \int_0^\infty \sum_{k=0}^\infty k e^{-\lambda t} \frac{(\lambda t)^k}{k!}\, dB(t)$$

$$= E[\,C\,] \int_0^\infty \lambda t\, dB(t)$$

$$= \lambda E[\,C\,] E[\,T\,].$$

If the steady-state distribution $\{\pi_i\}^{12}$ is to exist for the chain, then it is the solution of the system

$$\sum_{i=0}^{\infty} p_{ij}\pi_i = \pi_j \qquad (j \geqslant 0)$$

and

$$\sum_{i=0}^{\infty} \pi_i = 1.$$

From the transition matrix (as in Section 5.1.2),

$$\pi_j = \sum_{i=0}^{\infty} p_{ij}\pi_i$$

$$= c_j\pi_0 + \sum_{i=1}^{j+1} k_{j-i+1}\pi_i \qquad (c_0 \equiv 0).$$

If the foregoing stationary equation is multiplied by z^j and then summed on j, it is found that

$$\sum_{j=0}^{\infty} \pi_j z^j = \sum_{j=0}^{\infty} c_j\pi_0 z^j + \sum_{j=0}^{\infty}\sum_{i=1}^{j+1} k_{j-i+1}\pi_i z^j.$$

If the usual generating functions are defined as

$$\Pi(z) = \sum_{j=0}^{\infty} \pi_j z^j$$

and

$$K(z) = \sum_{j=0}^{\infty} k_j z^j,$$

then we find after reversing the order of summation in the final term that

$$\Pi(z) = \pi_0 C(z) + \frac{K(z)}{z}\sum_{i=1}^{\infty} \pi_i z^i$$

$$= \pi_0 C(z) + \frac{K(z)}{z}[\Pi(z) - \pi_0],$$

---

[12] Keep in mind that these $\{\pi_i\}$ are slightly different than those that would have resulted had we restricted ourselves to departure points only, as we did in $M/G/1$.

or

$$\Pi(z) = \frac{\pi_0[K(z) - zC(z)]}{K(z) - z}.$$

Furthermore, it can be shown that

$$K(z) = \sum_{j=0}^{\infty} \int_0^{\infty} \sum_{k=0}^{j} e^{-\lambda t} \frac{(\lambda t)^k}{k!} c_j^{(k)} \, dB(t) z^j$$

$$= \int_0^{\infty} e^{-\lambda t} \sum_{k=0}^{\infty} \frac{(\lambda t)^k}{k!} [C(z)]^k \, dB(t)$$

$$= \int_0^{\infty} e^{-\lambda t + \lambda t C(z)} \, dB(t)$$

$$= B^*[\lambda - \lambda C(z)] \qquad (|z| \le 1).$$

These results have all been derived without much difficulty in a manner similar to the approach for $M/G/1$. However, there now exists a problem that we have not faced before, namely, that the results derived for the imbedded Markov chain do not directly apply to the total general-time stochastic process, $\{X(t), t \ge 0 | X(t) = \text{number in the system at time } t\}$. In order to relate the general-time steady-state probabilities, say $\{p_n\}$, to $\{\pi_n\}$, we must appeal to some results from semi-Markov processes, which are presented in Chapter 6, Section 6.3. References on this subject for anyone with further interest in this material are Fabens (1961) and Gross et al. (1971).

5.1.10   Departure-Point State Dependence

In Section 3.7 we treated queues with state dependencies; that is, the mean service rate was a function of the number of customers in the system. Whenever the number in the system would change (arrival or departure), the mean service rate would itself change accordingly. But in many situations, it might not be possible to change the service rate at any time a new arrival may come, but rather only upon the initiation of a service (or, almost equivalently, at the conclusion of a service time). For example, in many cases where the service is a man-machine operation and the machine is capable of running at various speeds, the operator would set the speed only prior to the actual commencement of service. Once service had begun, he would not change speed until that service was completed for to do otherwise would necessitate stopping work to alter the speed setting, and then restarting and/or repositioning the work. Further, stopping the operation prior to completion may damage the unit. This type of situation where mean service rate can be adjusted only prior to commencing service, or at a customer departure point, is what we refer to as departure-point state dependency and is considered in this section.

We assume the state-dependent service mechanism is as follows. Let $B_i(t)$ be the service-time CDF of a customer who enters service with $i-1$ customers behind him (total of i in the system), and

$$k_{ni} = \Pr\{ n \text{ arrivals during a service time, given}$$
$$i \text{ in the system when service began} \}$$

$$= \int_0^\infty \frac{e^{-\lambda t}(\lambda t)^n}{n!} dB_i(t). \qquad (5.43)$$

Then the transition matrix, $P=[p_{ij}]$, is given as

$$P=[p_{ij}]=\begin{bmatrix} k_{01} & k_{11} & k_{21} & k_{31} & \cdots \\ k_{01} & k_{11} & k_{21} & k_{31} & \cdots \\ 0 & k_{02} & k_{12} & k_{22} & \cdots \\ 0 & 0 & k_{03} & k_{13} & \cdots \\ \cdot & \cdot & \cdot & \cdot & \cdots \\ \cdot & \cdot & \cdot & \cdot & \cdots \\ \cdot & \cdot & \cdot & \cdot & \cdots \end{bmatrix}. \qquad (5.44)$$

A sufficient condition for the existence of a steady-state solution [see Crabill (1968)] is[13]

$$\limsup\{\rho_i\} < 1, \qquad (5.45)$$

where

$$\rho_i \equiv \lambda E[S_i].$$

Thus, assuming this condition is met, we can find the steady-state prob-

---

[13]This condition says that all but a finite number of the $\{\rho_i, (i \geqslant 1)\}$ must be less than 1.

ability distribution by solving the stationary Equation (5.15),

$$\pi P = \pi.$$

Although this gives the departure-point state probabilities, we have shown in Section 5.1.3 for non-state-dependent service that these are equivalent to the general-time probabilities. The addition of state dependency of service does not alter the proof in any way, so that here, also, the departure-point and general-time state probabilities are equivalent.

Thus the use of Equation (5.15) results in

$$p_j = \pi_j = \pi_0 k_{j,1} + \pi_1 k_{j,1} + \pi_2 k_{j-1,2} + \pi_3 k_{j-2,3} + \pi_4 k_{j-3,4}$$

$$+ \cdots + \pi_{j+1} k_{0,j+1} \qquad (j \geqslant 0). \tag{5.46}$$

Again, we define the generating functions

$$\Pi(z) = \sum_{j=0}^{\infty} \pi_j z^j$$

and

$$K_i(z) = \sum_{j=0}^{\infty} k_{ji} z^j.$$

Then multiplying both sides of (5.46) by z^j and summing over all j, we get

$$\Pi(z) = \pi_0 K_1(z) + \pi_1 K_1(z) + \pi_2 z K_2(z) + \cdots + \pi_{j+1} z^j K_{j+1}(z) + \cdots.$$

This is as far as we are able to proceed in general. No closed-form expression for $\Pi(z)$ in terms of the $K_i(z)$ is obtainable, so we now present a specific case for $B_i(t)$ to illustrate the computational procedure and give another specific case as an exercise (see Problem 5.21).

Consider the case where the mean service rate is directly proportional to the number of customers at a departure point and where for any given mean, the service times are exponential. Thus

$$B_i(t) = 1 - e^{-i\mu t}, \tag{5.47}$$

where i is the system size when the to-be-served customer enters service. From (5.43) we have

$$k_{ni} = \int_0^{\infty} \frac{e^{-\lambda t}(\lambda t)^n}{n!} i\mu e^{-i\mu t} dt$$

$$= \frac{\lambda^n \mu i}{(\lambda + \mu i)^{n+1}}. \tag{5.48}$$

Now using (5.46) we thus have that

$$p_0 = p_0 k_{01} + p_1 k_{01}$$

or

$$p_1 = \frac{(1 - k_{01}) p_0}{k_{01}}.$$

But, from (5.48),

$$k_{01} = \frac{\mu}{\lambda + \mu};$$

hence

$$p_1 = \frac{\lambda}{\mu} p_0 \equiv \rho p_0. \qquad (5.49)$$

Now repeated use of (5.46) and (5.48) gives

$$
\begin{cases}
p_2 = \dfrac{\rho^2}{2!}\, (\rho + 2) \left[\dfrac{1}{\rho + 1} \right] p_0 \\[3mm]
p_3 = \dfrac{\rho^3}{3!}\, (\rho + 3) \left[\dfrac{2}{\rho + 2} + \dfrac{2}{(\rho + 1)^2} - \dfrac{1}{\rho + 1} \right] p_0 \\[3mm]
p_4 = \dfrac{\rho^4}{4!}\, (\rho + 4) \left[\dfrac{3}{\rho + 3} + \dfrac{6}{(\rho + 2)^2} - \dfrac{1}{\rho + 2} + \dfrac{6}{(\rho + 1)^3} \right. \\[3mm]
\qquad\qquad \left. - \dfrac{1}{(\rho + 1)^2} - \dfrac{1}{\rho + 1} \right] p_0 \\[3mm]
p_5 = \dfrac{\rho^5}{5!}\, (\rho + 5) \left[\dfrac{4}{\rho + 4} + \dfrac{12}{(\rho + 3)^2} + \dfrac{2}{\rho + 3} + \dfrac{24}{(\rho + 2)^3} + \dfrac{6}{(\rho + 2)^2} \right. \\[3mm]
\qquad\qquad \left. + \dfrac{3}{\rho + 2} + \dfrac{24}{(\rho + 1)^4} - \dfrac{2}{(\rho + 1)^3} + \dfrac{1}{(\rho + 1)^2} - \dfrac{8}{\rho + 1} \right] p_0
\end{cases}
$$

Unfortunately, it does not appear possible to determine a general expression for p_n and one must keep utilizing (5.46). Thus it is also not possible to determine p_0 in closed form, so that numerical methods must be employed. That is, we can write

$$p_n = f_n p_0, \qquad (5.50)$$

where f_n must be determined utilizing (5.46); hence summing all p_n to one yields

$$p_0 = \frac{1}{\sum\limits_{n=0}^{\infty} f_n}. \tag{5.51}$$

A problem exists, however, of when to stop the summation of f_n for sufficient accuracy; that is, numerically we can calculate $\hat{p}_0 = \sum_{n=0}^{M} f_n$ and would like M large enough for desired accuracy. Ideally we wish to get a bound on $\sum_{n=M+1}^{\infty} f_n$. However, this is not always possible, and one must essentially guess at a sufficient magnitude of M by successive trials. For a more detailed discussion the reader is referred to Gross and Harris (1971), particularly Section III. We present below an illustration patterned after one found in Harris (1967).

Example 5.4

The Hugh Borum Company has a production problem which involves drilling holes into castings. The interarrival times of the castings to the drill press were found to be governed closely by the exponential distribution, and there were such a variety of holes with different depths that service times were also closely approximated by an exponential distribution. The utilization factor ρ was found to be three-fourths, thus yielding an average of three castings at the drill press $(M/M/1)$.

Production had been falling behind demand so it was decided to increase the cutting rate μ, which could be done rather easily since the drill press had an adjustable speed mechanism. However, it was found that costly drills were wearing out prematurely because of the increased cutting rate as well as a deterioration of the quality of the holes. A compromise was considered by setting a policy of variable service rate such that the mean service rate would be $n\mu$, where n was the number of castings in the system at the beginning of service. It was felt that quality would not suffer greatly since only when there were large numbers of castings waiting to be drilled would the machine be operating at very high speeds. Management wished to know what the effect on average system size would be.

It is necessary, then, to solve for enough of the p_n's so that we can determine

$$L = \sum_{n=0}^{\infty} n p_n$$

with reasonable accuracy. Here ρ is still three-fourths yielding

$$
\begin{aligned}
k_{ni} &= \frac{\lambda^n \mu i}{(\lambda + \mu i)^{n+1}} \\
&= \frac{\rho^n i}{(\rho + i)^{n+1}} \\
&= \frac{(3/4)^n i}{(3/4 + i)^{n+1}}.
\end{aligned}
$$

A computer program was written to solve for p_n using (5.46) and the results are presented in Table 5.3. It appears that the estimate of L stabilizes at three decimal places for $M \geqslant 11$. Thus we see L is approximately 1, a considerable reduction in average system size.

5.2 MULTISERVER QUEUES WITH POISSON INPUT AND GENERAL SERVICE

We begin here with an immediate disadvantage from the point of view of being able to derive necessary results, since $M/G/c/\infty$ and $M/G/c/c$ do not possess imbedded Markov chains in the usual sense. This is so at

TABLE 5.3

M	$\hat{p}_0 = 1/\sum_{n=0}^{M} f_n$	$\hat{L} = \sum_{n=0}^{M} n\hat{p}_n$
1	0.571429	0.429
2	0.456211	0.745
3	0.415757	0.945
4	0.400619	1.066
5	0.394849	1.113
6	0.392612	1.141
7	0.391724	1.154
8	0.391364	1.161
9	0.391215	1.164
10	0.391153	1.165
11	0.391126	1.166
12	0.391115	1.166
13	0.391111	1.166

departure points because the number of arrivals during any interdeparture period is dependent upon more than just the system size at the immediate departure predecessor due to the presence of multiple servers. There are, however, some special $M/G/c$'s which do possess enough structure to get fairly complete results, including, of course, $M/M/c$. One such example is the $M/D/c$, which will be taken up in some detail in Chapter 6. What we wish then to do in this section is to try to get some general results for both $M/G/c/\infty$ and $M/G/c/c$ which could easily be applied by merely specifying G.

5.2.1 Some Results for $M/G/c/\infty$

For $M/G/c/\infty$, the main general result that may be found is a line version of the relationship between the kth factorial moment of system size and the regular kth moment of system delay given by (5.38), namely,

$$L_{(k)} = \lambda^k W_k.$$

The proof of this result follows.

To begin, we know from our earlier work on waiting times for $M/G/1$ (Section 5.1.6) that

$$\pi_n \equiv \Pr\{\, n \text{ in system just after a point of departure}\,\}$$

$$= \frac{1}{n!} \int_0^\infty (\lambda t)^n e^{-\lambda t}\, dW(t).$$

But this equation is also valid in modified form for $M/G/c$ if we consider everything in terms of the queue and not the system. Then it is true that

$$\pi_n^q \equiv \Pr\{\, n \text{ in } queue \text{ just after a departure}\,\}$$

$$= \frac{1}{n!} \int_0^\infty (\lambda t)^n e^{-\lambda t}\, dW_q(t).$$

Then the mean queue length at departure points, say $L_q^{(D)}$, is given by

$$L_q^{(D)} = \sum_{n=1}^\infty n \pi_n^q$$

$$= \int_0^\infty \lambda t\, dW_q(t)$$

$$= \lambda W_q,$$

which is Little's formula. Now suppose we use $L_{q(k)}^{(D)}$ to denote the kth factorial moment of the departure-point queue size. Then

$$L_{q(k)}^{(D)} = \sum_{n=1}^{\infty} n(n-1) \cdots (n-k+1) \pi_n^q$$

$$= \int_0^{\infty} dW_q(t) \sum_{n=1}^{\infty} \frac{n(n-1) \cdots (n-k+1)}{n!} (\lambda t)^n e^{-\lambda t}.$$

But the term inside the summation symbol is nothing more than the kth factorial moment of the Poisson, which can be shown to be $(\lambda t)^k$. Hence

$$\boxed{L_{q(k)}^{(D)} = \lambda^k W_{q,k}} \quad , \tag{5.52}$$

where $W_{q,k}$ is the ordinary kth moment of the line waiting time. This is now a generalization of Equation (5.38) to line waits for an $M/G/c$.

5.2.2 Some Results for $M/G/c/c$

Next, for $M/G/c/c$, we wish to go back to a result that we quoted earlier in Section 3.4. It is the surprising fact that the system-size distribution given by (3.31), namely, the truncated Poisson

$$p_n = \frac{(\lambda/\mu)^n / n!}{\sum_{i=0}^{c} (\lambda/\mu)^i / i!} \qquad (0 \leqslant n \leqslant c),$$

is valid for *any* $M/G/c/c$ independent of the form of G. Let us then consider the following proof of this assertion.

We begin by locating a Markov process within this generally non-Markovian queue, and then use this process via the Chapman-Kolmogorov equations to prove the required result for the single-channel case first, followed by the proof for $c=2$, which will suffice with no loss of generality for the general multichannel situation $c>1$. The Markov process we refer to in the foregoing is defined over all time $t>0$ with states $\{(n, u_1, \ldots, u_c) | n$ = number in system, u_i = amount of service already expended on the customer in the ith channel at time $t\}$. This is a technique which essentially goes back to the work of Cox (1955) and is called solution by supplementary variables. For the single-channel case then, the states are bivariate vectors (n,u), with joint probability function $p_n(u,t) \equiv \Pr\{$system has n at time t, and current customer in service has already been served for u units of time$\}$. That this process is Markovian can be seen since its future state is clearly a function of only its current position. (Satisfy yourself that this assertion is correct.)

The usual difference-type approach of the C-K equations allows us to write for this $M/G/1/1$ that

$$\Pr\{ n=1 \text{ at time } t+\Delta t \text{ with } u+\Delta t \text{ units of service}$$
$$\text{already expended}\}$$

$$= \Pr\{ n=1 \text{ at time } t, \text{ with service age} = u\} \cdot$$

$$\Pr\{ \text{no arrivals and no service completion}\},$$

or

$$p_1(u+\Delta t, t+\Delta t) = p_1(u,t)[1 - h(u)\Delta t] + o(\Delta t)$$

$$[\text{since any arrival is turned away}],$$

where $h(u)\,du$ is the probability that a service of age u will be completed in $(u, u+du)$ and is equal to $B'(u)\,du/[1-B(u)]$ (see Problem 5.23). The function $h(u)$ is common in reliability and is usually called the hazard or instantaneous failure rate. After a little algebra, it is found that

$$\frac{p_1(u+\Delta t, t+\Delta t) - p_1(u,t)}{\Delta t} = -h(u)p_1(u,t) + \frac{o(\Delta t)}{\Delta t},$$

or

$$\frac{p_1(u+\Delta t, t+\Delta t) - p_1(u+\Delta t, t)}{\Delta t} + \frac{p_1(u+\Delta t, t) - p_1(u,t)}{\Delta t}$$

$$= -h(u)p_1(u,t) + \frac{o(\Delta t)}{\Delta t} \quad .$$

When the limits of both sides of this equation are taken it is found that

$$\frac{\partial p_1(u,t)}{\partial t} + \frac{\partial p_1(u,t)}{\partial u} = -h(u)p_1(u,t).$$

However, we are in equilibrium; hence the foregoing equation becomes

$$\frac{dp_1(u)}{du} = -h(u)p_1(u),$$

where, after solving for $p_1(u)$, our final desired probability will be given by

$$p_1 = \int_0^\infty p_1(u)\,du.$$

To solve for $p_1(u)$ we simply integrate the differential equation and find that

$$p_1(u) = C\exp\left(-\int_0^u h(t)\,dt\right).$$

But it is a well-known result from probability that the CDF $B(u)$ may be written as (also see Problem 5.23)

$$B(u) = 1 - \exp\left(-\int_0^u h(t)\,dt\right). \tag{5.53}$$

Hence

$$p_1(u) = C[1 - B(u)]$$

and

$$p_1 = C\int_0^\infty [1 - B(u)]\,du$$

$$= \frac{C}{\mu},$$

since the integral of the complementary CDF is the expected value [see, for example, Parzen (1960)].Thus we see that p_1 is only dependent upon the mean service time and therefore must have the same value for any two service distributions with the same mean. Since we already know this result for the case of $M/M/1/1$ to be $p_1 = \lambda/(\lambda+\mu)$, this formula will now hold for any $M/G/1/1$ with the same service expectation, and will be independent of the form of G.

We now assume that $c = 2$ and the approach here will, in fact, carry over very nicely to cases for $c > 2$. The state vector is now (n, u_1, u_2), where u_1 thus indicates the service age of the customer in channel 1, while u_2 does likewise for channel 2. The difference equations are

$\Pr\{\, n = 2$ at time $t + \Delta t$ with $u_1 + \Delta t$ and

$\qquad u_2 + \Delta t$ units of service already expended

$\qquad$ in channels 1 and 2, respectively$\}$

$\qquad = \Pr\{\, n = 2$ at time t with u_1 and u_2 expended$\} \cdot$

$\qquad \Pr\{\,$no occurrences$\}$,

$\Pr\{\, n = 1$ at time $t + \Delta t$ with $u_1 + \Delta t$ and 0 or u_2

$\qquad + \Delta t$ and 0 units of service already expended$\}$

$\qquad = \Pr\{\, n = 1$ at time t with u_1 and 0 or u_2 and 0 expended$\} \cdot$

$\qquad \Pr\{\,$no occurrences$\}$

$\qquad + \Pr\{\, n = 2$ at time t with u_1 and u_2 expended, followed

$\qquad$ by 1 completion and 0 arrivals$\} + o(\Delta t)$,

and

$\Pr\{\, n = 0$ at time $t + \Delta t\} = \Pr\{\, n = 0$ at time $t\} \cdot \Pr\{\,$no arrivals$\}$

$\qquad + \Pr\{\, n = 1$ at time t with u_1 and 0 or

$\qquad u_2$ and 0 expended, followed

$\qquad$ by 1 completion $+ 0$ arrivals$\} + o(\Delta t)$.

These, in turn, lead to the difference equations

$$p_2(u_1 + \Delta t, u_2 + \Delta t, t + \Delta t) = p_2(u_1, u_2, t)\{1 - [h(u_1) + h(u_2)]\Delta t\} + o(\Delta t),$$

$$p_1(u_1 + \Delta t, 0, t + \Delta t) + p_1(0, u_2 + \Delta t, t + \Delta t) = p_1(u_1, 0, t)\{1 - [\lambda + h(u_1)]\Delta t\}$$

$$+ p_1(0, u_2, t)\{1 - [\lambda + h(u_2)]\Delta t\}$$

$$+ \Delta t \int_0^\infty p_2(u_1, u_2, t)[h(u_1)\,du_1 + h(u_2)\,du_2] + o(\Delta t),$$

$$p_0(0,0,t+\Delta t) = p_0(0,0,t)(1-\lambda\Delta t) + \Delta t \int_0^\infty p_1(u_1,0,t)h(u_1)\,du_1$$

$$+ \Delta t \int_0^\infty p_1(0,u_2,t)h(u_2)\,du_2 + o(\Delta t).$$

Therefore

$$\frac{p_2(u_1+\Delta t,u_2+\Delta t,t+\Delta t) - p_2(u_1,u_2,t)}{\Delta t}$$

$$= -[h(u_1)+h(u_2)]p_2(u_1,u_2,t) + \frac{o(\Delta t)}{\Delta t},$$

or

$$\frac{p_2(u_1+\Delta t,u_2+\Delta t,t+\Delta t) - p_2(u_1+\Delta t,u_2+\Delta t,t)}{\Delta t}$$

$$+ \frac{p_2(u_1+\Delta t,u_2+\Delta t,t) - p_2(u_1,u_2+\Delta t,t)}{\Delta t}$$

$$+ \frac{p_2(u_1,u_2+\Delta t,t) - p_2(u_1,u_2,t)}{\Delta t}$$

$$= -[h(u_1)+h(u_2)]p_2(u_1,u_2,t) + \frac{o(\Delta t)}{\Delta t}.$$

Taking limits as $\Delta t \to 0$ gives

$$\frac{\partial p_2(u_1,u_2,t)}{\partial t} + \frac{\partial p_2(u_1,u_2,t)}{\partial u_1} + \frac{\partial p_2(u_1,u_2,t)}{\partial u_2}$$

$$= -[h(u_1)+h(u_2)]p_2(u_1,u_2,t).$$

But we are in equilibrium; hence

$$\frac{\partial p_2(u_1,u_2)}{\partial u_1} + \frac{\partial p_2(u_1,u_2)}{\partial u_2} = -[h(u_1)+h(u_2)]p_2(u_1,u_2).$$

This partial differential equation is of a type we have seen before, namely, the planar form of $M/M/\infty$ transient analysis (Section 3.5.2). By exactly the same argument as used earlier, this PDE may be converted to the two ordinary differential equations

$$\frac{du_1}{1} = \frac{du_2}{1} = \frac{-dp_2(u_1,u_2)}{[h(u_1)+h(u_2)]p_2(u_1,u_2)}.$$

The solution of the first equation, $du_1 = du_2$, is clearly $u_1 = u_2 + C_1$, while the second gives (utilizing $du_1 = du_2$)

$$C_2 \exp\left\{ -\left[\int_0^{u_1} h(t)\,dt + \int_0^{u_2} h(t)\,dt \right] \right\} = p_2(u_1, u_2).$$

By an argument identical to that given for $M/M/\infty$, the foregoing equation leads to

$$p_2(u_1, u_2) = f(u_1 - u_2) \exp\left\{ -\left[\int_0^{u_1} h(t)\,dt + \int_0^{u_2} h(t)\,dt \right] \right\},$$

and using (5.53) gives

$$p_2(u_1, u_2) = f(u_1 - u_2)[1 - B(u_1)][1 - B(u_2)].$$

We know that

$$p_2 = \int_0^\infty \int_0^\infty p_2(u_1, u_2)\,du_1\,du_2,$$

and since $f(u_1 - u_2) = f(C_1)$, we therefore have

$$p_2 = \frac{f(C_1)}{\mu^2}.$$

Since $f(C_1)$ is a constant, we may use the same argument as that used for $M/G/1/1$, which tells us that p_2 must be the same for all $M/G/2/2$ with the same service mean, which, from (3.31), is

$$p_2 = \frac{\lambda^2}{2\mu^2 + 2\lambda\mu + \lambda^2}.$$

Next, for p_1, simplification yields

$$\frac{p_1(u_1 + \Delta t, 0, t + \Delta t) - p_1(u_1, 0, t)}{\Delta t} + \frac{p_1(0, u_2 + \Delta t, t + \Delta t) - p_1(0, u_2, t)}{\Delta t}$$

$$= -[\lambda + h(u_1)]p_1(u_1, 0, t) - [\lambda + h(u_2)]p_1(0, u_2, t)$$

$$+ \int_0^\infty p_2(u_1, u_2, t)[h(u_1)\,du_1 + h(u_2)\,du_2] + \frac{o(\Delta t)}{\Delta t},$$

which in turn leads to

$$\frac{\partial p_1(u_1,0,t)}{\partial t} + \frac{\partial p_1(u_1,0,t)}{\partial u_1} + \frac{\partial p_1(0,u_2,t)}{\partial t} + \frac{\partial p_1(0,u_2,t)}{\partial u_2}$$

$$= -[\lambda + h(u_1)]p_1(u_1,0,t) - [\lambda + h(u_2)]p_1(0,u_2,t)$$

$$+ \int_0^\infty p_2(u_1,u_2,t)[h(u_1)\,du_1 + h(u_2)\,du_2]\,,$$

or, in the steady state,

$$\frac{\partial p_1(u_1,0)}{\partial u_1} + \frac{\partial p_1(0,u_2)}{\partial u_2} = -[\lambda + h(u_1)]p_1(u_1,0) - [\lambda + h(u_2)]p_2(0,u_2)$$

$$+ \int_0^\infty p_2(u_1,u_2)[h(u_1)\,du_1 + h(u_2)\,du_2].$$

For p_0 we get

$$0 = -\lambda p_0(0,0) + \int_0^\infty [h(u_1)p_1(u_1,0)\,du_1 + h(u_2)p_1(0,u_2)\,du_2].$$

We shall not proceed any further at this point, but instead ask the reader to provide in Problem 5.24 the remaining steps of the proof to show that the entire probability distribution of system sizes for $M/G/2/2$ is identical to that of the $M/M/2/2$ with the same service-time expectation.

5.2.3   Some Further Results for $M/G/\infty/\infty$

The stationary system-size result derived in Section 5.2.2 does clearly allow c to be ∞. In other words, the stationary distribution for the ample-server, general service-time case is given by

$$p_n = \frac{e^{-\rho}\rho^n}{n!},$$

where $\rho \equiv \lambda E[T]$. In this section it is our intention to derive two further results for this model, namely, the transient distribution for the number of customers in the system at time t (as we did for $M/M/\infty$), and the transient distribution for the number of customers who have completed service by time t, that is, the departure counting process.

Let the system-size process be called $N(t)$, the departure process $Y(t)$, and the input process $X(t) = Y(t) + N(t)$. By the laws of conditional probability, we find that

$$\Pr\{N(t)=n\} = \sum_{i=0}^{\infty} \Pr\{N(t)=n|X(t)=i\} \frac{e^{-\lambda t}(\lambda t)^i}{i!}$$

since the input is Poisson. The probability that a customer who arrives at time x will still be present at time t is given by $1 - B(t - x)$, $B(u)$ being the service CDF. Hence it follows that the probability of an arbitrary one of these customers still being in service is given by

$$q(t) \equiv q = \int_0^t \Pr\{\text{service time exceeds } t - x| \text{ an arrival} \\ \text{at time } x\} \cdot \Pr\{\text{arrival at } x\} \, dx.$$

Since the input is Poisson, $\Pr\{\text{arrival at } x\}$ is uniform on $(0,t)$; hence it is $1/t$, from Equation (1.21). Thus

$$q = \frac{\int_0^t [1 - B(t - x)] \, dx}{t}$$

$$= \frac{\int_0^t [1 - B(x)] \, dx}{t},$$

and is independent of any other arrival. Thus by the binomial law,

$$\Pr\{N(t)=n|X(t)=i\} = \binom{i}{n} q^n (1-q)^{i-n} \qquad (n \geqslant 0).$$

Hence

$$\Pr\{N(t)=n\} = \sum_{i=n}^{\infty} \frac{\binom{i}{n} q^n (1-q)^{i-n} e^{-\lambda t}(\lambda t)^i}{i!}$$

$$= \frac{(\lambda q t)^n e^{-\lambda t}}{n!} \sum_{i=n}^{\infty} \frac{[\lambda t(1-q)]^{i-n}}{(i-n)!}$$

$$= \frac{(\lambda q t)^n e^{-\lambda t} e^{\lambda t - \lambda q t}}{n!}$$

$$= \frac{(\lambda q t)^n e^{-\lambda q t}}{n!},$$

namely, non homogeneous Poisson with mean $\lambda q t$.

As a check on our equilibrium solution, take the limit as $t \to \infty$ of this transient answer. It is thereby found that

$$\lim_{t \to \infty} (\lambda q t) = \lambda \int_0^\infty [1 - B(x)] \, dx$$

$$= \frac{\lambda}{\mu},$$

and hence the solution is indeed Poisson with mean λ/μ.

The distribution of the departure-counting process $Y(t)$ can be found by exactly the same argument as the foregoing using, instead of q, $1 - q = \int_0^t B(x) dx / t$. The result as expected is

$$\Pr\{ Y(t) = n \} = \frac{[\lambda(1-q)t]^n e^{-\lambda(1-q)t}}{n!}.$$

5.3 GENERAL INDEPENDENT INPUT AND EXPONENTIAL SERVICE

In this section we investigate the queueing situation where service times are assumed to be exponential and no specific assumption is made concerning the arrival pattern other than that successive interarrival times are independent and identically distributed. For this case, results can be obtained for c parallel servers using an analysis similar to that for the $c = 1$ case with a slight increase in complexity in certain probability calculations. So we first consider $c = 1$ and then generalize to c servers.

5.3.1 Arrival-Point Steady-State System-Size Probabilities for $GI/M/1$

We examine first a single server with mean service rate μ and exponential service times. We assume that the mean arrival rate is λ, that arrivals come singly, and, as mentioned above, that successive interarrival times are independent and identically distributed. An imbedded-Markov-chain approach is also utilized here to obtain our results.

Following the type of analysis used in Section 5.1 for $M/G/1$, we now consider the system just prior to an arrival so that the system state, X_i, represents the number in the system that the ith arrival sees upon joining the system ($X_i = 0, 1, 2, \ldots$). We then have

$$X_{n+1} = X_n + 1 - B_n \qquad (B_n \leqslant X_n + 1; \; X_n \geqslant 0),$$

where B_n is the number of customers served during the interarrival time $T^{(n)}$ between the nth and $(n+1)$st arrivals. Since the interarrival times are assumed independent, the random variable $T^{(n)}$ can be denoted by T, and we denote its CDF by $A(t)$. Since service is exponential, the random variable B_n depends on only the length of the interval and not on the extent of the service the present customer has already received. Thus we can drop the time-dependent subscript and henceforth denote B_n as B. We have, then, that

$$\Pr\{B = b\} = \int_0^\infty \Pr\{B = b \mid T = t\} \, dA(t)$$

$$= \int_0^\infty \frac{e^{-\mu t}(\mu t)^b}{b!} \, dA(t), \qquad (5.54)$$

so that[14]

$$p_{ij} \equiv \Pr\{X_{n+1} = j \mid X_n = i\} = \begin{cases} \Pr\{B = i+1-j\} & (i+1 \geqslant j \geqslant 1) \\ 0 & (i+1 < j) \end{cases}$$

$$= \begin{cases} \displaystyle\int_0^\infty \frac{e^{-\mu t}(\mu t)^{i+1-j}}{(i+1-j)!} \, dA(t) & (i+1 \geqslant j \geqslant 1) \\ 0 & (i+1 < j) \end{cases}. \qquad (5.55)$$

Since p_{i0} must be $1 - \sum_{j=1}^{i+1} p_{ij}$, we see from (5.55) that all p_{ij} depend only on i and j; thus we have a Markov chain.

Continuing on now in a manner similar to Section 5.1.2, we introduce the following simplifying notation:

$$b_n = \Pr\{n \text{ services during an interarrival time}\}$$

$$= \Pr\{B = n\}$$

$$= \int_0^\infty \frac{e^{-\mu t}(\mu t)^n}{n!} \, dA(t), \qquad (5.56)$$

[14]The case $j = 0$ must be treated separately since idle time results over a portion of the interarrival time T and it is not sufficient to say that $i + 1 - j$ are served during T as they could have been served in a time less than T.

so that we can obtain from (5.55) and (5.56)

$$P = [p_{ij}] = \begin{bmatrix} 1-b_0 & b_0 & 0 & 0 & 0 & \cdots \\ 1-\sum\limits_{k=0}^{1} b_k & b_1 & b_0 & 0 & 0 & \cdots \\ 1-\sum\limits_{k=0}^{2} b_k & b_2 & b_1 & b_0 & 0 & \cdots \\ \cdot & \cdot & \cdot & \cdot & \cdot & \cdot \\ \cdot & \cdot & \cdot & \cdot & \cdot & \cdot \\ \cdot & \cdot & \cdot & \cdot & \cdot & \cdot \end{bmatrix}. \tag{5.57}$$

Assuming a steady-state solution exists (this will be taken up later) and denoting by $q = \{q_n\}$ the probability vector that an arrival finds n in the system ($n = 0, 1, 2, \ldots$), we have the usual stationary equation

$$qP = q, \tag{5.58}$$

which yields

$$\begin{cases} q_i = \sum\limits_{k=0}^{\infty} q_{i+k-1} b_k & (i \geqslant 1) \\ q_0 = \sum\limits_{l=0}^{\infty} \left[q_l \left(1 - \sum\limits_{k=0}^{l} b_k \right) \right] \end{cases}. \tag{5.59}$$

We note here that a major difference between (5.59) and its counterpart for the $M/G/1$, Equation (5.16), is that the equations of (5.59) have an infinite summation, whereas each equation had a finite summation in (5.16). It turns out that this works to our advantage and we now employ the method of operators on (5.59).[15] Letting

$$Dq_i = q_{i+1},$$

we find for $i \geqslant 1$ that (5.59) can be written as

$$q_i - [q_{i-1}b_0 + q_i b_1 + q_{i+1}b_2 + \cdots] = 0.$$

[15]Problem 5.27 asks the reader to derive the same results using generating functions.

Hence

$$q_{i-1}[D - b_0 - Db_1 - D^2b_2 - D^3b_3 - \cdots] = 0.$$

Thus for a nontrivial solution,

$$D - \sum_{n=0}^{\infty} b_n D^n = 0. \tag{5.60}$$

But note that since b_n is a probability, the second term on the left is merely the probability generating function of b_n [call it $\beta(z)$], so that (5.60) becomes

$$\beta(z) = z. \tag{5.61}$$

Thus letting r_j denote the jth root $(0 < r_j < 1)$ of Equation (5.61), the solution for q_i is

$$q_i = \sum_j C_j r_j^i \qquad (i \geqslant 0),$$

where the $\{C_j\}$ are constants. But we shall prove shortly that for Equation (5.61) there is only a single admissible root between 0 and 1. Denoting this root by r_0, we thus have that

$$q_i = C r_0^i \qquad (i \geqslant 0). \tag{5.62}$$

The constant C as usual is to be determined from the boundary condition $\sum_{i=0}^{\infty} q_i = 1$.

To show that there is only one root of (5.61) between 0 and 1 we consider the two sides of the equation separately, that is,

$$y = \beta(z) \tag{5.63}$$

and

$$y = z. \tag{5.64}$$

We observe that

$$0 < \beta(0) = b_0 < 1$$

and

$$\beta(1) = \sum_{n=0}^{\infty} b_n = 1.$$

Also consider

$$\beta'(z) = \sum_{n=1}^{\infty} nb_n z^{n-1} \geqslant 0$$

and

$$\beta''(z) = \sum_{n=2}^{\infty} n(n-1)b_n z^{n-2} \geqslant 0,$$

which implies $\beta(z)$ is monotone nondecreasing and convex.[16] Since service times are exponential, all the b_n, $n>2$, are >0; hence $\beta(z)$ is strictly convex. We graph Equations (5.63) and (5.64) in Figure 5.1 to observe their intersections. There are two possible cases, either no roots between 0 and 1 or exactly one root between 0 and 1. The left-hand case (no roots between 0 and 1) occurs if

$$\beta'(1) = E[\text{number served during an interarrival time}] = \mu \frac{1}{\lambda} \leqslant 1.$$

This implies that $\mu/\lambda > 1$ (or equivalently $\rho = \lambda/\mu < 1$) is necessary and sufficient for ergodicity since otherwise $\sum q_i = \infty$. Hence when a steady-state solution exists, there is exactly one root (r_0) of Equation (5.62) between 0 and 1. Finding this root generally involves numerical procedures but it is readily obtainable. For example, the well-known Newton-Raphson

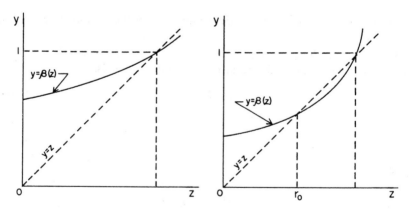

Fig. 5.1 Plot of Equations (5.63) and (5.64).

[16]A convex function is one for which a line segment joining any two points on the curve lies on or above the curve. Strict convexity requires that the line segment be strictly above the curve.

approximation method[17] is guaranteed to converge because of the convexity of $\beta(z)$.

It remains now to evaluate the constant C. This can be done in the usual manner by virtue of the fact that the probabilities must sum to one, which yields

$$C = 1 - r_0, \tag{5.65}$$

so that

$$\boxed{q_n = (1 - r_0)r_0^n \qquad (n \geq 0; \ \rho < 1)} \ . \tag{5.66}$$

It is interesting to note the analogy between (5.66) and the $M/M/1$ steady-state probability given by

$$p_n = (1 - \rho)\rho^n.$$

One can use all the expected-value measures of effectiveness results obtained for $M/M/1$ by merely replacing ρ by r_0. However, it must be pointed out that q_n is the steady-state probability of n in the system just prior to an arrival and not the general-time steady-state probability p_n, so that the expected-value measures apply only at arrival points. Unlike the $M/G/1$ model it is not true that $q_n = p_n$. In fact, it turns out that this is true if, and only if, the arrivals are Poisson; that is, $q_n = p_n$ for $GI/M/1$ if, and only if, $GI = M$. However, q_n and p_n can be related; more will be said about this in Chapter 6, Section 6.3.

The waiting-time distribution, $W_q(t)$, however, can be obtained from the $M/M/1$ with r_0 replacing ρ, that is,

$$\boxed{W_q(t) = \begin{cases} 1 - r_0 & (t = 0) \\ 1 - r_0 e^{-\mu(1 - r_0)t} & (t > 0) \end{cases}} \ .$$

This result is exact, since if one refers back to the development of Section 2.3, it can be seen that the steady-state probabilities used in the development leading to Equation (2.31) were in reality the q_n. In that case it was necessary to justify using p_n, thereby noting the two were equivalent because of Poisson arrivals.[18]

---

[17]See Hildebrand (1949), for example.

[18]Note here, as in all previous waiting-time considerations thus far, we are interested in actual waiting time, that is, the time an arriving customer would have to wait as opposed to virtual waiting time or the time a fictitious customer arriving at an arbitrary point of time would have to wait (see discussion at the beginning of Section 2.3).

5.3.2 Arrival-Point Steady-State System-Size Probabilities for $GI/M/c$

To generalize to c servers, everything remains the same except for calculating b_n. Equation (5.56) no longer applies since the system mean service rate is either $n\mu$ or $c\mu$, depending on its state; that is, b_n depends on i and j. This will lead to a somewhat different $P=[p_{ij}]$ matrix because of this dependence.

For a single server, we had [from (5.55)]

$$
p_{ij} = \begin{cases} 0 & (i+1<j) \\ b_{i+1-j} & (i+1 \geqslant j \geqslant 1) \\ 1 - \displaystyle\sum_{k=0}^{i} b_k & (j=0) \end{cases}.
$$

Considering c servers now, it is still true that

$$
p_{ij} = 0 \qquad (i+1<j).
$$

When $j \geqslant c$, the system is serving at mean rate $c\mu$, since all servers are busy, so that

$$
p_{ij} = b_{i+1-j} \qquad (i+1 \geqslant j \geqslant c),
$$

where now

$$
b_n = \int_0^\infty \frac{e^{-c\mu t}(c\mu t)^n}{n!} \, dA(t). \tag{5.67}
$$

The two cases which cause some difficulties are when $i \geqslant c$ and $j<c$, for which the system service rate varies from $c\mu$ down to $j\mu$, and when $i<c$ and $j<c$, for which the system service rate varies from $i\mu$ down to $j\mu$. Therefore, except for the first $c-1$ columns, the transition matrix will be identical to that given in (5.57), where b_n now is defined by (5.67).

Consider now the case when $j \leqslant i+1 \leqslant c$. Everyone is being served and the probability that anyone has completed service by a time t is the CDF of an individual server, namely, $1-e^{-\mu t}$. To go from i to j, there must be $i+1-j$ service completions by time t; hence using the binomial distribution, we have

$$
p_{ij} = \int_0^\infty \binom{i+1}{i+1-j} (1-e^{-\mu t})^{i+1-j} e^{-\mu t j} \, dA(t) \qquad (j \leqslant i+1 \leqslant c).
$$

$$
\tag{5.68}
$$

Lastly, it remains to consider the case $i+1>c>j$. The system starts out with all servers busy since $i \geqslant c$, and sometime during the interarrival time, T, servers start to become idle until finally only j servers are busy. Let us assume that at a time V after the arrival comes $(0<V<T)$, he goes into service (all prior customers have left), with $H(v)$ the CDF of V. Thus to get from state i to j in time T, we must have $c-j$ service completions from V to T, or during a time interval of $T-V$. Using the binomial distribution again and realizing that service time is memoryless we have

$$p_{ij} = \int_0^\infty \int_0^t \binom{c}{c-j} [1 - e^{-\mu(t-v)}]^{c-j} e^{-\mu(t-v)j} \, dH(v) \, dA(t).$$

The random variable V is merely the time until $i-c+1$ people have been served with all c servers working, which is the $(i-c+1)$-fold convolution of the exponential distribution with parameters $c\mu$. Hence $h(v) \equiv dH(v)/dv$ is Erlang type $(i-c+1)$, namely,

$$h(v) = \frac{(c\mu)^{i-c+1}}{(i-c)!} v^{i-c} e^{-c\mu v}.$$

Substituting for $dH(v)$ results in

$$p_{ij} = \binom{c}{c-j} \frac{(c\mu)^{i-c+1}}{(i-c)!} \int_0^\infty \int_0^t [1 - e^{-\mu(t-v)}]^{c-j} e^{-\mu(t-v)j} v^{i-c} e^{-c\mu v} dv \, dA(t)$$

$$(i+1>c>j).$$

Summarizing, then, for c servers we have

$$p_{ij} = \begin{cases} 0 & (i+1<j) \\[2mm] \int_0^\infty \binom{i+1}{i+1-j} (1-e^{-\mu t})^{i+1-j} e^{-\mu j} dA(t) & (j \leqslant i+1 \leqslant c) \\[2mm] \binom{c}{c-j} \frac{(c\mu)^{i-c+1}}{(i-c)!} \int_0^\infty \int_0^t [1-e^{-\mu(t-v)}]^{c-j} e^{-\mu(t-v)j} v^{i-c} e^{-c\mu v} \, dv \, dA(t) & (i+1>c>j) \\[2mm] b_{i+1-j} = \int_0^\infty \frac{e^{-c\mu t}(c\mu t)^{i+1-j}}{(i+1-j)!} dA(t) & (i+1 \geqslant j \geqslant c) \end{cases} \qquad .(5.69)$$

The stationary equation

$$qP = q$$

yields

$$q_j = \sum_{i=0}^{\infty} p_{ij} q_i \qquad (j \geqslant 0), \qquad (5.70)$$

where p_{ij} is given by (5.69). However, for $j \geqslant c$, we have

$$q_j = \sum_{i=0}^{j-2} 0 q_i + \sum_{i=j-1}^{\infty} b_{i+1-j} q_i$$

$$= \sum_{k=0}^{\infty} b_k q_{j+k-1} \qquad (j \geqslant c). \qquad (5.71)$$

Equation (5.71) is identical to the first line of (5.59), and hence using similar analyses as for $c = 1$, we have

$$q_j = Cr_0^j \qquad (j \geqslant c), \qquad (5.72)$$

where r_0 is the root of

$$\beta(z) = z$$

and

$$\beta(z) = \sum_{n=0}^{\infty} b_n z^n,$$

for b_n given by (5.67).

The constant C and q_j $(j = 0, 1, \ldots, c-1)$ must be determined from the boundary condition $\sum_{i=0}^{\infty} q_j = 1$ and the first $c - 1$ equations of the stationary equation, using the $\{p_{ij}\}$ given by (5.69). This is not particularly easy to do since the $c + 1$ equations in $c + 1$ unknowns are all infinite summations. We can, however, get an expression for C in terms of $q_1, q_2, \ldots, q_c$ and r_0, and then develop a recursive relation among the q_j $(j = 1, 2, \ldots, c)$. The procedure is as follows. The boundary condition yields

$$1 = \sum_{j=0}^{\infty} q_j = \sum_{j=0}^{c-1} q_j + \sum_{j=c}^{\infty} Cr_0^j.$$

Hence

$$C = \frac{1 - \sum_{j=0}^{c-1} q_j}{\sum_{j=c}^{\infty} r_0^j}$$

$$= \frac{1 - \sum_{j=0}^{c-1} q_j}{r_0^c (1 - r_0)^{-1}}. \qquad (5.73)$$

Now, a recursive relation for q_j $(j < c)$ can be obtained since

$$q_j = \sum_{i=0}^{\infty} p_{ij} q_i$$

$$= \sum_{i=0}^{c-1} p_{ij} q_i + \sum_{i=c}^{\infty} p_{ij} C r_0^i \, ,$$

which when using the first line of (5.69) gives

$$q_j = \sum_{i=j-1}^{c-1} p_{ij} q_i + C \sum_{i=c}^{\infty} p_{ij} r_0^i \qquad (1 \leqslant j \leqslant c-1).$$

Rewriting we have

$$q_{j-1} = \frac{q_j - \sum_{i=j}^{c-1} p_{ij} q_i - C \sum_{i=c}^{\infty} p_{ij} r_0^i}{p_{j-1,j}} \qquad (1 \leqslant j \leqslant c-1).$$

Dividing through by C and letting $q_j' = q_j / C$ gives

$$q_{j-1}' = \frac{q_j' - \sum_{i=j}^{c-1} p_{ij} q_i' - \sum_{i=c}^{\infty} p_{ij} r_0^i}{p_{j-1,j}} \qquad (1 \leqslant j \leqslant c-1). \qquad (5.74)$$

From (5.70) we can also write

$$q_c = \sum_{i=0}^{\infty} p_{ic} q_i$$

$$= \sum_{i=c-1}^{c} p_{ic} q_i + \sum_{i=c+1}^{\infty} b_{i+1-c} C r_0^i \, .$$

Hence

$$q_{c-1} = \frac{(1-p_{cc}) q_c - C \sum_{i=c+1}^{\infty} b_{i+1-c} r_0^i}{p_{c-1,c}}$$

and

$$q_{c-1}' = \frac{(1-p_{cc}) q_c' - \sum_{i=c+1}^{\infty} b_{i+1-c} r_0^i}{p_{c-1,c}}$$

$$= \frac{(1-b_1) q_c' - \sum_{i=c+1}^{\infty} b_{i+1-c} r_0^i}{b_0} \, .$$

But we also have

$$q_c' = \frac{C r_0^c}{C} = r_0^c,$$

so that q_{c-1}' can be determined using (5.69) or (5.67) for b_n. Then from repeated use of (5.74), $q_{c-1}', q_{c-2}', \ldots, q_0'$ can be obtained. Now writing (5.73) in terms of q_i' gives

$$C = \frac{1 - C \sum_{j=0}^{c-1} q_j'}{r_0^c (1 - r_0)^{-1}}$$

or

$$C = \left[\sum_{j=0}^{c-1} q_j' + r_0^c \frac{1}{1 - r_0} \right]^{-1}. \tag{5.75}$$

Thus we are able to obtain results without too much difficulty for the $GI/M/c$ model. It is somewhat ironic that while we can get fairly "tidy" results for $GI/M/c$, we cannot for $M/G/c$, which is, however, a more realistic model since it is far more common in practice to encounter Poisson input than it is to encounter exponential service.

To determine the waiting-time-in-queue distribution, $W_q(t)$, we observe first that

$$W_q(0) = \sum_{i=0}^{c-1} q_i$$

$$= C \sum_{i=0}^{c-1} q_i' = C \left(\frac{1}{C} - \sum_{i=c}^{\infty} q_i' \right)$$

$$= C \left(\frac{1}{C} - \frac{r_0^c}{1 - r_0} \right).$$

For $W_q(t)$, $t > 0$, we have c servers working when an arrival comes, so that the system is putting out at a mean rate $c\mu$. The time between completions of service is exponential with mean $c\mu$, so that if there are $n \geqslant c$ customers in the system upon arrival, there are $n - c$ waiting and the current arrival does not get into service until $n - c + 1$ customers have been serviced. Thus his waiting time is the $(n - c + 1)$-fold convolution of exponentials with mean $c\mu$, which is Erlang. Unconditioning upon n finally gives

$$\Pr\{W_q \le t\} = W_q(t) = \sum_{n=c}^{\infty} q_n \int_0^t \frac{(c\mu)^{n-c+1} v^{n-c} e^{-\mu c v}}{(n-c)!} \, dv + W_q(0)$$

$$= \sum_{n=c}^{\infty} \int_0^t \frac{(c\mu)^{n-c+1} v^{n-c} e^{-\mu c v}}{(n-c)!} Cr_0^n \, dv + W_q(0)$$

$$= Cr_0^c \int_0^t \sum_{n=c}^{\infty} \frac{(c\mu v r_0)^{n-c}}{(n-c)!} \mu c e^{-\mu c v} \, dv + W_q(0)$$

$$= Cr_0^c \int_0^t \mu c e^{-\mu c v(1-r_0)} \, dv + W_q(0)$$

$$= \frac{Cr_0^c}{1-r_0} [1 - e^{-\mu c(1-r_0)t}] + C\left(\frac{1}{C} - \frac{r_0^c}{1-r_0}\right)$$

$$= C\left[\frac{1}{C} - \frac{r_0^c}{1-r_0} e^{-\mu c(1-r_0)t}\right] \qquad (t > 0).$$

Example 5.5

Suppose that in a single-server queueing situation, we know the service time is exponential with mean μ but have no theoretical basis for assuming the input to be Poisson or Erlang. From past history, we have determined a k-point probability distribution for interarrival times, so that

$$\Pr\{\text{interarrival time} = t_i\} = a(t_i) = a_i \qquad (1 \le i \le k).$$

We must first determine the root r_0 from (5.61), namely,

$$\sum_{n=0}^{\infty} b_n z^n = z.$$

Equation (5.56) allows us to evaluate b_n as

$$b_n = \sum_{i=1}^{k} \frac{e^{-\mu t_i} (\mu t_i)^n}{n!} a_i.$$

TABLE 5.4   INTERARRIVAL-
TIME PROBABILITY DISTRIBUTION

t (min)	$a(t)$	$A(t)$
2	0.2	0.2
3	0.7	0.9
4	0.1	1.0

$1/\lambda = 2.9$ min

Then we must solve

$$\sum_{n=0}^{\infty} \sum_{i=1}^{k} \frac{a_i e^{-\mu t_i} \left(\mu t_i z \right)^n}{n!} = z,$$

or

$$\sum_{i=1}^{k} a_i e^{-\mu t_i (1-z)} = z. \tag{5.76}$$

This is a relatively easy equation to solve numerically. In fact, we can plot the left-hand side as a function of z, and its intersection with a 45° line from the origin yields r_0. Then Equation (5.66) can be used to find q_n, and the $M/M/1$ expected-value equation can be used (with r_0 replacing ρ) to yield $L^{(A)}, L_q^{(A)}$, and W and W_q, where the superscript (A) denotes expected queue length and system size at arrival points.

TABLE 5.5

z	$\beta(z)$
0	0.24
0.2	0.32
0.4	0.42
0.6	0.56
0.8	0.75
1.0	1.00

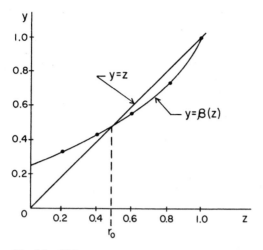

Fig. 5.2 $\beta(z)$ versus z.

To illustrate numerically consider the case where the interarrival-time distribution is given in Table 5.4 and $1/\mu = 2$ min. We must solve for the root of

$$\beta(z) = 0.2e^{-(1-z)} + 0.7e^{-1.5(1-z)} + 0.1e^{-2(1-z)} = z,$$

so the left-hand side is tabulated in Table 5.5 and the results graphed in Figure 5.2. It appears that

$$r_0 \doteq 0.46,$$

so that

$$q_n \doteq 0.54(0.46)^n, \qquad (n \geqslant 0)$$

$$L^{(A)} = \frac{r_0}{1-r_0} \doteq 0.85,$$

$$L_q^{(A)} = \frac{r_0^2}{1-r_0} \doteq 0.39,$$

$$W_q = \frac{r_0}{\mu(1-r_0)} \doteq 1.70 \text{ min,}[19]$$

and

$$W = W_q + \frac{1}{\mu} \doteq 3.70 \text{ min.}[19]$$

[19]Note that Little's formula using $L^{(A)}$ (that is, $W_q = L_q^{(A)}/\lambda$ and $W = L^{(A)}/\lambda$) does not hold.

We see that it is not difficult to obtain results for empirical distributions. This was true also for the $M/G/1$ illustrated by Example 5.2. It turns out that this is quite a useful result since *any* probability distribution can be approximated by a finite discrete distribution of k points, through the use of an approximating histogram.

Just as in $M/G/1$, there are numerous other $GI/M/1$-type problems we might want to consider, such as busy periods, $GI/M/1/K$, impatience, priorities, output, and transience. Due to space limitations we are not able to pursue any of these topics at length, and shall instead make a few comments on each and indicate a number of references.

Cohen (1969) is probably the most comprehensive reference for nearly all of these problems. Of course, specific references in the open literature may be better for particular subjects. For example, it is not a terribly difficult matter to get the expected length of the busy period for any $GI/M/1$ queue, and there is a very nice approach to this in Ross (1970) using renewal theory. We supply some of the details of this argument in the next chapter in the context of the $GI/G/1$ queue.

The approach to the truncation of the queue would be very similar to that described early in Section 5.1.8 for the imbedded chain of $M/G/1$, while impatience could be nicely introduced by permitting some departures from the system to occur before customers reach service. This can essentially be accomplished by changing the parameter of the exponential service to $\mu + r$, r now being the probability of such a renege. If, in addition, we desire to make reneges functions of queue size as they probably should be, then we would essentially have a problem with state-dependent departures, such that

$$b_{mn} = \Pr\{\, m \text{ services during an interarrival time} | n \text{ in system at}$$
$$\text{latest departure}\}$$

$$= \int_0^\infty \frac{e^{-[\mu+r(n)]t}\{[\mu+r(n)]t\}^m}{m!}\, dA(t),$$

where $r(n)$ is defined to be the reneging rate during interarrival periods which began with n in the system. The analysis would then proceed on in a way similar to the departure-point state dependence of Section 5.1.10.

As far as priorities are concerned, when the assumption of Poisson inputs is relaxed it becomes very difficult to obtain any results. One possible way of approaching the problem, suggested by Jaiswal (1968), is to use the technique of supplementary variables to gain some Markovianness. However, one supplementary variable would be required for each priority, hence at least two for any such problem. Even the near-Markovian assumption of Erlang input is messy, though some work has appeared on this problem for a small number of priorities.

For output, we have already indirectly obtained some results. It was noted in our discussion of series queues that the limiting output of an $M/G/1$ is Poisson if, and only if, G is exponential. Likewise, it can be shown that the limiting output of $GI/M/1$ is Poisson if, and only if, GI is exponential (see Problem 5.34).

Bulk services $(GI/M^{[Y]}/1)$ can be handled in a way comparable to the manner in which we solve the $M^{[X]}/G/1$ problem in Section 5.1.9.

Some results are also possible for these extensions of the basic $GI/M/1$ models when considering c-channels.

Finally, we close with a few comments about transient analysis. As for $M/G/1$, we have to again appeal to the Chapman-Kolmogorov equation

$$p_j^{(m)} = \sum_k p_k^{(0)} p_{kj}^{(m)},$$

where $p_j^{(m)}$ is the probability that the system state is j just before the mth customer has arrived. The necessary matrix multiplications must be done with some caution since we are still dealing with an $\infty \times \infty$ matrix when there is unlimited waiting room. But this can be done (albeit carefully) by truncating the transition matrix at an appropriate point [see Neuts (1973)].

PROBLEMS

5.1   Calculate the imbedded transition probabilities for an $M/G/1$ where service is uniformly distributed on (a,b).

5.2   Do Problem 4.16 without making any assumptions concerning the service-time distribution, that is, utilizing only the mean and variance of the service-time data.

5.3   (a) Find L, L_q, W_q, and W for an $M/G/1$ with service times which are beta distributed.

(b) Arrivals to an $M/G/1$ queue occur at the mean rate of one-third per unit time. Find the average steady-state waiting times in system and queue when the distribution function of the service time is an Erlang type 2.

5.4   For Example 5.1, find the σ_S^2 necessary to yield $L=5$ if the mean service time after training increases to 5.2 min.

5.5   Show that the expected amount of time required to finish the item in service encountered by an arbitrary arrival is given by $(\lambda/2)E[S^2]$ for any $M/G/1$ when S is the total service time of an item.

5.6   Verify Equation (5.16) for $i = 0, 1, 2, 3$.

5.7   Derive $\Pi(z)$ for $M/G/1$ as given by Equation (5.19).

5.8   From Equation (5.21), derive the P-K formula of Equation (5.11) by using $L = \Pi'(1)$ (*Hint*: Use L'Hôpital's rule twice.)

5.9   Use Equation (5.21) for the $M/G/1$ queue with G assumed exponential and show that it reduces to the generating function of $M/M/1$ given by Equation (2.16).

5.10   Calculate L for Example 5.2.

5.11   A certain assembly-line operation is assumed to be of the $M/G/1$ type, with input rate 5/hr and service times with mean 9 min and variance 90 (min)^2. Find L, L_q, W, and W_q. Is the operation improved or otherwise if the service times are forced to be exponential with the same mean?

5.12   Derive Equations (5.25) to (5.27) for the $M/G/1$ k-point service-time model following the analysis of Example 5.2.

5.13   Consider an $M/D/1$ queue with service time equal to b time units. Suppose further that one is only able to determine the system size when time is a multiple of b. Show that the stochastic process $\{X_n, n = 0, 1, 2, \ldots | X_n$ equal to the system size at time $t = nb\}$ is a Markov chain and find its transition matrix.

5.14   Expand Equation (5.33) in a manner similar to that used for Equation (5.34) in Section 5.1.5 by first writing $1/\{1 - z[1 + r(1 - z)]^k\}$ as a geometric series and then using the binomial expansion as needed to obtain the results of Equation (5.36).

5.15   Find the Laplace-Stieltjes transform of the waiting-time-in-line distributions for $M/D/1$ and $M/E_k/1$, and then invert the transforms with the use of Equation (5.41).

5.16   Verify the computation for Example 5.3 that $P''(1) \doteq 14.50$.

5.17   Find the third and fourth ordinary moments of the system delay for the Bearing Straight problem, given that the third and fourth moments of the system size are 149.2 and 1670.6, respectively.

5.18   Find the variance of the $M/G/1$ busy period from the Laplace-Stieltjes transform of its CDF.

5.19   Consider a single-server queue to which customers arrive according to a Poisson process with parameter $\lambda = 0.04/\text{min}$ and where the service times of all customers are fixed at 10 min. When there are three units on line, the system becomes saturated and all additional arrivals are turned away. The instants of departure give rise to an imbedded Markov chain with states 0, 1, 2, and 3. Find the one-step transition matrix of this chain and the resultant stationary distribution. Then compare this answer to the result you would have gotten without truncation.

5.20 Given the two-point service distribution of Table 5.2, find the output CDF for that $M/G/1$ queue.

5.21 Find the steady-state probabilities for an $M/G/1$ state-dependent queue where

$$B_i(t) = \begin{cases} 1 - e^{-\mu_1 t} & (i=1) \\ 1 - e^{-\mu t} & (i>1) \end{cases}$$

that is, where there are two possible mean service rates, μ_1 if there is no queue and μ if there is a queue. [*Hint*: In the transition probability matrix analogous to (5.44), after the first two rows it is only necessary to carry a single subscript for k.] Obtain the stationary equations for $\{\pi_n\}$ from $\pi P = \pi$ and the generating function $\Pi(z)$ in terms of $K_1(z)$ and $K(z)$. Use $\lim_{z \to 1} \Pi(z)$ and the appropriate equation from $\pi P = \pi$ to obtain two equations in the two unknowns π_0 and π_1 and solve. Using $\pi P = \pi$, find π_2, π_3, π_4 and show that they satisfy

$$\pi_n = \left[\frac{\rho_1^n}{(\rho_1+1)^{n-1}} + \sum_{k=0}^{n-2} \frac{\rho_1^{n-k} \rho^{k+1}}{(\rho_1+1)^{n-k-1}} \right] \pi_0,$$

which is, for this case, the closed-form general expression for π_n.

5.22 Prove that the distribution of system sizes just prior to arrivals is equal to that after departures in any $GI/G/c$.

5.23 Under the assumption of a continuous service CDF $B(u)$, show that the instantaneous failure rate $h(u)$ is found as

$$h(u) = \lim_{\Delta u \to 0} \left[\begin{array}{c} \Pr\{\text{service of age } u \text{ will be} \\ \text{completed in } (u, u+\Delta u)\} /\Delta u \end{array} \right]$$

$$= \frac{B'(u)}{1 - B(u)}.$$

Then show that this equation can be inverted to give $B(u)$ in terms of $h(u)$ as

$$B(u) = 1 - \exp\left(-\int_0^u h(t)\, dt \right).$$

5.24   Finish the proof begun in Section 5.2.2 for $M/G/2/2$ that

$$p_n = \frac{(\lambda/\mu)^n/n!}{\sum_{n=0}^{2} (\lambda/\mu)^n/n!} \qquad (0 \leqslant n \leqslant 2).$$

5.25   The Mutual Exclusive Life Insurance Company (MELIC) is building a new headquarters in downtown Burbank. The telephone company wishes to determine the number of lines to feed into the building to assure MELIC of no more than 5% loss in calls due to busy circuits. Find the number of lines when it is estimated that the calling stream is Poisson with mean 100/hr throughout the day and that the mean call duration is 2 min.

5.26   Derive the equivalent to Erlang's first formula (3.31) for the case in which the input source is *finite* with rate proportional to the remaining source size, service times are general, and there are two channels. (The resultant answer is an example of a so-called Engset formula.)

5.27   Derive Equation (5.61) using the method of generating functions on Equation (5.59), $i \geqslant 1$. Then show that $\beta(z) = A^*[\mu(1-z)]$.

5.28   Find the solution, r_0, of $\beta(z) = z$, the generating function equation for a $GI/M/1$ queue, when the interarrival times follow the hyperexponential (mixture of two exponentials) probability distribution, that is,

$$a(t) = 2q^2\lambda e^{-2q\lambda t} + 2(1-q)^2\lambda e^{-2(1-q)\lambda t},$$

$$(t > 0; 0 < q \leqslant \tfrac{1}{2}; \lambda > 0).$$

5.29   Show that if $GI = M$, the root r_0 of Equation (5.61) is ρ and that Equation (5.66) yields the $M/M/1$ results.

5.30   Prove that r_0 is always greater than $e^{-1/\rho}$.

5.31   Find $\beta(z)$ for $GI/M/1$ in the event that GI is (a) deterministic and (b) Erlang type 2. Then, under the assumption that the mean input rate λ is 3 and that the mean service time is $1/5$, find the steady-state arrival-point distribution for each of the two cases.

5.32   Suppose in Example 5.5 that there are two servers, each working at a mean rate of 4 minutes with their times being exponential. Find the steady-state system-size probabilities, the expected system size and queue length at arrival points, and the expected time in queue and in system.

5.33   Suppose it is known in a $D/M/1$ queue that some customers are reneging before reaching service and that the reneging pattern is such that

Pr{renege in $(t, t + \Delta t)$} = $r\Delta t + o(\Delta t)$. Find the resultant imbedded Markov chain.

5.34 Show (a) for $M/G/1$ queues, stationary system waiting times are exponential if, and only if, G is M; (b) stationary system waiting times are exponential for all $GI/M/1$; and (c) the stationary output of a $GI/M/1$ queue is Poisson if, and only if, GI is exponential. [*Hint*: Note for (c) that the idle time the queue undergoes in an arbitrary interdeparture period (called the virtual idle time) has a CDF given by $F(u) = A(u) + \int_u^\infty e^{-\mu(1 - r_0)(t - u)} dA(t)$ (See Problem 6.28). Then use the fact that each departure time is a sum of a virtual idle time and a service time.]

Chapter 6
ADDITIONAL MODELS
AND TOPICS

In this chapter we provide some assorted additional results. As a general rule these results were not included earlier because the models were either not Markovian or inappropriate for the discussions in Chapter 5 dealing with $M/G/1$ and $GI/M/c$. Most of this material follows in a logical way from the previous material in the sense that it ties up some "loose theoretical ends" and should also help to provide a more complete picture of the kinds of models which may occur in real life.

6.1 GENERAL INDEPENDENT INPUT, GENERAL SERVICE ($GI/G/1$)

Although nearly completely void of specific structure, we are nonetheless able to get some results for general input and service, and it is our intention to devote this section to a discussion of two important ones. The first is an integral equation of the Wiener-Hopf type for the stationary distribution of the waiting time in queue of an arbitrary customer. This equation is largely due to Lindley (1952) and goes under his name. The second result is one for the expected length of the busy period, which follows in Section 6.1.2. Additional results may be found in the literature [Cohen (1969) for example], but most are beyond the level of this text.

We begin by observing that the relationship between the line waiting times, say $W_q^{(n)}$ and $W_q^{(n+1)}$, of the nth and $(n+1)$st customers, which we noted earlier in the text as Equation (1.5), is just as valid for the arbitrary $GI/G/1$ as it is for $D/D/1$. This recurrence is given by [see (1.5)]

$$W_q^{(n+1)} = \begin{cases} W_q^{(n)} + S^{(n)} - T^{(n)} & \left(W_q^{(n)} + S^{(n)} - T^{(n)} > 0 \right) \\ 0 & \left(W_q^{(n)} + S^{(n)} - T^{(n)} \leqslant 0 \right) \end{cases},$$

or

$$W_q^{(n+1)} = \max\left(0, W_q^{(n)} + S^{(n)} - T^{(n)} \right),$$

where $S^{(n)}$ is the service time of the nth customer and $T^{(n)}$ is the time between the arrivals of the two customers. We can immediately note that the stochastic process $\{W_q^{(n)}, n = 0, 1, 2, \dots\}$ is a discrete-time Markov process, since the behavior of $W_q^{(n+1)}$ is only a function of the stochastically determined value of $W_q^{(n)}$ and is independent of prior waiting-time history.

Now, from basic probability arguments, we have

$$W_q^{(n+1)}(t) \equiv \Pr\left\{ \text{line delay } W_q^{(n+1)} \text{ of } (n+1)\text{st customer} \leqslant t \right\}$$

$$= \Pr\left\{ W_q^{(n+1)} = 0 \right\} + \Pr\left\{ 0 < W_q^{(n+1)} \leqslant t \right\}$$

$$= \Pr\left\{ W_q^{(n)} + S^{(n)} - T^{(n)} \leqslant 0 \right\} + \Pr\left\{ 0 < W_q^{(n)} + S^{(n)} - T^{(n)} \leqslant t \right\}$$

$$= \Pr\left\{ W_q^{(n)} + S^{(n)} - T^{(n)} \leqslant t \right\}.$$

If we now define the random variable $U^{(n)} \equiv S^{(n)} - T^{(n)}$ with CDF $U^{(n)}(x)$, then, making use of the convolution formula,

$$W_q^{(n+1)}(t) = \int_{-\infty}^{t} W_q^{(n)}(t-x)\,dU^{(n)}(x) \qquad (0 \leqslant t < \infty).$$

In the steady state[1] the two waiting-time CDFs must be identical; hence using $W_q(t)$ to denote the stationary delay distribution, we find that

$$\boxed{W_q(t) = \begin{cases} \int_{-\infty}^{t} W_q(t-x)\,dU(x) & (0 \leqslant t < \infty) \\ 0 & (t < 0) \end{cases}} \qquad (6.1a)$$

or[2]

$$\boxed{W_q(t) = -\int_{0}^{\infty} W_q(x)\,dU(t-x) \qquad (0 \leqslant t < \infty)}, \qquad (6.1b)$$

where $U(x)$ is the equilibrium $U^{(n)}(x)$ and is given by the convolution of S and $(-T)$, namely,

$$U(x) = \int_{x}^{\infty} B(y)\,dA(y-x). \qquad (6.2)$$

---

[1]Lindley showed in his 1952 paper that the limiting CDF, $W_q(t) = \lim_{n \to \infty} W_q^{(n)}(t)$, exists if, and only if, the utilization factor ρ, defined as the quotient of mean service time and interarrival time, is less than one, as we would intuitively expect.

[2]Both Equations (6.1a) and (6.1b) are often referred to as Lindley's equation.

The usual approach to the solution of a Wiener-Hopf integral equation such as (6.1a) or (6.1b) [see Feller (1966)] begins with the definition of a new function, $W_q^-(t)$, as

$$W_q^-(t) \equiv \begin{cases} 0 & (t \geqslant 0) \\ \int_{-\infty}^t W_q(t-x)\,dU(x) & (t<0) \end{cases}. \qquad (6.3)$$

Hence it follows from (6.1a) that

$$W_q^-(t) + W_q(t) = \int_{-\infty}^t W_q(t-x)\,dU(x) \qquad (-\infty < t < \infty). \quad (6.4)$$

Note that $W_q^-(t)$ is the virtual idle time introduced in Problem 5.34, part (c).

It turns out to be easiest to try to obtain $W_q(t)$, $t>0$ (that is, the positive part), since $W_q(t)$ is not continuous at 0, but has a jump of q_0, so that $W_q(0)=q_0$.[3] Denote the two-sided Laplace transforms (LT) of $W_q(t)$ and $W_q^-(t)$ as

$$\overline{W}_q(s) = \int_{-\infty}^{\infty} e^{-st} W_q(t)\,dt$$

$$= \int_0^{\infty} e^{-st} W_q(t)\,dt,$$

and

$$\overline{W}_q^-(s) = \int_{-\infty}^{\infty} e^{-st} W_q^-(t)\,dt = \int_{-\infty}^0 e^{-st} W_q^-(t)\,dt.$$

In addition, we shall use $U^*(s)$ as the (two-sided) Laplace-Stieltjes transform (LST) of $U(t)$.

We then take the two-sided Laplace transform of both sides of (6.4). The transform of the right-hand side is found to be

$$\mathcal{L}_2 \left\{ \int_{-\infty}^t W_q(t-x)\,dU(x) \right\} = \int_{-\infty}^{\infty} \int_{-\infty}^t e^{-(t-x)s} W_q(t-x) e^{-sx}\,dU(x)\,dt.$$

Since $W_q(t-x)=0$ for $x \geqslant t$ we can write

$$\mathcal{L}_2 = \left[\int_{-\infty}^{\infty} e^{-us} W_q(u)\,du \right]\left[\int_{-\infty}^{\infty} e^{-sx}\,dU(x) \right]$$

$$= \overline{W}_q(s) U^*(s).$$

---

[3]Recall that for all single-channel infinite capacity queues, $p_0 = 1-\lambda/\mu$ (see Table 2.1); however, the arrival-point q_0 is not always equal to p_0.

But U is the CDF of the difference of the interarrival and service times and hence by the convolution property must have (two-sided) LST equal to the product of the interarrival transform $A^*(s)$ evaluated at $-s$ and the service transform $B^*(s)$, since $A(t)$ and $B(t)$ are both zero for $t < 0$. Hence

$$U^*(s) = A^*(-s)B^*(s),$$

and thus from (6.4),

$$\overline{W}_q^-(s) + \overline{W}_q(s) = \overline{W}_q(s)A^*(-s)B^*(s),$$

or

$$\overline{W}_q(s) = \frac{\overline{W}_q^-(s)}{A^*(-s)B^*(s) - 1}. \qquad (6.5)$$

Thus given any $A(t)$ and $B(t)$ for an arbitrary $GI/G/1$, we can theoretically find the Laplace transform of the line delay. The determination of $\overline{W}_q^-(s)$ is the primary difficulty in this computation, often requiring advanced concepts from the theory of complex variables.

To partially verify the validity of Equation (6.5), let us compute $\overline{W}_q(s)$ from this formula for $M/M/1$ and then check the answer against our earlier result from Chapter 2. Here

$$B(t) = 1 - e^{-\mu t},$$

with

$$B^*(s) = \frac{\mu}{\mu + s} \quad \text{(see Appendix 3, Example A3.3)}$$

and

$$A(t) = 1 - e^{-\lambda t},$$

with

$$A^*(-s) = \frac{\lambda}{\lambda - s}.$$

In this case,

$$W_q^-(t) = \int_{-\infty}^t W_q(t - x)\,dU(x) \qquad (t < 0),$$

where, from Equation (6.2),

$$U(x) = \begin{cases} \int_0^\infty (1 - e^{-\mu y})\lambda e^{-\lambda(y-x)}\,dy & (x < 0) \\ \int_x^\infty (1 - e^{-\mu y})\lambda e^{-\lambda(y-x)}\,dy & (x \geqslant 0) \end{cases}$$

$$= \begin{cases} \dfrac{\mu e^{\lambda x}}{\lambda + \mu} & (x < 0) \\ 1 - \dfrac{\lambda e^{-\mu x}}{\lambda + \mu} & (x \geqslant 0) \end{cases} \tag{6.6}$$

Thus

$$W_q^-(t) = \frac{\lambda \mu}{\lambda + \mu} \int_{-\infty}^{t} W_q(t - x)e^{\lambda x}\,dx \qquad (t < 0).$$

Letting $u = t - x$ yields

$$W_q^-(t) = \frac{\lambda \mu}{\lambda + \mu} \int_0^\infty W_q(u)e^{-\lambda(u-t)}\,du$$

$$= \frac{\lambda \mu e^{\lambda t}}{\lambda + \mu} \int_0^\infty W_q(u)e^{-\lambda u}\,du$$

$$= \frac{\lambda \mu e^{\lambda t}\overline{W}_q(\lambda)}{\lambda + \mu}.$$

Now $\overline{W}_q(\lambda)$ may be easily found from some of our earlier work. In Section 5.2.1 we showed that

$$\pi_n^q = \Pr\{\,n \text{ in } \textit{queue} \text{ just after a departure}\,\}$$

$$= \frac{1}{n!} \int_0^\infty (\lambda t)^n e^{-\lambda t}\,dW_q(t)$$

for any $M/G/c$. Hence if $G = M$ and $c = 1$, we find that

$$\pi_0^q = \int_0^\infty e^{-\lambda t}\,dW_q(t).$$

Integration by parts then gives

$$\pi_0^q = e^{-\lambda t} W_q(t)\Big|_0^\infty + \lambda \int_0^\infty W_q(t) e^{-\lambda t}\, dt.$$

But $\lim_{t\to\infty} e^{-\lambda t} = 0$, and since we are only concerned computationally with $W_q(t)$ for $t>0$, let us make $W_q(0)=0$ for the time being to simplify the analysis henceforth. In the end, we shall simply set $W_q(0)=p_0$, since it is true for all $M/G/1$ (which always satisfy Little's formula) that $W_q(0)=p_0$ $= 1-\lambda/\mu$. Therefore

$$\pi_0^q = \lambda \overline{W}_q(\lambda).$$

We also know that π_0^q must be equal to $p_0 + p_1 = (1-\rho)(1+\rho)$, since this is an $M/M/1$. Hence

$$\overline{W}_q(\lambda) = \frac{(1-\rho)(1+\rho)}{\lambda},$$

and, from (6.5),

$$W_q^-(t) = \frac{e^{\lambda t}(1-\rho)(1+\rho)}{1+\rho}$$

$$= e^{\lambda t}(1-\rho),$$

with

$$\overline{W}_q^-(s) = \frac{1-\rho}{\lambda - s}.$$

Putting everything together we find, using (6.5), that

$$\overline{W}_q(s) = \frac{(1-\rho)/(\lambda - s)}{\lambda\mu/[(\lambda - s)(\mu + s)] - 1}$$

$$= \frac{(1-\rho)(\mu + s)}{s(\mu - \lambda + s)}$$

$$= \frac{1-\rho}{s} + \frac{\lambda(1-\rho)}{s(\mu - \lambda + s)},$$

which, from Table A3.1, inverts to

$$W_q(t) = 1 - \rho + \frac{\lambda(1-\rho)[1 - e^{-(\mu-\lambda)t}]}{\mu - \lambda}$$

$$= 1 - \rho e^{-\mu(1-\rho)t} \qquad (t > 0).$$

Now realizing that $W_q(0)$ actually equals $p_0 = 1 - \rho$, the result is the same as obtained previously in Chapter 2.

We illustrate the use of Lindley's equation by the following example:

Example 6.1.

Our friend the barber, H. R. Cutt, has decided that he would like to find the distribution of the line waits his customers undergo. He realizes that under the new priority rules he recently designated, service times may be either of two possibilities: exponential with mean 5 min (for the trims, used one-third of the time), and exponential with mean 12.5 min (for the others). The arrival stream remains Poisson with parameter $\lambda = 5$/hr. If we assume that there are no priorities, then Cutt's system is an $M/G/1$ queue, with service times given to be the mixed exponential, that is, the weighted average of two exponentials,[4]

$$B(t) = 1 - \left[\frac{1}{3} e^{-t/5} + \frac{2}{3} e^{-2t/25} \right],$$

with

$$b(t) = \frac{1}{15} e^{-t/5} + \frac{4}{75} e^{-2t/25}$$

and

$$B^*(s) = \frac{1/15}{s+1/5} + \frac{4/75}{s+2/25}$$

$$= \frac{1}{15s+3} + \frac{4}{75s+6}.$$

---

[4] Because the standard deviation of a mixture of exponentials exceeds the mean, it is referred to as hyper-exponential. Limited results are available for Poisson input, mixed-exponential service and mixed-exponential interarrival times, exponential service [see Morse (1958)].

So, from (6.3),

$$W_q^-(t) = \int_{-\infty}^t W_q(t-x)\,dU(x) \qquad (t<0),$$

where, from (6.2),

$$U(x) = \begin{cases} \displaystyle \int_0^\infty \left(1 - \frac{e^{-y/5} + 2e^{-2y/25}}{3}\right) \frac{e^{-(y-x)/12}}{12}\,dy & (x<0) \\[4mm] \displaystyle \int_x^\infty \left(1 - \frac{e^{-y/5} + 2e^{-2y/25}}{3}\right) \frac{e^{-(y-x)/12}}{12}\,dy & (x\geqslant 0) \end{cases}$$

$$= \begin{cases} \displaystyle e^{x/12}\left(1 - \frac{1/36}{1/12+1/5} - \frac{1/18}{1/12+2/25}\right) & (x<0) \\[4mm] \displaystyle 1 - \left[\frac{(1/36)e^{-x/5}}{1/12+1/5} + \frac{(1/18)e^{-2x/25}}{1/12+2/25}\right] & (x\geqslant 0) \end{cases}.$$

Thus

$$W_q^-(t) = \frac{1}{12}\left(\frac{468}{833}\right)\int_{-\infty}^t W_q(t-x)e^{x/12}\,dx \qquad (t<0)$$

$$= \frac{39}{833}e^{t/12}\overline{W}_q(\tfrac{1}{12}).$$

We must therefore next find $\overline{W}_q(\lambda)$, $\lambda = \tfrac{1}{12}$, for this model. Since the result we quoted earlier in this section from Section 5.2.1 regarding π_n^q was valid for any $M/G/1$, it is certainly true in this current case. Hence again,

$$\overline{W}_q(\lambda) = \frac{\pi_0^q}{\lambda}.$$

But here π_0^q will clearly be different from its value for $M/M/1$. We know that $\pi_0^q = \pi_0 + \pi_1$, where π_0 and π_1 refer to the departure-point probabilities of 0 and 1 in the system, respectively. Recalling the analysis of Chapter 5, we have $\pi_0 = 1-\rho$, and, from Equation (5.16), $\pi_1 = \pi_0(1-k_0)/k_0$. But

$$k_0 = \int_0^\infty e^{-\lambda t}\,dB(t)$$

$$= \int_0^\infty e^{-(t/12)}\left[\frac{e^{-(t/5)}}{15} + \frac{4e^{-(2t/25)}}{75}\right]dt$$

$$= \frac{468}{833}.$$

Since

$$\rho = \frac{1}{12}\left[\frac{1}{3}(5) + \frac{2}{3}\left(\frac{25}{2}\right)\right] = \frac{5}{6},$$

we get

$$\pi_0 = \frac{1}{6}$$

and

$$\pi_1 = \frac{1}{6}\left(\frac{365}{468}\right) = \frac{365}{2808}.$$

So

$$\pi_0^q \doteq 0.297$$

and

$$\overline{W}_q(\lambda) \doteq \frac{0.297}{1/12} = 3.564.$$

Therefore

$$W_q^-(t) = \frac{39}{833}(3.564)e^{t/12},$$

and thus

$$\overline{W}_q^-(s) \doteq \frac{(39/833)(3.564)}{1/12 - s} \doteq \frac{2.00}{1 - 12s}.$$

Since $A^*(-s) = 1/(1 - 12s)$, we now have, from (6.5), that

$$\overline{W}_q(s) \doteq \frac{(2.00)/(1 - 12s)}{[1/(1 - 12s)][1/(3 + 15s) + 4/(6 + 75s)] - 1}$$

$$= \frac{2.00(3 + 15s)(6 + 75s)}{6 + 75s + 12 + 60s - (1 - 12s)(3 + 15s)(6 + 75s)}$$

$$= \frac{2.00(18 + 315s + 1125s^2)}{36s + 2655s^2 + 13,500s^3}$$

$$\doteq \frac{1}{s} - \frac{0.010}{s + 0.18} - \frac{0.82}{s + 0.015}.$$

Therefore inversion of the LT yields

$$W_q(t) \doteq 1 - 0.01e^{-0.18t} - 0.82e^{-0.015t}.$$

As a partial check, $W_q(0)$ should equal $p_0 = \pi_0 = 1 - \rho = 1/6$. From the above, $W_q(0) \doteq 1 - 0.01 - 0.82 = 0.17$, which is the two-decimal approximation of $1/6$.

From the foregoing, it should be readily apparent that actual results for waiting times using Lindley's equation when interarrival and service times are continuous are very difficult to obtain. This is so primarily because of the complexity in obtaining $W_q^-(t)$. However, if the interarrival and service times are discrete distributions (recall that any continuous distribution can be approximated by a k-point distribution), then Equations (6.1a) or (6.1b) can be used iteratively (since their right-hand sides become sums and we know that $\Sigma W_q(t_i) = 1$) to obtain the values of $W_q(t)$ at all realizable values of the discrete random variable t.

This whole $GI/G/1$ problem can be greatly simplified if it can be assumed that the interarrival and service distributions can be expressed as a convolution of independent and not-necessarily-identical exponential random variables.[4a] This is not a particularly strong restriction since it can be shown that any CDF can be approximated to almost any degree of accuracy by such a convolution using an argument similar to that used to show the completeness of the polynomials in function space.

So we may write that

$$A^*(s) = \prod_{i=1}^{m} \frac{\lambda_i}{\lambda_i + s}$$

and

$$B^*(s) = \prod_{i=1}^{n} \frac{\mu_i}{\mu_i + s}.$$

Thus from (6.5),

$$\overline{W}_q(s) = \frac{\overline{W}_q^-(s)}{\prod_{i=1}^{m} [\lambda_i/(\lambda_i - s)] \prod_{i=1}^{n} [\mu_i/(\mu_i + s)] - 1}$$

$$= \frac{\overline{W}_q^-(s) \prod_{i=1}^{m} (\lambda_i - s) \prod_{i=1}^{n} (\mu_i + s)}{\prod_{i=1}^{m} \lambda_i \prod_{i=1}^{n} \mu_i - \prod_{i=1}^{m} (\lambda_i - s) \prod_{i=1}^{n} (\mu_i + s)}.$$

The denominator of $\overline{W}_q(s)$ is clearly a polynomial of degree $m + n \equiv k$ whose roots shall be denoted by $s_1, \ldots, s_k$, where s_1 is easily seen to be 0.

[4a]These results therefore apply to $E_k/E_l/1$.

Furthermore it can be shown from the form of the polynomial and Rouché's theorem that there are exactly $m-1$ roots, $s_2,\ldots,s_m$, whose real parts are positive and thus n, $s_{m+1},\ldots,s_k$, with negative real parts. Thus we must be able to write

$$\prod_{i=1}^{m} \lambda_i \prod_{i=1}^{n} \mu_i - \prod_{i=1}^{m} (\lambda_i - s) \prod_{i=1}^{n} (\mu_i + s)$$

$$= C_1 s(s - s_2) \cdots (s - s_m)(s - s_{m+1}) \cdots (s - s_k),$$

where C_1 is a constant. Hence letting $z_i = s_{m+i}$

$$\overline{W}_q(s) = \frac{\overline{W}_q^-(s) \prod_{i=1}^{m} (\lambda_i - s) \prod_{i=1}^{n} (\mu_i + s)}{C_1 s \prod_{i=2}^{m} (s - s_i) \prod_{i=1}^{n} (s - z_i)} .$$

But the numerator must also have the roots $s_2,\ldots,s_m$ to preserve analyticity for $\mathrm{Re}(s) > 0$ and hence may be rewritten as $C_2 f(s)\prod_{i=2}^{m}(s - s_i)$. So now after cancellation

$$\overline{W}_q(s) = \frac{C_3 f(s)}{s \prod_{i=1}^{n} (s - z_i)},$$

where C_3 is the constant equal to C_2/C_1. The final key step is to show that $f(s)$ is itself also a polynomial. This can, in fact, be done again using concepts from the theory of complex variables. The final form of $\overline{W}_q(s)$ is therefore determined by finding $C_3 f(s)$. It turns out that $f(s)$ cannot have any roots with positive real parts and, in fact, has the roots $\mu_1,\ldots,\mu_n$. Hence $\overline{W}_q(s)$ may be written as

$$\overline{W}_q(s) = \frac{C_3 \prod_{i=1}^{n} (s + \mu_i)}{s \prod_{i=1}^{n} (s - z_i)} .$$

To get C_3 now we note that

$$\mathscr{L}\{ W_q'(t) \} = s\overline{W}_q(s) - W_q(0)$$

$$= C_3 \prod_{i=1}^{n} \frac{s + \mu_i}{s - z_i} - q_0.$$

But the transform of $W_q'(t)$ evaluated at $s=0$ is equal to $1-q_0$ and thus

$$1-q_0 = C_3 \prod_{i=1}^{n} \frac{-\mu_i}{z_i} - q_0$$

and

$$C_3 = \prod_{i=1}^{n} \frac{-z_i}{\mu_i}.$$

So

$$\overline{W}_q(s) = \frac{\prod_{i=1}^{n} [(-z_i/\mu_i)(s+\mu_i)]}{s \prod_{i=1}^{n} (s-z_i)}.$$

The foregoing result is an extreme simplification since it now puts $\overline{W}_q(s)$ into an easily invertible form. A partial fraction expansion is then performed to give a result similar to that for $GI/M/1$, namely,

$$\overline{W}_q(s) = \frac{k_0}{s} + \sum_{i=1}^{n} \frac{k_i}{s-z_i}$$

and

$$W_q(t) = 1 + \sum_{i=1}^{n} k_i e^{z_i t},$$

where the $\{k_i\}$ would be determined in the usual way.

While conceptually, this is a simplification, in practice it is difficult to estimate the parameters λ_i $(i=1,2,\ldots,m)$ and μ_i $(i=1,2,\ldots,n)$ since m and n generally must be large in order for this method to be accurate.

Again, as a check, let us verify these results for $M/M/1$. To do so we need the roots of

$$\lambda\mu - (\lambda-s)(\mu+s) = s^2 - (\lambda-\mu)s$$

with negative real parts. There is clearly one and it is $z_1 = \lambda - \mu$. Hence

$$
\begin{aligned}
\overline{W}_q(s) &= -\frac{[(\lambda - \mu)/\mu][s + \mu]}{s(s - \lambda + \mu)} \\
&= \frac{(1 - \rho)(s + \mu)}{s(s - \lambda + \mu)}
\end{aligned}
$$

which completely agrees with the result obtained earlier.

6.1.1 Busy Period

We would now like to derive an expression for the expected length of the busy period in a $GI/G/1$ queue in terms of the as yet unspecified interarrival and service distributions, $A(t)$ and $B(t)$, respectively. The argument to be used here is very similar to that for $M/G/1$ in Section 5.1.7, except that we find now that the Poisson input assumption is no longer valid thus making the results to follow all approximate.

Let us begin by rewriting Equation (5.42) in a slightly different form, keeping in mind that we are still assuming a memoryless input.

$$
\begin{aligned}
G(x) = E_S[\,\Pr\{ \text{given first service time} = S, \\
\text{busy period generated by all arrivals} \\
\text{occurring during } S \leqslant x - S\}\,],
\end{aligned}
$$

where $G(x)$ is the CDF of the busy period. In Equation (5.42) the integral ranged from 0 to x, but since $S \leqslant x$, the effective range of S is then 0 to x; hence the expectation on S does coincide with the limits of the integral of (5.42). But the foregoing probability statement may be written as

$$
\Pr\{\quad\} = \sum_{n=0}^{\infty} \Pr\{N(S) = n\} G^{(n)}(x - S),
$$

where $N(S)$ is used here to denote the number of arrivals to the queue in $[0, S]$ and $G^{(n)}(x)$ is the n-fold convolution of $G(x)$. We therefore now have

$$
\begin{aligned}
G(x) &= E_S\left[\sum_{n=0}^{\infty} \Pr\{N(S) = n\} G^{(n)}(x - S)\right] \\
&= E_S\{E_N[G^{(N)}(x - S)]\},
\end{aligned} \tag{6.7}
$$

from which we are able to find a nice, neat expression for the expected length of the busy period. When expectations are taken on both sides of (6.7), we find that

$$E[X] = \int_0^\infty x \, dG(x)$$

$$= \int_0^\infty x \, dE_S \{ E_N [G^{(N)}(x - S)] \}$$

$$= E_S \left\{ E_N \left[\int_S^\infty x \, dG^{(N)}(x - S) \right] \right\} \qquad \text{(since } S \leqslant x)$$

$$= E_S \left\{ E_N \left[\int_S^\infty (x - S) \, dG^{(N)}(x - S) + S \int_S^\infty dG^{(N)}(x - S) \right] \right\}.$$

Because $\int_S^\infty (x - S) dG^{(N)}(x - S)$ is the expected value of an N-fold convolution, which equals N times the expectation of the original random variable, we can write

$$E[X] = E_S \{ E_N [E[X]N + S] \}$$

$$= E_S \{ E[X]m(S) + S \}$$

since the expected number of occurrences in $[0, S]$ is simply the arrival process mean-value function, say $m(t)$, evaluated at $t = S$. So, finally,

$$E[X] = E[X] \int_0^\infty m(t) \, dB(t) + E[S],$$

or

$$E[X] = \frac{E[S]}{1 - \int_0^\infty m(t) \, dB(t)}.$$

As a check let us verify this result for $M/G/1$. There $m(t) = \lambda t$; hence

$$E[X] = \frac{E[S]}{1 - \lambda \int_0^\infty t \, dB(t)}$$

$$= \frac{1/\mu}{1 - \lambda/\mu}$$

$$= \frac{1}{\mu - \lambda},$$

which checks with Section 5.1.7.

If we had desired the Laplace-Stieltjes transform $G^*(s)$ of $G(x)$, then

$$G^*(s) = \int_0^\infty \int_0^x \sum_{n=0}^\infty e^{-xs} p_n(t) G^{(n)}(x - t) \, dB(t) \, dx,$$

where $p_n(t) \equiv \Pr\{N(t) = n\}$. Now by changing the order of integration it is found that

$$G^*(s) = \int_0^\infty \sum_{n=0}^\infty p_n(t) \int_t^\infty e^{-xs} G^{(n)}(x - t) \, dx \, dB(t).$$

Using the convolution property of the transform we finally obtain

$$G^*(s) = \int_0^\infty \sum_{n=0}^\infty p_n(t) e^{-st} [G^*(s)]^n \, dB(t).$$

If it is possible to specify $p_n(t)$, then we should (theoretically) be able to obtain this approximate $G^*(s)$.

6.2 MULTIPLE-CHANNEL QUEUES WITH POISSON INPUT AND CONSTANT SERVICE ($M/D/c$)

In the event that an $M/G/c$ is found to have deterministic service, there is sufficient special structure available to permit the obtaining of the stationary probability generating function for the distribution of the queue size, something which is not possible generally for $M/G/c$. The approach we use is fairly typical, and a similar one may be found in Saaty (1961) which is essentially originally due to Crommelin (1932).

We begin by reordering time so that the constant service time [say b $(=1/\mu)$] is now the basic unit of time; then λ becomes λ/b and μ becomes 1. For purposes of ease of notation, let us henceforth use P_n to denote the CDF of the system size in the steady state. We are then able to observe (in the spirit of Problem 5.13) that the queueing process is indeed Markovian and therefore that

$$p_0 = \text{Pr}\{\, c \text{ or less in system at arbitrary instant}$$
$$\text{of time in the steady state}\,\} \cdot$$
$$\text{Pr}\{\, 0 \text{ arrivals in subsequent}$$
$$\text{unit of time}\,\}$$

$$p_1 = \text{Pr}\{\, c \text{ or less in system}\,\} \cdot \text{Pr}\{\, 1 \text{ arrival}\,\}$$
$$+ \text{Pr}\{\, c+1 \text{ in system}\,\} \cdot \text{Pr}\{\, 0 \text{ arrivals}\,\}$$

$$p_2 = \text{Pr}\{\, c \text{ or less in system}\,\} \cdot \text{Pr}\{\, 2 \text{ arrivals}\,\}$$
$$+ \text{Pr}\{\, c+1 \text{ in system}\,\} \cdot \text{Pr}\{\, 1 \text{ arrival}\,\}$$
$$+ \text{Pr}\{\, c+2 \text{ in system}\,\} \cdot \text{Pr}\{\, 0 \text{ arrivals}\,\}$$

$$\vdots$$

$$p_n = \text{Pr}\{\, c \text{ or less in system}\,\} \cdot \text{Pr}\{\, n \text{ arrivals}\,\}$$
$$+ \text{Pr}\{\, c+1 \text{ in system}\,\} \cdot \text{Pr}\{\, n-1 \text{ arrivals}\,\}$$
$$+ \cdots + \text{Pr}\{\, c+n \text{ in system}\,\} \cdot \text{Pr}\{\, 0 \text{ arrivals}\,\},$$

or

$$\begin{cases} p_0 = P_c e^{-\lambda} \\[2mm] p_1 = P_c \lambda e^{-\lambda} + p_{c+1} e^{-\lambda} \\[2mm] p_2 = \dfrac{P_c \lambda^2 e^{-\lambda}}{2} + p_{c+1} \lambda e^{-\lambda} + p_{c+2} e^{-\lambda} \\[2mm] \vdots \\[2mm] p_n = \dfrac{P_c \lambda^n e^{-\lambda}}{n!} + \dfrac{p_{c+1} \lambda^{n-1} e^{-\lambda}}{(n-1)!} + \cdots + p_{c+n} e^{-\lambda} \end{cases} \qquad (6.8)$$

When we define the usual generating function $P(z) \equiv \sum_{n=0}^{\infty} P_n z^n$, Equation

(6.8) leads, after multiplying the ith row by z^i and then summing over all rows, to (see Problem 6.9)

$$P(z) = \frac{\sum_{n=0}^{c} p_n z^n - P_c z^c}{1 - z^c e^{\lambda(1-z)}}. \tag{6.9}$$

Now to get rid of the $c+1$ unknown probabilities in the numerator of (6.9), we invoke the usual arguments, employing Rouché's theorem and the fact that $P(1)=1$. Since $P(z)$ is analytic and bounded within the unit circle, all the zeros of the denominator within and on the unit circle must also make the numerator vanish. Rouché's theorem will tell us that there are $c-1$ zeros of the numerator inside $|z| \leqslant 1$, while the cth is clearly $z=1$. Since the numerator is a polynomial of degree c, it may be written as

$$N(z) = K(z-1)(z-z_1) \cdots (z-z_{c-1}),$$

where $1, z_1, z_2, \ldots, z_{c-1}$ are the coincident roots of the denominator and numerator. To get K, we use the fact that $P(1)=1$. By L'Hôpital's rule,

$$1 = \lim_{z \to 1} P(z) = \frac{K(1-z_1) \cdots (1-z_{c-1})}{\lambda - c},$$

or

$$K = \frac{\lambda - c}{(1-z_1) \cdots (1-z_{c-1})}.$$

Hence

$$P(z) = \frac{\lambda - c}{(1-z_1) \cdots (1-z_{c-1})} \frac{(z-1)(z-z_1) \cdots (z-z_{c-1})}{1 - z^c e^{\lambda(1-z)}}. \tag{6.10}$$

The $\{p_n\}$ may now be theoretically obtained by the successive differentiation of $P(z)$, though this frankly gets to be rather difficult as c gets large, witness our experience with $M/D/1$ in Chapter 5. In addition, all moments of the queue size may be obtained from (6.10), which, in turn, enables us to find the moments of the line wait via the modification of Little's formula given in Section 5.2.

Various other measures of effectiveness may also be obtained for this model, though we shall not pursue it any further. For these the reader is referred to Saaty and/or Crommelin (op. cit.), where one may find such computations as the probability of no delay and the waiting-time distribution, as well as some approximate methods for computing measures of effectiveness which are useful when extreme accuracy is not required.

6.3   SEMI-MARKOV PROCESSES IN QUEUEING

In this section we treat a discrete-valued stochastic process which transits from state to state according to a Markov chain, but in which the time required to make each transition is a random variable which is a function of both the *to* and *from* states. Let us henceforth refer notationally to this semi-Markov process (SMP) as $\{X(t)|X(t)=\text{state of the process at time } t \geqslant 0\}$, with the state space to be of course the nonnegative integers. There are numerous references for semi-Markov processes that the authors could suggest as further reading, two of which are Pyke (1961) and Ross (1970).

For purposes of notation we shall define $Q_{ij}(t)$ to be the joint conditional probability that given we begin in state i the next transition will be to state j in an amount of time less than or equal to t. Then it should be clear that the (imbedded) Markov chain which underlies the SMP has transition probilities which are given by

$$p_{ij} = Q_{ij}(\infty).$$

Thus the conditional transition-time distribution function for the time to transit from i to j given i and j specified can be found from the definition of conditional probability to be

$$F_{ij}(t) = \frac{Q_{ij}(t)}{p_{ij}} \tag{6.11}$$

whenever $p_{ij} > 0$, and could be arbitrary otherwise. In addition, we shall define the joint marginal distribution of i and t, namely, $\sum_{j=0}^{\infty} Q_{ij}(t)$, as $G_i(t)$, which is thus the CDF of the time to the next transition given we started at state i. Let us also use $\{X_n|n=0,1,2,\dots\}$ to denote the imbedded Markov chain, $\{R_i(t)|t>0\}$ to denote the number of transitions into state i occurring in $(0,t]$, and $\{R(t)\}$ to denote the vector whose ith component is $R_i(t)$.

The stochastic process $\{R(t)\}$ is called a Markov renewal process (MRP), and though arising from the same probabilistic situation as the semi-Markov process, the MRP should be carefully distinguished from the SMP. The MRP is a counting process which keeps track of the total number of visits to each state, while the SMP only records at each point in time the single state at which the process finds itself. But for all intents and purposes, the one always determines the other.

To illustrate these concepts, let us display three well-known examples. The first of these is the $M/M/1/\infty$ queue, which we have seen to be a Markov chain. Here the queueing process $\{N(t)=\text{number in system at}$

time t} is clearly also semi-Markovian, with

$$Q_{01}(t) = 1 - e^{-\lambda t}$$

$$Q_{i,i-1}(t) = \Pr\{\text{one transition (service or arrival) by } t\} \cdot$$
$$\Pr\{\text{transition is a service}\}$$

$$= \frac{\mu[1 - e^{-(\lambda+\mu)t}]}{\lambda+\mu} \qquad (i \geqslant 1)$$

$$Q_{i,i+1}(t) = \Pr\{\text{one transition by } t\} \cdot$$
$$\Pr\{\text{transition is an arrival}\}$$

$$= \frac{\lambda[1 - e^{-(\lambda+\mu)t}]}{\lambda+\mu} \qquad (i \geqslant 1).$$

So

$$\begin{cases} p_{01} = 1 \\[2mm] p_{i,i-1} = \dfrac{\mu}{\lambda+\mu} \qquad (i \geqslant 1) \\[2mm] p_{i,i+1} = \dfrac{\lambda}{\lambda+\mu} \qquad (i \geqslant 1) \end{cases}$$

and

$$\begin{cases} F_{01}(t) = 1 - e^{-\lambda t} \\[2mm] F_{i,i-1}(t) = 1 - e^{-(\lambda+\mu)t} \\[2mm] \qquad\quad = F_{i,i+1}(t) \qquad (i \geqslant 1) \end{cases}$$

In fact, it should also be clear to us that any Markov chain, whether in continuous time or not, is indeed an SMP. The Markov chain in discrete time can be interpreted as an SMP whose transition times are the same constant, while the chain in continuous time has transition times which must all be exponential to maintain the memoryless property. So a simple second example of an SMP in queueing would be any birth–death model, such as the $M/M/c/\infty$ queue. There

$$\begin{cases} F_{01}(t) = 1 - e^{-\lambda t} \\[2mm] F_{i,i-1}(t) = \begin{cases} 1 - e^{-(\lambda+i\mu)t} & (1 \leqslant i \leqslant c) \\ 1 - e^{-(\lambda+c\mu)t} & (c < i < \infty) \end{cases} \\[4mm] \qquad\quad = F_{i,i+1}(t) \end{cases}$$

For a third example of the appearance of an SMP in queueing we turn to $M/G/1/\infty$. Here the system size is, in fact, neither Markovian nor semi-Markovian. But if we define a new stochastic process $\{X(t)\}$ as the number of customers who were left behind in the system by the most recent departure (this is not to be confused with the number in the system at the instant a customer departs, which is an imbedded Markov chain), then this new process is indeed semi-Markovian, and because it lies within the total general-time process it is often called an imbedded SMP.

The main use of SMPs in queueing is as a means to appropriately quantify systems which are non-Markovian, but which are either semi-Markovian or possess an imbedded SMP. The existence of semi-Markovian structure is quite advantageous and often easily permits the obtaining of some of the fundamental relationships required in queueing. To derive some of the key SMP results, we are going to need the following notation:

$H_{ij}(t) = $ CDF of time until first transition into j beginning from i;

$m_{ij} = $ mean first passage time from i to j

$$= \int_0^\infty t\, dH_{ij}(t), \qquad (m_{ii} \text{ is the mean recurrence time of state } i);$$

$m_i = $ mean time spent in state i during each visit

$$= \int_0^\infty t\, dG_i(t);$$

$\eta_{ij} = $ mean time spent in state i before going to j

$$= \int_0^\infty t\, dF_{ij}(t).$$

From the definition of $G_i(t)$ given previously and Equation (6.11) we are able to immediately deduce the intuitive result that

$$m_i = \sum_{j=0}^\infty p_{ij}\eta_{ij}.$$

The key results that we desire are those which determine the limiting probabilities of the SMP and then relate these to the general-time queueing process in the event that the SMP is imbedded and does not possess the same distribution as the general process. These relationships are all fairly intuitive but would of course require proof. A reference for the direct SMP results is Ross (op. cit.), while Fabens (1961) gives a presentation of the relationship between the imbedded SMP and the general-time process. The key points of interest to us are presented as follows.

(i) If state i communicates with state j (see Appendix 4) in the imbedded Markov chain of an SMP, $F_{ij}(t)$ is not lattice,[5] and $m_{ij} < \infty$, then

$$v_j \equiv \text{the steady-state probability of the SMP being}$$
$$\text{in state } j \text{ given it starts in state } i$$

$$= \lim_{t \to \infty} \Pr\{X(t) = j | X(0) = i\}$$

$$= \frac{m_j}{m_{jj}}.$$

(ii) Under the assumption that the underlying Markov chain of an SMP is irreducible, positive recurrent, and aperiodic (see Appendix 4), and that $m_j < \infty$ for all j, if π_j is the stationary probability of j in the imbedded Markov chain, then

$$v_j = \frac{\pi_j m_j}{\sum_{i=0}^{\infty} \pi_i m_i}.$$

(iii) If $\{X(t)\}$ is an aperiodic SMP in continuous time with $m_i < \infty$, then for $\delta(t)$ defined to be the time back to the most recent transition looking from t,

$$\lim_{t \to \infty} \Pr\{\delta(t) \leqslant \delta | X(t) = i\} = \int_0^\delta \frac{1 - G_i(x)}{m_i} dx \equiv U_i(\delta).$$

(iv) The general-time equilibrium probability of n in the system is found to be

$$p_n = \sum_i \left[v_i \int_0^\infty \Pr\{\text{appropriate changes in } t \text{ to bring} \right.$$

$$\left. \text{state from } i \text{ to } n\} dU_i(t) \right] \qquad (n \geqslant 0).$$

---

[5]A random variable is said to be lattice if it is discrete and all its possible outcomes are multiples of one of the values it takes.

We now proceed to use the foregoing results on those models that we have already developed which are or contain semi-Markov processes and for which further analysis was necessary at the time of their introduction. There are three such models, namely, $M^{[X]}/G/1$, $GI/M/1$, and $GI/M/c$. But $GI/M/1$ and $GI/M/c$ are so similar that we restrict ourselves to looking at the first two, with the third put in as a problem.

To obtain the general-time probabilities for an $M^{[X]}/G/1$, it is first necessary to derive the steady-state probabilities of the imbedded semi-Markov process. To do this we begin by noting that the service-time CDF $B(t)$ is also the distribution function of the unconditional waits, T_i, of the SMP in states $i > 0$, with means $m_i = E[T_i] = 1/\mu$. It also follows from the manner in which the regeneration points[6] were defined and from the Poisson input assumption that the unconditional wait in state 0 has CDF

$$B_0(t) = 1 - e^{-\lambda t},$$

with

$$\mu_0 = \frac{1}{\lambda}.$$

Since

$$E[T_0] = \frac{1}{\lambda}$$

and

$$E[T_i] = \frac{1}{\mu} \qquad (i > 0),$$

it then follows from result (ii) that

$$v_0 = \frac{\pi_0/\lambda}{\pi_0/\lambda + (1 - \pi_0)/\mu}$$

and, for $i > 0$, that

$$v_i = \frac{\pi_i/\mu}{\pi_0/\lambda + (1 - \pi_0)/\mu},$$

where $\{\pi_i\}$ are the stationary probabilities of the imbedded chain.

The $\{v_n\}$ can then be nicely related to the $\{p_n\}$ using results (iii) and (iv). We have here that the CDFs of the times back to the most recent

[6]A regeneration point is a point at which the process probabilistically restarts itself.

transition in the steady state are given by

$$U_i(t) = \int_0^t \frac{1 - G_i(x)}{m_i} \, dx$$

$$= \begin{cases} \int_0^t \lambda e^{-\lambda x} \, dx = 1 - e^{-\lambda t} & (i = 0) \\ \int_0^t \mu[1 - B(x)] \, dx & (i > 0) \end{cases}.$$

Therefore since the idle periods of the SMP and the general time process exactly coincide,

$$p_0 = v_0 = \frac{\pi_0/\lambda}{\pi_0/\lambda + (1 - \pi_0)/\mu}$$

and from Result (iv),

$$p_n = \sum_{i=1}^n \left[v_i \int_0^\infty \Pr\{ n - i \text{ arrivals in } t \} \, dU_i(t) \right]$$

$$= \left[\frac{\pi_0}{\lambda} + \frac{1 - \pi_0}{\mu} \right]^{-1} \sum_{i=1}^n \pi_i \int_0^\infty \sum_{k=0}^{n-i} \frac{e^{-\lambda t}(\lambda t)^k}{k!} c_{n-i}^{(k)} [1 - B(t)] \, dt,$$

where $\{c_n^{(k)}\}$ is again the k-fold convolution of the batch-size distribution with itself.

Next, with regard to $GI/M/1$, we recall that the regeneration points are now the arrivals, so that the unconditional wait T_i of the SMP in state i has the same CDF as the interarrival distribution, namely, $A(t)$, with mean $1/\lambda$. Hence it follows that the imbedded SMP has probabilities

$$v_i = \frac{q_i/\lambda}{\sum_{i=0}^\infty q_i/\lambda} \qquad (i \geqslant 0)$$

$$= q_i = (1 - r_0)r_0^i \quad,$$

which is expected since the interoccurrence times between the regeneration points of the SMP are independent and identically distributed.

Now the times back to the latest transition have CDFs $\{U_i(t)\}$ which must be

$$U_i(t) = \lambda \int_0^t [1 - A(x)] \, dx \qquad \text{(for all } i\text{)}$$

$$\equiv U(t).$$

So, for $n > 0$,

$$p_n = \sum_{i=n-1}^{\infty} \left\{ v_i \int_0^{\infty} \Pr\{ i-n+1 \text{ departures in } t \} \lambda [1 - A(t)] \, dt \right\},$$

with

$$p_0 = \sum_{i=0}^{\infty} \left\{ v_i \int_0^{\infty} \Pr\{ \text{at least } i+1 \text{ departures in } t \} \lambda [1 - A(t)] \, dt \right\}.$$

Therefore

$$p_n = \lambda \sum_{i=n-1}^{\infty} (1-r_0) r_0^i \int_0^{\infty} \frac{e^{-\mu t} (\mu t)^{i-n+1}}{(i-n+1)!} [1 - A(t)] \, dt.$$

Letting $j = i - n + 1$ we have

$$p_n = \lambda (1-r_0) r_0^{n-1} \int_0^{\infty} \left\{ e^{-\mu t} [1 - A(t)] \sum_{j=0}^{\infty} \frac{(r_0 \mu t)^j}{j!} \right\} dt$$

$$= \lambda (1-r_0) r_0^{n-1} \int_0^{\infty} e^{-\mu t(1-r_0)} [1 - A(t)] \, dt.$$

Now integration by parts yields

$$p_n = \frac{\lambda}{\mu} r_0^{n-1} \left[1 - \int_0^{\infty} e^{-\mu t(1-r_0)} \, dA(t) \right].$$

But, from Equation (5.61) and Problem 5.27,

$$\int_0^{\infty} e^{\mu t(1-r_0)} \, dA(t) = \beta(r_0)$$

$$= r_0 ;$$

hence for $n > 0$,

$$p_n = \frac{\lambda}{\mu} r_0^{n-1} (1 - r_0).$$

Returning now to p_0, we have

$$p_0 = \lambda \sum_{i=0}^{\infty} (1 - r_0) r_0^i \int_0^{\infty} \int_0^t \frac{\mu e^{-\mu x} (\mu x)^i}{i!} [1 - A(t)] \, dx \, dt$$

[since the sum of exponentials is Erlang]

$$= (1 - r_0) \lambda \mu \int_0^{\infty} \int_0^t e^{-\mu x (1 - r_0)} [1 - A(t)] \, dx \, dt$$

$$= \lambda \int_0^{\infty} [1 - e^{-\mu t (1 - r_0)}] [1 - A(t)] \, dt$$

$$= \lambda \left\{ \int_0^{\infty} [1 - A(t)] \, dt - \frac{1 - \int_0^{\infty} e^{-\mu t (1 - r_0)} dA(t)}{\mu (1 - r_0)} \right\}$$

$$= \lambda \left[\frac{1}{\lambda} - \frac{1 - r_0}{\mu (1 - r_0)} \right]$$

$$= 1 - \frac{\lambda}{\mu}$$

$$= 1 - \rho \qquad (\rho \equiv \lambda / \mu),$$

as we should have expected.

6.4   OTHER QUEUE DISCIPLINES

As was mentioned early in Chapter 1, there are many possible approaches to the selection from the queue of customers to be served, and certainly first come, first served is not the only choice available. We have already considered some priority models in Chapter 4, and mentioned in Chapter 1 at least two other important possibilities, namely, random selection for service (SIRO) and last come, first served (LIFO). We might even consider a third to be general discipline (GD), that is, no particular pattern specified. We shall obtain results in this section for some of our earlier FIFO models under the first two of these three possible variations. To do any more would be quite time consuming, and it would not really be in the best intetest of the reader to be bogged down in such detail.

We should note that in case of the modification of discipline from FIFO to another, the system state probabilities do not change, since discipline considerations never enter their calculation. In addition, the proof of Little's formula remains unchanged, and since the average system size is unaltered, the average waiting time is likewise. But there will indeed be changes in the waiting-time distribution, and, of course, this implies that any higher moment generalization of Little's formula is, in general, not going to be applicable any further. For the sake of illustration now we shall derive the waiting-time distribution for the $M/M/c$ under the SIRO discipline and for the $M/G/1$ under LIFO, two results which are thought to be quite typical.

We begin with a discussion of the waiting times for the $M/M/c$ when service is in random order. In the usual way let us define $W_q(t)$ as the CDF of the line delay, and then it may be written that

$$W_q(t) = 1 - \sum_{j=0}^{\infty} p_{c+j} \tilde{W}_q(t|j) \qquad (t \geqslant 0),$$

where $\tilde{W}_q(t|j)$ now represents the probability that the delay undergone by an arbitrary arrival who joined when $c+j$ were in the system is more than t. But, from the results of Chapter 3, this may also be written as

$$W_q(t) = 1 - p_c \sum_{j=0}^{\infty} \rho^j \tilde{W}_q(t|j).$$

To calculate $\tilde{W}_q(t|j)$, we observe that the waiting times depend not only on the number of customers found in line by an arbitrary arrival, but also on the number who arrive afterward. We shall then derive a differential-difference equation in $\tilde{W}_q(t|j)$ by considering this Markov process over the time intervals $(0, \Delta t)$ and $(\Delta t, t + \Delta t)$, and evaluating the appropriate Chapman-Kolmogorov equation. There are three possible ways to have a waiting time greater than $t + \Delta t$, given $c+j$ were in the system upon arrival: (1) the interval $(\Delta t, t + \Delta t)$ passes without any change; (2) there is another arrival in $(0, \Delta t)$ [thus bringing the system size, not counting the first arrival, up to $c+j+1$] and then the remaining waiting time is greater than t, given $c+j+1$ were in the system at the instant of arrival; and (3) there is a service completion in $(0, \Delta t)$ [thus leaving $c+j-1$ of the originals], the subject customer is not the one selected for service, and the line wait now exceeds t, given $c+j-1$ were in the system. Thus

$$\tilde{W}_q(t+\Delta t|j) = [1 - (\lambda + c\mu)\Delta t]\tilde{W}_q(t|j) + \lambda\Delta t\tilde{W}_q(t|j+1)$$

$$+ \frac{j}{j+1}c\mu\Delta t\tilde{W}_q(t|j-1) + o(\Delta t) \qquad \left(j \geqslant 0, \tilde{W}_q(t|-1) \equiv 0\right). \quad (6.12)$$

The usual algebra on Equation (6.12) leads to

$$\frac{d\tilde{W}_q(t|j)}{dt} = -(\lambda + c\mu)\tilde{W}_q(t|j) + \lambda\tilde{W}_q(t|j+1)$$

$$+ \frac{j}{j+1}c\mu\tilde{W}_q(t|j-1) \qquad (j \geqslant 0) \qquad (6.13)$$

with $\tilde{W}_q(t|-1) \equiv 0$ and $\tilde{W}_q(0|j) = 1$ for all j. Equation (6.13) is somewhat complicated, but can be examined using the following approach. Assume that $\tilde{W}_q(t|j)$ has the Maclaurin series representation

$$\tilde{W}_q(t|j) = \sum_{n=0}^{\infty} \frac{\tilde{W}_q^{(n)}(0|j)t^n}{n!} \qquad (j = 0, 1, \dots),$$

with $\tilde{W}_q^{(0)}(0|j) \equiv 1$. We thus get

$$W_q(t) = 1 - p_c \sum_{j=0}^{\infty} \rho^j \sum_{n=0}^{\infty} \frac{\tilde{W}_q^{(n)}(0|j)t^n}{n!}. \qquad (6.14)$$

Then the derivatives can be directly determined by the successive differentiation of the original recurrence relation given by Equation (6.13). For example,

$$\tilde{W}_q^{(1)}(0|j) = -(\lambda + c\mu)\tilde{W}_q^{(0)}(0|j) + \lambda\tilde{W}_q^{(0)}(0|j+1) + \frac{j}{j+1}c\mu\tilde{W}_q^{(0)}(0|j-1)$$

$$= -(\lambda + c\mu) + \lambda + \frac{jc\mu}{j+1}$$

$$= \frac{-c\mu}{j+1},$$

$$\tilde{W}_q^{(2)}(0|j) = -(\lambda + c\mu)\tilde{W}_q^{(1)}(0|j) + \lambda\tilde{W}_q^{(1)}(0|j+1) + \frac{j}{j+1}c\mu\tilde{W}_q^{(1)}(0|j-1)$$

$$= \frac{(\lambda + c\mu)c\mu}{j+1} - \frac{\lambda c\mu}{j+2} - \frac{(c\mu)^2}{j+1},$$

and so on. Putting everything together into (6.14) gives a final series representation for $W_q(t)$. The ordinary moments of the line delay may be found by the appropriate manipulation of $W_q(t)$ and its complement [this manipulation is well-detailed by Parzen (1960)].

For the second and last model of this section, let us now consider $M/G/1/\infty/\text{LIFO}$. In this case, the waiting time of an arriving customer who comes when the server is busy is the sum of the duration of time from his instant of arrival, T_A, to the first subsequent service completion, at T_S, and the length of the total busy period generated by the $n \geqslant 0$ other customers who arrive in (T_A, T_S).

To find $W_q(t)$, let us first consider the joint probability that n customers arrive in (T_A, T_S) and $T_S - T_A \leqslant x$. Then, for $R(t)$ equal to the CDF of the remaining service time (see Section 5.1.6),

$$\Pr\{ n \text{ arrivals in } (T_A, T_S) \text{ and } T_S - T_A \leqslant x \} = \int_0^x \frac{(\lambda t)^n}{n!} e^{-\lambda t} dR(t),$$

$$R(t) = \mu \int_0^t [1 - B(x)] dx.$$

Since $\pi_0 = 1 - \rho$, it is found by an argument similar to that used for the $M/G/1$ busy period that

$$W_q(t) = 1 - \rho + \Pr\{ \text{system busy} \} \cdot$$

$$\sum_n \Pr\{ n \text{ customers arrive in } (T_A, T_S), T_S - T_A \leqslant t - x,$$

$$\text{and total busy period generated by these } n \text{ arrivals is } x \}$$

$$(6.15)$$

because under the last-in, first-out discipline the most recent arrival in the system is the one chosen for service when the server has just had a completion. But the busy-period distribution is exactly the same as that derived in the FIFO case in Section 5.1.7 since the sum total of customer service times is unchanged by the discipline, and it will be denoted henceforth by $G(x)$, with $G^{(n)}(x)$ used as its n-fold convolution. Therefore (6.15) may be rewritten as

$$W_q(t) = 1 - \rho + \rho \sum_{n=0}^{\infty} \int_0^t \int_0^{t-x} \frac{(\lambda u)^n}{n!} e^{-\lambda u} dR(u) dG^{(n)}(x).$$

A change of the order of integration then gives

$$W_q(t) = 1 - \rho + \rho \sum_{n=0}^{\infty} \int_0^t \frac{(\lambda u)^n}{n!} e^{-\lambda u} dR(u) \int_0^{t-u} dG^{(n)}(x)$$

$$= 1 - \rho + \rho \sum_{n=0}^{\infty} \int_0^t \frac{(\lambda u)^n}{n!} e^{-\lambda u} G^{(n)}(t-u) dR(u). \qquad (6.16)$$

Though this form of $W_q(t)$ is adequate for most purposes, it turns out to be possible to still refine the result to become free of $G(t)$. The final version as given in Cooper (1972) is

$$W_q(t) = 1 - \rho + \lambda \sum_{n=1}^{\infty} \int_0^t \frac{(\lambda u)^{n-1}}{n!} e^{-\lambda u} \left[1 - B^{(n)}(u) \right] du \qquad (6.17)$$

where $B^{(n)}(t)$ is the n-fold convolution of the service-time CDF.

There are, of course, numerous other models with varying disciplines but, as mentioned earlier, it is felt that this discussion should suffice for this text. It is to be emphasized, however, that each different discipline must be approached in a unique manner and any interested reader is referred to the literature, one specific reference on this subject again being Cooper (1972).

6.5   HEAVY-TRAFFIC AND NONSTATIONARY QUEUES

It is our intent in this section to indicate some interesting limit results and inequalities for $GI/G/1$ models in which the traffic intensity is just barely less than one ($1 - \epsilon < \rho < 1$) or is equal to or greater than one ($\rho \geq 1$). The former will be said to be saturated or in heavy traffic, while the latter will be described as nonstationary, divergent, or unstable. Of course, any transient results derived earlier in the text are valid for heavy-traffic and divergent queues since no assumptions are made on the size of ρ.

For limit results, the virtual wait, $V(t) =$ wait a customer would undergo if that customer arrives at time t, and the actual wait, $W_n =$ actual waiting time of the nth customer, will be used to construct random sequences which are then shown to stochastically converge. These convergence theorems will then be applied to obtain approximate distributions of the actual and virtual waits in queue of a customer and their averages. In addition, bounds on the average waiting times will be obtained, some of these directly from the limit results. While we shall present some nice theoretical results, the practicality of these is subject to question since the results depend on allowing time to go to ∞. This is unrealistic as it is unlikely that systems with congestion would be permitted to run for very long before remedial action was taken.

If we were to go back to Lindley's approach to waiting times for the $GI/G/1$, or even to the $M/M/1$ case, and try to get results for $\rho \geqslant 1$, we would fail. The convergence theory which exists for stationary queues cannot be directly extended in either case, generally because the relevant quantities get excessively large as time increases, so that a completely new approach is required. This new approach is to appropriately scale and shift the random variable to permit some form of convergence. For example, when we study $W^{(n)}$, the appropriate quantity to consider in the event that $\rho \geqslant 1$ is $(W^{(n)} - an)/(b\sqrt{n})$, where a and b will be suitably chosen constants. It will be shown that the distribution of this new random variable will converge to a normal or a truncated normal distribution for $\rho > 1$ and $\rho = 1$, respectively. Two references for the following discussion are contained in Smith and Wilkinson (1965), specifically, the paper on heavy traffic queues (the $\rho < 1$ case) by J. F. C. Kingman and the paper on divergent queues ($\rho \geqslant 1$) by C. R. Heathcote.

We shall begin by first discussing in some detail the heavy-traffic problem in the context of $GI/G/1$. Here the system is completely specified by the sequences of interarrival and service times. If each of these sequences contains independent and identically distributed random variables which are independent of each other, then the main result which will be shown is that the line waiting times in heavy traffic are approximately exponentially distributed with mean

$$W_q^{(H)} = \frac{1}{2} \frac{Var[\text{interarrival times}] + Var[\text{service times}]}{1/\lambda - 1/\mu}. \quad (6.18)$$

Recall from Section 6.1 that we defined the difference of the nth service and interarrival times to be

$$U^{(n)} \equiv S^{(n)} - T^{(n)}.$$

Let us also now define the nth partial sum of the $\{U^{(n)}\}$ to be

$$P^{(n)} = \sum_{i=0}^{n-1} U^{(i)},$$

and let the mean and variance of the IID $\{U^{(n)}\}$ be denoted by $-\alpha$ and β^2, respectively. Then the expectation of $P^{(n)}$ is $-n\alpha$, while its variance is $n\beta^2$. Note since we are in heavy traffic α should be small. Now when the central limit theorem for IID random variables is applied to $\{P^{(n)}\}$, it is found that [6a]

$$Y_n \equiv \frac{P^{(n)} + n\alpha}{\sqrt{n}\ \beta} \overset{df}{\to} N(0, 1).$$

Without going through the details of the proof, suffice it to say that by using the asymptotic normality of Y_n it can be shown that the Laplace-Stieltjes transform of $W_q^{(n)}$ for large n is given approximately by

$$W_q^*(s) \doteq \left(1 + \frac{\beta^2 s}{2\alpha}\right)^{-1}$$

when α/β is small. From this (see Example A3.3 of Appendix 3) we therefore deduce that the line delay is negative exponential with mean

$$\frac{\beta^2}{2\alpha} = -\frac{1}{2}\frac{Var[U]}{E[U]}$$

$$= \frac{1}{2}\frac{Var[\text{service times}] + Var[\text{interarrival times}]}{1/\lambda - 1/\mu}, \qquad (6.19)$$

which is equivalent to Equation (6.18).

The foregoing limiting results are rather loose, and it thus becomes important to have some inequalities from which it can be determined just how close any particular queue is to a heavy-traffic situation for which the above exponential waiting distribution is valid. To do so, we proceed as follows. We have from (6.19) the approximation that

$$W_q^{(H)} \doteq -\frac{1}{2}\frac{Var[U]}{E[U]}.$$

For $M/G/1$, the P-K formula [see Equation (5.11)] gives the exact result as

$$W_q = \frac{\rho^2 + \lambda^2\sigma^2}{2\lambda(1 - \rho)} = \frac{\lambda(1/\mu^2 + \sigma^2)}{2(1 - \lambda/\mu)} = \frac{E[S^2]}{-2E[U]}.$$

Hence for $M/G/1$,

$$W_q - W_q^{(H)} \doteq \frac{E[S^2] - Var[U]}{-2E[U]} = \frac{E[S^2] - Var[S] - Var[T]}{-2E[U]}.$$

[6a]By df is meant "in distribution function."

But $Var[T] = E^2[T]$ for Poisson streams; hence

$$W_q - W_q^{(H)} \doteq \frac{E[S^2] - E[S^2] + E^2[S] - E^2[T]}{-2(E[S] - E[T])} = -\frac{E[S] + E[T]}{2}.$$

So the heavy-traffic approximation would be too large by an amount of the order of one mean interarrival time since the expectations of S and T are arbitrarily close to each other. The magnitude of the error here appears to be fairly typical of arbitrary $GI/G/1$'s, and it has, in fact, been shown by Kingman (1962) that the $W_q^{(H)}$ is always an upper bound for W_q, asymptotically improving as ρ approaches 1.

To appropriately estimate the error in the approximation we need to find a lower bound for W_q, but this problem is somewhat more difficult than that of obtaining the upper bound, and the arguments leading to it are too complex for the treatment intended here. Work has also been done on finding bounds on the variance of the delays and of the CDF itself. But these, too, would draw us into great detail and will therefore not be pursued. The interested reader is referred to the aforementioned references. We can extend the $GI/G/1$ bound given by (6.18) to $GI/G/c$ by noting that the system is essentially always operating as a single-server queue with service rate $c\mu$. Hence it would be fair to expect exponential delays in heavy traffic with

$$W_q^{(H)} \doteq \frac{1}{2} \frac{Var[\text{interarrival times}] + (1/c^2) Var[\text{service times}]}{1/\lambda - 1/c\mu}.$$

We now move to $\rho \geqslant 1$ and a brief study of divergent queues. Once again we make use of

$$W_q^{(n+1)} = \max\left(0, W_q^{(n)} + U^{(n)}\right) \tag{6.20}$$

where

$$U^{(n)} = S^{(n)} - T^{(n)},$$

but we shall now operate under the assumption that $E[U] \geqslant 0$.

First let $1/\mu - 1/\lambda = E[U] > 0$, that is, $\rho > 1$. We would then expect the difference $\Delta W_q^{(n)} \equiv W_q^{(n+1)} - W_q^{(n)}$ to behave like $U^{(n)}$ when n gets large, since by observation of (6.20) we see that it is unlikely that $W_q^{(n+1)}$ will ever be zero again when $\rho > 1$. But we have already stated earlier in this section that $\sum U^{(i)}$ properly normalized as

$$Y_n = \frac{\sum U^{(i)} + n\alpha}{\sqrt{n}\,\beta} \qquad (\alpha < 0)$$

converges in distribution function to the unit normal. Since $W_q^{(0)} = 0$,

$$\sum_{i=0}^{n-1} \Delta W_q^{(i)} = W_q^{(n)} = \sum_{i=0}^{n-1} U^{(i)} = P^{(n)},$$

and hence

$$\Pr\left\{ \frac{W_q^{(n)} + n\alpha}{\sqrt{n}\,\beta} \leqslant x \right\} \to \int_{-\infty}^{x} \frac{e^{-t^2/2}}{\sqrt{2\pi}}\,dt. \tag{6.21}$$

One immediate result of (6.21) is the determination of an estimate for the probability of no wait for n large. This is clearly given by

$$\Pr\left\{ W_q^{(n)} = 0 \right\} = \Pr\left\{ \frac{W_q^{(n)} + n\alpha}{\sqrt{n}\,\beta} = \frac{n\alpha}{\sqrt{n}\,\beta} \right\}$$

$$= \int_{-\infty}^{n\alpha/(\sqrt{n}\,\beta)} \frac{e^{-t^2/2}}{\sqrt{2\pi}}\,dt.$$

We can get an interesting alternative approximation to this probability which does not use the normal distribution by using Chebyshev's inequality, which says that

$$\Pr\left\{ |X - \mu| \geqslant k\sigma \right\} \leqslant \frac{1}{k^2}$$

for any variable X possessing two moments. Applying this result here it is found that

$$\Pr\left\{ W_q^{(n)} = 0 \right\} = \Pr\left\{ X + n\alpha \leqslant n\alpha \right\} = \frac{\Pr\left\{ |X + n\alpha| \geqslant n\alpha \right\}}{2} \leqslant \frac{\beta^2}{2n\alpha^2},$$

which provides a fairly reasonable upper bound for the probability that an arbitrary arrival encounters an idle system. Notice that this number does go to zero as n approaches ∞ as required.

There are some additional limit theorems for the special $GI/G/1$ queue with $GI = M$ and $\rho = 1$. These results shall be stated without proof, and if the reader is interested in further pursuing the matter he is referred to Cohen (1969).

For $V^m(t)$ defined as the maximum of the virtual wait on the interval $(0, t)$, $W_q^{(k)m}$ as the largest value of $W_q^{(n)}$ between $n = 1$ and k, and $X^m(t)$ as the maximum system size on $(0, t)$, we have that $\mu V^m(t)/t$, $\mu W_q^{(k)m}/k$, and

$X^m(t)/t$ all converge in distribution as t (or k) go to ∞ to a random variable with CDF

$$G(x) = \begin{cases} e^{-1/x} & (x > 0) \\ 0 & (x \leqslant 0) \end{cases}. \tag{6.22}$$

Similar results were obtained by Iglehart (1972) for arbitrary $GI/G/1$ queues with both $\rho = 1$ and $\rho > 1$.

Comparable results are available for $GI/M/1$ queues which say that for $\rho = 1$ and T_2, the second moment of the interarrival time, $2V^m(t)/(\lambda T_2)$, $2W_q^{(n)m}/(\lambda n T_2)$, and $2X^m(t)/(\lambda^2 t T_2)$ all converge in distribution function to the $G(x)$ of (6.22).

The final result we wish to state is that the limit of the time-averaged virtual wait at t almost surely (with probability one, often abbreviated by as) goes to $\lambda - 1$ as t goes to ∞, that is,

$$\lim_{t \to \infty} \left[\frac{V(t)}{t} \right] \overset{as}{=} \lambda - 1.$$

6.6 MORE ON BOUNDS AND INEQUALITIES

This section is concerned with a discussion of the use of bounds and inequalities for arbitrary stationary $GI/G/1$ queues, and as such extends some of the work of the previous section. Much of the following development parallels the 1962 paper of Kingman and one by Marshall (1968).

We begin by finding some fairly simple relationships between the moments of the interarrival times, service times, idle periods, and waiting times for an arbitrary $GI/G/1$ queue with $\rho < 1$. The starting point again is the iterative equation for the line delays,

$$W_q^{(n+1)} = \max\left(0, W_q^{(n)} + U^{(n)}\right),$$

$$U^{(n)} \equiv S^{(n)} - T^{(n)}.$$

If we now let $X^{(n)}$ be the negative slack in the event of idle time, that is,

$$X^{(n)} = \min\left(0, W_q^{(n)} + U^{(n)}\right),$$

then[6b]

$$W_q^{(n+1)} + X^{(n)} = W_q^{(n)} + U^{(n)}. \tag{6.23}$$

[6b]Note that $-X^{(n)}$ can be interpreted as virtual idle time.

Since the queue is stationary and $E[W_q^{(n+1)}] = E[W_q^{(n)}]$, when expectations are taken of both sides of (6.23) it is found that

$$0 = E[U] - E[X]$$

or

$$E[U] = E[X].$$

But

$$E[U] = \frac{1}{\mu} - \frac{1}{\lambda}$$

and

$$E[X] = -\Pr\{\text{system found empty by an arrival}\} \cdot E[\text{length of idle period}].$$

If "length of idle period" is henceforth denoted by I and arrival-point probabilities as $\{q_n\}$, then

$$E[I] = -\frac{E[U]}{q_0} = \frac{1/\lambda - 1/\mu}{q_0}. \tag{6.24}$$

The next result we find is a formula for the expected wait for a stable $GI/G/1$ in terms of the first and second moments of U and I. First square both sides of (6.23) to get

$$\left[W_q^{(n+1)}\right]^2 + 2W_q^{(n+1)}X^{(n)} + [X^{(n)}]^2 = \left[W_q^{(n)}\right]^2 + 2W_q^{(n)}U^{(n)} + [U^{(n)}]^2.$$

$$\tag{6.25}$$

We see from the definitions of $W_q^{(n+1)}$ and $X^{(n)}$ that

$$2W_q^{(n+1)}X^{(n)} = 0$$

since one or the other of the two terms must always be zero. Since $W_q^{(n)}$ and $U^{(n)}$ are independent, when expectations are taken of both sides of (6.25) and n goes to ∞, we find that

$$0 = E[U^2] - E[X^2] + 2E[U]W_q,$$

or

$$W_q = \frac{E[X^2] - E[U^2]}{2E[U]}.$$

But

$$E[X^2] = \Pr\{\text{system found empty by an arrival}\} \cdot E\left[(\text{length of idle period})^2\right].$$

Hence

$$W_q = \frac{q_0 E[I^2] - E[U^2]}{2E[U]},$$

and, from Equation (6.24),

$$\boxed{W_q = \frac{-E[I^2]}{2E[I]} - \frac{E[U^2]}{2E[U]}} \qquad (6.26)$$

For Poisson arrivals, $E[I] = 1/\lambda$ and $E[I^2] = 2/\lambda^2$, and (6.26) checks out correctly to give the $P\text{-}K$ expression

$$W_q = \frac{\rho^2 + \lambda^2\sigma^2}{2\lambda(1-\rho)}.$$

A similar expression can then be found for the variance by cubing Equation (6.23). The final result is given by Marshall (op. cit.).

The foregoing results now lead to some bounds valid for all $GI/G/1$ queues. The first is a lower bound on the mean idle time. From Equation (6.24), we see that since $q_0 \leqslant 1$,

$$E[I] \geqslant \frac{1}{\lambda} - \frac{1}{\mu},$$

with the inequality binding for $D/D/1$. Note that we have already obtained an upper bound for the mean delay in Section 6.5 as Equation (6.18), namely,

$$W_q \leqslant \frac{1}{2} \frac{Var[\text{interarrival times}] + Var[\text{service times}]}{1/\lambda - 1/\mu},$$

which can be slightly rewritten as

$$\boxed{W_q \leqslant \frac{\lambda(\sigma_A^2 + \sigma_B^2)}{2(1-\rho)}} \qquad (6.27)$$

If now it is assumed that the interarrival and service distributions are both known and given by $A(t)$ and $B(t)$, respectively, then a lower bound on the stationary line wait may be found to be $W_q \geqslant r_0$, where r_0 is the unique nonnegative root when $\rho < 1$ of

$$f(z) = z - \int_{-z}^{\infty} [1 - U(t)]\,dt = 0,$$

where, as before, $U(t)$ is the CDF of $U = S - T$.

To prove this assertion, we begin by observing that $f(z)$ does indeed have a unique nonnegative root. It is easily seen that $f(z)$ is monotonically increasing for $z \geqslant 0$, since

$$f'(z) = 1 - [1 - U(-z)]$$
$$= U(-z) \geqslant 0.$$

When $z = 0$,

$$f(z) = -\int_0^\infty [1 - U(t)] \, dt < 0$$

since the integrand is always positive, while when z is large, say M,

$$f(z) = M - \int_{-M}^\infty [1 - U(t)] \, dt$$

$$= M - \int_{-M}^\infty \int_t^\infty dU(x) \, dt$$

$$= M - \int_{-M}^\infty \int_{-M}^x dt \, dU(x)$$

$$= M - \int_{-M}^\infty (x + M) \, dU(x)$$

$$\geqslant M - \int_{-\infty}^\infty (x + M) \, dU(x)$$

$$= M - \left(\frac{1}{\mu} - \frac{1}{\lambda}\right) - M$$

$$= -\left(\frac{1}{\mu} - \frac{1}{\lambda}\right)$$

$$= \frac{1 - \rho}{\lambda} > 0.$$

Since $f(z)$ is monotonic and goes from $f(0) < 0$ to $f(M) > 0$, we have thus shown there is a unique nonnegative root when $\rho < 1$, which shall be henceforth known as r_0. It therefore remains for us to show that $W_q \geqslant r_0$.

Now rewrite $f(z)$ as

$$f(z) = z - f_1(z),$$

where

$$f_1(z) = \int_{-z}^{\infty} [1 - U(t)] \, dt.$$

Then we have that

$$f_1(z) = \int_{-z}^{\infty} [1 - U(t)] \, dt \quad \begin{cases} > z & (z < r_0) \\ \leqslant z & (z \geqslant r_0) \end{cases}. \quad (6.28)$$

The function $f_1(z)$ is, in fact, continuous and convex, and we are going to apply Jensen's inequality for the expected value of a convex function of a nonnegative random variable [see Parzen (1960), for example] to get a relationship between W_q and f_1. Given that $W_q^{(n)}$ is (say) x, we see that

$$E[W_q^{(n+1)} | W_q^{(n)} = x] = E[\max(0, x + U^{(n)})]$$

$$= \int_{-x}^{\infty} (x + t) \, dU^{(n)}(t)$$

$$= \int_{-x}^{\infty} t \, dU^{(n)}(t) + x[1 - U^{(n)}(-x)]$$

$$= \int_{-x}^{\infty} \int_{0}^{t} dv \, dU^{(n)}(t) + x[1 - U^{(n)}(-x)]$$

$$= \int_{0}^{\infty} \int_{v}^{\infty} dU^{(n)}(t) \, dv - \int_{-x}^{0} \int_{-x}^{v} dU^{(n)}(t) \, dv + x[1 - U^{(n)}(-x)]$$

$$= \int_{-x}^{\infty} [1 - U^{(n)}(v)] \, dv$$

$$= f_1(x).$$

Hence by the law of total probability,

$$E[W_q^{(n+1)}] = \int_{0}^{\infty} f_1(x) \, dW_q^{(n)}(x).$$

Jensen's inequality then tells us that

$$E[W_q^{(n+1)}] \geqslant f_1(E[W_q^{(n)}]),$$

or in the steady state that

$$W_q \geqslant f_1(W_q) = \int_{-W_q}^{\infty} [1 - U(t)] \, dt. \tag{6.29}$$

We now assume that $W_q < r_0$ and proceed to prove the result by contradiction.

Equation (6.28) says that

$$\int_{-W_q}^{\infty} [1 - U(t)] \, dt > W_q$$

for $W_q < r_0$. But this is a contradiction of (6.29); hence the result is shown.

When the inequalities are finally put together, we get

$$\boxed{r_0 \leqslant W_q \leqslant \frac{\lambda(\sigma_A^2 + \sigma_B^2)}{2(1-\rho)}} \quad .$$

To illustrate, let us look at $M/M/1$ to see what the bounds look like. Going back to Section 6.1, Equation (6.6), we have

$$U(t) = \begin{cases} \dfrac{\mu e^{\lambda t}}{\lambda + \mu} & (t < 0) \\[2mm] 1 - \dfrac{\lambda e^{-\mu t}}{\lambda + \mu} & (t \geqslant 0) \end{cases} \quad .$$

The lower bound r_0 is then found by solving $f(z) = 0$; that is,

$$0 = r_0 - \int_{-r_0}^{\infty} [1 - U(t)] \, dt$$

$$= r_0 - \int_{-r_0}^{0} \left(1 - \frac{\mu e^{\lambda t}}{\lambda + \mu}\right) dt - \int_{0}^{\infty} \frac{\lambda e^{-\mu t}}{\lambda + \mu} \, dt$$

$$= r_0 - r_0 + \frac{\mu(1 - e^{-\lambda r_0})}{\lambda(\lambda + \mu) - \lambda / [\mu(\lambda + \mu)]}$$

$$= \frac{\mu^2 - \lambda^2 - \mu^2 e^{-\lambda r_0}}{\lambda \mu(\lambda + \mu)}$$

$$= \frac{\mu - \lambda}{\lambda \mu} - \frac{\mu e^{-\lambda r_0}}{\lambda(\lambda + \mu)} \, .$$

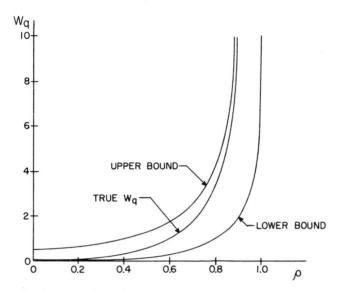

Fig. 6.1   Bounds on the expected wait in the $M/M/1$ queue.

So

$$\mu^2 - \lambda^2 = \mu^2 e^{-\lambda r_0},$$

or

$$1 - \rho^2 = e^{-\lambda r_0}.$$

Hence finally,

$$r_0 = -\frac{1}{\lambda} \ln(1 - \rho^2).$$

The upper bound here is

$$\frac{\lambda(1/\lambda^2 + 1/\mu^2)}{2(1 - \lambda/\mu)} = \frac{1}{\lambda} \frac{1 + \rho^2}{2(1 - \rho)}.$$

Note that both r_0 and the upper bound go to ∞ as ρ goes to 1, as expected. To get a better idea of the behavior of the bounds, we supply a graph in Figure 6.1 of the bounds and the actual mean for all ρ between zero and one.

In his paper, Marshall goes on to get some more results for special classes of possible arrival distributions, but it is not felt necessary to go into them here. The reader can pursue them in the quoted reference.

Example 6.2.

To see how these results might apply to a real problem, let us examine their application to a $GI/G/1$ queue in which both the interarrival times and service times are empirical. Assume that the service times are as given by Table 5.2, namely, $b(9)=2/3$ and $b(12)=1/3$, and that the interarrival-time probabilities are given by $a(10)=2/5$ and $a(15)=3/5$.

A little bit of calculation gives $\sigma_B^2=2$ min^2 and $\sigma_A^2=6$ min^2, with $\mu=1/10$, $\lambda=1/13$, and $\rho=10/13$. So we can immediately calculate the upper bound from (6.27) as

$$W_q \leqslant \frac{(1/13)(8)}{2(3/13)} = \frac{4}{3} \text{ min.}$$

The obtaining of the lower bound is slightly more difficult since we must find the root of the nonlinear equation

$$f(r_0) = 0 = r_0 - \int_{-r_0}^{\infty} [1 - U(t)]\, dt.$$

We begin by first calculating $U(t)$ directly from the empirical forms of the interarrival and service distributions. It is thus found that the only possible values of the random variable $U = S - T$ are $-6 = 9 - 15$ [with probability $(2/3)(3/5)=2/5$], $-3 = 12 - 15$ (with probability $1/5$), $-1 = 9 - 10$ (with probability $4/15$), and 2 (with probability $2/15$). Therefore

$$U(t) = \begin{cases} 0 & (t < -6) \\ \dfrac{2}{5} & (-6 \leqslant t < -3) \\ \dfrac{3}{5} & (-3 \leqslant t < -1) \\ \dfrac{13}{15} & (-1 \leqslant t < 2) \\ 1 & (t \geqslant 2) \end{cases}$$

and

$$1 - U(t) = \begin{cases} 1 & (t < -6) \\ \dfrac{3}{5} & (-6 \leqslant t < -3) \\ \dfrac{2}{5} & (-3 \leqslant t < -1) \\ \dfrac{2}{15} & (-1 \leqslant t < 2) \\ 0 & (t \geqslant 2) \end{cases}.$$

So

$$\int_{-r_0}^{\infty}[1-U(t)]\,dt = \int_{-r_0}^{2}[1-U(t)]\,dt$$

$$= \begin{cases} \dfrac{2r_0}{15}+\dfrac{4}{15} & (0 \leqslant r_0 \leqslant 1) \\[2mm] \dfrac{4r_0}{15}+\dfrac{2}{15} & (1 < r_0 \leqslant 3) \\[2mm] \dfrac{3r_0}{15}+\dfrac{5}{15} & (3 < r_0 \leqslant 6) \\[2mm] \dfrac{6r_0}{15}-\dfrac{13}{15} & (r_0 > 6) \end{cases}.$$

Since the upper bound is just over one, we surmise that the lower bound should probably be less than one; that is to say, it is the solution to

$$r_0 = \frac{2r_0}{15}+\frac{4}{15},$$

or

$$r_0 = \frac{4}{13} \text{ min},$$

which must be correct, since r_0 is the unique nonnegative solution.
Therefore,

$$\frac{4}{13} \leqslant W_q \leqslant \frac{4}{3}.$$

6.7 DIFFUSION APPROXIMATIONS

Since we have already seen in Section 6.5 that heavy-traffic $GI/G/1$ queues have exponential waiting times, it might therefore seem quite logical that these queues are birth–death and of the $M/M/1$ type[7] when $\rho = 1 - \epsilon$, independent of the form of the arrival and service distributions. However, this is not correct since we know from Equation (6.19) that the interarrival and service-time variances must be involved. But we can obtain a fairly simple approximate result for the transient queue distribution of the $M/M/1$ in heavy traffic by using a continuous approximation of the diffusion type. After that we shall use another, but similar, approach

---

[7]Remember that $M/G/1$ queues have exponential system waits if $G = M$, and that $GI/M/1$ queues also have exponential waits if $GI = M$ (see Problem 5.34).

to get a diffusion approximation for the transient distribution of the line delay for the $M/G/1$ in heavy traffic.

The $M/M/1$ diffusion result is found by first considering the elementary random walk approximation of a birth–death process which changes every Δt time units and has transition probabilities given by

$$
\begin{cases}
\Pr\{\text{state goes up by one over "unit" interval}\} = \lambda \Delta t \\
\Pr\{\text{state goes down by one over "unit" interval}\} = \mu \Delta t \quad, \\
\Pr\{\text{no change}\} = 1 - (\lambda + \mu)\Delta t
\end{cases}
$$

assuming for the moment that since the queue is in heavy traffic the walk is unrestricted with no impenetrable barrier at 0.

Now the elementary random walk is clearly a Markov chain since its future behavior is only a probabilistic function of its current position and is independent of past history. More specifically, we may write that the probability $p_k(n)$ of finding the walk in position k after n steps is given by the Chapman-Kolmogorov equation

$$
p_k(n+1) = p_k(n)[1 - (\lambda + \mu)\Delta t] + p_{k+1}(n)\mu\Delta t + p_{k-1}(n)\lambda\Delta t,
$$

which may be rewritten after a bit of algebra as

$$
\frac{p_k(n+1) - p_k(n)}{\Delta t} = \frac{\mu + \lambda}{2}[p_{k+1}(n) - 2p_k(n) + p_{k-1}(n)]
$$

$$
+ \frac{\mu - \lambda}{2}[p_{k+1}(n) - p_k(n)]
$$

$$
+ \frac{\mu - \lambda}{2}[p_k(n) - p_{k-1}(n)]. \tag{6.30}
$$

We then observe that Equation (6.30) has been written in terms of the discrete version of derivatives, the left-hand side with respect to step (time) and the right-hand with respect to the state variable.[8] If we then appropriately take limits of (6.30) so that the time between transitions shrinks to zero, while simultaneously the size of the state steps goes to zero, we find ourselves ending up with a partial differential equation of the diffusion type.

[8]See Appendix 2, Section A2.2.

To be more exact, let the length of a "unit" state change be denoted by θ and the step time by Δt, and then (6.30) leads to

$$\frac{p_{k\theta}(t+\Delta t)-p_{k\theta}(t)}{\Delta t} = \frac{\mu+\lambda}{2}\left[p_{(k+1)\theta}(t)-2p_{k\theta}(t)+p_{(k-1)\theta}(t)\right]$$

$$+\frac{\mu-\lambda}{2}\left[p_{(k+1)\theta}(t)-p_{k\theta}(t)\right]$$

$$+\frac{\mu-\lambda}{2}\left[p_{k\theta}(t)-p_{(k-1)\theta}(t)\right]. \qquad (6.31)$$

Now let both Δt and θ go to 0, preserving the relationship that $\Delta t = \theta^2$ (which guarantees that the state variance is meaningful) and at the same time letting k increase to ∞ such that $k\theta \to x$. Then $p_{k\theta}(t) \to p(x,t|X_0 = x_0)$, which is now the probability density for the system state X, given that the queueing system began operation at a size of x_0. Utilizing the definitions of first and second derivatives, Equation (6.31) becomes

$$\frac{\partial p(x,t|x_0)}{\partial t} = \left(\frac{\mu+\lambda}{2}\right)\frac{\partial^2 p(x,t|x_0)}{\partial x^2} + (\mu-\lambda)\frac{\partial p(x,t|x_0)}{\partial x}, \qquad (6.32)$$

one form of the well-known diffusion equation, which among other things describes the movement of a particle under Brownian motion [see Prabhu (1965a), for example]. This particular form of the diffusion equation often goes under the name of Fokker-Planck, and, in addition, turns out to be the version of the forward Kolmogorov equation that is found for the continuous-state Markov process.

So now we would like to solve (6.32) under the boundary conditions that

$$\begin{cases} p(x,t|x_0) \geqslant 0 \\ \int_{-\infty}^{\infty} p(x,t|x_0)\,dx = 1 \\ \lim_{t\to 0} p(x,t|x_0) = 0 \qquad (x \neq x_0) \end{cases},$$

the first two being usual properties of densities, while the third is essentially a continuity requirement at time 0. It can then be shown (see Prabhu, op. cit.) that the solution to (6.32) is given by

$$p(x,t|x_0) = \frac{e^{-[x-x_0+(\mu-\lambda)t]^2/2(\mu+\lambda)t}}{\sqrt{2\pi t(\mu+\lambda)}},$$

which is a Wiener process starting from x_0 with drift $-(\mu-\lambda)t$ and variance $(\mu+\lambda)t$. That is to say, the random process $X(t)$ is normal with mean $x_0-(\mu-\lambda)t$ and variance $(\mu+\lambda)t$, so that

$$\Pr\left\{n-\tfrac{1}{2}<X(t)<n+\tfrac{1}{2}|X_0=x_0\right\}\doteq\int_{n-1/2}^{n+1/2}N(x_0-[\mu-\lambda]t,[\mu+\lambda]t)\,dx.$$

But we observe that this result is not too meaningful since $\lambda<\mu$ and the drift is therefore negative, thus bringing the process eventually to one with a negative mean.[9] In order to counter this possibility we must impose an impenetrable barrier upon the walk at $x=0$. This is added to the problem in the form of the additional boundary condition that[10]

$$\lim_{x\to0}\frac{\partial p(x,t|x_0)}{\partial t}=0\qquad(\text{for all }t).$$

The new solution is then given by

$$p(x,t|x_0)=\frac{1}{\sqrt{2\pi(\lambda+\mu)t}}\left\{e^{-[x-x_0-(\lambda-\mu)t]^2/2(\mu+\lambda)t}+\left(e^{-2x(\mu-\lambda)/(\mu+\lambda)}\right).\right.$$

$$\left.\left(e^{-[x+x_0+(\lambda-\mu)t]^2/2(\mu+\lambda)t}+\frac{2(\mu-\lambda)}{\mu+\lambda}\int_x^\infty e^{-[y+x_0+(\lambda-\mu)t]^2/2(\mu+\lambda)t}dy\right)\right\}.$$

$$(6.33)$$

This solution is then valid as an approximation for any $M/M/1$ provided $\rho=1-\epsilon$.

It is particularly interesting to make two additional computations. The first is to allow $\lambda=\mu$ throughout the derivation, while the second is to look at the limiting behavior of $p(x,t|x_0)$ as t goes to ∞. When $\lambda=\mu$, Equation (6.32) is quite simplified and becomes

$$\frac{\partial p(x,t|x_0)}{\partial t}=\lambda\frac{\partial^2 p(x,t|x_0)}{\partial x^2}.$$

Under the same augmented boundary conditions as led to (6.33), this

[9]This is due to neglecting the impenetrable barrier at 0. However, the approximation would be valid if x_0 were large since it is then unlikely for the queue to empty.

[10]This is used because the process cannot move beyond zero to the negatives and therefore $p(x,t|x_0)=0$ for $x<0$ and $\lim_{x\to0}\Delta p(x,t|x_0)=0$.

differential equation has the solution

$$p(x,t|x_0) = \frac{1}{\sqrt{4\pi\lambda t}} \left\{ e^{-(x-x_0)^2/4\lambda t} + e^{-(x+x_0)^2/4\lambda t} \right\},$$

again a Wiener process, but one with no drift. This one may be used to approximate the transient solution for any $M/M/1$ with $\rho = 1$.

On the other hand, when we let $t \to \infty$ in (6.32), it is found that

$$0 = \left(\frac{\mu + \lambda}{2} \right) \frac{d^2 p(x)}{dx^2} + (\mu - \lambda) \frac{dp(x)}{dx},$$

which is a homogeneous, second-order linear differential equation with solution

$$p(x) = C_1 + C_2 e^{-[2(\mu-\lambda)/(\mu+\lambda)]x}.$$

Since $p(x)$ must integrate to one, we see that $C_1 = 0$ and $C_2 = (\mu + \lambda)/[2(\mu - \lambda)]$. Thus $p(x)$ is an exponential density, as might have been expected, since $M/M/1$ lengths are geometric in distribution, which is just the discrete analog of the exponential. That this mean of $(\mu + \lambda)/[2(\mu - \lambda)]$ $= (1 + \rho)/[2(1 - \rho)]$ makes sense can be seen by noting that for ρ nearly 1, $(1 + \rho)/2 \doteq \rho$; hence $(1 + \rho)/[2(1 - \rho)] \doteq \rho/(1 - \rho)$, the usual $M/M/1$ result.

Another approach to this approximation which leads to the same result is the use of the central limit theorem on the independent and identically distributed random variables making up the random walk. One then uses the same kind of limiting argument to get the results in terms of the same variables as before.

The $M/G/1$ heavy-traffic diffusion approximation is due to Gaver (1968). In order to approximate the conditional "density function,"[11] say $w_q(x,t|x_0)$, of the virtual line delay[12] $V(t)$, its mean μ and variance σ^2 are approximated by

$$\mu\Delta t = E[V(t+\Delta t) - V(t)|V(t)] = (\lambda E[S] - 1)\Delta t + o(\Delta t)$$

and

$$\sigma^2\Delta t = Var[V(t+\Delta t) - V(t)|V(t)] = \lambda E[S^2]\Delta t + o(\Delta t)$$

---

[11]Recall from Chapter 2 that $w_q(x,t|x_0)$ is not a true density since there is a nonzero probability of a zero wait.

[12]It turns out to be easier to use virtual delay then actual in the derivation here, but the two are identically distributed random variables in the $M/G/1$.

since the change in the virtual wait over the time increment Δt assuming a loaded system is the total service time needed to serve all arrivals over Δt, minus Δt. But it is known that $V(t)$ is a continuous-parameter continuous-state Markov process and hence will satisfy the Fokker-Planck equation for its conditional "density" $w_q(x, t | x_0)$ given by [see Newell (1972)]

$$\frac{\partial w_q(x, t | x_0)}{\partial t} = -\mu \frac{\partial w_q(x, t | x_0)}{\partial x} + \frac{\sigma^2}{2} \frac{\partial^2 w_q(x, t | x_0)}{\partial x^2},$$

subject to the boundary conditions

$$\left\{ \begin{array}{l} w_q(x, t | x_0) \geqslant 0 \\ \displaystyle\int_0^\infty w_q(x, t | x_0)\, dx = \frac{\lambda}{\mu} \\ \displaystyle\lim_{t \to 0} w_q(x, t | x_0) = 0 \qquad (x \neq x_0) \end{array} \right. .$$

From the earlier discussion of heavy traffic, it is to be expected that if

$$\frac{-\mu}{\sigma^2} = \frac{1 - \lambda E[S]}{\lambda E[S^2]}$$

is positive and small, then the diffusion solution should provide a good approximation. The final expression for $w_q(x, t | x_0)$ is found to be

$$w_q(x, t | x_0) = \frac{1}{\sqrt{2\pi t \sigma^2}} \left\{ e^{-(x - x_0 - \mu t)^2 / 2\sigma^2 t} + e^{2x\mu/\sigma^2} \left[e^{-(x + x_0 + \mu t)^2 / 2\sigma^2 t} \right. \right.$$

$$\left. \left. + \frac{2\mu}{\sigma^2} \int_x^\infty e^{-(y + x_0 + \mu t)^2 / 2\sigma^2 t}\, dy \right] \right\}.$$

This section was by no means meant to exhaust the subject of diffusion approximations in queueing, but rather to give the reader a brief introduction to the area. The interested reader will find a limited amount of literature on the subject, including in addition to Prabhu and Newell (op. cit.) such references as Feller (1966) and Karlin (1966).

PROBLEMS

6.1   Find $W_q(t)$ for: (a) $D/M/1$; (b) $D/E_k/1$; (c) $E_k/D/1$, and (d) $E_k/E_l/1$.

6.2   The Bearing Straight Corporation of Example 5.2 now finds that its machines do not break down as a Poisson process, but rather as the two-point distribution given below.

Interarrival Time	Probability	$A(t)$
9	2/3	2/3
18	1/3	1

Given that the service has the same two-point distribution as it did earlier, use Lindley's equation to find the line-delay CDF.

6.3   Use the approach indicated for $M/G/1$ waiting times in Section 5.1.6 to verify the waiting-time result of Example 6.1.

6.4   Use Equation (6.5) to verify Equation (6.26).

6.5   Find the probability mass functions for the arrival counts associated with deterministic and Erlang input streams.

6.6   Using the results of Problem 6.5, find the mean lengths of the busy period for: (a) $D/E_k/1$; (b) $E_2/D/1$; (c) $E_2/E_l/1$.

6.7   Verify the formula given in Section 6.1.1 for the mean length of the busy period by the use of the expression given there for $G^*(s)$.

6.8   Use the results of Problem 6.5 to find the Stieltjes transform of the busy period for $E_2/M/1$.

6.9   Verify the derivation of Equation (6.9) from (6.8).

6.10   Given an $M/D/2$ with $\lambda = 1$ and $\mu = 1$, fully specify the form of its generating function, including the necessary values of the roots of the denominator.

6.11   Use the result of Problem 6.10 to find the variance of the line wait under the same assumptions.

6.12   Consider an $M^{[X]}/M/1$ queue in which if there are two or more customers waiting when the server is free, then he serves the next two together. The service time is assumed exponential with mean $1/\mu$ regardless of the number being served. Find all the single-step transition probabilities of the imbedded Markov chain. Then use the result of Section 6.3 to obtain the chain's steady-state distribution and finally the general-time stationary probabilities via Equation (6.11).

6.13   Do as in Problem 6.12 for the two-channel exponential-exponential machine repair problem.

6.14   Use the theory of semi-Markov processes on $GI/M/c$ to find the general-time probabilities.

6.15 Find the general-time probability distribution associated with the queue in Example 5.5, and then from this find the mean system and queue sizes.

6.16 Equation (6.13) may be solved in another manner somewhat different from that indicated in the text. This new approach requires the obtaining of the generating function of $\tilde{W}_q(t|j)$, say

$$\tilde{W}(t,z) = \sum_{j=0}^{\infty} \tilde{W}_q(t|j) z^n.$$

Obtain the partial differential equation in $\tilde{W}(t,z)$ that results from the proper multiplication and summing of (6.13).

6.17 Take the Laplace transform of the PDE obtained in Problem 6.16 with respect to t.

6.18 Find the Laplace-Stieltjes transform of the line wait for the $M/G/1/LIFO$ model of Section 6.4.

6.19 Use the result of Problem 6.18 to find the variance of the line delay for the $M/G/1/LIFO$ model.

6.20 Remember the Bearing Straight Corporation (of Example 5.2 and Problem 6.2 fame)—they are now in bad straits because they are finding that an estimated machine-breakdown rate of as low as 5/hr is much too optimistic and that a more realistic estimate would be 6/hr. It is then observed that their service rate is also 6/hr based on a time of 9 min two out of three times and 12 min one-third of the time. Use the results of Section 6.5 to determine what kind of a decrease below the 9 min figure would guarantee Bearing machines of an average wait twice what it was before, namely, now equal to 72 min.

6.21 Considering Problem 6.20, what kind of a decrease below 9 min would guarantee Bearing machines waits of more than 400 min less than 5% of the time?

6.22 Our moonlighting graduate assistant of Problem 2.11 now finds in the new semester that the arrival rate to his counter has increased to 20/hr. If his service times remain exponential with mean 4 min, use the results of Section 6.5 to find the approximate probability that his Nth customer (N large) will have to wait. Then compare this result to what you would have obtained via the Chebyshev approximation.

6.23 Use Equation (6.26) to derive the mean line wait for a stationary $E_k/E_l/1$ queue.

6.24 Derive the inequality given by (6.27) from Equation (6.26).

6.25 Employ the bounds on W_q derived in Section 6.6 for $D/M/1$.

6.26   Show that the density function $p(x,t|x_0)$ given by (6.33) satisfies the diffusion partial differential equation (6.32).

6.27   Apply the central limit theorem to the one-dimensional random walk which moves left with probability q, right with probability p, and stands still with probability r. Then use the limiting procedures of Section 6.7 which led to Equation (6.33) to show that the continuous problem is a Wiener process, and that this Wiener density satisfies an equation of the same form as (6.33).

6.28   Show that the virtual idle time for a $GI/M/1$ has CDF

$$F(u) = A(u) + \int_u^\infty e^{-\mu(1-r_0)(t-u)} dA(t).$$

Chapter 7
STATISTICAL INFERENCE
AND DESIGN AND CONTROL

In this chapter we present two distinct but important topics in queueing theory. The first section deals with statistical problems in queueing, with an emphasis on the estimation of input and service parameters and/or distributions in various queueing contexts. Since one must, in practice, often use observable data to decide on what input and service patterns of a queueing system actually are, it is extremely important to utilize the data to the fullest extent possible.

Known statistical procedures can help in making the best use of existing data, or in determining what and how much new data should be taken. That this is an important facet of real queueing studies is readily seen, since the output from a queueing model can be no better than its input. Furthermore, when statistical procedures are used to estimate input parameters (say from data), the output measures of effectiveness actually become random variables, and it is often of interest to obtain confidence statements concerning them.

The second topic discussed in this chapter involves the use of queueing models in the making of operating decisions concerning the queueing system which the model describes. Often one desires to know such things as the optimal number of servers to have, the proper queue discipline to choose, the proper amount of waiting space to have, when to turn on (and off) automatic service devices, and so on. These types of questions fall under the category of design and control problems in queueing and are discussed in Section 7.2.

7.1. STATISTICAL INFERENCE IN QUEUEING

We normally think of queueing as the applied probabilistic analysis of waiting lines, but it should be clear that statistical inference must play an extremely vital role in any use of queueing as an aid to decision-making. This point becomes even clearer in Section 7.2 when we talk about design

344

of queues and in Chapter 8 when we discuss simulation of queues. Most probabilists generally divide the statistical problems into two types, which we shall call "parameter estimation" and "distribution selection." In the parameter estimation case, a particular type of probabilistic model is prespecified, and we wish to estimate its parameters, while for the distribution selection case, appropriate data are to be examined as a basis for determining a choice of model. This second type of problem is the more important of the two, but usually also the more difficult. It is for this reason that the overwhelming portion of the literature has been concerned with parameter estimation. In the following we go into each of these situations in moderate detail to give the reader a feel for the current state of statistics in queueing. The primary references to be noted are a survey paper by D. R. Cox entitled "Some Problems of Statistical Analysis Connected with Congestion," which appeared in Smith and Wilkinson (1965), as well as Clarke (1957), Billingsley (1961), and Wolff (1965).

7.1.1.   Parameter Estimation Problems

As introduced, it is our intention in this section to describe the problems of statistical inference which arise when it is assumed that a specific model form is indeed applicable. This in turn leads to a further subdivision into problems of estimation, hypothesis testing, and confidence statements, though all of course are clearly related.

The initial step in any statistical procedure is a determination of the availability of sample information. The method that is chosen for estimation and the form of the estimators are very much a function of the completeness of the monitoring process. It is one thing if one can observe fully a system over a certain period and is thus able to record the instants of service for each customer, but quite a different problem to have such incomplete information as only the queue size at each departure.

The earliest work on the statistics of queues did, in fact, assume that the subject queue was fully observed over a period of time and therefore that complete information was available in the form of the arrival instants, and points of the beginning and end of the service of each customer. As would then be expected, the queue was assumed to be a Markov chain in continuous time. Clarke (op. cit.) began a sequence of papers on this and related topics by obtaining the maximum-likelihood estimators for the arrival and service parameters of an $M/M/1$ queue, in addition to the variance-covariance matrix for the two statistics. He was followed shortly thereafter by a similar exposition for $M/M/\infty$ by Beneš (1957). Clarke's work was pioneering and therefore a brief discussion of it follows. Prior to discussing Clarke's work, however, we first develop some classical maximum-likelihood estimators based solely on individual observations in the form of interarrival or service-time data.

The class of estimators known as maximum-likelihood estimates (MLE) have some nice statistical properties and can be obtained as follows. First consider the case where we believe our underlying distribution is exponential with parameter θ and we wish to estimate θ from the sample data. We form the likelihood function (joint density function of the sample), assuming the observations are *independent*, as

$$L(\theta) = \prod_{i=1}^{n} \theta e^{-\theta t_i} = \theta^n e^{-\theta \sum_{i=1}^{n} t_i}$$

where t_i is the ith sample observation (say interarrival or service time). The MLE of θ is the value which maximizes L. It is often more convenient to find the maximum of $\ln L = \mathcal{L}$. Thus $\hat{\theta}$, the MLE of θ, is the value for which

$$\max_{\theta} \mathcal{L}(\theta) = \max_{\theta} \left[n \ln \theta - \theta \sum_{i=1}^{n} t_i \right].$$

Taking $d\mathcal{L}/d\theta$ and setting it to zero yields the maximizing value as

$$\frac{n}{\hat{\theta}} - \sum_{i=1}^{n} t_i = 0,$$

or

$$\hat{\theta} = \frac{n}{\sum_{i=1}^{n} t_i}.$$

Note that the MLE of the mean of the exponential is

$$\frac{1}{\hat{\theta}} = \frac{\sum_{i=1}^{n} t_i}{n},$$

the sample arithmetic average.

Another important problem is that of determining maximum-likelihood estimators for an Erlangian density. For the density

$$f(t) = \frac{\phi(\phi t)^{k-1} e^{-\phi t}}{(k-1)!},$$

the likelihood function may be written as

$$L(\phi,k) = \prod_{i=1}^{n} \frac{\phi(\phi t_i)^{k-1} e^{-\phi t_i}}{(k-1)!}$$

$$= \frac{\phi^{nk} e^{-\phi \sum_{i=1}^{n} t_i} \left(\prod_{i=1}^{n} t_i \right)^{k-1}}{\left[(k-1)! \right]^n}$$

Thus the log-likelihood is

$$\mathcal{L}(\phi,k) = nk \ln \phi - \phi \sum_{i=1}^{n} t_i + (k-1) \sum_{i=1}^{n} \ln t_i - n \ln (k-1)!.$$

Therefore

$$\frac{\partial \mathcal{L}}{\partial \phi} = \frac{nk}{\phi} - \sum_{i=1}^{n} t_i$$

and

$$\hat{\phi} = \frac{\hat{k}}{\bar{t}} \qquad \left(\bar{t} = \sum_{i=1}^{n} t_i/n \right).$$

To now get the complete pair $(\hat{\phi}, \hat{k})$, consider k to be a continuous variable (say x) and proceed in the usual way to obtain the MLE $\hat{x}$ of x as the numerical solution to

$$\frac{\partial \mathcal{L}}{\partial x} = 0 = n \ln \hat{\phi} + \sum_{i=1}^{n} \ln t_i - n \psi (\hat{x})$$

$$= n (\ln \hat{x} - \ln \bar{t}) + \sum_{i=1}^{n} \ln t_i - n \psi (\hat{x}),$$

where $\psi(x)$ is the logarithmic derivative of the Γ function, that is,

$$\psi(x) \equiv \frac{d \ln \Gamma(x)}{dx}.$$

The function $\psi(x)$ is tabulated in Abramowitz and Stegun (1964), and, if needed, a good approximation to $\psi(x)$ when x is not too small (say $\geqslant 3$) is

$$\psi(x) \doteq \ln(x - 1/2) + \frac{1}{24(x - 1/2)^2}.$$

Hence the MLE $\hat{k}$ of k is either $[\hat{x}]$ or $[\hat{x}] + 1$ (where $[x]$ is the greatest integer in x), depending on which pair, $([\hat{x}]/\bar{l}, [\hat{x}])$ or $(([\hat{x}] + 1)/\bar{l}, [\hat{x}] + 1)$, gives a higher value to the log-likelihood.

Returning now to Clarke's work, assume it is known that a stationary $M/M/1$ queue is being observed, with unknown mean arrival rate λ and mean service rate μ ($\lambda/\mu \equiv \rho$). We suppose that the queue begins operation with n_0 customers present and then consider both the conditional like-lihood given $N(0) = n_0$ and the likelihood ignoring the initial condition. It is clear that times between transitions are exponential, with mean $1/(\lambda + \mu)$ when the zero state is not occupied and mean $1/\lambda$ when the zero state is occupied. All jumps are upward from zero, while jumps upward from nonzero states occur with probability $\lambda/(\lambda + \mu)$ and jumps downward with probability $\mu/(\lambda + \mu)$, all independent of previous queue history.

Let us assume that we are observing the system for a fixed amount of time t, where t is sufficiently large to "guarantee" some appropriate number of observations. Note that t must be chosen independently of the arrival and service processes so that the sampling interval is independent of λ and μ. Let us then use t_e and t_b to denote respectively the amount of time the system is empty and busy ($t_b = t - t_e$). In addition, let n_c denote the number of customers who have been serviced, n_{ae} the number of arrivals to an empty system, and n_{ab} the number to a busy system, with the total number of arrivals $n_a = n_{ae} + n_{ab}$. This is essentially Clarke's notation, and it is most convenient to use in this situation.

The likelihood function is then made up of components which are formed from the following kinds of information:

(i) Intervals of length x_b spent in a nonzero state and ending in an arrival or departure;

(ii) Intervals of length x_e spent in the zero state and ending in an arrival;

(iii) The very last (unended) interval (length x_l) of observation;

(iv) Arrivals to a busy system;

(v) Departures;

(vi) The initial number of customers, n_0.

The contributions to the likelihood of each of these are:

(i) $(\lambda + \mu)e^{-(\lambda + \mu)x_b}$;

(ii) $\lambda e^{-\lambda x_e}$;

(iii) $e^{-\lambda x_l}$ or $e^{-(\lambda + \mu)x_l}$;

(iv) $\lambda/(\lambda + \mu)$;

(v) $\mu/(\lambda+\mu)$;

(vi) $\Pr\{N(0)=n_0\}$.

Since $\sum x_b = t_b$, $\sum x_e = t_e$, and $n_{ae} + n_{ab} = n_a$, the likelihood is found to be (with the unended time x_l properly included in either t_b or t_e)

$$L(\lambda,\mu) = (\lambda+\mu)^{n_c+n_{ab}} e^{-(\lambda+\mu)\sum x_b} \lambda^{n_{ae}} e^{-\lambda\sum x_e} \left(\frac{\lambda}{\lambda+\mu}\right)^{n_{ab}} \left(\frac{\mu}{\lambda+\mu}\right)^{n_c} \Pr\{n_0\}$$

$$= e^{-(\lambda+\mu)t_b} \lambda^{n_a} e^{-\lambda t_e} \mu^{n_c} \Pr\{n_0\}.$$

The log-likelihood function $\mathcal{L}$ corresponding to the foregoing L is clearly given by

$$\mathcal{L}(\lambda,\mu) = -(\lambda+\mu)t_b + n_a\ln\lambda - \lambda t_e + n_c\ln\mu + \ln\Pr\{n_0\}$$

$$= -\lambda t - \mu t_b + n_a\ln\lambda + n_c\ln\mu + \ln\Pr\{n_0\}. \qquad (7.1)$$

In the event that the queue is in equilibrium, the initial size may be ignored, and then the maximum-likelihood estimates, $\hat{\lambda}$ and $\hat{\mu}$ ($\hat{\rho}=\hat{\lambda}/\hat{\mu}$), would be given by the solution to

$$\begin{cases} \dfrac{\partial\mathcal{L}}{\partial\lambda}=0 \\ \dfrac{\partial\mathcal{L}}{\partial\mu}=0 \end{cases}.$$

So, in this case,

$$\frac{\partial\mathcal{L}}{\partial\lambda} = -t + \frac{n_a}{\lambda}$$

and

$$\frac{\partial\mathcal{L}}{\partial\mu} = -t_b + \frac{n_c}{\mu}.$$

Thus[1]

$$\begin{cases} \hat{\lambda} = \dfrac{n_a}{t} \\ \\ \hat{\mu} = \dfrac{n_c}{t_b} \end{cases}. \qquad (7.2)$$

[1]Note that this is what one would obtain by observing individual interarrival and service times and taking their sample averages (ignoring the last unended interval.), that is, the previously derived classical MLE s.

The ratio $\hat{\lambda}/\hat{\mu}$ must be less than one since equilibrium has been assumed; but if this condition is violated, then we must assume that $\hat{\lambda} \doteq \hat{\mu}$ and would minimize $\mathcal{L}(\lambda,\mu) + \theta(\lambda,\mu)$, where θ is a Lagrange multiplier, and obtain as the common estimator, $\hat{\lambda} = \hat{\mu} = (n_a + n_c)/(t + t_b)$.

If we now instead assume that $N(0)$ cannot be ignored, then in order to obtain any meaningful results, some assumption must be made regarding the distribution of this initial size. If ρ is known to be less than one, then by choosing $\Pr\{N(0) = n_0\}$ to be $\rho^{n_0}(1 - \rho)$ we would immediately place ourselves into the steady state. On the other hand, we may want to do the estimation under the assumption that ρ could indeed be greater than one. But the suggested geometric distribution for system size would not be appropriate when $\rho > 1$. In this case, an alternative approach must be tried, and the choice is then somewhat arbitrary.

Let us, now, for illustrative purposes use $\Pr\{n_0\} = \rho^{n_0}(1 - \rho)$, in which case Equation (7.1) becomes

$$\mathcal{L}(\lambda,\mu) = -\lambda t - \mu t_b + n_a \ln\lambda + n_c \ln\mu + n_0(\ln\lambda - \ln\mu) + \ln\left(1 - \frac{\lambda}{\mu}\right).$$

Then

$$\frac{\partial\mathcal{L}}{\partial\lambda} = -t + \frac{n_a}{\lambda} + \frac{n_0}{\lambda} - \frac{1}{\mu - \lambda}$$

and

$$\frac{\partial\mathcal{L}}{\partial\mu} = -t_b + \frac{n_c}{\mu} - \frac{n_0}{\mu} + \frac{\lambda}{\mu(\mu - \lambda)}.$$

The estimators $\hat{\lambda}$ and $\hat{\mu}$ are thus the solution to

$$\begin{cases} 0 = -t + \dfrac{n_a + n_0}{\hat{\lambda}} - \dfrac{1}{\hat{\mu} - \hat{\lambda}} \\ 0 = -t_b + \dfrac{n_c - n_0}{\hat{\mu}} + \dfrac{\lambda}{\hat{\mu}(\hat{\mu} - \hat{\lambda})} \end{cases} \tag{7.3}$$

The first of these equations simplifies to $\hat{\mu} - \hat{\lambda} = \hat{\lambda}/[(n_a + n_0) - \hat{\lambda}t]$. Then eliminating $\hat{\mu}$ from the second equation gives a quadratic in $\hat{\lambda}$, which would then be used to obtain two values of $\hat{\lambda}$. Any negative value obtained is rejected and for the remaining values of $\hat{\lambda}$, the corresponding value of $\hat{\mu}$ is obtained. In addition, one would reject any $(\hat{\lambda}, \hat{\mu})$ pair for which

$$(i) \quad \hat{\mu} \leqslant 0$$

or

$$(ii) \quad \frac{\hat{\lambda}}{\hat{\mu}} > 1.$$

If both solutions are valid, then the one which maximizes the likelihood function is kept. If neither solution is valid and it is positiveness that is violated, let the violating parameter be equal to a small positive ϵ; otherwise, let $\hat{\lambda} = \hat{\mu}$.

An alternative approach of incorporating the initial state would be to adjust (7.2) by subtracting the initial system size minus the estimated mean equilibrium system size divided by the observing time from $\hat{\mu}$, since the effect of the difference of n_0 from the steady-state mean will then be removed. The reverse is true of $\hat{\lambda}$; that is, this quantity is added. This then gives the approximations

$$\begin{cases} \hat{\lambda} \doteq \dfrac{n_a}{t} + \dfrac{n_0 - (n_a t_b / n_c t)/(1 - n_a t_b / n_c t)}{t} \\[4mm] \hat{\mu} \doteq \dfrac{n_c}{t_b} - \dfrac{n_0 - (n_a t_b / n_c t)/(1 - n_a t_b / n_c t)}{t_b} \end{cases} \qquad (7.4)$$

where $(n_a t_b / n_c t)/(1 - n_a t_b / n_0 t)$ is an estimate of L, namely, $\hat{L} = \hat{\rho}/(1 - \hat{\rho})$.

Lilliefors (1966) has considered the problem of finding confidence intervals for the actual $M/M/1$ traffic intensity from the maximum-likelihood estimators given by Equation (7.2). Since the individual interar-rival times are independent, identically distributed exponential random variables, the quantity t is Erlang type n_a with mean n_a/λ; hence λt is an Erlang type n_a with mean n_a. Likewise, the quantity μt_b is Erlang type n_c with mean n_c. Therefore, as pointed out by Lilliefors, the distribution for the ratio $t_b/t\rho$ is given as $F_{2n_c, 2n_a}(t_b/t\rho)$,[2] where $F_{a,b}(x)$ would be the usual F distribution with degrees of freedom a and b. But, from (7.2), $t_b/t\rho = (n_c/n_a)\hat{\rho}/\rho$, and thus confidence intervals can be readily found for ρ by the direct use of the F distribution, with the upper $1 - \alpha$ confidence limit, say ρ_u, found from the F table by the equation

$$\frac{n_c \hat{\rho}}{n_a \rho_u} = F_{2n_c, 2n_a}\left(\frac{\alpha}{2}\right)$$

and the lower $1 - \alpha$ limit, say ρ_l, from

$$\frac{n_c \hat{\rho}}{n_a \rho_l} = F_{2n_c, 2n_a}\left(1 - \frac{\alpha}{2}\right).$$

In addition, confidence intervals can be found for any of the usual measures of effectiveness which are functions of ρ.

---

[2]The twos in the degrees of freedom enter when the Erlang distributions are converted to χ^2 distributions (see Table 3), the ratio of which then yield the F distribution.

Example 7.1

As a simple illustration, let us consider the following problem. Observations are made of an $M/M/1$ queue, and it is noted at time $t = 400$ hr that all of 60 arrivals have been duly served and departed. Of the 400 hr of observation, the server was actually busy for a total of 300 hr. Let us then find a 95% confidence interval for the traffic intensity ρ.

By our previous discussion,

$$\hat{\lambda} = \frac{60}{400} = \frac{3}{20}$$

and

$$\hat{\mu} = \frac{60}{300} = \frac{1}{5},$$

so that $\hat{\rho} = 3/4$. Furthermore, we are interested in confidence intervals at a level $\alpha = 0.05$. Then the appropriate upper and lower limits for degrees of freedom of 120 and 120 approximately are (from Table A5.7, Appendix 5)

$$\frac{n_c}{n_a} \frac{\hat{\rho}}{\rho_u} \doteq 0.70$$

and

$$\frac{n_c}{n_a} \frac{\hat{\rho}}{\rho_l} \doteq 1.43.$$

Therefore

$$\rho_u \doteq \frac{\hat{\rho}}{0.70} \doteq 1.07$$

and

$$\rho_l \doteq \frac{\hat{\rho}}{1.43} \doteq 0.52,$$

and we conclude with 95% confidence that ρ will fall in the interval (0.52, 1.07).

These same kinds of ideas can be nicely extended to exponential queues with many servers, and to cases with Erlang input and/or service. In addition, Billingsley (op.cit.) has made a detailed study of likelihood estimation for Markov chains in continuous time, including limit theory and hypothesis testing, and then these results were used and extended to obtain results for birth–death queueing models by Wolff (op.cit.).

Suppose that we now wish to apply the likelihood procedure to an

$M/G/1$ queue. The approach is similar except that the loss of memory-lessness is going to alter the likelihood function since no use can be made of data which distinguish between empty and busy intervals and there are now four components of the likelihood, namely:

(i)   Interarrival intervals of length x which are exponential, with contribution $\lambda e^{-\lambda x}$;

(ii)   Service times of duration x for the n_c completed customers with contribution $b(x)$;

(iii)   Time spent in service (say x_l) by the very last customer, with contribution $1 - B(x_l)$;

(iv)   The initial number of customers.

Hence the likelihood may be written as

$$L(\lambda,\mu) = \lambda^{n_a} e^{-\lambda(t-x_l)} \left[\prod_{i=1}^{n_c} b(x_i) \right] [1 - B(x_l)] \Pr\{n_0\},$$

and the log-likelihood as

$$\mathcal{L}(\lambda,\mu) = n_a \ln \lambda - \lambda(t - x_l) + \sum_{i=1}^{n_c} \ln b(x_i) + \ln[1 - B(x_l)] + \ln \Pr\{n_0\}.$$

Then derivatives are taken in the usual way and the procedure follows that of the $M/M/1$ thereafter. To illustrate this method, the following example is presented.

Example 7.2

Let us find the maximum-likelihood estimators for an $M/E_2/1$ queue with mean arrival rate λ and mean service time $2/\mu$. From the aforementioned, since $b(t) = \mu^2 t e^{-\mu t}$, the log-likelihood may be written as

$$\mathcal{L}(\lambda,\mu) = n_a \ln \lambda - \lambda(t - x_l) + \sum_{i=1}^{n_c} [2\ln \mu + \ln x_i - \mu x_i]$$

$$+ \ln \left[1 - \int_0^{x_l} \mu^2 t e^{-\mu t} \, dt \right] + \ln \Pr\{n_0\}.$$

But the integral of an Erlang may be rewritten in terms of a Poisson sum [see Equation (1.20)] as

$$\int_0^{x_l} \mu^2 t e^{-\mu t}\, dt = 1 - e^{-\mu x_l} \sum_{i=0}^{1} \frac{(\mu x_l)^i}{i!};$$

hence

$$\mathcal{L}(\lambda, \mu) = n_a \ln \lambda - \lambda(t - x_l) + 2n_c \ln \mu + \sum_{i=1}^{n_c} \ln x_i - \mu \sum_{i=1}^{n_c} x_i$$

$$+ \ln \left[e^{-\mu x_l} + \mu x_l e^{-\mu x_l} \right] + \ln \Pr \{ n_0 \}.$$

The partial derivatives can be computed to be

$$\left\{ \begin{array}{l} \dfrac{\partial \mathcal{L}}{\partial \lambda} = \dfrac{n_a}{\lambda} - (t - x_l) + \dfrac{\partial \ln \Pr \{ n_0 \}}{\partial \lambda} \\[3mm] \dfrac{\partial \mathcal{L}}{\partial \mu} = \dfrac{2n_c}{\mu} - \displaystyle\sum_{i=1}^{n_c} x_i - x_l + \dfrac{x_l}{1 + \mu x_l} + \dfrac{\partial \ln \Pr \{ n_0 \}}{\partial \mu} \end{array} \right.$$

If now the initial state is assumed to be chosen free of λ and μ, then the MLE's for λ and μ are found by equating the partial derivatives to zero as

$$\hat{\lambda} = \frac{n_a}{t - x_l},$$

and the appropriate solution $\hat{\mu}$ to the quadratic equation

$$x_l \left(x_l + \sum_{i=2}^{n_c} x_i \right) \hat{\mu}^2 + \left(\sum_{i=1}^{n_c} x_i - 2n_c x_l \right) \hat{\mu} - 2n_c = 0.$$

Thus far we have been in the position of assuming that we were dealing with simple processes subject to complete information. But suppose now that certain kinds of information are just not available. For example, suppose we observe only the stationary output of an $M/G/1$ (G known) queue and are then asked to estimate the mean service and interarrival times. Under the assumption that the stream is in equilibrium, the mean interarrival time $1/\lambda$ must equal the long-term arithmetic mean of the interdeparture times. If the mean of the departure process is denoted by $\bar{d}$, then the maximum-likelihood estimator of λ is $\hat{\lambda} = 1/\bar{d}$.

If service is exponential, then we know that the limiting distribution of output is the same as that of the input. Hence no inference is possible about the mean service time. But if service is assumed to be other than exponential, then the estimation is possible. The CDF of the interdeparture process of any $M/G/1$ queue is given by [recall from (5.20) that $p_0 = 1 - \lambda/\mu$]

$$C(t) = \frac{\lambda}{\mu} B(t) + \left(1 - \frac{\lambda}{\mu}\right) \int_0^t B(t-x)\lambda e^{-\lambda x}\,dx, \qquad (7.5)$$

where the last term is the convolution of service and arrival-time CDFs. In the special case where the service time is constant, (7.5) reduces to

$$C(t) = \begin{cases} 0 & (t < 1/\mu) \\[2mm] \dfrac{\lambda}{\mu} & (t = 1/\mu). \\[3mm] \dfrac{\lambda}{\mu} + \left(1 - \dfrac{\lambda}{\mu}\right)[1 - e^{-\lambda(t - 1/\mu)}] & (t > 1/\mu) \end{cases}$$

Since there is nonzero probability associated with the point $t = 1/\mu$, we may directly obtain our estimate by equating $1/\mu$ with the minimum observed interdeparture time.

On the other hand, if the distribution is other than exponential or deterministic, the approach is to take Laplace-Stieltjes transforms of Equation (7.5), yielding

$$C^*(s) = \frac{(1 + s/\mu)B^*(s)}{1 + s/\lambda}, \qquad (7.6)$$

where $B^*(s)$ is the Laplace-Stieltjes transform of the service-time CDF whose form is known but not the values of its parameters. Then the moments of $C(t)$ and $B(t)$ may be directly related by the successive differentiation of Equation (7.6) using enough equations to determine all parameters of $B(t)$. However, there still exists a small problem since successive interdeparture times will be correlated, and this correlation must be considered when calculating moments from data. For example, if enough data are present, data spread "sufficiently" far apart may be considered to form an approximately random sample so that formulas based on uncorrelated observations can be used. It would be advisable nevertheless to test for lack of correlation by computing the sample correlation coefficient between successive observations before making any definitive statements. We have asked the reader to solve a problem of this type at the end of the chapter as Problem 7.6. There is also a discussion of this and related questions in Cox (op. cit.).

Suppose we slightly modify the previous problem and instead now consider an $M/G/1$ with the form of G known and observations now made on both the input and output. The analysis might then proceed in a very similar manner, except that since we are observing ordered instants of arrival and departure, the successive customer waiting times are now available. The relevant relationship between the transforms of $B(t)$ and the system-wait CDF $W(t)$ [see Equation (5.40)] is

$$W^*(s) = \frac{(1-\lambda/\mu)sB^*(s)}{s-\lambda+\lambda B^*(s)}. \qquad (7.7)$$

Again, problems of autocorrelation surface. But assuming that these are taken into account as previously discussed, we know that in the event $B(t)$ is exponential,

$$\left. \frac{dW^*(s)}{ds} \right|_{s=0} = -\frac{1}{\mu-\lambda}.$$

Hence

$$\hat{W} = \frac{1}{\hat{\mu}-\hat{\lambda}},$$

or

$$\hat{\mu} = \hat{\lambda} + \frac{1}{\hat{W}}.$$

In fact, for any one-parameter service distribution, we may directly appeal to the Pollaczek-Khintchine formula (found via the first derivatives of Equation (7.7) or from the results of Section 5.1), namely,

$$W = \frac{1}{\mu} + \frac{(\lambda/\mu)^2 + \lambda^2 \sigma_s^2}{2\lambda(1-\lambda/\mu)}.$$

In the deterministic case, for example, we find that

$$W = \frac{1}{\mu} + \frac{\lambda/\mu^2}{2(1-\lambda/\mu)} = \frac{(1/\mu)(2-\lambda/\mu)}{2(1-\lambda/\mu)}.$$

Therefore we may write that

$$\hat{W} = \frac{(1/\hat{\mu})(2-\hat{\lambda}/\hat{\mu})}{2(1-\hat{\lambda}/\hat{\mu})},$$

or that $\hat{\mu}$ is the appropriate solution to the quadratic

$$2\hat{W}\hat{\mu}^2 - 2(\hat{W}\hat{\lambda} + 1)\hat{\mu} + \hat{\lambda} = 0,$$

where $\hat{\lambda}$ is found as before from $\bar{d}$. Again, we leave the Erlang case as a problem (see Problem 7.7).

As for $M/G/\infty$, we have shown previously in Section 5.2.3 that both output and system size are nonhomogeneous Poisson streams. Hence we should be able to estimate parameters by converting all the observable processes to Poisson processes from which it is easy to obtain appropriate estimators.

7.1.2 Distribution Selection Problems

Here we are going to be concerned with the separate and marginal analysis of interarrival times, service times, waiting times, and so on to permit the selection of an appropriate queueing model. There are three key questions that should be addressed when attempting to find the interarrival or service distribution from data. These are as follows:

(i) Is the process exponential?

(ii) Is there homogeneity in the data (or the lack thereof) with respect to time of day?

(iii) Is there autocorrelation in the data being used?

Question (i) is relevant not only because Poisson/exponential processes are so important in queueing, but maybe even more so because their absence may create nearly insurmountable obstacles to closed-form analytical modeling. A negative answer to Question (ii) means that extra caution must be taken in the analysis since parameters and/or distributions may be changing with time. For example, there may be a systematic dependence of parameters on time, such as in the nonhomogeneous Poisson, or a parameter might merely be slowly changing throughout a day. Question (iii) is effectively asking the analyst to be very careful in treating data and not to assume that he has a truly random sample. This is especially true when analyzing output and waiting data.

Much of the foregoing essentially amounts to frequency curve fitting. Curve fitting is itself a fairly large subject area, and it is felt not to be in the reader's best interest to give an exhaustive survey since there appears a great volume of material on the subject in the open literature. There is, however, one specific problem of this nature which is extremely important to the queueing analyst, namely, the "optimal" manner in which to test for a Poisson/exponential stream. Since this problem does appear so often, we shall make some suggestions for its solution.

Generally, the easiest and most familiar way to test for Poisson/exponential character is to use a χ^2 goodness-of-fit test on the data presented in block histogram form against a theoretical distribution with the parameter replaced by its maximum-likelihood estimator, which is $\hat{\lambda} = n/\sum_{i=1}^{n} t_i$ for both the exponential $\lambda e^{-\lambda t}$ and Poisson $(\lambda t)^n e^{-\lambda t}/n!$, where t_i is the time between the $(i-1)$st and ith occurrences. The resulting statistic is then[3]

$$\chi_k^2 = \sum_{i=1}^{n} \frac{(o_i - e_i)^2}{e_i},$$

where o_i is the number observed in the ith frequency class (out of a total of between 10 and 20 classes), e_i the number expected in the ith frequency class if the hypothesized distribution were correct, and k the number of degrees of freedom, always equal to the total number of classes less one and then minus one for each parameter estimated. Of course, the usual precautions must be taken to keep the number in any case from being too small (a rule of thumb being less than five).

Great care should always be exercised in doing χ^2 goodness-of-fit tests, and the analyst would, of course, be well advised to search for a definitive exposition on the subject in the statistical literature. In Chapter 9, Section 9.3, we do run some χ^2 tests with the details presented. The basic weaknesses of the χ^2 test are its requirement for large samples, its heavy dependence upon the choice of the number and position of the time-axis intervals, and its possibly very high type II error[4] for some feasible alternative distributions.

There are, of course, numerous other approaches to goodness-of-fit (for example, graphical procedures), but since this is a small (though certainly important) aspect of statistical analysis in queueing, we shall not spend an excessive amount of time on the subject. However, since the exponential distribution is so important in queueing, we would like to point out three specific tests for exponentiality which are reasonably powerful (power meaning ability to discern false hypotheses) against almost any alternative hypothesis and will often outperform χ^2. They are the F test, the Kolmogorov-Smirnov (K-S) test [see Fisz (1963)] and the Anderson-Darling (A-D) test [or, as it is also referred to, the Weighted Cramer-Smirnov-Von Mises test; see Anderson and Darling (1952)].

First considering the F test, if the first $r(\doteq n/2)$ and last $(n-r)$ of a set of n hypothesized exponential interoccurrence times $\{t_i\}$ are grouped and

---

[3] χ^2 critical values can be found in Appendix 5, Table A5.6.

[4] This is stated in terms of the probability of accepting a false hypothesis.

if S_i is used to denote the ith normalized spacing, that is,

$$S_i = (n - i + 1)(t_i - t_{i-1}) \qquad (t_0 \equiv 0),$$

then it should be clear that the $\{S_i\}$ are independent and identically distributed exponentials with exactly the same mean as the underlying distribution. Thus it follows as in Section 7.1.1 that the quantity

$$F = \frac{\displaystyle\sum_{i=1}^{r} S_i / r}{\displaystyle\sum_{i=r+1}^{n} S_i / (n - r)}$$

is the ratio of two Erlangs and is distributed as an F distribution with $2r$ and $2(n - r)$ degrees of freedom when the hypothesis of exponentiality is true. Therefore a two-tailed F test would be performed on the F calculated from a set of data in order to determine whether the stream is indeed truly exponential. Tables of critical points for the F distribution, $\alpha = 0.01$, 0.025, and 0.05, may be found in Appendix 5, Table A5.7.

The K-S test compares deviations of the empirical CDF from the theoretical CDF, and uses as its test statistic a modified maximum absolute deviation, namely,

$$K = \max_j \left[\max \left\{ \left| \frac{j}{n} - F(t_j) \right|, \left| \frac{j-1}{n} - F(t_j) \right| \right\} \right],$$

where t_j is the jth ordered (ascending) observation, and $F(t_j)$ is the exponential CDF, $1 - e^{-t_j/\bar{t}}$, $\bar{t}$ being the sample arithmetic average. Critical values for the K-S statistic when testing for exponentiality with unknown population mean are given in Table 7.1A [see Lilliefors (1969a)].

Now, for the A-D test, the test statistic with t_j the jth ordered observation and $F(t_j) = 1 - e^{-t_j/\bar{t}}$ is[5]

$$A = -n - \frac{1}{n} \sum_{j=1}^{n} \left\{ (2j - 1) \ln F(t_j) + (2n - 2j + 1) \ln \tilde{F}(t_j) \right\}.$$

[5]Note that the F, K-S, and A-D statistics require all the data points, while for χ^2, data can come already grouped. Also, the A-D test is similar in type to the K-S test except that the A-D test statistic is a type of weighted deviation of theoretical from actual CDF while the K-S test statistic is based only on the maximum deviation. Also, the whole argument for the F can be extended easily to the case in which there are randomly occurring incomplete interoccurrence periods, such as is common in repair-like problems.

TABLE 7.1 CRITICAL VALUES FOR KOLMOGOROV-SMIRNOV AND ANDERSON-DARLING, EXPONENTIAL WITH UNKNOWN PARA-METER

Sample Size N	A. Kolmogorov-Smirnov[a] Level of Significance α				
	0.20	0.15	0.10	0.05	0.01
3	0.451	0.479	0.511	0.551	0.600
4	0.396	0.422	0.449	0.487	0.548
5	0.359	0.382	0.406	0.442	0.504
6	0.331	0.351	0.375	0.408	0.470
7	0.309	0.327	0.350	0.382	0.442
8	0.291	0.308	0.329	0.360	0.419
9	0.277	0.291	0.311	0.341	0.399
10	0.263	0.277	0.295	0.325	0.380
11	0.251	0.264	0.283	0.311	0.365
12	0.241	0.254	0.271	0.298	0.351
13	0.232	0.245	0.261	0.287	0.338
14	0.224	0.237	0.252	0.277	0.326
15	0.217	0.229	0.244	0.269	0.315
16	0.211	0.222	0.236	0.261	0.306
17	0.204	0.215	0.229	0.253	0.297
18	0.199	0.210	0.223	0.246	0.289
19	0.193	0.204	0.218	0.239	0.283
20	0.188	0.199	0.212	0.234	0.278
25	0.170	0.180	0.191	0.210	0.247
30	0.155	0.164	0.174	0.192	0.226
>30	$\dfrac{0.86}{\sqrt{N}}$	$\dfrac{0.91}{\sqrt{N}}$	$\dfrac{0.96}{\sqrt{N}}$	$\dfrac{1.06}{\sqrt{N}}$	$\dfrac{1.25}{\sqrt{N}}$

Sample Size N	B. Anderson-Darling[b] Level of Significance α				
	0.20	0.15	0.10	0.05	0.01
3	0.736	0.812	0.951	1.092	1.63
5	0.766	0.854	0.991	1.224	1.88
7	0.781	0.873	1.024	1.260	1.90
10	0.788	0.889	1.028	1.280	1.91
15	0.801	0.896	1.033	1.302	1.93
$\geqslant 20$	0.806	0.903	1.044	1.305	1.94

[a]Reprinted with permission of the *Journal of the American Statistical Association*.
[b]Reprinted with permission of the author, H. W. Lilliefors.

Power comparisons are made between this statistic and others in common usage, including $K\text{-}S$, for a number of typical alternatives, in a paper by Lilliefors (1969b). A table of critical values for the $A\text{-}D$ statistic is given as Table 7.1B. These three tests (F, $K\text{-}S$, and $A\text{-}D$) are illustrated in Example 7.3.

Example 7.3.

The following sequence of 25 service times was observed for an $M/G/1$ queue. We test these data for exponentiality using the F, $K\text{-}S$, and $A\text{-}D$ tests, respectively.

$$
\begin{bmatrix}
27.6 & 28.9 & 3.8 & 16.6 & 13.3 & 3.3 & 7.8 \\
55.3 & 12.6 & 1.8 & 12.9 & 4.8 & 12.6 & 8.8 \\
3.3 & 2.7 & 0.6 & 1.3 & 1.1 & 21.3 & 11.3 \\
14.9 & 15.7 & 8.6 & 9.6 & & &
\end{bmatrix}
$$

For the F test, we arrange the observations in ascending order and divide the 25 points into two sets of 13 and 12, respectively.

$$
\begin{bmatrix}
13: & 0.6, 1.1, 1.3, 1.8, 2.7, 3.3, 3.3, 3.8, 4.8, 7.8, 8.6, 8.8, 9.6 \\
12: & 11.3, 12.6, 12.6, 12.9, 13.3, 14.9, 15.7, 16.6, 21.3, 27.6, 28.9, 55.3.
\end{bmatrix}
$$

Thus, after computing the S_i's,

(for example, $\quad S_1 = (25 - 1 + 1)(0.6 - 0) = 15.0$

$$S_2 = (25 - 2 + 1)(1.1 - 0.6) = 12.0, \text{ etc.})$$

we find

$$
F = \frac{143.9/13}{127.7/12} = 1.04.
$$

The 95% critical values for $F_{26,24}$ are $1/2.23 \doteq 0.45$ and 2.26, respectively, thus we accept the hypothesis that the data are exponential.

We now proceed to illustrate the $K\text{-}S$ test. Ordering the entire set of data in ascending order we must compare the empirical CDF with a theoretical exponential CDF, with mean $\bar{t}$ of 12.02 (our sample mean). Letting t_j be the jth ordered observation, we must compare $j/25$ (empirical) with $1 - e^{-t_j/12.02}$, for all j, and choose the largest absolute difference; that is, we desire

$$
K = \max_j \left[\max \left\{ \left| \frac{j}{25} - 1 + e^{-t_j/12.02} \right|, \left| \frac{j-1}{25} - 1 + e^{-t_j/12.02} \right| \right\} \right].
$$

To illustrate the computation, the maximum for $j = 1$ is

$$\max\left\{\left|\frac{1}{25} - 1 + e^{-0.6/12.02}\right|, 1 - e^{-0.6/12.02}\right\} \doteq 0.05.$$

Computing the 25 deviations, the largest one turns out to be for $j = 10$, $t_j = 7.8$, and equals 0.12. From Table 7.1A, the critical K-S value for $\alpha = 0.05$ and $n = 25$ is 0.21; hence we again accept the hypothesis that the data are exponential.

Next, the A-D test was performed, the calculations following the previously given formula for A, which in this case reduces to

$$A = -25 - \frac{1}{25}\sum_{j=1}^{25}\left\{(2j-1)\ln[1 - e^{-t_j/12.02}] + (50 - 2j + 1)\frac{-t_j}{12.02}\right\}.$$

The calculated A value from our data is 0.300, and when compared to the critical value from Table 7.1B for $n = 25$ and $\alpha = 0.05$ of 1.305, we see that once again the data are accepted as being exponential.

The χ^2 test was not performed here since 25 data points are too few for breaking the data up into intervals to obtain the required frequency classes. Whenever the raw data exist, one is generally better off, particularly for small sample sizes, using F, K-S, or A-D. In Chapter 9 we resort to χ^2 since the data are presented in histogram form and it is impossible to reconstruct the actual values of the observations.

It should also be noted at this point that tests for exponentiality can be made directly on output data for an $M/G/1$ $(GI/M/1)$ to test for Markovian character since $M/G/1$ $(GI/M/1)$ has exponential output if, and only if, $G = M$ $(GI = M)$. This is a very important use of the goodness-of-fit tests, since queueing systems are often observed as black boxes with input and output the only available data.

One final problem we wish to discuss in this section is the determination of the service distribution for an arbitrary $M/G/1$ from its system-waiting-time data. Let us assume that the data have essentially been randomized through an appropriate scheme and are then presented to us for analysis. In addition, let us suppose that the input rate λ is known and that the service rate μ is known for the more than one parameter, nonexponential service-time distribution. From Equation (7.7), which relates the waiting-time and service Laplace-Stieltjes transforms, we can obtain a relationship for each moment of the service times in terms of those of the waiting times. Then, for example, the first four service moments can be used to estimate the parameters of the Pearson differential equation for the density $b(t)$,

$$b'(t) = \frac{(t+a)b(t)}{b_0 + b_1 t + b_2 t^2} \qquad \text{[see Fisz (op. cit.), for example]}.$$

It is well-known that the moment estimators of a, b_0, b_1, and b_2 are respectively

$$\begin{cases} \hat{a} = \dfrac{-T_3(T_4 + 3T_2^2)}{A} \\[2ex] \hat{b}_0 = \dfrac{-T_2(4T_2 T_4 - 3T_3^2)}{A} \\[2ex] \hat{b}_1 = \dfrac{-T_3(T_4 + 3T_2^2)}{A} \\[2ex] \hat{b}_2 = \dfrac{-2T_2 T_4 - 3T_3^2 - 6T_2^2}{A} \end{cases},$$

where T_i, $i = 1, 2, 3, 4$, denotes the ith waiting-time moment around the mean T_1, and $A = 10 T_2 T_4 - 18 T_2^3 - 12 T_3^2$. The form of density is then determined by the values of these estimates and is placed into one of seven categories or types accordingly. The primary types of concern to the queueing analyst are I (betas), III (gammas), VI (F's), and VII (Student t's). Notice that this is not a typical frequency estimation problem since the estimator must be derived from moments and not raw data.[6]

To close now, we note that though the distribution selection problem is extremely essential to any queueing analysis, we have devoted somewhat less space to it here than to the parameter estimation problem. This is often so because the analyst may have prior information on arrival and service processes thus eliminating the need for any additional detailed statistical work. It is also true that often the entire model is determined by the presence of exponential streams and we have devoted considerable space to that problem here. Finally, as we noted earlier, the analyst is often going to have to use graphical or seat-of-the-pants approaches and there is not much that can be said about these in any formal way.

---

[6]Another possible approach to this problem is to express the service distribution as the convolution of a number of independent, nonidentical exponentials, using as many as the number of available moments warrants (see Ch. 6, end of Sec. 6.1).

7.2   DESIGN AND CONTROL OF QUEUES

Models have on occasion been classified into two general types—descriptive or prescriptive. Descriptive models are models which *describe* some current "real-world" situation, while prescriptive models (also at times called normative) are models which *prescribe* what the real-world situation should be, that is, the "optimal" behavior at which to aim.

Most of the queueing models presented thus far are descriptive in that, for given types of arrival and service patterns, and specified queue discipline and configuration, the state probabilities and expected-value measures of effectiveness which describe the system are obtained. This type of model does not attempt to prescribe any action (such as put on another server, change from FIFO to priority, etc.), but merely represents the current state of affairs.

On the other hand, consider a resource allocation model, for example, a linear programming model for determining how much of each type of product to make for a certain sales period. This model indicates the best setting of the variables, that is, how much of each product *should be* made, within the limitations of the resources needed to produce the products. This, then, is a prescriptive model—it prescribes the optimal course of action to follow. There has been some (albeit not very much) work on prescriptive queueing models, and this effort is generally referred to under the title of *design and control of queues*.

Design models are those dealing with what the optimal system parameters (in this context, they are actually variables) should be (for example, the optimal mean service rate, the optimal number of channels, and so on). Just which and how many parameters are to be optimized depends entirely on the system being modeled, that is, the particular set of parameters which are actually subject to control.

Generally, the controllable parameters are the service pattern (μ and/or its distribution), number of channels, and queue discipline, or some combination of these. Occasionally, one may even have control over arrivals in that certain arrivals can be shunted to certain servers, truncation can be instituted, removed, increased or decreased, and so on. In many cases, however, some "design" parameters are beyond control or perhaps limited to a few possibilities. For example, physical space may prevent increasing the number of channels, and the union may prevent any proposed decreases. Similar types of effects could fix or limit other potentially controllable parameters.

Models dealing strictly with control of queues (as opposed to design of queues) are not meant to indicate optimal (or better) parameter settings directly, but are intended to detect changes in parameters. Bhat and Rao (1972) develop a control-chart approach to determining when changes in the traffic intensity ρ for $M/G/1$ and $GI/M/1$ queues take place. It may be important to know when ρ is changing since remedial action of some sort (opening another channel or increasing μ) might be called for. The literaure on these types of control models is even scantier than that on design models.

It is not always easy to make a distinction between design and control models. As we comment later, there is a class of models which deals with optimal control of service rates, and whether these are properly classified as design or control models is not immediately obvious. It often depends on the choice of the author and is not really that important. The meaningful thing is that these are prescriptive rather than descriptive models.

Although it was not directly pointed out, we have already in this text introduced the ideas of prescriptive models. The reader is specifically referred to Examples 3.6, 4.1, and 5.3, and Problems 2.12, 2.13, 2.17, 3.12c, 3.13, 3.17d, 3.18, and 3.32 for some illustrations of prescriptive type analyses. Because of the importance of the subject, we feel it warrants the added visibility provided by a chapter section, and thus devote this section to a discussion of the topic in some detail.

The work to date on prescriptive queueing models (design and control of queues) can be conveniently classified into four categories as:

(i) Cost or profit functions superimposed on classical descriptive models so that λ, μ, c, or some combination of these parameters can be optimized (these types of models are sometimes referred to as *static* models);

(ii) Rate-control policies (often referred to as optimal control policies) which deal with when and how arrival and/or service rates should be changed to optimize some objective function, generally costs or profits (these types of models are sometimes referred to as *dynamic* models, since the arrival or service rate policy specifies how these should change with system state;

(iii) Models dealing with optimization of queue discipline;

(iv) Scheduling rule (mostly in a job-shop context) selection and optimization.

Not specifically categorized here is the aforementioned control-chart approach of Bhat and Rao, as it is the only endeavor of its kind to date.

Traversing the list above in reverse order, the work on scheduling (category iv) has generally taken the form of comparing different operating rules (the shortest-processing-time (SPT) rule, most-imminent-due-date rule, etc.) against criteria such as average lateness, average time in system, maximum lateness, and other such similar measures of effectiveness. In addition, a great deal of this work, especially for job-shop scheduling which can be a rather complicated network of queues in series, has been via simulation. We refer the interested reader to the following references and their respective bibliographies: Crabill and Maxwell (1969), and Conway et al. (1967).

The models in category iii are somewhat related to category iv in that they, too, deal with determining "best" queue discipline. However, they are not scheduling oriented per se and their criteria of effectiveness involve costs. A typical type of problem illustrating this category would be a single-server queue with two priority classes. The operating policy is such that a low priority customer in service will be preempted only if his elapsed service time is less than some value, say z. Given a certain cost function, the optimal value of z is desired. These types of problems can be found in Jaiswal (1968).

Bhat (1969b) discusses the types of problems above in slightly more detail and mentions other works dealing with optimization of queue discipline of the following nature. Different arrival streams come to a server and are assigned priorities. Cost functions are imposed on expected queue size for each stream and it is desired to find the optimal priority ordering which minimizes total cost.

For models falling in categories i and ii, we present more detailed discussions which follow.

7.2.1. Rate-Control Models

Considerably more effort has been put forth on the rate-control models mentioned in category ii. These generally result in a Markovian analysis and are structured as what have been referred to as Markov decision processes. Most of the effort has concentrated on service rates with a few exceptions such as Miller (1969). He considers a c-server no-waiting-space queueing system with m customer classes, each class yielding a different reward. At first glance, Miller's model may seem quite similar to the category iii model mentioned earlier where an optimal priority scheme was desired for different arrival streams. However, in this case, the decision to be made is not a priority ordering but rather when to accept and reject customers, since no queue is allowed to form here. This, in effect, amounts to changing the arrival rate. For example, for a two-class two-server case, a possible policy might be, "if both servers are idle, allow an arrival of either class to enter service; if only one server is available, save him for a class 2 arrival; that is, if a class 1 arrival comes with only one free server, turn him away." The queueing model is assumed to be $M/M/c/c$, and the problem is formulated as an infinite-horizon, continuous-time Markov decision problem. The objective function (criterion of effectiveness) here is the expected reward rate over an infinite planning horizon, and it is desired to find the policy which maximizes this. "Qualitative" results which characterize the form of the optimal policy are given, as well as a comparison, via simulation, of some approximate policies deduced from the analysis, for arbitrary service-time distributions.

There have been several other studies on controlling the arrival process, such as charging customers an entrance fee based on system size to cut down congestion at peak periods. More detailed discussion of these types of models appears in Crabill et al. (1973).

A series of models concerned with varying the service rate of the system have been treated in the literature. We have actually introduced this concept in Section 3.7 dealing with state-dependent service, specifically Example 3.6 and Problem 3.32. The earliest model of this type in the literature appears to be that of Romani (1957). He considers a policy where, if the queue builds up to a certain critical value, additional servers are added as new arrivals come, thus preventing the queue from ever exceeding the critical value. Servers are then removed when they finish service and no one is waiting in the queue.

Moder and Phillips (1962) modify the Romani model in that there are a certain number of servers, say c_1, always available. If the number in the queue exceeds a critical value, say M_1, additional servers are added as arrivals come in a manner similar to the Romani model, except here there is a limit, say c_2, to how many servers can be added. Furthermore, in the Moder-Phillips model, the added servers are removed when they finish serving and the queue falls below another critical value, say M_2.

The usual measures of effectiveness, such as idle time, mean queue lengths, and mean wait, are derived as well as the mean number of channel starts. No explicit cost functions or optimizations are considered so that these models are in reality more descriptive than prescriptive, although the measures of effectiveness are compared for various values of certain parameters such as the "switch points" (critical values of queue size for which servers are added and removed).

Yadin and Naor (1967) further generalize the Moder-Phillips model by assuming that the service rate can be varied at any time and is under the control of the decision maker. The class of policies considered is of the following form. Denoting the feasible service capacities by $\mu_0, \mu_1, \ldots, \mu_k, \ldots$, where $\mu_{k+1} > \mu_k$ and $\mu_0 = 0$, the policy is stated as, "whenever system size reaches a value R_k (from below) and service capacity equals μ_{k-1}, it is increased to μ_k; whenever system size drops to S_k (from above) and service capacity is μ_{k+1}, it is decreased to μ_k." The $\{R_k\} = \{R_1, R_2, \ldots, R_k, \ldots\}$ and $\{S_k\} = \{S_0, S_1, \ldots, S_k, \ldots\}$ are vectors of integers, ordered by $R_{k+1} > R_k$, $S_{k+1} > S_k$, $R_{k+1} > S_k$, $S_0 = 0$, and are the policy parameters; that is, a specific set of values for $\{R_k\}$ and $\{S_k\}$ yields a specific decision rule (policy) from the class of policies described above within the quotation marks. Input is assumed to be Poisson and service exponential so the queueing model is essentially $M/M/1$ with state-dependent service.[7]

Given the class of policies above, the authors derive the steady-state probabilities, the expected system size, and expected number of rate switches per unit time. While they do not specifically superimpose any cost functions in order to compare various policies of the class studied, they do discuss the problem in general terms. Given a set of feasible $\{\mu_k\}$, a cost structure made up of costs proportional to customer wait, service rate, and number of rate switches, and sets of $\{R_k\}$ and $\{S_k\}$ which represent a particular policy, the expected cost of the policy can be computed. Thus various policies (different sets of $\{R_k\}, \{S_k\}$) can be compared. They further point out the extreme difficulty, however, of trying to find the optimal policy, that is, finding the particular sets $\{R_k\}$ and $\{S_k\}$ which minimize expected cost.

Gebhard (1967) considers two particular service-rate-switching policies for situations with two possible service rates, also under the Poisson input exponential service assumption (state-dependent $M/M/1$). He refers to the first as single level control, and the policy is, "whenever the system size is $\leqslant N_1$, use rate μ_1; otherwise use rate μ_2." The second policy he considers, called bilevel hysteretic control, can be stated as, "when the system size reaches a value N_2 from below, switch to rate μ_2; when the system size drops to a value N_1 from above, switch back to rate μ_1." The term *hysteretic* (which was also used by Yadin and Naor) stems from the "control loop" which can be seen in a plot of system size versus service rate as shown in Figure 7.1. Gebhard actually compares his two policies for specific cost functions (including both service and queueing costs) after deriving the steady-state probabilities, expected system size, and expected rate of service switching.

As the Yadin-Naor and Gebhard papers appeared simultaneously, it was not specifically pointed out that the Gebhard policies were special cases of the Yadin-Naor policies. The single level control policy is the case for which, in the Yadin-Naor notation, $R_1 = 1$, $R_2 = N_1 + 1$, $S_1 = N_1$, and the bilevel hysteretic control is the case for which $R_1 = 1$, $R_2 = N_2$, and $S_1 = N_1$.

Heyman (1968) considers an $M/G/1$ state-dependent model where there are two possible service rates, one being zero (the server is "turned off") and the other being μ (the server is "turned on"). He considers a server start-up cost, a server shut-down cost, a cost per unit time when the server is running, and a customer waiting cost. He proves that the form of

---

[7]Note that this type of $M/M/1$ model with state-dependent service can also represent a situation where additional channels are added and subtracted as the state of the system changes.

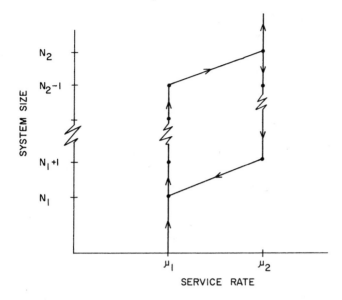

Fig. 7.1   Bilevel hysteretic control

the optimal policy is, "turn the server on when there are n customers in the system, and turn the server off when the system is empty." Heyman considers various combinations of cases involving discounting or not discounting costs over time and a finite or infinite planning horizon.

This paper represents a sort of turning point in emphasis, since rather than choose specific policies or classes of policies as was done in the aforementioned works on rate switching, Heyman's major thrust is determining what the optimal form of the class of policies should be. He does, however, also consider the problems of determining the optimal n for the various combinations of cases mentioned above (infinite horizon with and without discounting and finite horizon), although this is not the prime emphasis of his work.

Sobel (1969) considers the same problem as did Heyman, namely, starting and stopping service but generalizes it to $GI/G/1$, as well as assuming a more general cost structure. He considers only the criterion, average cost rate (undiscounted) over an infinite horizon, and shows that almost any type of stationary policy is equivalent to one which is a slight generalization of Heyman's policy and further, that it is optimal under a wide class of cost functions. The policy form is, "provide no service (server turned off) if system size is m or less; when the system size increases to M ($M > m$), turn the server on and continue serving until the system size again drops to m." He refers to these as (M, m) policies and one can readily see that Heyman's class of policies are a proper subset of Sobel's, namely, (M,m) policies where $m = 0$. Sobel's results are strictly "qualitative" in that he is interested in showing that almost any type of policy imaginable for this kind of problem falls into his class of (M,m) policies. He also shows conditions under which (M,m) policies are optimal but does not deal with determining, for specific costs, the optimum values of M and m.

The reader with some familiarity in inventory theory will notice the similarity of the (M,m) policies to the classical inventory (S,s) policies. Sobel does, in one section of his paper, show the applicability of his (M,m) policies to a production inventory system, namely, that the rule becomes, "if inventory is at least as high as M, do not produce until inventory drops to m; at that time, start to produce and continue until the inventory level reaches M."

Blackburn (1972) also treats an extension of Heyman's model in that he incorporates balking and reneging into the $M/G/1$ queue. Now, the longer the server is in the off position, the more chance there is of a balk or a renege. He shows that the stationary optimal policy which maximizes discounted reward over an infinite horizon can also be characterized by a simple pair of critical values, (M,m), as in the Sobel case, and furthermore that the critical value m will be either 0 or -1, that is, either the server is turned off when the system is empty ($m = 0$) or the server is never turned off ($m = -1$), depending on the particular costs involved, especially the magnitude of the shut-down (or start-up) cost. Blackburn analyzes the problem as a Markov renewal decision process.

Brosh (1970) considers a two-service-rate model, with a limit on queue size. The system is observed at short, equally spaced intervals of time, and at each interval, after observing the system size, a decision on which service rate to use is made for the next interval.

A Markovian analysis is used, since the assumption is made that in any interval at most one arrival occurs with the probability of an arrival being λ. The probability of a service completion is μ_k ($k = 1$, or 2, depending on which service rate was chosen the previous period) and is independent of how long the present customer has been in service. Thus Brosh is essentially dealing with a state-dependent $M/M/1/K$ model.

Considering different costs for each service rate and a cost for lost customers, the optimal policy which minimizes the long-run average cost rate is desired, that is, the optimal partition of the possible state space $\{0, 1, 2, \ldots, K\}$ into two portions, those states for which μ_1 is to be used and those for which μ_2 is to be used. Brosh categorizes the structure of the policy space, eliminating from consideration those policies (partitions) which are dominated by better ones, and from this structuring of the space develops an algorithm which searches over the admissible policies. He also gives references to linear programming algorithms for Markov chains which can solve the problem, but which, he claims, may present greater computational difficulties for K large.

Crabill (1972) treats a problem similar to that of Yadin and Naor described above, namely, an $M/M/1$ queue with k possible service rates, where now k is restricted to be finite. Crabill, however, explicitly considers a cost structure, comprised of costs associated with each service rate and a cost associated with the state of the system. He assumes no particular class of policies, as his goal is to find the characterstics of the policy class which will minimize the long-run average cost rate. He proves that the optimal policy is characterized completely by $k-1$ numbers and that the optimal service rate is nondecreasing in the state of the system. He further briefly discusses computations for the two-service-rate problem ($k = 2$). Crabill has a series of papers in process which consider variations and extensions of this work. The interested reader is referred to his Ph.D. dissertation, Crabill (1969), and Crabill et al. (op. cit.).

7.2.2 Economic (Cost or Profit) Models

We now consider category i models, which are generally descriptive and rather standard types of queues, with a superimposed cost or profit function. Parameters such as λ, μ, or c (number of channels) are considered to be controllable, and it is desired to optimize the cost or profit function (objective function) with respect to these. Most frequently, the optimization has been with respect to μ or c. It is to problems of this type that we have given the most attention throughout this text in the examples and problems specifically pointed out earlier in Section 7.2.

One of the earliest efforts on this type of queueing design model in the literature is described in Brigham (1955). He was concerned with the optimum number of clerks to place behind tool-crib service counters in plants belonging to the Boeing Airplane Company. From observed data, Brigham justifies Poisson arrivals and exponential service so that an $M/M/c$ model can be used. Costing both clerk idle time and customer waiting time, he presents curves showing the optimal number of clerks (optimal c) as a function of λ/μ and the ratio of customer waiting cost to clerk idle cost.[8]

Morse (1958) treats several economic models. He considers optimizing the mean service rate μ for an $M/M/1$ problem framed in terms of ships arriving at a harbor with a single dock (op. cit., p. 22). He desires the μ which will minimize costs proportional to it and the mean customer wait W. This is easily done by taking the derivative of the total cost expression

---

[8]This study is presented in detail in Chapter 9.

with respect to μ and equating it to zero. Morse also treats an $M/M/1/K$ problem where he considers a service cost proportional to μ and a cost due to lost customers (op. cit., p. 19). The problem is actually framed as a profit model (a lost customer detracts from the profit) and again differential calculus is employed to find the optimal μ.

Yet another model considered by Morse deals with finding an optimal μ when there is customer impatience (op. cit., p. 27). The greater μ, the less impatience on the part of a customer. Again, a profit per customer and a service cost proportional to μ is considered (see Problem 7.14). Morse also considers a model dealing with optimizing the number of channels in an $M/M/c/c$ situation (no queue allowed to form). The cost of service is now assumed proportional to c rather than μ, and again lost customers are accounted for economically by a profit per customer term. His solution is in the form of a graph where, for a given λ, μ, and ratio of the service cost to profit per customer, the optimal c can be found (op. cit., p. 33).

Finally, Morse considers several machine-repair problems, with a cost proportional to μ and a machine profitability dependent on the machine's "up-time" (op. cit., p. 159 and 165). In the latter model, he also builds in a preventive maintenance function.

A second reference on economic models in queueing of the category i type discussed here is Hillier and Lieberman (1967), Chapter 11. Much of this material is taken from the earlier works of Hillier. They consider three general classes of models, the first dealing with optimizing c, the second dealing with optimizing both μ and c, and the third considering optimization of λ and c.[9] Mention is also made of optimization with respect to λ, c, and μ. Many of the previously mentioned models of Morse turn out to be special cases of one of the classes above. We present here a few of the examples from Hillier and Lieberman, and refer the reader to this source for further cases.

We first consider an $M/M/c$ model with unknown c and μ. Assume that there is a cost per server per unit time of C_s and a cost of waiting per unit time for each customer of C_w. Then the expected cost rate (expected costs per unit time) is given by

$$E[C] = cC_s\mu + C_w L. \tag{7.8}$$

Considering first the optimal value of c, we see that as far as the first term is concerned, it is independent of c for a fixed value of ρ. Morse (op. cit., p.103) shows that for a given value of ρ, increasing c increases L, so in order to minimize $E[C]$ we would like $c = 1$. Thus the optimal c is

[9]Optimization with respect to λ deals with how to allocate arrivals to servers.

$c = 1$ and we can now consider finding the optimal value of μ for an $M/M/1$ queue.

Using (7.8) we have

$$E[C] = C_s\mu + C_w L$$

$$= C_s\mu + C_w \frac{\lambda}{\mu - \lambda}. \qquad (7.9)$$

Taking $dE[C]/d\mu$ and setting it equal to zero yields

$$0 = C_s - \frac{\lambda C_w}{(\mu^* - \lambda)^2},$$

so that

$$\mu^* = \lambda + \sqrt{\frac{\lambda C_w}{C_s}}.$$

Looking at the sign of the second derivative shows μ^* to be the value which minimizes (7.9).

Hillier and Lieberman point out an interesting interpretation of the results. If a service channel consists of a crew, such that the mean service rate is proportional to crew size, it is better to have one large crew ($M/M/1$) whose size corresponds to μ^* than several smaller crews ($M/M/c$).[10] This, of course, is true only as long as the assumptions of crew service rate being proportional to crew size and cost functions being linear are valid. Stidham (1970) extends this model by relaxing the exponential assumptions on interarrival and service times and also considers nonlinear cost functions. He shows that even for these relaxed assumptions, $c = 1$ is generally optimal.

Another interesting model treated by Hillier and Lieberman deals with finding λ and c for a given μ. The situation considered is one in which a population (such as employees of a building) must be provided with a certain service facility (such as a rest room). The problem is to determine what proportion of the population to assign to each facility (or equivalently the number of facilities) and the optimal number of channels for each facility (in the rest room example, a channel would correspond to a

[10]Hillier and Lieberman correctly point out that given an individual service rate of μ, using μ^* to obtain crew size may not yield an integer value, that is, μ^*/μ may not be integer. However, $E[C]$ is convex (the second derivative is everywhere positive) so that checking the integer value on either side of μ^*/μ will give the optimum crew size.

"stall"). To simplify the analysis, it is assumed that λ and c are the same for all facilities. Thus if λ_p is the mean arrival rate for the entire population, we need to find the optimal λ, say λ^* (optimal mean number of arrivals to assign to a facility). The optimal number of facilities then would be λ_p/λ^*.

Using C_s to denote the marginal cost of a server per unit time and C_f to be a fixed cost per facility per unit time, we desire to minimize $E[C]$ with respect to c and λ (or c and n), where

$$E[C] = (C_f + cC_s)n + nC_wL$$

$$\text{subject to: } n = \frac{\lambda_p}{\lambda}.$$

Hillier and Lieberman show under quite general conditions ($C_f \geqslant 0$, $C_s \geqslant 0$, and $L = \lambda W$) that the optimal solution is to make $\lambda^* = \lambda_p$, or equivalently $n = 1$, that is, to provide only a single service facility. The problem then reduces to finding the optimal value for c which minimizes

$$E[C] = cC_s + C_wL,$$

a problem previously discussed by Hillier and Lieberman (op. cit., p. 333) and Brigham (op. cit.).

It is interesting to consider further that $n = 1$ was shown to be optimal, regardless of the particular situation. For example, in the Empire State Building, this model would say to have only one (although it might be gigantic) rest room. This is obviously absurd and the reason, as Hillier and Lieberman point out, is that travel time to the facility is not considered in the model. If travel time is insignificant (which might be the case in certain situations), then a single facility would be optimal. When travel time is explicitly considered, this is not always the case, depending on the amount of travel time involved. Hillier and Lieberman present a variety of travel time models with examples, and again, it is recommended that the interested reader consult this source.

We present below two examples from Hillier and Lieberman which illustrate the economic design models discussed in this section.

Example 7.4 [11]

Consider a tool crib, where the average interarrival times of mechanics are Poisson with mean 50 sec and a clerk can service a mechanic according to

---

[11]Hillier and Lieberman, op. cit., p. 349. Reprinted with permission of Holden-Day, Inc.

an exponential distribution with mean 60 sec. The cost of a mechanic is $7/hr and of a clerk is $3/hr. What is the optimal number of clerks?

The cost function to be minimized is

$$E[C] = cC_s + C_w L$$

$$= 3c + 7L \ (\$/hr).$$

Using Equations (3.11) and (3.18) for an $M/M/c$ model we have

$$
\begin{cases}
r = \dfrac{\lambda}{\mu} = \dfrac{60}{50} = 1.2 \quad (c \text{ must be at least } 2) \\[12pt]
\lambda = \dfrac{3600}{50} = 72/\text{hr} \\[12pt]
\mu = \dfrac{3600}{60} = 60/\text{hr} \\[12pt]
P_0 = \left[\displaystyle\sum_{n=0}^{c-1} \dfrac{1}{n!}(1.2)^n + \dfrac{1}{c!}(1.2)^c \left(\dfrac{60c}{60c - 72} \right) \right]^{-1} \\[20pt]
L = 1.2 + \dfrac{P_0(1.2)^{c+1}/c}{c![1 - (1.2)/c]^2}
\end{cases}
$$

This can be tabulated by iterating on c as

c	cC_s	P_0	L	$C_w L$	$E[C]$
2	$6	0.250	1.875	$13.12	$19.12
3	$9	0.294	1.294	$9.06	$18.06
4	$12	0.300	1.216	$8.51	$20.51

The optimum number of clerks is therefore three.

Example 7.5 [12]

A machine shop manager desires to determine the optimal number of machines to assign to a repairman. Once the assignment is made, the

[12]Hillier and Lieberman, op. cit., p. 348. Reprinted with permission of Holden-Day, Inc.

repairman of a group is the only one who can service the machines of his group; he cannot give or receive help from other repairmen.

The time between breakdowns of a machine is exponential with mean 120 min. The service time is also exponential with a mean of 6 min. The net cost to the company of each operator is \$3/hr. It is estimated that the detraction from profit of a machine, while it is "down" (waiting for or being serviced) is \$30/hr.

The queueing model required here is that of Section 3.6, dealing with finite source queues. It is necessary to develop an appropriate cost model. We desire to minimize the expected cost of service and down time for all machines. We cannot minimize the expected cost per hour per operator, since the number of operators depends on the value we finally decide upon for our decision variable (the number of machines to be assigned to an operator). What we desire, then, is to minimize the expected cost per hour per machine, as the total number of machines is invariant. Thus we wish to find the M that minimizes

$$E[C] = \frac{C_s + C_w L}{M}$$

$$= \frac{3 + 30L}{M} \ (\$/\text{hr}/\text{machine}),$$

where M is the number of machines assigned to an operator. From Equations (3.42) and (3.43) we have (here $c = 1$)

$$p_0 = \left[\sum_{n=0}^{M} \frac{M!}{(M-n)!} \left(\frac{\lambda}{\mu} \right)^n \right]^{-1}$$

and

$$L = p_0 \left[\sum_{n=1}^{M} \frac{nM!}{(M-n)!} \left(\frac{\lambda}{\mu} \right)^n \right].$$

The expected cost rate per machine, $E[C]$, is tabulated below and we see that the optimal number of machines to assign to a repairman is six.

M	p_0	L	$E[C]$
4	0.81093	0.2186	$2.390
5	0.76436	0.2872	$2.323
6	0.71814	0.3628	$2.314
7	0.67233	0.4466	$2.343

It is interesting to observe the relative insensitivity of $E[C]$ to M near the optimal value.

A considerably different approach to these types of models is taken by Evans (1971). He considers optimizing the steady-state probabilities of a Markov chain with respect to an unknown parameter, say μ. He assumes a general cost function and a rather general form of the transition matrix of the chain as follows. Denoting the steady-state probability vector as a function of the unknown parameter μ by $p(\mu)$ and assuming the matrix of transition probabilities is of the form $Q + \mu W$, Evans formulates the problem as:

$$\text{optimize } c(\mu) + [f \cdot p(\mu)]$$

$$\text{subject to: } p(\mu)(Q + \mu W) = p(\mu),$$

where $c(\mu)$ is a cost associated with the parameter value μ, f is a vector of costs or profits, and $f \cdot p(\mu)$ is the dot product of the vectors f and $p(\mu)$.

Evans concentrates on obtaining expressions for the derivatives of the objective function with respect to μ. These expressions lead to iterative schemes for calculation which in turn suggest gradient algorithms for finding locally optimal chains. This model can be interpreted in the following way. In the absence of any action ($\mu = 0$) the Markov transition matrix is Q. This may be modified in fixed proportions by adding $\mu \cdot W$ to Q, at a cost of $c(\mu)$. The problem defines the modification μ^* which minimizes the cost of the control $c(\mu)$ and a cost dependent on the steady-state probabilities $f \cdot p(\mu)$.

To conclude this entire section on design and control of queues, we make one final comment. While the number of articles dealing with design and control models is relatively small when compared to the voluminous literature in queueing theory as a whole, this area has had a great deal of interest lately, and it is our conjecture that much of the future work in queueing theory will be along these lines.

PROBLEMS

7.1   Find the quadratic in $\hat{\lambda}$ that arises in the simultaneous solution of Equation (7.3), and then solve.

7.2   Get maximum-likelihood estimators for λ and μ in an $M/M/1$, first ignoring the initial state and then with an initial state of 4 in the steady state, from the following data:

$$\begin{cases} t = 150 & n_a = 16 \\ t_b = 100 & n_c = 12 \end{cases}.$$

7.3   Under the assumption that n_0 is always fixed at 0, find the maximum-likelihood estimators for λ, μ, and ρ in an $M/M/1$ from the following data:

$$\begin{cases} t = 150 & n_a = 16 \\ t_b = 100 & n_c = 8 \end{cases}.$$

7.4   Give a 95% confidence interval for the ρ in Problem 7.3.

7.5   Find the formulas for the maximum-likelihood estimates of λ and μ in an $M/E_3/1$.

7.6   Use Equation (7.6) to find the estimators of the service parameter from the output of an $M/E_2/1$ queue.

7.7   Use Equation (7.7) to find the estimators of the service parameter from the system waiting times of the $M/E_2/1$ queue.

7.8   Test the following data for "Poissonianness":

Number of Arrivals n	Number of 3-min Intervals Observed with n
0	21
1	23
2	10
3	4
4	1
5	0
6	0
7	1

7.9   The observed output (interdeparture time) of an $M/G/1$ queue is as follows:

0.6,	2.4,	1.0,	1.1,	0.2,	0.2,	0.2,	2.7,	1.5,	0.3,
0.6,	0.9,	0.5,	0.2,	0.5,	3.8,	0.4,	0.1,	1.5,	1.4,
0.7,	0.5,	0.1,	0.6,	1.6,	1.5,	0.5,	0.8,	1.3,	0.4,
0.7,	2.4,	2.4,	0.3,	0.8,	0.9,	1.5,	0.3,	1.2,	1.0,
0.6,	0.1,	0.4,	0.3,	2.5,	3.5,	0.8,	0.6,	9.5,	1.6.

Test to determine whether this stream is exponential and therefore whether the queue is $M/M/1$.

7.10   Draw figures comparable to Figure 7.1 for the following rate-switching policies discussed in Section 7.2.1:

(a)   Romani.
(b)   Moder-Phillips.
(c)   Yadin-Naor.
(d)   Gebhard Single Control.
(e)   Heyman.
(f)   Sobel.

7.11   Redo Problem 3.33 for $C_1 = \$4$, $C_2 = \$23$.

7.12   Consider a two-server system, where each server is capable of working at two speeds. The service times are exponential with mean rate μ_1 or μ_2. When there are $k(>2)$ customers in the system, the mean rate of both servers switches from μ_1 to μ_2. Suppose the cost per operating hour at low speed is $\$5$ and at high speed is $\$22$, and the cost of waiting time (time spent in the system) per customer is $\$1/\text{hr}$. The arrival rate is Poisson with mean $20/\text{hr}$. The service rates μ_1 and μ_2 are 7.5 and $15/\text{hr}$, respectively. Use the results of Problem 3.34 to find the optimal k. Compare the solution with that of Problem 3.33.

7.13   Consider an $M/M/1$ queue with a three-state service rate as explained in Problem 3.35. Suppose that management policy sets the upper switch point k_2 at 5. Using the results of Problem 3.35 find the optimal k_1, the lower switch point, for a system with mean arrival rate of 40, μ_1 of 20, μ_2 of 40, and μ of 50. Assume the waiting cost per customer is $\$1/\text{hr}$ and that the service costs per hour are $\$10$, $\$15$, and $\$25$, respectively.

7.14   Suppose we have an $M/M/1$ queue with customer balking. Furthermore, suppose the balking function b_n is given as $e^{-n/\mu}$, where μ is the mean service rate (see Section 3.8.1). It is known that the salary of the server depends on his skill, so that the marginal cost of providing service is proportional to the rate μ and, in fact, is estimated as 0.50μ/hr. Arrivals are Poisson with a mean rate of 10/hr. The profit per customer served is estimated at $25. Use the results of Problem 3.37 and find the optimal value of μ.

7.15 Suppose that in Example 7.4 the manager believes that costs are really only associated with idle time, that is, the time that the clerks are idle (waiting for mechanics to come) and the time that the mechanics are idle (waiting in the *queue* for a clerk to become available). Using the same set of costs ($3/hr idle per clerk and $7/hr idle per mechanic) and the same λ and μ, find the optimal number of clerks. Comment.

7.16 Consider an $M/M/1$ queue where the mean service rate μ is under the control of management. There is a customer waiting cost per unit time spent in the system of C_w and a service cost which is proportional to the *square* of the mean service rate, the constant of proportionality being C_s. Find the optimal value of μ in terms of λ, C_s, and C_w. Solve for $\lambda = 10$, $C_s = 2$, and $C_w = 20$.

Chapter 8

SIMULATION

It often turns out that it is not possible to develop analytical models for queueing systems. This can be due to the characteristics of the input or service mechanisms, the complexity of the system design, the nature of the queue discipline, or combinations of the above. For example, a multistation multiserver system with some recycling, where service times are (truncated) normally distributed and a complex priority system is in effect, would be impossible to model analytically. Furthermore, even some of the models treated previously in the text provided only steady-state results, and if one were interested in transient effects or if the probability distributions were to change with time, it would not be possible to develop analytical solutions in these cases. For such problems as these, it may be necessary to resort to analyses by simulation. It should be emphasized, however, that if analytical models are achievable, they should be used and that simulation should be relied upon only in cases where analytical models are either not achievable and approximations not acceptable or they are so complex that solution time is prohibitive.

While simulation may offer a "way out" for many analytically intractable models, it is not in itself a panacea. There are a considerable number of pitfalls one may encounter in using simulation. Since simulation is comparable to analysis by experimentation, one has all the usual problems associated with running experiments in order to make inferences concerning the real world, and must be concerned with such things as run length, number of replications, and statistical significance. However, the theory of statistics (including, of course, experimental design) can be of help here.

Another drawback to simulation analyses occurs if one is interested in optimal design of a queueing system. Suppose that it is desired to determine the optimal number of channels or the optimal service rate for a particular system where conflicting system costs are known. If an analytical model can be developed, the mathematics of optimization (differential calculus, mathematical programming, etc.) can be utilized. However, if it is necessary to study the system using simulation, then one must rely on the techniques for searching experimental output. These search techniques are often not as neat as the mathematics of optimization for analytical functions. Frequently the experimenter will merely try a few alternatives and simply choose the best among them. It might well be that none of the

381

alternatives tried is optimal nor even near optimal. How close one gets to optimality in a simulation study often depends on how clever the analyst is in considering the alternatives to be investigated. Because of this, simulation analysis has often been referred to as an *art*. Nevertheless, simulation can be an extremely important tool and is often the only procedure that can be used in analyzing many of the complex queueing systems encountered in practice.

8.1 ELEMENTS OF A SIMULATION MODEL

A simulation model can be considered as consisting of two basic phases: data generation and bookkeeping. Data generation involves the production of representative interarrival times and service times where needed throughout the queueing system. Generally this involves producing representative observations from prespecified probability distributions, and it is this aspect to which the term *Monte Carlo* has been applied. Thus a Monte Carlo simulation is one in which it is necessary to generate at least one stream of random observations from some specified probability distribution.

For illustration, consider a situation where it is desired to determine the number of toll booths to have at the entrance of the toll plaza to the Henry A. Smogchoke Memorial Expressway. Furthermore, use of manned and automatic booths is planned, and it is desired to find the number of each kind to have. It is known that the service time of an automatic toll booth (exact change lane) is deterministic, while that of a manned booth is (truncated) normally distributed. Drivers will choose generally the lane with the fewest number of cars waiting, although because of the exact change required no more than 40% of the cars will go to the exact change lane. Also, in about 2% of the cases, a car in an exact change lane will require special service because the driver may miss the coin slot, find his wife really does not have the exact change that she said she did, or encounter a shriek from one of the children in the back seat causing him to drop the coin between the seats. For these cases, the service-time distribution is highly skewed with a tail to the left, which perhaps can be fit by a beta distribution.

Even assuming that the interarrival times were exponential, this situation would be extremely difficult, if at all possible, to model analytically. One way of studying this system would be to build x manual and y automatic toll booths, observe the actual ongoing situation, then systematically add or remove toll booths of each type, observing the system every time a change is made. Obviously, the real-time experimentation would be ex-

tremely costly (building and tearing down toll booths), as well as potentially causing huge traffic jams whenever the cases of too few booths are
being observed. If representative observations of interarrival times (say
from an exponential distribution) and representative observations of service
times from truncated normal and beta distributions could be generated,
the process could be studied on paper (or what may be much more
convenient, on a computer). Changing the number of booths now would
require only the expense of more computer running time. The Monte Carlo
process which generates random variates from a specified probability
distribution (either a theoretical or empirical distribution) is discussed in
the next section.

   Not all simulations are Monte Carlo. Instead of generating random
variates, actual historical data could be used. If for an ongoing system
detailed data were accumulated over a previous period so that actual
interarrival times and service times were recorded, these past observations
could be used in a simulation to determine what would have happened to
the system had an extra server been put on or had an existing server been
taken away. However, for the vast majority of queueing problems encountered, either enough data do not exist, or the data are not in the proper
form to be utilized directly in the simulation; hence most queueing
simulations are of the Monte Carlo type.

   The bookkeeping phase of a simulation model deals with updating the
system when new events (arrivals and departures) occur, monitoring and
recording the system states as they change, and keeping track of quantities
of interest such as idle time and waiting time so that measures of effectiveness can be calculated. In the aforementioned example, the bookkeeping portion would involve allocating cars to lanes according to the smal-
lest-number-in-the-lane rule and the exact change choice probability of
0.4, determining when beta service at exact change lanes takes place,
keeping track of the increases and decreases of the number of cars in each
lane as arrivals and departures occur, recording the waiting and system
time for each car, and calculating such quantities as average queue sizes
for each type of booth, average system size, average waiting time, and so
on. One possible aggregate block diagram of a simulation model for this
example illustrating the data generation and bookkeeping phases is shown
in Figure 8.1. Data generation components are shown as dashed lines while
bookkeeping components are shown as solid. Notice that although we have
separated these two major phases for discussion purposes, in the simulation procedure shown in Figure 8.1, the process actually goes back and
forth between data generation and bookkeeping. This is generally the most
efficient procedure although one could generate all random variate streams
first, record these in some sort of table, and then select as needed during

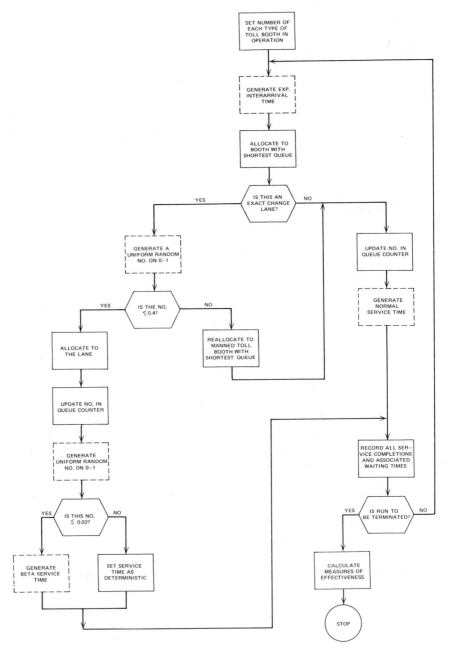

Fig. 8.1 Block diagram of simulation model.

384

the bookkeeping phase, Bookkeeping aspects of simulation queueing models are discussed further in a following section

8.2   MONTE-CARLO RANDOM-VARIATE GENERATION

We now consider the problem of generating a sequence of numbers which could be considered as being typical observations from a given probability distribution. For example, suppose it were desired to generate a sequence of Bernoulli random variates, with $p = \frac{1}{2}$. One could take a coin (presumed to be fair) and toss it in the air, record a 1 if it lands with heads showing and 0 if it lands with tails showing, repeating this process until the desired number of variates were generated. Assuming there is no bias in the coin tossing procedure, the resulting string of 1's and 0's should be a representative sequence of Bernoulli random variates with parameter $p = \frac{1}{2}$. A similar experimental procedure could be employed to obtain binomial random variates by repeating experiments consisting of tossing a coin n times and recording the number of successes. However, aside from being a time-consuming process, physical experimentation cannot be used for the vast majority of probability distributions, since they are generally not a direct result of a physical process.

Conceptually, one could take any distribution, approximate it by a finite number of discrete points, take a large bowl (or hat if desired), place into it chips designated with the value of the random variable in amounts proportional to the probabilities, and draw chips randomly and record values. Again, this is a time-consuming process and at best approximate, the degree of approximation depending on the discrete representation of the probability distribution (this could be exact for finite, discrete random variables) and how close to purely random the selection process can be made.

To avoid some of these problems, we concentrate on a procedure (which we refer to as Monte Carlo) which generates mathematically a stream of pseudorandom observations from any given probability distribution. These are referred to as pseudorandom because they are in reality a deterministic sequence (they are generated by a completely specified mathematical procedure and can be reproduced as desired) which "acts" as if it is random. By "acts" is meant that they pass a variety of statistical tests on their randomness and representativeness. It turns out for simulation purposes that it is often quite advantageous to have the ability to be able to duplicate exactly a stream of random variates. This is further discussed in a later section.

The Monte-Carlo procedure consists of two phases. First it is necessary to generate random variates from a uniform distribution on (0, 1) and then using these, random variates can be generated from any desired probability distribution.

8.2.1 Generation of Uniform-(0, 1) Random Numbers

Once again, one method for generating random numbers that are uniformly distributed is to use some physical procedure such as a spinner, or bowl of chips, or perhaps even some sort of electronic device which might yield random "beeps." We have already discussed the disadvantages of such procedures.

There are available tables of random numbers which have been generated by physical devices. One of the better known ones, which was generated on an analog computer, is available through the RAND Corporation (1955). While using tables such as these does provide a means for reproducing previous random number sequences and avoiding physical experimentation, a table must be read into the computer. Even though the RAND numbers are available on punch cards, it is often more convenient to generate random numbers as needed within a simulation program.

Most computer libraries contain a random-number-generator subroutine which generates pseudorandom numbers. These are pseudo in the sense mentioned previously in that they are completely reproducible by a mathematical algorithm but random in the sense that they have passed statistical tests, which basically test for equal probability of all values and statistical independence. Most computer routines are based on congruential methods which involve modulo arithmetic. The multiplication congruential procedure, one that is commonly employed, is a recursive algorithm of the form

$$r_{n+1} = kr_n (\operatorname{mod} m), \tag{8.1}$$

where k and m are positive integers ($k < m$). That is, r_{n+1} is the remainder when kr_n is divided by m. For example, if $k = 4$ and $m = 9$, and we choose r_0 initially as 1, we generate the numbers shown in Table 8.1. Since the smallest number[1] (remainder using mod 9) could be 0 and the largest number could be 8, the range is [0–8]. To normalize to [0–1], all numbers are divided by $m - 1 = 8$.[2] It is clear from the table even without utilizing

---

[1] Generating a zero would, of course, be disastrous since nothing but zeros would then follow. This can be prevented by not allowing m to be a multiple of k nor allowing r_0 to be 0.

[2] If the initial value is always chosen strictly between 0 and m, that is, $0 < r_0 < m$ and all numbers are divided by m instead of $m - 1$, then the normalized numbers will be uniform on the open interval (0, 1).

TABLE 8.1

r_n	Normalized Numbers $[0-1]=r_n/8$
1	0.125
4	0.500
7	0.875
1	0.125
4	0.500
7	0.875
1	0.125
4	0.500
7	0.875
1	0.125
.	.
.	.
.	.

statistical tests that this sequence would not be acceptable. First of all, only three of the possible nine numbers appear. Second we see that the sequence is cyclic with a cycle length of three. If one desired more than three random numbers, this sequence would be unusable. Changing k to 3 and m to 7 yields

$$1, 3, 2, 6, 4, 5, 1, 3, 2, 6, 4, 5, 1, \ldots.$$

Here all possible numbers [0–6], except 0, are generated but the cycle length is only 6 [the maximum cycle length of any stream using Equation (8.1) is $m - 1$].

Thus we see very careful consideration must be given to divising k and m (and to some extent r_0 also). One set that does work fairly well is to choose $k = 65539$, $m = 2^{31}$, and r_0 any odd integer less than 9999. Table A5.1 in the table section at the rear of the book presents 250 pseudorandom numbers generated using these values, with $r_0 = 1951$. The initial value, r_0, is often referred to as the *seed*. Since the largest number possible is $2^{31} - 1 = 2,147,483,647$, all generated numbers were divided by this to reduce the range to $(0, 1)$. This algorithm is an IBM library subroutine called RANDU.[3]

---

[3]There are other random number generators that perform significantly better but for our purposes here this will suffice.

It is convenient, for binary computers, to set $m = 2^b$, where b is the word length in bits. For machines with 32 bit words, one bit is used for sign leaving 31 bits to represent the number; hence $m = 2^{31}$. The reason for this convenience is that if a number exceeds 31 bits, only the last 31 bits are kept, thus automatically performing the modulo 2^{31} arithmetic. For example, if the word size were 4 bits, one bit used for sign leaves 3 for the number, so if the number 9 were generated, the machine would keep only the last 3 bits. The binary representation of 9 is 1001, but the number kept by the computer when throwing away all but the last 3 bits is 1. The number 26 in binary is 11010. The last three bits give $010 = 2$. Thus a 4-bit-word-size computer automatically performs modulo $2^3 = 8$ arithmetic. For a more detailed discussion of random-number generation the interested reader is referred to Naylor et al. (1966), Tocher (1963), and Fishman (1973b).

8.2.2 Generation of Random Variates Other Than Uniform-(0, 1)

We desire now to generate representative observations from any specified probability distribution, with CDF (say) $F(x)$. Again, there are several methods of achieving this. We present the most popular and refer the reader to the aforementioned references for further detail, if desired.

The method we present is sometimes referred to as the inverse or probability transformation method, or generation by inversion. It can best be described graphically by considering a plot of the CDF from which we desire to generate random variates. Such a plot is shown in Figure 8.2. The procedure is to first generate uniform-(0, 1) random variates, say $r_1, r_2, \ldots$.

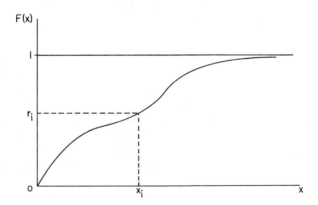

Fig. 8.2   Inversion technique for generating random deviates.

To obtain x_1, the first random variate corresponding to $F(x)$, we simply enter the ordinate with r_1, project over and down, as shown in Figure 8.2, and the resulting value from the abscissa is x_1. Repeating the procedure with $r_2, r_3, \ldots$ will yield $x_2, x_3, \ldots$.

To prove that this procedure works we would like to show that a random variate (say X_i) generated by this procedure obeys the relation

$$\Pr\{X_i \leqslant x\} = F(x).$$

We have, considering Figure 8.2, that

$$\Pr\{X_i \leqslant x\} = \Pr\{R_i \leqslant F(x)\}.$$

Since R_i is uniform-$(0,1)$, we can write

$$\Pr\{R_i \leqslant F(x)\} = F(x);$$

hence

$$\Pr\{X_i \leqslant x\} = F(x).$$

For some theoretical distributions, it is not necessary to plot $F(x)$ to obtain the x_i, since the inversion can be obtained analytically. For example, consider the exponential distribution with parameter θ. Its CDF is given by

$$F(x) = 1 - e^{-\theta x} \qquad (x \geqslant 0).$$

Entering the ordinate with a uniform-$(0,1)$ random number, r, and finding the resulting x after the projection procedure amounts to solving the following equation for x,

$$r = 1 - e^{-\theta x}.$$

Algebraic manipulation yields

$$e^{-\theta x} = 1 - r.$$

Since r is uniform-$(0,1)$, it is immaterial whether we use r or $1 - r$ as our random number and hence we can write

$$e^{-\theta x} = r.$$

Taking natural logarithms of both sides finally gives

$$x = \frac{-\ln r}{\theta}.$$

Unfortunately, analytical inversion is not possible for most probability distributions. If the distribution is discrete with frequency function $p(x)$, the following slightly different procedure can be used for computer inversion. The random variate x is the smallest value for which

$$\sum_{i=0}^{x} p(i) \geq r. \tag{8.2}$$

For example, to generate binomial random variates with parameters (n,p), we would desire the smallest x for which

$$\sum_{i=0}^{x} \binom{n}{i} p^{i} (1-p)^{n-i} \geq r.$$

If the distribution were empirical, then $p(i)$ would be in tabular form and denoting this by

$$p(i) = p_i ,$$

where p_i is the relative frequency for the value i, we would then desire the smallest x for which

$$\sum_{i=0}^{x} p_i \geq r.$$

This procedure can be somewhat time-consuming for cases where the parameters are such that there are sizable probabilities of encountering large values which would frequently require calculating sums with many terms before satisfying the inequality given in (8.2). For example, generating Poisson random variates with a mean of 20 would require sums of 25 or more terms in over 15% of the cases, since the probability of a Poisson random variable with mean 20 exceeding 24 is approximately 0.1567.

For continuous random variables for which analytical inversion is not possible, either numerical inversion techniques would have to be employed or a discrete approximation would be performed and the procedure above for empirical distributions utilized. However, the discrete approximation could present some problems in obtaining representative observations from the tails of the distribution.

For many random variables of interest, we can often make use of some statistical theory to aid in the random variate generation procedure. For example, if we wish to generate Erlang type k random variables, instead of attempting inversion on the Erlang CDF, we can merely take sums of k

exponential random variates which are quite easy to generate by inversion. This was actually done in generating the data for Problem 4.16 of Chapter 4. Thus if we desired to generate random deviates from an Erlang type k with mean $1/\mu$, we could obtain this type of random variate, say x, from the uniform-$(0,1)$ random variates $r_1, r_2, \ldots, r_k$ by

$$x = \sum_{i=1}^{k} \left(-\frac{\ln r_i}{ku} \right)$$

$$= -\frac{\ln \prod_{i=1}^{k} r_i}{k\mu}.$$

To generate normal random variates, use can be made of the central limit theorem. One common procedure is to take sums of twelve uniform-$(0,1)$ random numbers since these are approximately normally distributed with mean 6 and variance 1. Subtracting 6 yields standard normal variates (mean 0, variance 1). Thus to get any desired mean and variance (say μ and σ^2), multiply the resulting numbers by σ and add μ. Hence an approximately normal random variate with mean μ and variance σ^2 can be obtained from 12 uniform-$(0,1)$ random numbers as

$$x = \sigma \left(\sum_{i=1}^{12} r_i - 6 \right) + \mu.$$

A better degree of approximation can be accomplished by taking sums of more than twelve uniform-$(0,1)$ random numbers. Twelve is convenient to choose since the variance of a uniform-$(0,1)$ random variable is $\frac{1}{12}$ so that summing twelve values gives unit variance. Also, the sum of uniform random variables converges quite quickly to normal so that 12 is generally sufficient.

These sorts of procedures are quite helpful in alleviating some of the problems of generating random variates on a computer by inversion. As two last examples we mention that χ^2 random variates can be obtained from summing squares of standard normal variates, while variates from an F distribution can be gotten by taking the ratio of two χ^2 variates.

It is possible, therefore, to generate random variates from any probability distribution using the procedures above, although in some cases it may be time-consuming and/or approximate. Nevertheless, in most cases it can be done without too much difficulty. We next turn our attention to the consideration in more detail of the bookkeeping phase of a simulation analysis.

8.3   BOOKKEEPING ASPECTS OF SIMULATION ANALYSIS

As mentioned previously, the bookkeeping aspects have to do with updating the system status when events occur, recording items of interest, and calculating measures of effectiveness. There are two general methods of accomplishing the bookkeeping portion of the study: (1) time-oriented bookkeeping (sometimes called synchronous) and (2) event-oriented bookkeeping (sometimes called asynchronous).

Time-oriented bookkeeping involves setting a basic time unit (second, minute, etc.) and setting up in the computer a "master clock" which advances each basic time unit, updating the system states time unit by time unit. Event-oriented bookkeeping updates the system state only when events (arrivals or departures) occur. Since there is not necessarily an event every basic time unit, in event-oriented bookkeeping the master clock is increased by a variable amount each time rather than a fixed amount as in time-oriented bookkeeping. These methods can be best illustrated by an example.

Consider a single-channel queue with arrival and service data given in Table 8.2. Table 8.3 illustrates time-oriented bookkeeping, while Table 8.4 illustrates event-oriented bookkeeping. If the reader were to reconstruct these tables from the data of Table 8.2 he would see pros and cons for each method. For example, in comparing Tables 8.3 and 8.4 one sees that only 18 lines are required for Table 8.4, while 27 lines are necessary for Table 8.3. Furthermore, slightly more information is given in Table 8.4 in that waiting times are given for all 12 customers, while in Table 8.3 time in queue is available only for 10 customers and time in system available only for nine customers. To obtain waiting time for all 12 customers would require the master clock to reach time 31. Thus time-oriented bookkeeping,

TABLE 8.2

						i						
	1	2	3	4	5	6	7	8	9	10	11	12
Interarrival time between customers $i+1$ and i	2	1	3	1	1	4	2	5	1	4	2	—
Service time of customer i	1	3	6	2	1	1	4	2	5	1	1	3

TABLE 8.3 TIME-ORIENTED BOOKKEEPING

(1)	(2)	(3)	(4)	(5)	(6)	(7)	(8)
Master Clock Time (min)	Arrival (Customer ⓘᵃ)	Start of Service (Customer ⓘᵃ)	Service Completion (Customer ⓘᵃ)	No. in Queue	No. in System	Time in Queue (min)	Time in System (min)
0	①	①		0	1	0,①ᵃ	
1			①	0	0		1,①ᵃ
2	②	②		0	1	0,②	
3	③			1	2		
4				1	2		
5		③	②	0	1	2,③	3,②
6	④			1	2		
7	⑤			2	3		
8	⑥			3	4		
9				3	4		
10				3	4		
11		④	③	2	3	5,④	8,③
12	⑦			3	4		
13		⑤	④	2	3	6,⑤	7,④
14	⑧	⑥	⑤	2	3	6,⑥	7,⑤
15		⑦	⑥	1	2	3,⑦	7,⑥
16				1	2		
17				1	2		
18				1	2		
19	⑨	⑧	⑦	1	2	5,⑧	7,⑦
20	⑩			2	3		
21		⑨	⑧	1	2	2,⑨	7,⑧
22				1	2		
23				1	2		
24	⑪			2	3		
25				2	3		
26	⑫	⑩	⑨	2	3	6,⑩	7,⑨

ᵃCircled numbers denote customer number.

TABLE 8.4   EVENT-ORIENTED BOOKKEEPING

(1)	(2)	(3)	(4)	(5)	(6)	(7)	(8)
Master Clock Time	Arrival/Departure Customer i	Time Arrival(i) Enters Service	Time Arrival(i) Leaves Service	Time in Queue	Time in System	No. in Queue Just After Master Clock Time	No. in System Just After Master Clock Time
0	①-A	0	1	0	1	0	1
1	①-D					0	0
2	②-A	2	5	0	3	0	1
3	③-A	5	11	2	8	1	2
5	②-D					0	1
6	④-A	11	13	5	7	1	2
7	⑤-A	13	14	6	7	2	3
8	⑥-A	14	15	6	7	3	4
11	③-D					2	3
12	⑦-A	15	19	3	7	3	4
13	④-D					2	3
14	⑧-A;⑤-D	19	21	5	7	2	3
15	⑥-D					1	2
19	⑨-A;⑦-D	21	26	2	7	1	2
20	⑩-A	26	27	6	7	2	3
21	⑧-D					1	2
24	⑪-A	27	28	3	4	2	3
26	⑫-A; ⑨-D	28	31	2	5	2	3

since it records times when nothing happens, requires more space to record the same information as the event-oriented method.

In addition, it is more cumbersome to record individual waiting times using time-oriented bookkeeping since, for example, to determine the time in system of customer i, when i departs one must go back to find the time i arrived and subtract the two. In the event-oriented table, it is only a matter of subtracting the value in the first column from the value in the fourth column.

There is an easier approximate method of obtaining average waiting time per customer for time-oriented bookkeeping which is quite accurate for long run lengths. If, for example, average wait for service is desired, the computation would consist of summing column 5 of Table 8.3, which gives total unit-minutes of 'waiting, and dividing by the total number of customers. The error results in that the total from column 5 is somewhat deflated since only partial waiting times of customers still in the system are included. Using this procedure, an estimate of average wait for service yields $39/12 = 3.25$. Taking the average of the twelve actual waiting times from Table 8.4 gives 3.33.

To help illustrate the differences in carrying out the bookkeeping for the two methods, flow charts necessary to determine waiting time in queue are shown in Figures 8.3 and 8.4 for the time-oriented (Table 8.3) and event-oriented (Table 8.4) procedures, respectively. Problem 8.5 asks for more complete flow charts for the two procedures. Figure 8.5 shows an alternate (but less straightforward) way of obtaining the waiting time in queue for the event-oriented method given and illustrates how efficiencies can be achieved by clever programming. The advantage of the procedure in Figure 8.5 is that one need not work with cumulative sums as in Figure 8.4, so that the magnitudes of the numbers are not increasing as the simulation progresses in time. For long run lengths, the procedure of Figure 8.4 necessitates working with large numbers (NAT, TIS, TOS), since these numbers increase monotonically as simulation time progresses. The reader should work through three or four iterations for each of the procedures (Figures 8.3, 8.4, and 8.5), using the data in Table 8.2 to convince himself of the validity of each.

Turning now to consideration of average system size, one need only take the arithmetic average of the values in the sixth column of Table 8.3 (time-oriented method), while for the event-oriented method, it is necessary to multiply the values in column 8 of Table 8.4 by the elapsed time between the present and next event, then sum up the results and divide by total elapsed time. The reader should verify that both procedures give an L of $65/27 \doteq 2.4$ (note that since time started at 0 the actual elapsed time is 27 min, since we count the minute which begins at clock time 26 and ends just prior to the clock advancing to 27).

There is also an easier method for event-oriented bookkeeping in calculating the average system or queue size. Suppose average system size is desired. Instead of using column 8 in Table 8.4 and multiplying by elapsed time between successive events, one can use the time in system (column 6). To get results up to a clock time of 27 (through minute 26), one can add up the total time in system (total of column 6) and divide by total elapsed time, yielding $70/27 \doteq 2.6$. This answer is somewhat inflated (compare with previous answer of $L \doteq 2.4$), since the total time in system for the last two customers is included even though part of these times extends beyond clock time 26. However, for a long run length this error would be small.

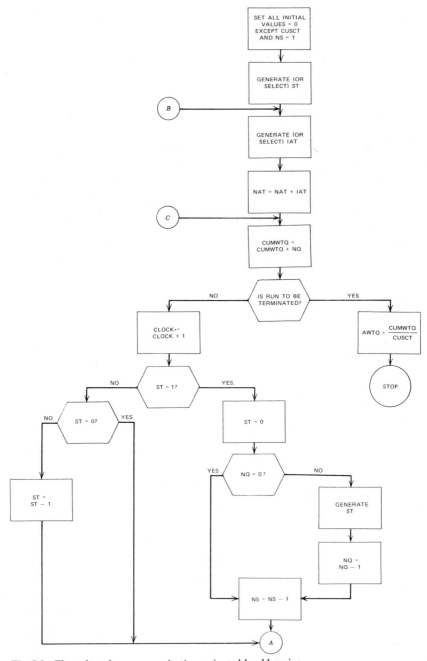

Fig. 8.3 Flow chart for queue wait, time-oriented bookkeeping.

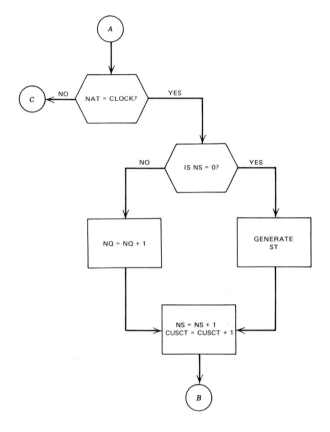

Fig. 8.3 (Continued).

Definition of Terms:

AWTQ = Average waiting time in queue per customer
CLOCK = Master clock time
CUMWTQ = Cumulative waiting time (all customers) in queue
CUSCT = Cumulative number of arrivals
IAT = Interarrival time
NAT = Clock time for next arrival
NQ = Number in queue
NS = Number in system
ST = Service time (also remaining service time)

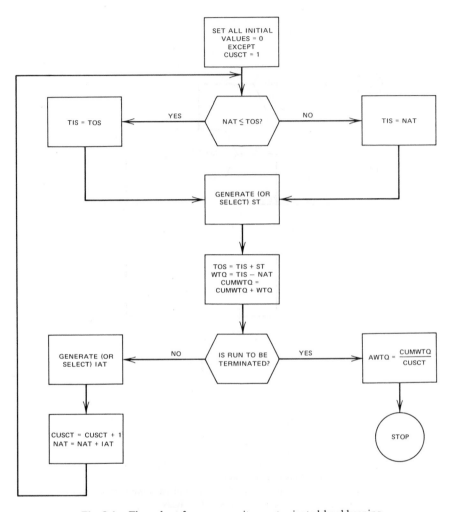

Fig. 8.4 Flow chart for queue wait, event-oriented bookkeeping.

Definition of Terms:

AWTQ = Average waiting time in queue per customer
CUMWTQ = Cumulative waiting time (all customers) in queue
CUSCT = Cumulative number of arrivals
IAT = Interarrival time
NAT = Time of next arrival (cumulative arrival time)
ST = Service time
TIS = Time into service
TOS = Time out of service

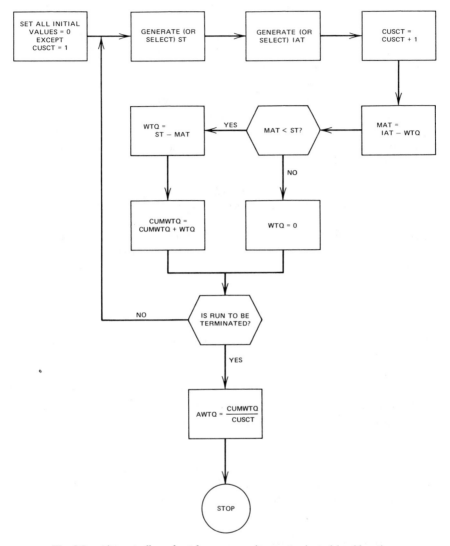

Fig. 8.5 Alternate flow chart for queue wait, event-oriented bookkeeping.

Definition of Terms:

AWTQ = Average waiting time in queue per customer
CUMWTQ = Cumulative waiting time (all customers) in queue
CUSCT = Cumulative number of arrivals
IAT = Interarrival time
MAT = Marginal arrival time (minutes present customer arrives *after* previous customer entered service; negative value indicates *before*)
ST = Service time
WTQ = Waiting time of a customer in queue

Both this latter method of obtaining L through use of system times (or similarly L_q by using waiting times) and the "easier approximate" method previously mentioned for time-oriented bookkeeping of obtaining W (or W_q) by using the number in system (or number in queue) column is a direct illustration of Little's formula (see Problem 2.9). Note that if we ignore the error introduced by partial waits when terminating a continuing system at an arbitrary clock time and then divide the numerator and denominator of the $70/27$ figure given above by 12 (the number of customers), we have

$$L \doteq \frac{70}{27} = \frac{70/12}{27/12} = \frac{W}{1/\lambda} = \lambda W.$$

Similarly, the value of W_q obtained for time-oriented bookkeeping, previously given as the total of column 5 of Table 8.3 divided by the number of customers, can be considered as

$$W_q \doteq \frac{39}{12} = \frac{39/27}{12/27} \doteq \frac{L_q}{\lambda}.$$

The errors due to partial waits when terminating a continuing system are very minor in any realistic simulation run, since run lengths are generally long enough to make these errors insignificant.

Problems 8.5 and 8.6 provide a feel for the respective difficulties in programming each of the bookkeeping methods (time and event-oriented). As to which method takes less time on a computer depends somewhat on what one desires for output, but mostly on how quickly events happen. If events occur very frequently, the time-oriented method could be more efficient. However, for most queueing simulation studies, event-oriented bookkeeping has generally been used and is probably the more efficient procedure when dealing with queueing situations.

8.4   SIMULATION PROGRAMMING LANGUAGES

The most well-known and widely used programming language for scientific-mathematical studies is undoubtedly FORTRAN. FORTRAN is a general-purpose mathematically oriented language which is extremely flexible, relatively easy to learn, and available for almost any digital computer manufactured today. There are other similar types of languages, such as ALGOL and PL1; however, these are not as popular or readily accessible as FORTRAN.

Since simulation studies have many elements in common, for example, the concept of a clock time, status updating, and event sequencing, special-purpose languages tailored for simulation have been developed. Among the most popular are GPSS, SIMSCRIPT, and DYNAMO. The first two are particularly well-suited to queueing-type problems. We describe briefly some of these simulation languages to give the reader a "flavor" of what these languages are like, their advantages and their disadvantages. For more detail, the reader is referred to IBM Corporation (1970), Kiviat et al. (1968), Krasnow and Merikallio (1964), Meir et al. (1969), Kiviat (1966), Naylor et al. (1966), and Fishman (1973b).

8.4.1 GPSS Simulation Language

GPSS stands for General Purpose Systems Simulator and was developed at IBM by G. Gordon and R. Efron.[4] It is a *block* diagrammatic language where the system structure must be defined in terms of a fixed set of predefined block types. Each block type has its own symbol (see Figure 8.6) so that the program is actually "written" by a block diagram. Input into the computer is by punch cards, with each block requiring one card to describe it. In addition to a card for each block, some definition and control cards are also required.

The blocks represent command-type operations (specific actions) such as those illustrated in Figure 8.6. Built into the language are certain automatic

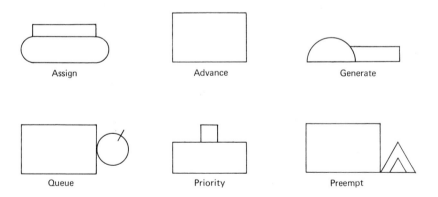

Assign   Advance   Generate

Queue   Priority   Preempt

Fig. 8.6   Some examples of GPSS block symbols.

[4]The most recent version for the IBM 360 is actually called the General Purpose Simulation System, but the original name is the better known and appears as such throughout most of the literature on simulation languages.

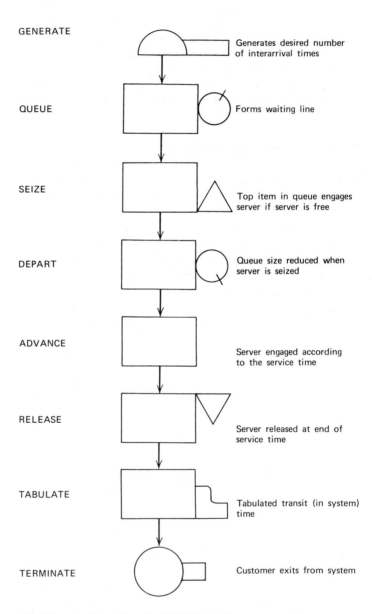

| BLOCK OPERATION | SYMBOL | EXPLANATION |

GENERATE — Generates desired number of interarrival times

QUEUE — Forms waiting line

SEIZE — Top item in queue engages server if server is free

DEPART — Queue size reduced when server is seized

ADVANCE — Server engaged according to the service time

RELEASE — Server released at end of service time

TABULATE — Tabulated transit (in system) time

TERMINATE — Customer exits from system

Fig. 8.7 An illustration of a GPSS block-diagram program.

calculations of interest; for example in the QUEUE block, average queue size and average time in queue are automatically measured. Also, if desired, the distribution of time spent in queue is provided. Figure 8.7 shows how a simple single-server queue may be "programmed" in GPSS.[5]

The block-diagram program, in addition to consisting of block symbols, would also have information contained within the block symbols. For example, the first two blocks (numbered 10 and 11) might look as shown in Figure 8.8. The GENERATE block shows that 1000 customers are to be created, arriving with a mean interarrival time of 10 from a function FN1. This function could be, if desired, the exponential CDF. The generated arrivals next enter a queue. The QUEUE block has two numbers in it representing the index number of the queue and the number of units which are to be added to the queue. In our example, since there is only a single queue and units enter one at a time, the number one appears in both places.

The punch cards describing these two blocks are illustrated in Figure 8.9. While the language does have uniform-(0, 1) random-number generators incorporated in it, to generate random deviates from other distributions it is necessary to define the function desired by reading in points from the desired CDF. This is required even though mathematical inversion may be possible. Shown in Figure 8.9 are 24 points from an exponential CDF with mean equal to one. The function argument, RN1, is an internal uniform-(0, 1) random-number generator, so that by using this argument and the CDF points, random variates from an approximately exponential distribution can be provided. The GENERATE block takes 1000 of these values and multiplies each by 10 to achieve exponential interarrival times with a mean of 10.

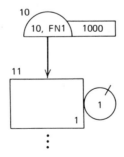

Fig. 8.8 An illustration of information required in block symbols.

[5]This example and those to follow which illustrate the simulation languages are not intended to be exact or complete in every detail but rather are intended to show the general nature of the languages.

| 1 | FUNCTION | RN1, C24 | | | | | | | | | |

0	0	.1	.104	.2	.222	.3	.355	.4	.509	.5	.69
.6	.915	.7	1.2	.75	1.38	.8	1.6	.84	1.83	.88	2.12
.9	2.3	.92	2.52	.94	2.81	.95	2.99	.96	3.2	.97	3.5
.98	3.9	.99	4.6	.995	5.3	.998	6.2	.999	7.	.9997	8.

| 10 | GENERATE | 10 | FN1 | 1000 |

| 11 | QUEUE | 1 | 1 |

⋮

Fig. 8.9   An illustration of punch card input describing blocks.

The bookkeeping for GPSS is event-oriented. The program keeps track of future events in a future-events chain and updates the clock time to the next future event to take place. During any current clock time, customers are pushed through as many blocks as possible. When no one can move any further, the clock is updated to the next future-event time. Suppose, for example, that the next event on the future-events chain is service completion. Clock time is updated and the customer in the ADVANCE block is now pushed to the RELEASE block freeing the server. Next, the served customer is pushed to the TABULATE block and then to the TERMINATE block where he exists from the system. The queue is then searched. Assuming there is a nonempty queue, the customer at the front is pushed to the SEIZE block, thus engaging the server. He is then pushed to the DEPART block, reducing queue size by one. Next, he is pushed to the ADVANCE block where he is given a service time. Assuming the service time is not zero, he can go no further during this clock time. The future-events chain is now searched and the clock is updated to the next event, which will either be a new arrival or the completion of service. If the next event were an arrival, the arriving customer would be pushed to the QUEUE block. He cannot be pushed further, since even if there is no queue, the server is busy. Thus this is as far as he can go during this clock time, and the future-events chain is searched once again. Had the server been idle when the new arrival came (implying also an empty queue), he could have been pushed as far as the ADVANCE block.

There are some 45 block types in the GPSS language so that the model can be quite sophisticated. For example, there are blocks which can examine, test, loop, alter, link, and logically switch. In addition to blocks, there are two other key elements to the language, which we have already been using in our illustrations, but have not pointed out explicitly. They are *transactions* and *system entities*. The transactions are the basic units flowing through the simulator or what we have been referring to as "customers" in our queueing models. Of course, these "customers" could represent people, automobiles, messages, jobs, and so on. Transactions are always nonpermanent, in that they are constantly created (arrive) and terminated (exit).

System entities are permanent elements of the simulation model and are categorized into three types: facilities, storages, and logical switches. As an example, for a traffic model a storage entity might be a road, a facility a toll booth, and a logical switch a traffic light. In our simple single-server queue example, the server is a facility and there are no logical switches or storage entities. Had we considered a finite waiting capacity, the queue would then have been replaced by a storage entity, requiring the replacement of the QUEUE block by an ENTER and a LEAVE block. The ENTER block would prevent a customer from joining the queue if it is at its capacity. As an illustration of the logical switch concept in our single-server queueing model example, suppose we wish to allow for periodic breakdowns in our service facility, so that when the server is down, the customers are shunted to an auxiliary queue. This could be accomplished with a logical switch and block types LOGIC and GATE. Depending on the state of the logical switch (logical switches are binary in that they have only two possible states), the customers could either flow to the server or the auxillary.

From the brief description above, one can get a feeling for the modeling ability of GPSS with respect to queueing models. It is capable of modeling quite complex queueing systems (as well as other types of systems which are not of particular interest to us here). It is relatively easy to learn and has good diagnostic capabilities for program debugging. It is not, however, available for all computers, since it requires a fairly large machine on which to operate. Also, it is not as flexible as the general purpose languages such as FORTRAN in that the types of calculations and output are fixed.

8.4.2 SIMSCRIPT Simulation Language

SIMSCRIPT, originally developed by the RAND Corporation and now marketed by Consolidated Analyses Centers, Inc., of California, is an "algebraic" type language more like FORTRAN than the block diagrammatic GPSS. In fact, the more recent version of SIMSCRIPT is being marketed as a general-purpose language competitor to FORTRAN as well as a simulation language. Its major features of interest here, however, are its simulation capabilities.

In observing any system, one must be cognizant of two things: (1) the instantaneous state of the system at any point in time and (2) the points in time at which the state of the system changes. SIMSCRIPTS's modeling concept (or "world view" as it is often referred to) is to describe the system status by means of *entities, attributes*, and *sets*, and to account for status changes by *event* routines.

An entity is an object associated with the system to be simulated, such as a customer or a server. Entities can be either permanent (remain with the system throughout) or temporary (come and go). Servers are generally permanent entities while customers are temporary. It is possible, of course, to wish to model a system where servers are phased in and out, in which case they would also be temporary entities.

Entities may have properties associated with them which are called attributes. Attributes take on values, for example, the entity "customer" may have an attribute "age." Or, if the customer happens to be a vehicle, its attributes may be volume and weight.

Similar entities are related through the use of sets. For example, customers waiting for a particular server are said to *belong* to a set which may be called a "queue." The set called "queue" is *owned* by the server, so that all sets have members and owners. In some cases, a set's owner may be the system itself, rather than an entity of the system. Entities may be filed in a set on a first-in, first-out (FIFO) basis, last-in, first-out (LIFO) basis or ranked according to some attribute value.

SIMSCRIPT requires that the system to be simulated be completely described in terms of entities, attributes, and sets. The user specifies all entities, along with a list of their attributes and possible set memberships and ownerships. Also required is a specification of the types of events the system is to encounter. These are all defined in the first section of the program which is called the PREAMBLE, and in some ways is analogous to the common portion of a FORTRAN program.[6] An example of a portion of a PREAMBLE is shown in Figure 8.10.

PREAMBLE

EVENT NOTICES INCLUDE ARRIVAL
    EVERY ENDSERV HAS A CUST
TEMPORARY ENTITIES
    EVERY CUSTOMER HAS AN ATIME,
    AND MAY BELONG TO THE QUEUE
THE SYSTEM OWNS THE QUEUE
DEFINE QUEUE AS A FIFO SET
DEFINE LAMDA, MU, IDLE AND ELAPSTIME AS VARIABLES
ACCUMULATE LQ AS THE AVG OF N. QUEUE
TALLY W AS THE AVG OF ELAPSTIME
END

Fig. 8.10 An illustration of a SIMSCRIPT PREAMBLE.

[6]Older versions of SIMSCRIPT do not have a PREAMBLE portion but instead require that all entities, attributes, and set relations be described on a prespecified definition form.

The preamble shown in Figure 8.10 is not complete in every detail (e.g., the mode—integer or floating point—of each variable must be designated), but nevertheless illustrates the basic ideas of a PREAMBLE required for, say, an $M/M/1$ queueing model. There are two types of events, arrivals and service completions, the latter designated by ENDSERV. Because ENDSERV is associated with a customer, it is necessary to give it an attribute which is designated as CUST, since an attribute cannot take the identical name of an entity.

CUSTOMER is a temporary entity, since customers will be created and destroyed as the simulation progresses. CUSTOMER is also given its arrival time as an attribute because it is needed in calculating the total time spent in the system which is required in computing W. Customers are also members of the set QUEUE, which is defined as a FIFO set. No definition of a permanent entity, say server, is required explicitly since the language always assumes that there is at least one permanent entity called SYSTEM. Since we have a single waiting line queue, the server and system are identical. Had we had a multiple-server system with separate queues allowed in front of each, we would have had to define the servers as permanent entities by including

PERMANENT ENTITIES
EVERY SERVER OWNS A QUEUE

in the PREAMBLE.

Variables to be used throughout the simulation are defined and include the mean arrival and service rates (which the user may wish to vary from run to run), a variable IDLE to be used later for checking if the server is busy, and the total time spent in the system by each customer (ELAPSTIME). It is not necessary to define a variable for the number in the queue as SIMSCRIPT defines a counter variable automatically for every set defined. This counter variable is designated by the set name preceded by N.; hence for the set QUEUE we automatically have the variable N. QUEUE.

The two lines prior to the END statement illustrate two of the many internal features of the language designed especially for simulation models (which are particularly useful in queueing simulations). The ACCUMULATE and TALLY statements automatically generate statistics of interest concerning measures of effectiveness. ACCUMULATE is used when desiring statistics concerning time-dependent phenomena while TALLY yields statistics involving time-independent phenomena. To obtain average queue size, for example, we have previously seen in Section 8.3 that for event-oriented bookkeeping (which SIMSCRIPT is) it is necessary not only to record the number in the queue but the time the queue spends in each state. On the other hand, obtaining average time spent in the system by a customer requires only a straight average of all the elapsed times of each customer. The user can, if he wishes, also obtain the variance and even a histogram of these measures of effectiveness.

Following the PREAMBLE is the MAIN program (similar to the MAIN program in FORTRAN) which reads in data, initializes counters, specifies the output and formatting (which, incidentally, is quite flexible and easy to do), schedules the first event to take place, and starts and stops the simulation. The statement START SIMULATION engages the simulator internal timing routine which keeps track of the schedule or "calendar" of future events much like the future-events chain in GPSS. As bookkeeping in SIMSCRIPT is also event-oriented, the simulated clock time advances to the next event scheduled on the calendar. What actually happens when events occur is programmed by the user in event routines (which are similar to subroutines in FORTRAN) so that control passes back and forth between the timing routine and the event routines.

Event routines are indicated by either the statement EVENT *name* or UPON *name* and appear after the MAIN program. Figure 8.11 illustrates an event routine dealing with what happens when the calendar indicates that the next event to take place is an arrival. The clock time (TIME.V) is advanced to the time the arrival is to take place as given on the calendar and control is passed to the event routine EVENT ARRIVAL (or UPON ARRIVAL if preferred).

The EVENT ARRIVAL subroutine of Figure 8.11 first creates a customer, setting the customer's arrival-time attribute to the current clock

```
EVENT ARRIVAL
CREATE CUSTOMER
LET ATIME = TIME.V
IF IDLE = 0, FILE CUSTOMER IN QUEUE
        GO TO SCHED.NEXT.ARVL
OTHERWISE
LET IDLE = 1
SCHEDULE AN ENDSERV(CUSTOMER) IN
        EXPONENTIAL.F(1./MU, - 1) HOURS
'SCHED.NEXT.ARVL'
        SCHEDULE AN ARRIVAL IN
        EXPONENTIAL.F(1./LAMBDA, - 1) HOURS
        RETURN
        END
```

Fig. 8.11   An illustration of a SIMSCRIPT event routine.

time. Next, a test is made to determine if the server is busy (IDLE = 0) and if so, the customer is placed in the QUEUE set and the GO TO command sends the program to the next *labeled* statement (a statement in single quotation marks). The next arrival is then scheduled according to an exponential distribution with mean $1/\lambda$ which is an internal subroutine of the language (unlike GPSS which requires an exponential CDF to be read in as data). Control then returns to the timing routine and the calendar is searched for the next event to occur. If the server is free (IDLE $\neq$ 0), an end of service is scheduled according to an exponential distribution with mean $1/\mu$. Control then returns to the timing routine.

Had the next event to take place on the calendar been the completion of a service, control would go to an end of service event routine which might appear as that illustrated in Figure 8.12. The first thing done after notice is received of an end of service event is to calculate the customer's elapsed time (time in system) which is needed to get W. Notice that all that is necessary is merely to calculate each customer's elapsed time, as the TALLY statement in the PREAMBLE takes care of providing any desired statistics concerning elapsed time; in our example all we ask for is the mean W. Next the customer is destroyed (leaves the system) and the queue is checked to see whether it is empty, and if it is, IDLE is set to so indicate this (IDLE = 1). If the queue is not empty, the first customer is removed and placed in service where an end of service time is generated for him. Control then returns to the timing routine. Notice that when a member is removed from a set, the set counter automatically reduces by one. It is not necessary to program this explicitly. Also notice that the SIMSCRIPT language accepts commands such as IF QUEUE IS EMPTY, CREATE, DESTROY, and so on, which illustrates its ability to accept a more English-oriented code than FORTRAN.


```
UPON ENDSERV(CUST)
LET ELAPSTIME = TIME.V − ATIME(CUST)
DESTROY CUSTOMER CALLED CUST
IF QUEUE IS EMPTY, LET IDLE = 1
        RETURN
OTHERWISE
REMOVE FIRST CUSTOMER FROM QUEUE
SCHEDULE AN ENDSERV(CUSTOMER) IN
    EXPONENTIAL.F(1./MU, − 1) HOURS
RETURN
END
```

Fig. 8.12 An illustration of a service completion event routine.

The language also abounds in synonyms. For example, in place of OTHERWISE one could use ELSE. For an arithmetic test, one may use the symbol <, or the word statement LT, LS, or LESS THAN.

There are many convenient logical statements in addition to DO loops which might require several lines of code to accomplish in FORTRAN. For example, there is an UNTIL statement which can be used with a FOR statement as illustrated by the following.

.

.

.

FOR I = 1 TO N, UNTIL ABS.F(YN − YO) < .000005.
DO LET Y(N) = YO + (X(I)/YO − YO)/2
LET Y(O) = Y(N)
LOOP

.

.

.

The illustration above shows the calculation portion for Newton's square root approximation, where for each $X(I)$, the square root is calculated iteratively until two successive values have an absolute difference less than 0.000005. In addition to UNTIL, there are statements such as WHILE, WITH, and UNLESS which allow looping to continue "while," "with," or "unless" something is satisfied. These statements can also be used without looping, of course.

The language also has many convenient library functions such as ABS.F (absolute value), SQRT.F (square root), LOG.E.F (natural logarithm), MAX.F (maximum value), to mention a few.

In comparing SIMSCRIPT to GPSS, SIMSCRIPT appears to have a great deal more power and flexibility than GPSS. The SIMSCRIPT user is always free to write his own routines in SIMSCRIPT source language, a capability not available for GPSS. Outputting and report generation are quite extensive and flexible in SIMSCRIPT. However, SIMSCRIPT is much more difficult to learn. One can learn to program queueing models of modest nature in a matter of hours using GPSS with no particular knowledge of any other programming language. On the other hand, weeks may be involved in learning SIMSCRIPT, even with prior knowledge of FORTRAN, which turns out to be quite helpful in understanding the SIMSCRIPT language.

For sophisticated models, SIMSCRIPT appears to be more efficient in running time and computer capacity, although statements of this nature must be guarded since in any particular situation such general statements may be violated.

8.4.3 Other Simulation Languages

While GPSS and SIMSCRIPT are the most popular of the simulation languages, there are a variety of other simulation languages. Meier et al. (1969) briefly describe some additional languages which we quote below.[7]

CSL   —Control and Simulation Language developed by Buxton and Laski of IBM United Kingdom, Ltd., and Esso Petroleum, respectively. Uses concepts of entities, set and activities. Based on FORTRAN.

SIMPAC   —Developed at Systems Development Corporation. Uses concepts of activities, transactions, queues, and resources. Requires knowledge of SCAT and is closer to machine language than most other simulation languages.

GASP   —Developed at United States Steel. Consists of a collection of about 30 FORTRAN subroutines. Uses concepts of elements, attributes, queues, and events which cause activities.

OPS − 3   —Developed at the Massachusetts Institute of Technology for on-line, time-sharing computer system. OPS is a general system of which simulation is a part. Uses concepts of activities placed on an agenda.

SIMULA   —Developed at Norwegian Computing Center as an extension of ALGOL. Uses concepts of processes which become active, suspended, passive, or terminated due to occurrence of events.

One rather well-known language which has not been mentioned thus far is DYNAMO. DYNAMO was developed at The Massachusetts Institute of Technology and is an outgrowth of the industrial dynamics concept of J. W. Forrester. It is not particularly suited for queueing models, as DYNAMO considers systems to be of the closed-loop information feedback type which are continual in time, whereas queueing systems are not continual in time, since they change their state only at discrete intervals. The DYNAMO world view of a system is that it can be described in terms of level and rate of change equations, where the level of a particular quantity of interest depends on its past level and its current rate of change. Although these equations would actually be integral equations, they are approximated by difference equations where the system status is updated every Δt (DYNAMO must use time-oriented bookkeeping, since it attempts to model continuous systems on a discrete computer). DYNAMO

[7]Robert C. Meier, William T. Newell, and Harold L. Pazer, *Simulation in Business and Economics*, © 1969, p. 243. Reprinted by permission of Prentice-Hall, Inc., Englewood Cliffs, N. J.

appears to be more applicable to "macro" simulations, that is, simulations of a firm or industry or even the economy, rather than "micro" simulations which involve smaller scale models of "pieces" of a firm such as queueing, production scheduling, and inventory models.

8.5 STATISTICAL CONSIDERATIONS

We have mentioned previously in the introduction to this chapter that simulation is comparable to experimentation. In addition to being concerned with such things as designing meaningful experiments and statistical significance of results, which are concerns of the physical experimenter as well, we must also validate our simulation model, that is, make sure we have the correct "laboratory" in which to conduct our experiments. Furthermore, in many queueing studies, steady-state results are of prime interest and it is desired to know when the transient effects are "washed out." We have also mentioned previously the problem of using simulation in determining optimal or near optimal conditions, a problem which the physical experimenter also faces. Therefore we consider in this section topics concerning model validation, the attainment of near steady-state conditions, experimental design and statistical analysis of output, and the utilization of simulation models to arrive at optimal system designs.

8.5.1 Validation

The problem of validating simulation models has two aspects. The first deals with internal correctness in that the model is programmed correctly and performs the calculations and logical sequences intended by the author. This is the easier of the two aspects of model validation and can be carried out by running simple cases which can be checked by hand calculations. Often, if the model is complex, several simple runs which are designed to test only one facet of the model's logic at a time may be necessary. Nevertheless, internal correctness of a model can be substantiated rather easily.

A far more formidable aspect of model validation lies in deciding whether the model is an adequate representation of the real world that it is intended to describe. Two procedures can be used in aiding this type of validation. First, if the model is a description of an actual on-going system for which historical data are available, these data can be used as input to the model and the subsequent model output checked with actual history. Second, under certain conditions of input, the simulation model may coincide with a known analytical model. Running the simulation model with these input conditions will allow for a comparison of the model output to known theoretical results. For example, the use of exponential interarrival and service times will often allow one to compare simulation results with previously developed analytical results, provided, of course, all the other assumptions leading to the analytical development are reasonably well met.

In comparing simulation results with either past history or known analytical results, statistical procedures must be employed, since the output from simulation models (assuming the models are Monte Carlo) is statistical in nature. More is mentioned on this in a later section.

For models which attempt to describe new systems where no historical data are available and which are of such a nature that there are no input conditions which correspond to known analytical results, the problem of model validation is, indeed, extremely difficult. About all one can do in this situation is to recheck the logic of the design, run the model over a range of different inputs to determine whether the outputs are within the realm of plausibility and, if one is so inclined, pray a lot.

8.5.2 Steady State

Many systems which we are interested in modeling are continuing systems. Studying these by simulation necessitates artificially starting and stopping the systems being modeled. Thus it is important to decide when the starting conditions are no longer significantly influencing the results. Of course, there are situations in which we may be directly interested in starting (and stopping) conditions. Examples of this could be almost any type of retail establishment such as a bank, barber shop, or department store, which opens and closes at specified times. However, even with these, peak periods are usually the major concern and although in some cases the transient effects are also of interest, generally we wish to obtain results on the steady state during peak loads.

One general procedure for eliminating transient effects in simulation studies that immediately comes to mind is to discount (or ignore completely) what happens during the early portion of a run so that any calculations involving measures of effectiveness would not commence until after the model has been running a "suitable" amount of time. The problem is determining what "suitable" is. Again, this is often somewhat of an art.

Prior to running the model for results, a few pilot runs may provide an indication of when the starting conditions are no longer a sizable effect.

Measures of effectiveness such as L or W can be calculated at successive intervals of time and when it appears that these values are becoming stable, the transient effects can be considered no longer important. One should be cautious, since cumulative measures may stabilize far faster than the system itself and may even "settle down" in cases where the system does not even possess a steady state. To illustrate the above procedure, consider a pilot run for T units of simulated time, in which the T units are broken up into k intervals. We calculate, say L_i, the average number in the system up to and including interval i, so that we have a sequence $L_1, L_2, \ldots, L_k$. Plotting L_i versus i should give us some idea as to when initial conditions may be "washed out."

If the last l of the L_i are relatively close (they will never be exactly the same, since there is variation even in the steady state), we can assume steady state occurred at time $t_s = (k - l)(T/k)$. One should repeat this procedure a few times, changing the seed of the random number generator and then choosing a conservative t_s. The width of the interval to use in this procedure should depend on the volume of traffic going through the simulated system. Statements such as, "the simulator was run for one year prior to assuming steady state" are meaningless, for the system might have encountered only two or three transactions in that time. If one uses a measure such as W, it is not necessary to break T into intervals, but merely to calculate W every so often, say after 100, 200, 300, and so on customers. Once again, as in so many aspects of simulation, subjective judgement must be used. Generally, this judgement improves through experience.

Although most simulations are started with the system empty, the time to steady state can be cut down by loading the system judiciously when starting. Ideally, the system should be loaded at something near its steady-state mean system size which, of course, if known, might obviate the need to simulate in the first place. However, almost any guess at steady-state conditions will do better than starting the system empty. One conservative procedure would be to start the system empty in the pilot run which is used to determine t_s. Then, using t_s as the stabilization period in future runs (no measure of effectiveness is calculated during this time), start the system loaded with the final value obtained from the pilot run.

Things become somewhat more complicated when the simulation is to be run for various alternatives (e.g., different numbers of servers or different values of μ). What might be good loading for one alternative might be poor for another. One could start at different loadings for each alternative studied, but this may introduce further variation into the system. What is generally done is to choose some compromise loading which is used to start all runs, regardless of the system alternative being studied, as long as these alternatives are going to be compared to each other. Conway (1963) and Conway et al. (1959) are two references which discuss problems such as model validation and stabilization periods.

8.5.3 Statistical Analyses of Output

The output from any Monte Carlo simulation model is stochastic, as are results from physical experiments. There is a major difference, however. The stochastic nature of physical experiments is introduced by effects which lie outside the experimenter's control—effects caused by imprecise recording instruments, weather conditions, soil variations, and so on. In simulation models, the statistical variation is directly due to the developer of the model; that is, statistical variation is deliberately built into the model through random number generators, and, in a sense, is under the control of the experimenter. Since most of the classical statistical procedures were developed for physical experiments where statistical variation is uncontrollable and successive observations are generally independent, they are not always applicable directly for the analysis of simulation output, much of which is correlated. Nevertheless, some existing statistical procedures can be utilized and we concentrate discussions on these, as well as mention some problems involved with the use of the more general procedures developed for physical experiments.

The major tenet of any experimenter is to reduce extraneous variation to the greatest degree possible. The physical experimenter goes to great lengths to design his experiment to remove the effects of uncontrollable variables which increase variability of results. For example, in an agricultural experiment to determine which of several types of fertilizer produces the greatest yield of a certain crop, the item of interest is the variation in crop yield caused by the different fertilizers and not the variation that may be caused by different soil conditions. The experimenter wishes, then, to *block* the effect due to soil condition and thus design his experiment so that each given type of fertilizer is used in different areas of the experimental field in order to average out the soil effect. The experimental design used may be that of randomized blocks or latin squares, depending on whether only a single direction soil gradient (north-south or east-west) is anticipated or soil gradation in both directions is suspected.

The simulation experimenter has the power to carry this blocking procedure to the limit, so to speak, since he can duplicate randomness exactly from run to run by using the same seed in starting his random number generation. Thus if he is interested in comparing two alternative system designs, he can compare them both on the same stream of random numbers and any variation resulting should depend only on the difference between designs, and the inherent variation due to the random number stream chosen. It is rather paradoxical that this ability of the simulation experimenter to control randomness renders inapplicable much of the classical theory of the analysis of variance due to the lack of independence of the results. Of course, the simulation experimenter could use a different seed for generating random numbers in comparing alternatives, thus achieving independence, but the added variation introduced may, in some cases, be too high a price to pay to use existing statistical techniques. Furthermore, there do exist some statistical procedures that can be legitimately used in analyzing nonindependent simulation output.

We first consider the problem of determining some measure of effectiveness (for purposes of illustration let us say the mean number in the system) from a simulation run. From ergodic theory, if we made one "very long" run of the simulation model and computed the time-weighted average of the number in the system, we know that this converges to the true mean number in the system L, assuming steady state is achievable (refer to the introduction to Chapter 2 dealing with ergodicity and the steady state). That is, if we run for a length of time T, and the resulting computation of average system size is L_T, then from Chapter 2, we see that

$$\lim_{T \to \infty} L_T = L.$$

Unfortunately, in practice we can never reach the limit, so for a run of any finite length T, we end up with L_T and not L. The fact that for T large L_T (which again depends on volume of traffic through the system) will be "close" to L is comforting, but does not allow us to make any kind of confidence interval estimate concerning L.

We saw from Chapter 2 that if we have ergodicity, L is also the expected value of the ensemble average. The ensemble processes can be approached by repeating runs of the simulation, keeping all design parameters constant, but using a different random number stream each time, that is, changing the starting seed used in the random number generator. Multiple runs of this nature are referred to as *replications*. Denoting the average system-size calculation for run i by L_i, the ensemble average (assuming steady state) will be

$$\lim_{n \to \infty} \sum_{i=1}^{n} \frac{L_i}{n} = L.$$

Of course, it is not possible to take an average over an infinite set either, but we can point estimate L by an finite average, and furthermore, we can calculate the variance of the point estimate and thereby obtain a confidence interval estimate.

We shall assume that we take n replications, and that the initialization period is such that transient conditions are washed out. We shall further assume that the random number streams used in the replications are not correlated with one another (this may not always be true if we randomly select the starting seed since, by chance, we may choose two seeds which give us streams of random numbers displaced by only a small number of values; however, the probability of this happening is small). The point estimate of L is then

$$\hat{L} = \sum_{i=1}^{n} \frac{L_i}{n}$$

and the estimate of the variance of $\hat{L}$ is

$$s_{\hat{L}}^2 = \frac{1}{n} \frac{\sum_{i=1}^{n} \left(L_i - \hat{L}\right)^2}{n-1}. \tag{8.3}$$

If n is large, we may employ the central limit theorem and obtain a confidence interval estimate of L as

$$\hat{L} \pm z_\alpha s_{\hat{L}},$$

where z_α is obtained from the standard normal tables and $1 - \alpha$ is the degree of confidence. For n small, one must replace z_α by t_α, the corresponding value from the Student t distribution; however, this requires the further assumption that the underlying distribution of the L_i is normal.[8]

Subjective judgment must still be used in determining the run size for each replication, for the longer the run length, the smaller $s_{\hat{L}}$ should be (the L_i are "closer" to L). Some pilot runs are generally necessary to determine the trade-off (here computer running time also enters the picture) between run length and number of replications, since increasing n also reduces $s_{\hat{L}}$. We desire, of course, the situation which gives us the smallest $s_{\hat{L}}$ so that we obtain narrow confidence intervals.

While large numbers of replications tend to reduce the variance $s_{\hat{L}}^2$, every time a replication is made the stabilization period must be run through, so

[8]This procedure can also help, in certain cases, in determining whether initial conditions are washed out. If we are simulating a case for which an analytical answer is available (e.g., $M/M/1$), we can, starting from time zero, make confidence interval estimates (or test hypotheses) based on n replications of runs of length T. If the confidence interval estimate contains the true (analytical) value, we have confidence that the initial condition influence has vanished.

that some of the running time is unproductive. One way to avoid this problem is to make one long run (say n times as long as the run size would be if replicating), break up the single run into n "blocks," and treat each block as a separate run (a replication). Thus the initial stabilization period is encountered only once. However, we might pay a price for this in that the replications may not be statistically independent, that is, the L_i may be correlated.

If the block interval is large, the correlation should be very small and can safely be neglected. However, one can explicitly account for correlation in calculating the estimate of the variance of L_i (i here refers to the particular block). If we assume that the autocorrelation among the L_i is stationary, that is, the correlation between L_i and L_{i+k} depends only on k and not i, then denoting the sample covariance of lag k by c_k we have

$$c_k = \frac{1}{n-k} \sum_{i=1}^{n-k} \left(L_i - \hat{L} \right) \left(L_{i+k} - \hat{L} \right) \qquad (k = 0, 1, 2, \ldots), \qquad (8.4)$$

where $\hat{L}$ is, as before, the sample mean of the L_i. Generally, c_k will be small for all but very small values of k (note that c_0 is the sample variance of the L_i, based on a divisor of n instead of $n-1$). The variance of $\hat{L}$ can then be estimated by

$$s_{\hat{L}}^2 = \frac{1}{n} \left[c_0 + 2 \sum_{k=1}^{m} \left(1 - \frac{k}{n} \right) c_k \right], \qquad (8.5)$$

where m must be considerably smaller than n. Hopefully, the c_k for $k \geqslant m$ will be negligible; if it turns out that c_k ($k \geqslant m$) is not small, larger block interval sizes are necessary to reduce the autocorrelation for large lags. A rule of thumb concerning the relationship between m and n is that m should not be larger than $n/4$. Again, for n large, normal theory can be employed for confidence interval estimation.[8a]

Conway (1963) presents an example showing the effect of ignoring autocorrelation. The simulation model is of a job shop, the measure of effectiveness of interest being the average time a job spends in the shop. A single long run was made (9300 jobs after stabilization), recording individual job times. These, of course, are highly autocorrelated, since if a particular job has a long wait, the one behind it is likely to also. No blocking was done (or equivalently, a block size of one was used), and the sample variance (ignoring autocorrelation) was computed. If the correlation between job times is ignored, then the variance of the mean of 100 jobs is simply $1/100$ of the originally computed variance. Conway, from

---

[8a]Note that this procedure can be used on any autocorrelated series such as successive waiting times and is an alternative to replication or blocking to achieve independence.

the same run, then actually computed the variance of blocks of size 100 and found that it was more than eight times larger than the estimate based on individual job variance divided by 100, that is, the estimate obtained from an equation similar to Equation (8.3). Thus if autocorrelation is to be ignored, it is important to have adequate block sizes when only a single run is going to be used.

Gordon (1969) presents a procedure for utilizing the variance of the mean of autocorrelated observations [Equation (8.5)] in determining when initial conditions are washed out. It can be shown [see Fishman (1967)] that for a process in steady state, the variance of the mean of autocorrelated observations goes down as run length increases, that is, $s^2 = C/T$, where C is a constant and T is the run length. When the observations are independent, C is merely σ^2, the variance of the underlying population [Equation (8.5) reduces to Equation (8.3)]. Since we have the inverse relationship even for autocorrelated observations, Gordon suggests plotting $\log s$ versus $\log T$, since

$$s = \frac{\sqrt{C}}{\sqrt{T}}$$

and

$$\log s = \tfrac{1}{2}\log C - \tfrac{1}{2}\log T$$

$$= C' - \tfrac{1}{2}\log T.$$

When the process reaches steady state, the points should line up roughly with a slope of $-\tfrac{1}{2}$. This procedure requires replicating so that s can be calculated for various T's.

We now turn to the problem of comparing two alternative system designs to determine which may be better. We shall again, for illustrative purposes, consider L as the measure of effectiveness. As mentioned previously, to keep extraneous variation at a minimum, it is desirable to compare the two designs on the same random number sequence. This can be easily done whether we replicate or use a single "blocked" run. Then, we obtain a set of paired values for each replication (or block) denoted by $L_{i,1}$ for design one and $L_{i,2}$ for design two. Since we are interested in any difference between designs, we can simply look at the paired differences $d_i = L_{i,i} - L_{i,2}$. Assuming independence of replication (or block), the mean and variance of the differences can be computed and a confidence interval formed on the mean difference as

$$\hat{d} \pm z_\alpha \left[\frac{1}{n} \frac{\sum\limits_{i=1}^{n} \left(d_i - \hat{d} \right)^2}{n-1} \right]^{1/2},$$

where

$$\hat{d} = \frac{\sum_{i=1}^{n} d_i}{n}.$$

If n is small, then z_α is replaced by t_α (again requiring the assumption that the underlying distribution of the d_i is normal). This procedure is referred to in the literature as the paired t test.

Problems are encountered when desiring to consider more than two alternatives. We could of course, look only at two at a time, always keeping the best to compare against another. However, if we had k alternatives of interest, and used this procedure with confidence of $1 - \alpha$ for each statement concerning a pair, we can only state that the overall confidence would be at least $(1 - \alpha)^k$. There are, in the statistical literature, a variety of multiple ranking procedures (Dunnett, Tukey, Cornfield, and Scheffé), which allow simultaneous confidence statements. These, however, have certain assumptions associated with them such as normality, common variance, and *independence*, conditions that the simulation experimenter often will have difficulty in meeting. For the same reasons, the classical analyses of variance cannot be used either. As previously mentioned, independence can be achieved by using different random number streams for each alternative within a given replication and in some cases this may be the best procedure to follow. Nevertheless more often than not, comparison by pairs is usually chosen.

There has been some relatively current work in studying and developing procedures that can be used on simulation output. Gafarian and Ancker (1966), Fishman (1967, 1968, 1971, 1972b), and Fishman and Kiviat (1967) employ time-series models and spectral-analysis methodology, which explicitly account for autocorrelation, to simulation output.[9] Other references on the subject of analyzing simulation output are Kleijnen and Naylor (1969), Hunter and Naylor (1970), and Naylor and Wonnacott (1971), these latter generally concerned with how to apply the classical theory of analysis of variance, design of experiments, and multiple ranking procedures to simulation output. One might quite naturally ask whether it

---

[9] An elementary use of time series analysis methodology was employed previously in calculating the variance of autocorrelated observations given by Equations (8.4) and (8.5). Fishman (op. cit.) makes use of this variance in developing confidence interval estimates, as well as considering its asymptotic properties for large sample sizes. The effect of initial conditions is also considered in his (1972b) paper where he shows that the procedure of removing an initial period to diminish initial condition influence, while accomplishing removal of initial condition bias, may also increase the resulting variance of the point estimators thus increasing the confidence interval.

is preferable to block (or replicate) to get rid of autocorrelation or not to block (or replicate) and simply account for autocorrelation in the variance computations [Equation (8.5)]. Also, in comparing alternatives, should one use the same random numbers for a given replication thus correlating observations, or use different random number streams so that observations can be considered independent? Questions such as these are also addressed somewhat in the references above, as well as in Chapter 8 of Emshoff and Sisson (1972).

We briefly mention here one further topic dealing with statistical considerations of running simulations, prior to taking up optimization through simulation in the next section. No discussion on statistical considerations in simulation analyses would be complete without mentioning the topic of *variance-reducing techniques*. We have already discussed ways of eliminating extraneous variance—for example, reusing the same random number stream when comparing alternatives—but variance-reducing techniques deal with specific ways of improving upon the Monte Carlo generation process itself. Techniques such as proportional sampling, importance sampling, fixed sequence sampling, control variates, and the use of concomitant information are the types of techniques considered. Although variance-reducing techniques have had little popularity or use in queueing simulations, Ehrenfield and Ben-Tuvia (1962) do illustrate some of these techniques applied to estimating the mean waiting time in queue. The added complexities and computations of using these techniques, even though they may be able to yield better estimates for smaller run lengths have kept them in relative obscurity, as far as queueing analyses are concerned (this is also true for other microsimulation models of a firm such as inventory, production scheduling, and so on, as well as many macroeconomic models). Their biggest use to date has been in simulation studies of statistical sampling procedures and mathematical calculations such as evaluating integrals through Monte Carlo analyses.

One variance-reducing technique that merits more than passing reference here is that of "virtual measures" in computer simulation [see Carter and Ignall (1972)]. This has to do with studies in which there is interest in observing (or at least including in the general observation of the process) events which occur only rarely. An example in a queueing framework might be a system which is near saturation (ρ near one), in which there is interest in what might occur during an idle period. Suppose several different events could occur during an idle period according to some probability distribution. To simulate the system, in addition to generating random interarrival and service times, when the system is idle, the possible idle period events would also be generated randomly. But even in a long run of the simulator, since the idle periods occur infrequently, there will be very few observations on the idle period events. The idea behind virtual measures is to use the simulation to determine the frequency of idle periods and then to either simulate separately what happens during an idle period so that more observations are obtained for the rare events or, when possible, analytically determine what transpires during the idle period. The usefulness of this technique to queueing simulation depends on whether there is interest in rare events which may be associated with the process. However, there are models of other processes (inventory being one of these) where rare events (in the case of an inventory model, a stock out) are always of interest.

One final comment on recent works by Fishman (1972a, c), (1973a), and Crane and Iglehart (1974) is in order prior to ending this section. These studies consider simultaneously the problems of calculating variances, variance reducing techniques, confidence interval estimates, and initial conditions.

The procedures for analyzing simulation output discussed thus far were general in that they apply to any stochastic system being simulated. The analysis of these works centers on the fact that the system being simulated is a queueing system and that a stable queueing system periodically becomes idle and thus can be looked at as a renewal process. This means

TABLE 8.5   MAJOR APPROACHES IN STATISTICAL
ANALYSIS OF OUTPUT

Initial Conditions [References]	Confidence Interval Estimates and Comparison of Alternatives [References]
Removal of initialization period [most general texts on simulation, for example, Gordon (1969)].	Replication or blocking; pair-wise comparisons of alternatives [most general texts on simulation, for example, Gordon (1969) or Emsoff and Sisson (1972)].
	Application of classical statistical procedures such as design of experiments, multiple rankings, and so on. [Kleijnen and Naylor (1969), Hunter and Naylor (1970), Naylor and Wonnacott (1971)].

Explicit consideration of initial condition bias; accountability for correlation of observations—time series/spectral density approach [Gafarian and Anker (1966), Fishman (1967, 1968, 1971, 1972b), Fishman and Kiviat (1967), Emsoff and Sisson (1972)].

Renewal theoretic approach making use of queueing process structure [Crane and Iglehart (1972), Fishman (1972a, c, 1973a)].

that successive periods between idle points (busy periods) are independent. If one measures run length in terms of number of busy periods and estimates, for example, the state probabilities $\{p_n\}$ by taking the average, over the total number of busy periods run, of the amount of time the system is observed in state n in each period divided by the length of that busy period, one has m independent observations if m busy periods are run. Since most quantities of interest (measures of effectiveness) turn out to be expected values of ratios of random variables (this was certainly true for p_n as both the time spent in state n during a busy period and the duration of a busy period are random), statistical theory developed for ratio estimators is employed to develop confidence interval estimates. Also, since a busy period begins with an arrival to an empty system, there is no initial starting condition problem—one merely starts the simulation with an arrival to an empty system.

Table 8.5 summarizes the major approaches in the statistical analysis of queueing simulation output.

8.5.4 Optimization

For situations where we are trying to determine which alternative system design to choose from among a finite number of possibilities, we can, in spite of some of the aforementioned problems involved with statistical significance of simulation output, usually find the best by simulating all of the possibilities, since generally the number of alternatives is small. Suppose, however, that we have a cost model where some costs increase with a certain design parameter (e.g., number of channels) and some costs decrease, and we wish to find the parameter setting which minimizes total cost.[10] Furthermore, the design parameter may be a continuous one such as μ, or we may even be attempting to find the optimum setting of several design parameters simultaneously. There is no longer, now, a finite set of alternatives to search, although we could (and this is often done) restrict ourselves to a finite subset of the infinite space. Earlier in this chapter (the introduction) we touched on this problem as one potential drawback of using simulation rather than analytical methods. Often, however, analytical analysis is not possible and we must simulate.

If the output from a simulation analysis was deterministic instead of random, we could employ a number of mathematical optimum search procedures such as Fibonacci search, golden section, and so on [see Wilde (1964) for a discussion of these and other methods]. The fact that simulation output is stochastic adds much greater complexity to the problem.

---

[10]Models like those treated in Section 7.2, but which are not amenable to analytical solution.

One procedure alluded to above is to a priori and subjectively choose a grid that is, a finite set of points (parameter settings), and then choose the best of those investigated, using whatever statistical analysis (probably paired comparisons) is available as discussed in the previous section. The degree of optimization achieved depends on the experience and cleverness of the experimeter.

Fortunately (or perhaps unfortunately) in practice, one usually does not have unlimited choices for parameter settings. Exogenous factors, such as management policy, labor force, and physical capacity limitations, generally limit the number of alternatives to a finite set. However, in some cases, limits do not exist and one is faced with the optimization problem mentioned previously.

There have been some techniques developed for this problem of finding optimality when output is stochastic, although, once again, many of the techniques assume independent observations, among other things. These techniques are sequential in nature and occasionally heuristic in that the results from one stage determine what might be done next, and in some of these procedures, there is a measure of subjectivity involved in the decision making. These types of techniques are often referred to as optimum seeking methods, response surface analysis, and/or evolutionary operation (some of these names are also applied to the deterministic procedures mentioned earlier).

We shall not consider specifically any of these here but supply ample references for the interested reader. Wilde (op. cit.) discusses response surfaces with experimental error in Chapter 6 of his book, although most of the text deals with deterministic situations. Meir et al. (1969) present one technique called simplex search (not to be confused with the completely different simplex linear programming procedure) which for an n dimensional search starts with $n+1$ points (hence the name simplex since a convex polyhedron formed from $n+1$ linearly independent points in n space is said to form a simplex) and sequentially adds and drops points as the procedure continues. Meir et al. actually describe a modification to the original procedure, where, for simulation studies, points are replicated to enable statistical analysis to be performed at each step.

Another procedure is that of Karr et al. (1965) which combines three procedures in phases. First, the decentralized gradient approach is utilized to get close to optimum quickly. Then, when the solution appears to be near optimum, two "finer" techniques, linear response surface analysis and quadratic response surface analysis, may be employed to improve on the solution found by the gradient approach. These latter two methods are closely related to response surface techniques developed by G. E. P. Box, a name intricately connected with stochastic response surface methodology [see Davies (1956), Chapter 11, and Box and Wilson (1951), Box (1952), Box (1954), Box and Youle (1955), Box and Hunter (1957), and Box (1957)].

A recent article on the desirable characteristics of a computer simulation optimization routine is Smith (1973). He discusses the problems of optimization on computer output and suggests several possible routines and how to incorporate them into the simulation program. Two recent textbooks on simulation, which also give coverage to simulation optimization as well as the other problems of statistical analysis previously discussed, are Emshoff and Sisson (op. cit.) and Fishman (1973b), with the latter, as the reader may have guessed from the references of this section, devoting a considerable portion of the book to these topics.

The actual success to date of these optimization techniques in simulation studies is, as yet, difficult to ascertain. Work is still continuing on improving this aspect of simulation analysis, although not at nearly the pace many simulation users would say is desirable.

8.6 SIMULATION MODEL EXAMPLE

The toll booth situation described in the introductory section of this chapter and flow charted in Figure 8.1 was programmed in FORTRAN IV and run on an IBM 370/145. The number of manned toll booths, c, and the number of automatic toll booths, c_a, are up to the discretion of the user. The Poisson input generator and normal, deterministic, and beta service generators are programmed as subroutines.

In validating the model, initial runs were made for $c = 1$ and $c_a = 0$, with exponential service times. These runs were also used for determining the effect of initial conditions. Cumulative measures of average waiting time in system were calculated at periodic intervals of running time for $\rho = 8/9$ and $4/9$ with $(1/\lambda, 1/\mu)$ of $(9, 8)$ and $(9, 4)$ sec, respectively. The theoretical answers obtained from Equation (2.34) are 72 and 7.2 sec, respectively. Figure 8.13 shows graphs of cumulative average waiting time versus run length for the two cases, three replications each. Note that for the high ρ, stabilization takes much longer. This is true in general, and is rather unfortunate, since many cases of interest deal with heavy "traffic." A fairly conservative estimate for stabilization appears to be 40 hr which is roughly equivalent to $(1/9)(3600)(40)$ or 16,000 customers observed. Figure 8.14 shows the same cases as those of Figure 8.13, but with the initial 40 periods removed from the calculations. The cumulative mean wait settles down somewhat more quickly here.

It was decided to make one long run and block, rather than actually to replicate, so that the stabilization period would be incurred only once. It was subjectively felt that block widths of 5 hr (approximately 2000 customers) or higher would be sufficient to allow neglecting autocorrelation. This was checked by computing the sample autocorrelation one period lag between block means for a pilot run of 40 blocks, which yielded a value of -0.045. Since this value is quite close to zero, we can be confident in the independence of block means for block sizes of 5 hr or greater.

To obtain an idea of what a good combination of block width and run length (number of blocks) might be, three cases were investigated for $\rho = \frac{8}{9}$ and the associated confidence interval estimates obtained using the methodology of assuming block means to be independent, previously described in Section 8.5.3. The results are presented in Table 8.6. For a total run length of 300 hr (approximately 120,000 customers), the best case appears to be the 30 blocks of size 10 (approximately 40,000 customers per block).

One final pilot run was performed, still assuming exponential service times but now increasing c to 2 (while continuing to keep $c_a = 0$), using an initialization period of 40 hr, block width of 10 hr, and running 30 blocks. The theoretical answer for an $M/M/2$ model with $\rho = \frac{8}{9}, \lambda = \frac{1}{9}, \mu = \frac{1}{16}$ is, from Equations (3.17) and (3.11), $W = 76.24$. A 95% confidence interval estimate of the simulation output gave (75.86, 90.14), with the point estimate being $\hat{W} = 83.00$.

We note that the toll booth problem is not strictly an $M/M/2$, since queues form in front of each booth. Nor is it two parallel independent $M/M/1$'s, each with a mean arrival rate of $\lambda/2$, since customers will choose the booth with the shortest queue. If neither has a queue, the program sends the customers to booth one. If there were complete jockeying between booths, the system would be quite close to an $M/M/2$, even though separate lines are allowed.[11] Even without complete jockeying as in

TABLE 8.6   BLOCK SIZE VERSUS NUMBER OF BLOCKS [a]

Block Width (hr)	Number of Blocks	Point Estimate	Confidence (95%) Interval Estimate
5	60	78.53	(68.91, 88.15)
10	30	78.75	(69.60, 87.91)
15	20	78.84	(69.43, 88.25)

[a] $\rho = \frac{8}{9}$; Total Run Length = 300 hr

[11] The reader may wish to refer back to a similar discussion on appropriate modeling at the end of Section 1.2.

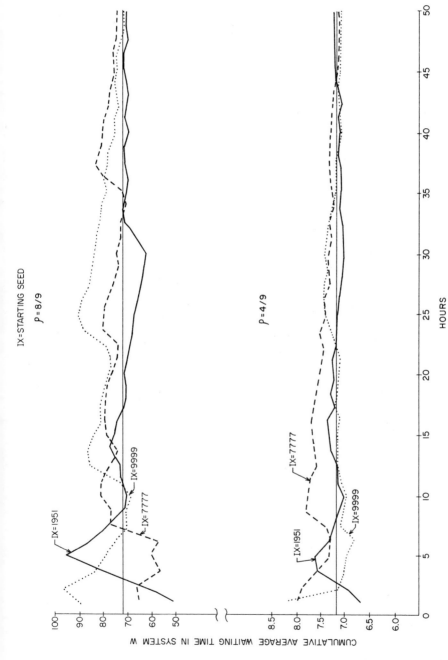

Fig. 8.13 Cumulative W vs. run length.

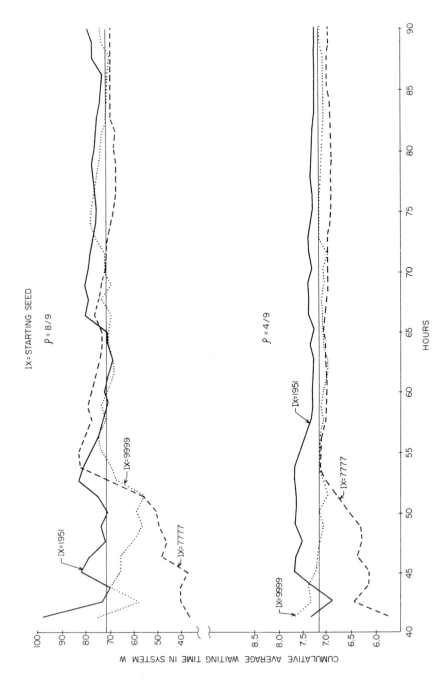

Fig. 8.14 Cumulative W vs. run length—initial period removed.

our case, an $M/M/2$ is the closest standard analytical model. We would, however, expect somewhat higher average waits in our toll booth system over a strict $M/M/2$ situation, which indeed turned out to be the case (83.00 versus 76.24). Furthermore, we would not expect waits as high as one would get using two independent $M/M/1$ models, each with a mean arrival rate of $(1/2)(1/9)$, which gives a W of 144. Thus it appears the simulation model is yielding realistic values.

With a fair degree of confidence in our procedure thus far, two alternatives to an assumed current situation of two manned booths $(c=2, c_a=0)$ were considered and compared by a paired t test. The "real" conditions simulated were Poisson arrivals with mean interarrival time of 9 sec, manned service normal with mean 16 sec and standard deviation of 1 sec, and automatic service (if there are automatic booths) deterministic at 10 sec, except when coins are dropped service time becomes beta, skewed to the left, with mean 20 sec on a range of 10 to 24 sec. The case of $(c=2, c_a=1)$ [adding an automatic booth] was compared to $(c=3, c_a=0)$ [adding another manned booth]. The results are presented in Table 8.7.

As a further check on validity, the current situation, case 1, should yield results close to an $M/\text{Normal}/2$ model. Since the normal distribution used for manned service time had a very small coefficient of variation $(\sigma/\mu = \frac{1}{16})$, this should be very close to an $M/D/2$ model. An $M/D/2$ model is difficult to solve analytically; however, an approximate W can be gotten by considering W_q for an $M/D/1$ with double the service rate[12] and using $W = W_q + 16$. For $M/D/1$ with $\lambda = \frac{1}{9}$ and service time of $\frac{16}{2} = 8$, W_q from Equations (4.24) or (5.11) turns out to be 32 so that W should be in the vicinity of 48. Similarly, for case 3, using $M/D/1$ with service time of $\frac{16}{3}$ to approximate the $M/\text{Normal}/3$, W_q is 3.67 yielding a W of 19.67. The simulation results for cases 1 and 3 from Table 8.7 are fairly close to these values, giving us added confidence in our simulation results.

TABLE 8.7   COMPARISON OF TWO ALTERNATIVES

Case	Point Estimate of W	95% Confidence Interval Estimate
1. $c=2, c_a=0$ (present)	49.66	(45.86, 53.48)
2. $c=2, c_a=1$	18.28	(18.14, 18.41)
3. $c=3, c_a=0$	18.96	(18.83, 19.09)

[12]W_q for $M/D/1$ will be different than W_q for the "corresponding" $M/D/2$ model but the two values should not differ by very much.

Comparing the two alternatives, cases 2 and 3, the mean waits are reduced to 18.28 when an automatic booth is added as opposed to 18.96 when adding another manned booth. A paired t test gave a t statistic of 8.83, which is highly significant. Recall 30 blocks were used (equivalent to 30 replications) and for 29 degrees of freedom, t values for levels of significance of 5 and 1%, respectively, are 2.045 and 2.756. However, while there is strong evidence of *statistical* significance, *practical* considerations in making the final choice would weigh heavily here, for the average percentage difference of the mean wait between the two cases is equal to $100([18.96-18.28]/18.28)$ or about 4%, a relatively small amount. Furthermore, because only 40% of the customers have exact change for the case $(c=2, c_a=1)$, the waits at the manned booths are much higher (W for the manned booths averaged 22.97, while W for the automatic booth averaged 10.85, yielding the overall W of 18.28). Thus unless the automatic booth is considerably cheaper to operate, it is quite likely that if another booth is to be added (and it certainly seems warranted since average waits can be reduced from around 50 sec to 18 or 19 sec although 50 sec is not unduly long), it would probably be a manned booth in spite of the fact that the addition of an automatic booth is slightly "better." "Better" here refers to average waits, and we would undoubtedly want to look at the histogram of waiting times prior to making any decisions. We might also be interested in queue lengths in front of the booths. Waiting time histograms are given in Figure 8.15 for the two alternatives. We note that although the mean wait for case $(c=2, c_a=1)$ was shown statistically to be significantly less than for the case $(c=3, c_a=0)$, there are vehicles waiting longer times (a greater tail on the histogram). This is due to the long queues which form from time to time in front of the manned booths, even though the automatic booth may be relatively uncongested. This again tends to favor adding a manned booth rather than an automatic booth.

In concluding this chapter, we point out that simulation can be an important device in analyzing the complex queueing situations one often encounters in real life. However, to achieve meaningful results, a great deal of care and thought must go into planning and running the simulator, especially in the areas of run-length determination and the interpretation of the output. These aspects in designing and performing simulation analyses have all too often been ignored, probably because it is now a rather routine matter actually to get a simulation model "working," especially with the advent of simulation languages. The success or failure, however, of simulation studies often lies not in the programming and coding of the model, but in how it is used and how the output is interpreted.

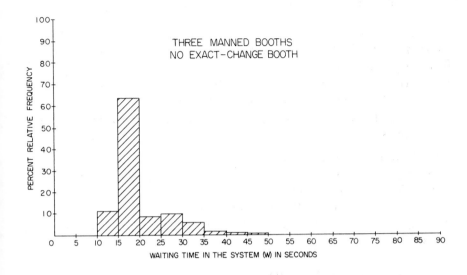

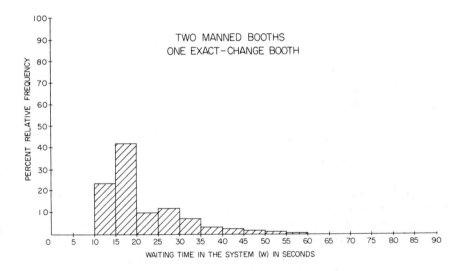

Fig. 8.15 Waiting-time histograms.

PROBLEMS

8.1 Generate by hand calculation the first three values of Table A5.1 of Appendix 5. (Since this table is printed directly from computer output the first three values are those of the first *row* not *column*.)

8.2 Write a computer program in FORTRAN to generate Table A5.1 (assuming the computer has a 32 bit word size).

8.3 Using Table A5.1, generate five observations from the following distributions.

(a) Uniform, between 5 and 15.
(b) Exponential, mean = 5.
(c) Erlang type 3, with mean = 5.
(d) Normal, mean = 5, standard deviation = 2.

8.4 Using Table A5.1, generate five observations from the following distributions.

(a) The triangular distribution, where

$$f(x) = \begin{cases} \dfrac{2x}{3} & (0 \leqslant x \leqslant 1) \\ 1 - \dfrac{x}{3} & (1 \leqslant x \leqslant 3) \end{cases}.$$

(b) Poisson, with mean = 2.
(c) The distribution given by

x:	0	1	2	3	4	5
$f(x)$:	0	$\frac{1}{10}$	0	$\frac{3}{10}$	$\frac{2}{10}$	$\frac{4}{10}$

8.5 Expand the flow-logic diagrams of Figures 8.3, 8.4, and 8.5 for programming a computer to carry out the bookkeeping methods illustrated by Tables 8.3 and 8.4 to include waiting time in system, idle time, queue size, and system size.

8.6 Write programs in FORTRAN for the flow diagrams of Problem 8.5 which will yield calculations for L, L_q, W, W_q, and percentage idle time. Check your results (either by hand calculation or computer) with Tables 8.3 and 8.4.

8.7 Suppose it were desired for the example leading to Tables 8.3 and 8.4 to obtain the probability distributions for system size and waiting time in queue as well as expected values. Indicate the modifications necessary to the program (or flow diagram) of the previous problems to accomplish this.

8.8   Write a simulation program in FORTRAN giving expected system size and waiting time in queue for a single-channel queueing model where interarrival times and service times are to be generated from subroutines to be specified by the user. Check out the program by using exponential interarrival and service times and comparing results with the known formula for $M/M/1/\infty$ given in Chapter 2.

8.9   Use GPSS to program the $M/M/1/\infty$ model of Problem 8.8. Compare programming effort and running time with the FORTRAN program.

8.10   Program an $M/M/1/\infty$ model in SIMSCRIPT. Compare programming effort and running time with FORTRAN and GPSS. Would these comparisons for a simple model also be typical for more complicated models? Discuss the relative advantages and disadvantages of each of the three languages.

8.11   To determine the length of the stabilization period, a particular queueing simulator was run until 2400 customers were observed, starting at time zero with no one in the system. The 2400 customers were broken up into 24 blocks of 100 each and the cumulative average waiting time per customer was calculated, and is given below. What would you consider to be a suitable stabilization period?

Block number:	1	2	3	4	5	6	7	8	9	10	11	12
Cumulative average waiting time:	1.3	9.4	7.9	2.8	4.4	6.8	5.5	4.3	5.1	5.3	5.3	5.4

Block number:	13	14	15	16	17	18	19	20	21	22	23	24
Cumulative average waiting time:	6.1	5.2	5.5	5.9	6.4	6.8	7.3	6.8	6.5	6.6	7.0	6.8

8.12   Suppose you have programmed a general single-channel queueing simulator which allows for any input and service patterns. To validate the model, you decide to make runs with exponential interarrival times, mean 10, and exponential service times, mean 8. The following are results of average system-size calculations for 20 replications.

$$5.21, \quad 3.63, \quad 4.18, \quad 2.10, \quad 4.05,$$
$$3.17, \quad 4.42, \quad 4.91, \quad 3.79, \quad 3.01,$$
$$3.71, \quad 2.98, \quad 4.31, \quad 3.27, \quad 3.82,$$
$$3.41, \quad 5.00, \quad 3.26, \quad 3.19, \quad 3.63.$$

Based on these values, what can you conclude?

8.13 In comparing two alternate system designs via a queueing simulation model, the following results on mean waiting times under each design for 15 replications are obtained. For each replication, both designs were compared on the same random number stream. Does it appear that one design may be preferable?

Replication Number	Mean Waiting Time Design 1	Mean Waiting Time Design 2
1	23.02	23.97
2	25.16	24.98
3	19.47	21.63
4	19.06	20.41
5	22.19	21.93
6	18.47	20.38
7	19.00	21.97
8	20.57	21.31
9	24.63	23.17
10	23.91	23.09
11	27.19	26.93
12	24.61	24.82
13	21.22	22.18
14	21.37	21.99
15	18.78	20.61

8.14 Suppose that in Problem 8.13, design 2 is the current design. Another alternate design, which we shall call design 3, is proposed. Results for 15 replications (using the same random number streams as before) are presented below, where replications 1–5 are given in the first row, 6–10 the second, and 11–15 the third. Is Design 3 any better than current practice? Comment from both a statistical and a practical point of view.

23.91,   24.95,   21.52,   20.37,   21.90,

20.17,   21.90,   21.26,   23.10,   23.02,

26.90,   24.67,   22.09,   21.91,   20.60.

8.15 The manager of your local supermarket is having problems in determining the number of checkers (and baggers) to have on Saturday morning, one of his busiest periods. Because your wife (husband) is always complaining about the wait and challenges you to put some of your queueing theory education to use, you undertake the task of aiding the manager.

You decide, because of the complexities of jockeying and reneging, the ability to add and subtract baggers (which decreases or increases the μ of a channel) and other complexities, to model the system by simulation. Develop the simulation model and program it in any language you desire. Try to estimate, by actual observation over several Saturday mornings, arrival and service patterns as well as queue discipline. Use part of the observations to develop empirical distributions for the simulation and validate the simulator using the remaining observations. Then attempt to determine the solutions to the manager's problems.

If you are successful, give your suggestions to the manager if (a) your wife (husband) agrees to stop nagging you, and (b) the manager gives you at least one month's free groceries.

8.16 Take a system which gives you great annoyance (registration for college courses, barber or beauty shop, local cafeterias, etc.), and build a simulation model of the system with an eye toward determining a better system design.

Chapter 9
APPLYING QUEUEING THEORY—
A CASE STUDY

In this chapter we present a case study from the literature which not only illustrates many of the topics we have presented in this text, but also shows the many facets of analysis that must be considered to apply the theory successfully to a real situation. Rather than present many examples of case studies in brief detail, we have decided to present one in-depth illustration. For other examples of applications and case studies, we refer the reader to Buffa (1966), Chapters 13, 14, and 15, Lee (1966) Chapters 9 through 15 inclusive, and Panico (1969), Chapter 4. These applications cover such diverse areas as materials handling, hospital administration, airline passenger check-in and terminal design, freight reservations, job vacancies, and many others.

The study we present here is one by Brigham (1955) which we mentioned briefly in Chapter 7, Section 7.2.2. Although in Chapter 7 we were interested only in the design and control aspects of the study, here we consider all phases of the problem, which include statistical inference on input and service, the descriptive queueing model and analysis, the optimal design aspects of the model, and implementation. While this is one of the earliest case studies in the literature, it nevertheless serves to illustrate the many facets and diverse aspects of a successful queueing study.

9.1 THE PROBLEM[1]

The problem studied by Brigham involves clerks who service tool crib counters throughout the Boeing Airplane Company's factory areas. The cribs store a variety of tools required by mechanics in the shops and in the assembly lines. There were about 60 cribs scattered throughout three plants, each employing from one to five clerks.

---

[1] Permission to present this study has been obtained from *Operations Research*. Portions of this material are quoted directly from the reference, although sizable amounts of rewriting and editorial commenting are also given.

The number of clerks that *should* be assigned to each crib had long been a topic of great discussion within the company. Foremen complained that there were not enough clerks and that mechanics were waiting in line far too long, thus tying up valuable mechanic time which resulted in either having to hire more mechanics (or scheduling more overtime) or lowering actual production output. Management, however, was under great pressure to reduce overhead and thus wished to reduce the number of clerks that were already employed. Management pressure, being greater than that exercised by the foremen, won out, and an order to reduce the number of clerks ensued. This led to a formal request by the tool department to determine if a reasonable criterion that could justify a certain level of personnel might be found. The results of this request led to the study reported here.

There was no question that the situation to be studied was one of a queueing nature. The problem then remained to determine the appropriate queueing model and exercise it in such a way that an "optimum" number of clerks at each crib could be determined.

At this point, all one really knew for sure was that the queueing model for a crib had c channels (c clerks), and that there was essentially no limit on waiting room. Information on the queue discipline, input pattern, and service pattern was needed to pin down the queueing model so that it could be useful in providing the information necessary to consider optimizing the number of clerks.

9.2 THE COST MODEL

It is usually a good idea in any optimization problem to determine first the cost (or profit) function to be optimized and any constraints that might be present. In this way, one would then know just exactly what information is needed from the queueing model which may influence how the queueing model is to be formulated.

The problem here was one of "balancing" the cost of the mechanics' idle time against that of the clerks. The ideal solution would be to have no idle time on the part of clerks and no waiting time on the part of mechanics. This, of course, is impossible to obtain, since for one thing, there is randomness in the system, and for another, clerks cannot be cut into pieces (i.e., the number of clerks must be an integer). However, the total variable costs due to mechanics waiting and clerks being idle is, ideally, to be minimized. Denoting expected clerk idle time by I, expected time in queue on the part of a mechanic by W_q, the marginal cost of a clerk per unit time by C_s, and the marginal cost of a mechanic per unit time by C_w, we wish to find the number of channels (c) which minimizes[2]

$$E[C] = C_s cI + (C_w W_q)\lambda \qquad (\$/\text{unit time}), \qquad (9.1)$$

where λ is the mean rate of arrivals of mechanics for service. Since we have a c-channel queue, we have previously shown (see Example 3.1 or Problem 3.8) that the average percentage idle time of any given clerk is $(1 - \lambda/cu)$ so that Equation (9.1) becomes

$$E[C] = C_s\left(c - \frac{\lambda}{\mu}\right) + \lambda C_w W_q. \qquad (9.2)$$

The problems now remaining are to find the parameters of the queueing model (λ and μ), to find W_q from the queueing model, to estimate the cost parameters C_s and C_w, and finally to find the optimum number of clerks [the c which minimizes Equation (9.2)].

9.3   THE QUEUEING MODEL—INPUT

Before being able to find W_q, it is necessary to determine the queueing model. Thus far, we have decided that the queueing model is $?/?/c/\infty/?$. It is now necessary to remove the question marks.

Two tool cribs were chosen for detailed study, one in final assembly and one in the machine shop. The first was chosen because it was one of the largest in the company and could furnish a great deal of data in a short period of time. The second was chosen because it serviced a shop with quite different work characteristics. It was believed that the other tool cribs operated in a similar manner, but with possibly different parameters. Although it was not economically feasible to study all cribs in the same detail as the two mentioned above, several others were checked and were found to behave either like the assembly crib or the machine-shop crib.

Considering first the queue discipline, by observation it was decided that service was essentially in random order. Although when only a few mechanics were waiting, a first-come, first-served discipline was followed, as the congestion increased, the order of service became random, since when a clerk became free, he would tend to service the mechanic standing closest to him, rather than walk the length of the counter to serve a mechanic who may have arrived earlier. The discipline is really not too

---

[2]An equivalent objective function would be to minimize $C_s c + \lambda C_w W$ [see Hillier and Liebermann (1969), p. 334 and 337, and/or Problem 7.15].

important here (except that mechanic complaint may be cut down if a strict FIFO discipline were followed), since the cost model requires an average unconditional waiting time over all mechanics which would be the same for SIRO or FIFO.

The next step was to obtain some data on arrival and service patterns. An observer with a stopwatch can rather easily check service time by sampling. When a clerk commences service on a customer, the observer can start the stopwatch, stopping it when the mechanic leaves the crib. After recording the service time, he can choose at random another customer to study. However, it is far more difficult getting interarrival time data, since the observer must record time for successive arrivals, and, if the system is busy, he can miss an arrival (or arrivals) while recording data.

To facilitate the taking of data, a box-mounted panel on which were

TOOL CRIB NO. I

DATA TAKEN ON 7-22-54

MEAN SERVING TIME: 71.55 SEC

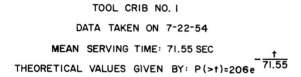

THEORETICAL VALUES GIVEN BY: $P(>t) = 206e^{-\frac{t}{71.55}}$

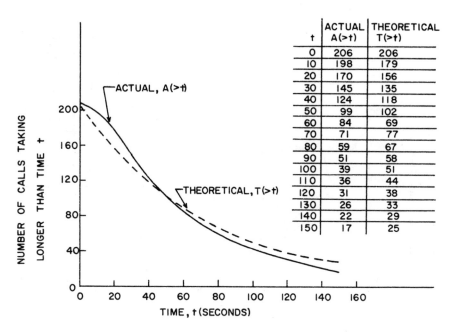

t	ACTUAL A(>t)	THEORETICAL T(>t)
O	206	206
10	198	179
20	170	156
30	145	135
40	124	118
50	99	102
60	84	69
70	71	77
80	59	67
90	51	58
100	39	51
110	36	44
120	31	38
130	26	33
140	22	29
150	17	25

Fig. 9.1   Distribution of serving times, machine-shop tool crib.

installed two small hand-operated switches and two signal lights was constructed. There was a 6-V battery inside and whenever one of the switches was depressed, a 6-V signal was fed to a two-channel brush recorder, one switch controlling one channel, and the corresponding signal light going on as a check for the operator's convenience. The recorder paper was run at a speed of 5 mm/sec, which provided sufficient accuracy for the purpose desired. The observer depressed the switch briefly each time a man arrived at the counter, thus recording easily and accurately any number of arrivals that occurred.

This machine was found more useful than a stopwatch in recording serving times. For such a record the switch was kept depressed for as long as the transaction was carried out and was released when it ended, thus

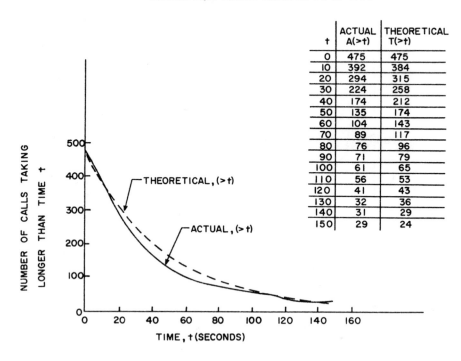

TOOL CRIB NO. 27

DATA TAKEN ON 7-2-54 & 7-12-72

MEAN SERVING TIME: 50.5 SEC.

THEORETICAL VALUES GIVEN BY: $P(>t) = 475\, e^{-\frac{t}{50.5}}$

t	ACTUAL A(>t)	THEORETICAL T(>t)
0	475	475
10	392	384
20	294	315
30	224	258
40	174	212
50	135	174
60	104	143
70	89	117
80	76	96
90	71	79
100	61	65
110	56	53
120	41	43
130	32	36
140	31	29
150	29	24

Fig. 9.2   Distribution of serving times, final assembly tool crib.

making a continuous trace on the paper. With a little practice the operator found that he could record the length of two transactions simultaneously, and hence was able to accumulate data faster.

Figures 9.1 and 9.2 show the resulting observations of service times for the machine shop and assembly cribs, respectively. Shown on the graphs is the actual complementary CDF of the observed service times along with a theoretical complementary CDF of the exponential distribution with its mean set equal to that of the observed mean.

Similarly, plots are presented in Figures 9.3 and 9.4 of the actual frequency of arrivals per mean service time versus a theoretical Poisson for the machine-shop and final assembly cribs, respectively.

Brigham does not mention specifically any application of statistical goodness-of-fit tests (see Chapter 7, Section 7.1.2) to these data but concludes, apparently on the basis of inspection, that service times are approximately exponentially distributed and arrivals are approximately Poisson. While the visual evidence does appear adequate, we nevertheless ran χ^2 goodness-of-fit tests on the data of Figures 9.1 through 9.4 with the

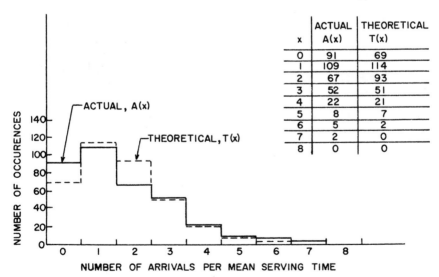

TOOL CRIB NO. I

DATA TAKEN ON 7-20-54

AVERAGE NUMBER OF ARRIVALS PER MEAN SERVING TIME = 1.64

THEORETICAL VALUES GIVEN BY: $T(x) = 357 \dfrac{e^{-1.64} \, 1.64^x}{x!}$

x	ACTUAL A(x)	THEORETICAL T(x)
0	91	69
1	109	114
2	67	93
3	52	51
4	22	21
5	8	7
6	5	2
7	2	0
8	0	0

Fig. 9.3   Distribution of arrivals, machine-shop tool crib.

results shown in Tables 9.1 through 9.4 and summarized in Table 9.5.

From the tables, it appears there is evidence to reject both the exponential service-time and the Poisson arrival-rate assumptions. Glancing at Table 9.5, the exponential-service assumption seems particularly suspect for the final assembly tool crib, while the Poisson arrival-rate assumption seems poorest for the machine-shop tool crib. It is quite possible to find other distributions offering better fits; however, a more complex model would result and perhaps not even be amenable to analytic treatment. We shall proceed, accepting Brigham's conjecture of Poisson arrivals and exponential service, realizing that the ensuing model is only an approximation to reality anyway, but shall comment further upon this matter in Section 9.7.

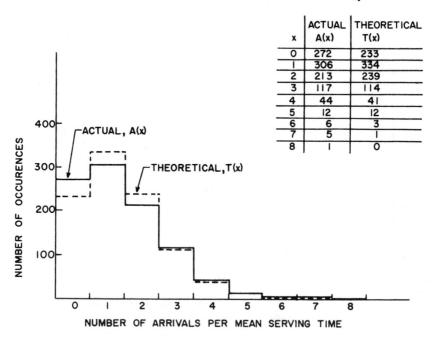

TOOL CRIB NO. 27

DATA TAKEN ON 7-6-54, 7-8-54, 7-9-54

AVERAGE NUMBER OF ARRIVALS PER MEAN SERVING TIME = 1.43

THEORETICAL VALUES GIVEN BY: $T(x) = 976 \dfrac{e^{-1.43} \, 1.43^x}{x!}$

x	ACTUAL A(x)	THEORETICAL T(x)
0	272	233
1	306	334
2	213	239
3	117	114
4	44	41
5	12	12
6	6	3
7	5	1
8	1	0

NUMBER OF ARRIVALS PER MEAN SERVING TIME

Fig. 9.4 Distribution of arrivals, final assembly tool crib.

TABLE 9.1 χ^2 TEST FOR DATA OF FIGURE 9.1

t	Actual	Theoretical	Difference	(Difference)^2	$\dfrac{\text{(Difference)}^2}{\text{Theoretical}}$
0–10	8	27	19	361	13.35
10–20	28	23	5	25	1.09
20–30	25	21	4	16	0.76
30–40	21	17	4	16	0.94
40–50	25	16	9	81	5.05
50–60	15	13	2	4	0.31
60–70	13	12	1	1	0.08
70–80	12	10	2	4	0.40
80–90	8	9	1	1	0.11
90–100	12	7	5	25	3.57
100–120	8	13	5	25	1.92
120–140	9	9	0	0	0.00
>140	22	29	7	49	1.69
	206	206			29.27

Degrees of freedom $= 13 - 2 = 11$; $\chi^2_{0.05} = 19.68$; $\chi^2_{0.01} = 24.73$

TABLE 9.2 χ^2 TEST FOR DATA OF FIGURE 9.2

t	Actual	Theoretical	Difference	(Difference)^2	$\dfrac{\text{(Difference)}^2}{\text{Theoretical}}$
0–10	83	91	8	64	0.70
10–20	98	69	29	841	12.20
20–30	70	57	13	169	2.96
30–40	50	46	4	16	0.35
40–50	39	38	1	1	0.03
50–60	31	31	0	0	0.00
60–70	15	26	11	121	4.65
70–80	13	21	8	64	3.05
80–90	5	17	12	144	8.45
90–100	10	14	4	16	1.14
100–120	20	22	2	4	0.18
120–140	10	14	4	16	1.14
>140	31	29	2	4	0.14
	475	475			34.99

Degrees of freedom $= 13 - 2 = 11$; $\chi^2_{0.05} = 19.68$; $\chi^2_{0.01} = 24.73$

TABLE 9.3 χ^2 TEST FOR DATA OF FIGURE 9.3

x	Actual	Theoretical	Difference	(Difference)^2	$\dfrac{(\text{Difference})^2}{\text{Theoretical}}$
0	91	69	22	484	7.02
1	109	114	5	25	0.22
2	67	93	26	676	7.27
3	52	51	1	1	0.02
4	22	21	1	1	0.05
≥ 5	15	9	6	36	4.00
					18.58

Degrees of freedom $= 6 - 2 = 4$; $\chi^2_{0.05} = 9.49$; $\chi^2_{0.01} = 13.28$

TABLE 9.4 χ^2 TEST FOR DATA OF FIGURE 9.4

x	Actual	Theoretical	Difference	(Difference)^2	$\dfrac{(\text{Difference})^2}{\text{Theoretical}}$
0	272	233	39	1521	6.53
1	306	334	28	784	2.35
2	213	239	26	676	2.83
3	117	114	3	9	0.08
4	44	41	3	9	0.22
≥ 5	24	16	8	64	4.00
					16.01

Degrees of freedom $= 6 - 2 = 4$; $\chi^2_{0.05} = 9.49$; $\chi^2_{0.01} = 13.28$

TABLE 9.5 SUMMARY OF χ^2 TESTS

Figure	χ^2 (Computed)	Degrees of Freedom	$\chi^2_{0.05}$	$\chi^2_{0.01}$	$\chi^2_{0.005}$
9.1	29.27	11	19.68	24.73	26.76
9.2	34.99	11	19.68	24.73	26.76
9.3	18.58	4	9.49	13.28	14.86
9.4	16.01	4	9.49	13.28	14.86

9.4 THE QUEUEING MODEL—OUTPUT

It has been decided, then, on the basis of observation and rationalization, to accept as an approximation to the real system the queueing model $M/M/c/\infty/\text{SIRO}$. We can now utilize Equations (3.11) and (3.15) to find an expression for W_q (or equivalently $L_q = \lambda W_q$) in Equation (9.2); that is, we desire the value of c which minimizes

$$E[C] = C_s(c - r) + C_w L_q \qquad (r = \lambda/\mu), \qquad (9.3)$$

where L_q is given by

$$L_q = \left[\frac{r^{c+1}/c}{c!(1 - r/c)^2} \right] p_0, \qquad (9.4)$$

$$p_0 = \left[\sum_{n=0}^{c-1} \frac{1}{n!} r^n + \frac{r^c}{c!} \left(\frac{c}{c - r} \right) \right]^{-1}. \qquad (9.5)$$

Thus for any values of r, C_s, and C_w, $E[C]$ is merely a function of c and can be computed for various values of c. Since the optimal value of c depends actually on the relative values of C_s and C_w (doubling both, for example, merely multiplies $E[C]$ by 2) we can look at

$$\frac{E[C]}{C_s} \equiv C' = (c - r) + R L_q, \qquad (9.6)$$

where $R \equiv C_w/C_s$.

9.5 RESULTS

Using Equations (9.4), (9.5), and (9.6), curves were obtained for C' as a function of the ratio of waiting-to-service cost R and the mean number of arrivals per mean service time r. These are reproduced in Figure 9.5.

For a fixed R and c, C' plots as a function of r into a J-shaped curve which starts at c on the ordinate when $r = 0$, reaches a minimum for some value $r < c$ and goes to infinity at $r = c$. Such a representative curve is shown in Figure 9.5, for $R = 5$ and $c = 2$. (C' is not defined, though in fact it is still infinite, for $r > c$. Recall that for steady-state conditions r must be $< c$.)

For various values of c, a series of J-shaped curves is obtained, but since

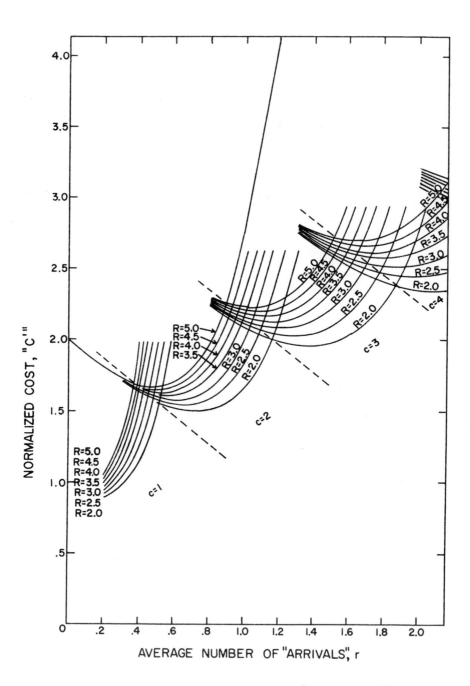

Fig. 9.5 Normalized expected cost curves.

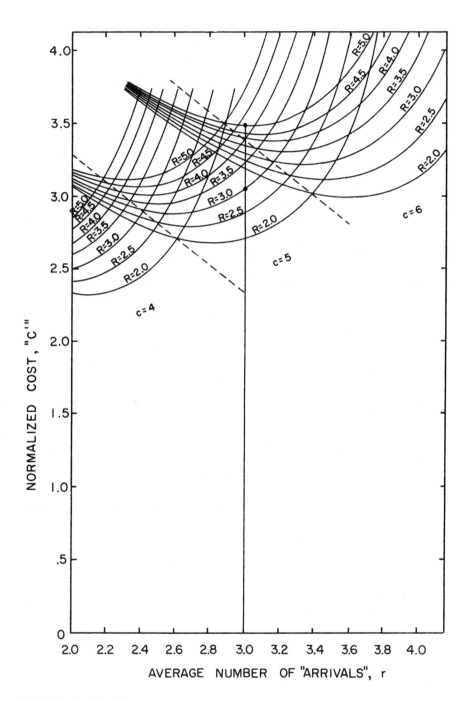

Fig. 9.5 (continued).

it is desired to minimize cost, it follows that what is wanted is the number of clerks which gives the least cost for a given value of r. Hence it is necessary to keep only the scalloped bottom of this series of curves, labeling each scallop with the corresponding number of clerks.

Finally R is varied. As R increases, the normalized cost increases, and the family of curves of Figure 9.5 is obtained. For a given number of arrivals per mean serving time (r) and a given ratio of costs (R), the graph will give the normalized cost C' and, according to the scallop, the number of clerks to be used. For example, if r is 3.0 and $R = 5.0$, the optimal c would be 6, whereas if r is 3.0 and $R = 3.0$, the optimal c is 5.

Values of C' were computed for c ranging from 1 to 6 in steps of 1, and R ranging from 2 to 5 in steps of 0.5. These ranges were decided upon from a consideration of the size of the counters being maintained in the company and the costs of clerks and mechanics likely to be encountered. It is interesting to note that in many cases the parameters may vary a good bit, yet the optimum number of clerks remains the same. Further investigation for the exact value of these parameters is needed only in those cases that are borderline. In practice, of course, the judgment of the foreman of the tool department still came into play when other factors had to be considered, such as the experience of the personnel.

The payoff for this study was rather indirect. As mentioned earlier, it was started at the same time as the tool department was instructed to cut down on personnel. The number of clerks was reduced from about 270 to 240, representing a savings of about $10,000 a month. As it turned out, the number 240 was within 1 or 2 of the optimum number calculated according to Figure 9.5. So while the credit for the savings could not be taken as a direct result of this study, it did provide the department with the means of justifying their final number, and of modifying it when required. This should have at least ensured that the number would not build back up to an uneconomical level, or for that matter, get forced down further, also to an uneconomical level.

9.6 MODEL REFINEMENT

The model was further refined to incorporate the fact that not all clerk idle time is wasted, since part of this time can be used productively in work behind the counter (sorting, straightening, etc.). This productive time during an idle period was limited by two factors: a minimum amount of idle time needed to do productive work (set-up time), and the maximum amount of time the productive work consumes, beyond which the idle time of a clerk really is wasted.

It was necessary, then, to find the probability distribution of clerk idle time. This proceeds as follows. Denoting by s_n the probability of n clerks being busy ($c - n$ clerks being idle) and letting $\tilde{I}_n(t)$ be the conditional probability that one of the $c - n$ idle clerks remains idle for a time greater than t, the unconditional probability of a clerk being idle for a period greater than t is

$$\tilde{I}(t) = \sum_{n=0}^{c-1} s_n \tilde{I}_n(t). \tag{9.7}$$

The probability that n clerks are busy ($c - n$ idle) is merely the probability that there are n customers in the system, $n < c$, which, from Equation (3.9), is

$$s_n = p_n = \frac{r^n}{n!} p_0, \tag{9.8}$$

with p_0 being given by Equation (9.5).

To find the conditional probability that a clerk remains idle for a time greater than t given $c - n$ clerks are presently idle, a birth–death analysis is used as follows. If one of the $c - n$ clerks is idle at t, he will remain idle at $t + \Delta t$, if (a) no customer arrives in Δt, or (b) a customer arrives and the clerk is not chosen. Assuming clerks are chosen at random by customers when several are idle, we have then

$$\tilde{I}_n(t + \Delta t) = \tilde{I}_n(t) \left[1 - \lambda \Delta t + \lambda \Delta t \frac{c - n - 1}{c - n} + o(\Delta t) \right]$$

$$= \tilde{I}_n(t) \left[1 - \frac{\lambda}{c - n} \Delta t + o(\Delta t) \right].$$

Dividing by Δt and taking the limit as Δt goes to zero in the usual fashion gives the differential equation

$$\frac{d\tilde{I}_n(t)}{dt} = - \frac{\lambda}{c - n} \tilde{I}_n(t),$$

whereupon integration of both sides yields

$$\tilde{I}_n(t) = C e^{-\lambda t / (c - n)}.$$

To find C, use is made of the boundary condition, $\tilde{I}_n(0) = 1$, which gives

$$\tilde{I}_n(t) = e^{-\lambda t / (c - n)}. \tag{9.9}$$

Using Equations (9.8) and (9.9) in Equation (9.7) produces finally

$$\tilde{I}(t) = \sum_{n=0}^{c-1} \frac{r^n}{n!} p_0 e^{-\lambda t/(c-n)}.$$

The density function of the complementary CDF above is

$$i(t) = -\frac{d\tilde{I}(t)}{dt} = \lambda p_0 \sum_{n=0}^{c-1} \frac{r^n}{n!(c-n)} e^{-\lambda t/(c-n)}. \tag{9.10}$$

The productive time that can be utilized by the clerks when they are idle can now be determined. We let τ_{min} represent the minimum time a clerk would need (set-up time) to accomplish any behind-the-counter productive work and τ_{max} be the maximum productive work time required (these are to be determined from the specific situation prevailing at a crib and are parameters just as C_s and C_w are). Furthermore, denoting the total expected free (idle) time that clerks have available in periods longer than τ_{min} as τ, the useful expected clerk idle time (call I_u) can be written as

$$I_u = \min(\tau, \tau_{max}),$$

where

$$\tau = (\text{number of waits longer than } \tau_{min}) \cdot (\text{average length}$$
$$\text{of waits longer than } \tau_{min}).$$

The number of waits longer than τ_{min} is the product of the number of waits times the probability of a wait longer than τ_{min}. Since a potential wait occurs every time a customer departs (albeit it may be for a time zero if more than c customers are present), we have, for a time T, λT departures; hence

$$\tau = \left[\lambda T \int_{\tau_{min}}^{\infty} i(t) \, dt \right] \frac{\int_{\tau_{min}}^{\infty} ti(t) \, dt}{\int_{\tau_{min}}^{\infty} i(t) \, dt}$$

$$= \lambda T \int_{\tau_{min}}^{\infty} ti(t) \, dt,$$

where $i(t)$ is given by Equation (9.10).

Thus the cost equation, Equation (9.2), would be adjusted by subtracting the expected useful idle time and becomes

$$E[C] = C_s\left(C - \frac{\lambda}{\mu} - \frac{I_u}{T}\right) + \lambda C_w W_q,$$

where

$$\frac{I_u}{T} = \min\left[\lambda \int_{\tau_{\min}}^{\infty} t i(t)\, dt, \tau'_{\max}\right],$$

and $\tau'_{\max}$ is now normalized by dividing $\tau_{\max}$ by T, the period of time over which the operation is being considered; that is, $\tau'_{\max}$ is the maximum behind-the-counter productive time per unit time.

We point out here that if it were not for $\tau_{\min}$, the minimum time which is needed for productive work during an idle period, the analysis would have been quite simple in that I_u would have been merely the minimum of $\tau_{\max}$ and the immediately calculable expected clerk idle time, $(c - r)T$.

Curves were presented showing τ/T as a function of r and $\tau_{\min}$ for various values of c. We will not reproduce these here; the interested reader is referred to the orignal source (Brigham, op. cit.).

9.7 DISCUSSION

This study has shown the many facets of a complete queueing analysis. First, the problem was formulated as a multichannel queueing design model with an appropriate objective (cost) function. Next, it was necessary to determine the basic characteristics of the queueing system (input, service, queue discipline) by observation and data gathering, the latter requiring some ingenuity for efficient and accurate recording of information.

From the data-gathering effort, a decision was made to use, as an approximation to the "real world," an $M/M/c$ model. Computer computations resulted in graphs of the expected cost as a function of the number of servers, so that, for any specific situation (λ, μ, C_s, C_w), the optimal number of servers could be obtained.

Implementation took the form of validating management's decision[3] to reduce the number of clerks and the establishment of a permanent decision-making aid to help in making any decisions of this nature which may be required in the future.

[3]Recall that it was the tool department, which was unhappy with management's decision in the first place, that initiated the study.

Finally, the model was further refined to include the fact that not all clerk idle time is really wasted.

We return now to the assumptions of Poisson input and exponential service that we questioned earlier in this chapter. The χ^2 goodness-of-fit test gave evidence that these assumptions were suspect. This, then, leaves three alternatives: (1) to use the Poisson-exponential model regardless as an approximation; (2) to search for other theoretical distributions and models which may better accommodate the data; and (3) to use the actual empirical distributions.

Alternative 2 appears to be the least desirable in this case, since for a multichannel optimization problem, it would be nearly impossible to derive any analytical results without the Poisson-exponential assumptions. Perhaps an Erlang-Erlang model ($E_k/E_l/c$) might be workable (see the final comments in Section 4.3.2), but the resulting complexities are enormous. Furthermore, quickly checking the service time data of Figure 9.2, for example, by calculating the sample variance shows the standard deviation to be larger than the mean, indicating no Erlang fit possible (Erlang $k < 1$). Another workable possibility is a $GI/M/c$ model (see Section 5.3.2), but this still restricts us to exponential service. What we would ideally like are results for $GI/G/c$ which would yield L_q; but alas, no such results are available for multichannel problems.

Alternative 3 would, of course, be the most compatible with the data, but again, any hope of analytical treatment is practically nonexistent. Almost surely, simulation would have to be employed.

Thus the decision really boils down to using the $M/M/c$ model, even though statistically we are on "thin ice," or performing a Monte Carlo simulation. To produce graphs equivalent to those given in Figures 9.5 and 9.6 using simulation would require a great deal of computer running time.

The best course of action here might be to develop a simulation model to check, for one or two sets of conditions, the results of the $M/M/c$ analytical model. Hopefully, the costs will not be too sensitive to the exact shape of the arrival and service distributions so that the $M/M/c$ model can be justifiably used. If it turns out this is not the case, there is no alternative but to use a complete Monte Carlo simulation analysis.

BIBLIOGRAPHY

References

Abramowitz, M., and Stegun, I. A. (1964). *Handbook of Mathematical Functions*. National Bureau of Standards, Applied Mathematics Series 55.

Agnew, R. P. (1942). *Differential Equations*. New York: McGraw-Hill.

Anderson, T. W., and Darling, D. A. (1952). "Asymptotic Theory of Certain 'Goodness of Fit' Criteria Based on Stochastic Processes." *Ann. Math. Statist.* **23**, 193–212.

Bailey, N. T. J. (1952). "Study of Queues and Appointment Systems in Out-Patient Departments with Special Reference to Waiting Times." *J. Roy. Statist. Soc. Ser. B.* **14**, 185–199.

Bailey, N. T. J. (1954). "A Continuous Time Treatment of a Simple Queue Using Generating Functions." *J. Roy. Statist. Soc. Ser. B.* **16**, 288–291.

Barrer, D. Y. (1957a). "Queuing with Impatient Customers and Indifferent Clerks." *Oper. Res.* **5**, 644–649.

Barrer, D. Y. (1957b). "Queuing with Impatient Customers and Ordered Service." *Oper. Res.* **5**, 650–656.

Beneš, V. E. (1957). "A Sufficient Set of Statistics for a Simple Telephone Exchange Model." *Bell Syst. Tech. J.* **36**, 939–964.

Beneš, V. E. (1963). *General Stochastic Processes in the Theory of Queues*. Reading, Mass.: Addison-Wesley.

Beneš, V. E. (1965). *Mathematical Theory of Connecting Networks and Telephone Traffic*. New York: Academic Press.

Betz, H., Burcham, P. B., and Ewing, G. M. (1954). *Differential Equations with Applications*. New York: Harper and Row.

Bhat, U. N. (1964). "Imbedded Markov Chain Analysis of Single-Server Bulk Queues." *J. Aust. Math. Soc.* **4**, 244–263.

Bhat, U. N. (1967). "Some Explicit Results for the Queue $GI/M/1$ with Group Service." *Sankhya Ser. A.* **29**, 199–206.

Bhat, U. N. (1968a). "Transient Behavior of Multi-Server Queues with Recurrent Input and Exponential Service Times." *J. Appl. Prob.* **5**, 158–168.

Bhat, U. N. (1968b). *A Study of the Queueing Systems $M/G/1$ and $GI/M/1$*. Lecture Notes in Operations Research and Mathematical Economics, No. 2. Berlin: Springer-Verlag.

Bhat, U. N. (1969a). "Queueing Systems with First-Order Dependence." *Opsearch* **6**, 1–24.

Bhat, U. N. (1969b). "Sixty Years of Queueing Theory." *Manage. Sci.* **15**, B280–B292.

Bhat, U. N., and Rao, S. S. (1972). "A Statistical Technique for the Control of Traffic Intensity in the Queuing Systems $M/G/1$ and $GI/M/1$." *Oper. Res.* **20**, 955–966.

Billingsley, P. (1961). *Statistical Inference for Markov Processes*. Chicago, Ill.: University of Chicago Press.

Blackburn, J. D. (1972). "Optimal Control of a Single-Server Queue with Balking and Reneging." *Manage. Sci.* **19**, 297–313.

Borofsky, S. (1950). *Elementary Theory of Equations*. New York: Macmillan.

Box, G. E. P. (1952). "Multi-Factor Design of First Order." *Biometrika* **39**, 49–57.

Box, G. E. P. (1954). "The Exploration and Exploitation of Response Surfaces; Some General Considerations and Examples." *Biometrics* **10**, 16–60.

Box, G. E. P. (1957). "Evolutionary Operation: A Method for Increasing Industrial Productivity." *Appl. Statist.* **6**, 81–101.

Box, G. E. P. and Hunter, J. S. (1957). "Multifactor Experimental Design for Exploring Response Surfaces." *Ann. Math. Statist.* **28**, 195–241.

Box, G. E. P., and Wilson, K. B. (1951). "On the Experimental Attainment of Optimum Conditions." *J. Roy. Statist. Soc. Ser. B.* **13**, 1–38.

Box, G. E. P., and Youle, P. U. (1955). "The Exploration and Exploitation of Response Surfaces; an Example of the Link Between the Filled Surface and the Basic Mechanism of the System." *Biometrics* **11**, 287–323.

Brigham, G. (1955). "On a Congestion Problem in an Aircraft Factory." *J. Oper. Res. Soc. Amer.* **3**, 412–428.

Brosh, I. (1970). "The Policy Space Structure of Markovian Systems with Two Types of Service." *Manage. Sci.* **16**, 607–621.

Buffa, E. S. (1966). *Readings in Production and Operations Management*. New York: Wiley.

Carter, G., and Ignall, E. (1972). "Virtual Measures for Computer Simulation Experiments." P-4817, The Rand Corporation, Santa Monica, California.

Champernowne, D. G. (1956). "An Elementary Method of Solution of the Queueing Problem with a Single Server and a Constant Parameter." *J. Roy. Statist. Soc. Ser. B.* **18**, 125–128.

Churchill, R. V. (1958). *Operational Mathematics*, 2nd ed.. New York: McGraw-Hill.

Churchill, R. V. (1960). *Complex Variables and Applications*, 2nd ed.. New York: McGraw-Hill.

Clarke, A. B. (1957). "Maximum Likelihood Estimates in a Simple Queue." *Ann. Math. Statist.* **28**, 1036–1040.

Cobham, A. (1954). "Priority Assignment in Waiting Line Problems." *Oper. Res.* **2**, 70–76; correction *ibid*. **3**, 547.

Cohen, J. W. (1969). *The Single Server Queue*. New York: Wiley.

Conway, R. W. (1963). "Some Tactical Problems in Digital Simulation." *Manage. Sci.* **10**, 47–61.

Conway, R. W., Johnson, B. M., and Maxwell, W. L. (1959). "Some Problems of Digital Systems Simulation." *Manage. Sci.* **6**, 92–110.

Conway, R. W., and Maxwell, W. L. (1962). "A Queueing Model with State Dependent Service Rates." *J. Ind. Eng.* **12**, 132–136.

Conway, R. W., Maxwell, W. L., and Miller, L. W. (1967). *Theory of Scheduling*. Reading, Mass.: Addison-Wesley.

Cooper, R. B. (1972). *Introduction to Queueing Theory*. New York: Macmillan.

Cox, D. R. (1955). "The Analysis of Non-Markovian Stochastic Processes by the Inclusion of Supplementary Variables." *Proc. Camb. Phil. Soc.* **51**, 443–441.

Crabill, T. B. (1969). "Optimal Control of a Queue with Variable Service Rates." Technical Report No. 75. Department of Operations Research, Cornell University, Ithaca, N. Y..

Crabill, T. B. (1972). "Optimal Control of a Service Facility with Variable Exponential Service Time and Constant Arrival Rate." *Manage. Sci.* **18**, 560–566.

Crabill, T. B., Gross, D., and Magazine, M. J. (1973). "A Survey of Research on Optimal Design and Control of Queues." Technical Paper T-280. Institute for Management Science and Engineering, The George Washington University, Washington, D. C.

Crabill, T. B., and Maxwell, W. L. (1969). "Single Machine Sequences with Random Processing Times and Random Due-Dates." *Nav. Res. Log. Quart.* **16**, 549–554.

Crane, M. A., and Iglehart, D. L. (1974). "Simulating Stable Stochastic Systems I: General Multi-Server Queues. *J. Assoc. Comput. Mach.* **21**, 103–113.

Crommelin, C. D. (1932). "Delay Probability Formulae When the Holding Times are Constant." *P.O. Elec. Eng. J.* **25**, 41–50.

Davies, O. L. (1956). *The Design and Analysis of Industrial Experiments.* London: Oliver and Boyd.

Doeh, G. (1960). "A Graphic Tool for the No-Queue Model." *Oper. Res.* **8**, 143–145.

Ehrenfield, S., and Ben-Tuvia, S. (1962). "The Efficiency of Statistical Simulation Procedures." *Technometrics* **4**, 257–275.

Eilon, S. (1969). "A Simpler Proof of $L = \lambda W$." *Oper. Res.* **17**, 915–917.

Emshoff, J. R., and Sisson, R. L. (1970). *Design and Use of Computer Simulation Models.* New York: Macmillan.

Erlang, A. K. (1909). "The Theory of Probabilities and Telephone Conversations." *Nyt Tidsskrift Matematik, B.* **20**, 33–39.

Erlang, A. K. (1917). "Solution of Some Problems in the Theory of Probabilities of Significance in Automatic Telephone Exchanges." *Electroteknikeren* (Danish) **13**, 5–13. [English translation in the *P.O. Elec. Eng. J.* **10**, 189–197 (1917–1918).]

Evans, R. V. (1971). "Programming Problems and Changes in the Stable Behavior of a Class of Markov Chains." *J. App. Probl.* **8**, 543–550.

Fabens, A. T. (1961). "The Solution of Queueing and Inventory Models by Semi-Markov Processes." *J. Roy. Statist. Soc. Ser. B.* **23**, 113–127.

Feller, W. (1966). *An Introduction to Probability Theory and Its Applications,* Vol. 2. New York: Wiley.

Fishman, G. S. (1967). "Problems in the Statistical Analysis of Simulation Experiments: The Comparison of Means and the Length of Sample Records." *Comm. ACM* **10**, 94–99.

Fishman, G. S. (1968). "The Allocation of Computer Time in Comparing Simulation Experiments." *Oper. Res.* **16**, 280–295.

Fishman, G. S. (1971). "Estimating Sample Size in Computer Simulation Experiments." *Manage. Sci.* **18**, 21–38.

Fishman, G. S. (1972a). "Output Analysis for Queueing Simulations." Technical Report 56. Department of Administrative Sciences, Yale University, New Haven, Conn.

Fishman, G. S. (1972b). "Bias Considerations in Simulation Experiments." *Oper. Res.* **20**, 785–790.

Fishman, G. S. (1972c). "Estimation in Multiserver Queueing Simulation." Technical Report 58. Department of Administrative Sciences, Yale University, New Haven, Conn.

Fishman, G. S. (1973a). "Statistical Analysis of Multiserver Queueing Simulations." Technical Report 64. Department of Administrative Sciences, Yale University, New Haven, Conn.

Fishman, G. S. (1973b). *Concepts and Methods in Discrete Event Digital Simulation.* New York: Wiley.

Fishman, G. S., and Kiviat, P. J. (1967). "The Analysis of Simulation-Generated Time Series." *Manage. Sci.* **13**, 525–557.

Fisz, M. (1963). *Probability Theory and Mathematical Statistics.* New York: Wiley.

Foster, F. G. (1953). "On Stochastic Matrices Associated with Certain Queuing Processes." *Ann. Math. Statist.* **24**, 355–360.

Fry, T. C. (1928). *Probability and Its Engineering Uses.* Princeton, N. J.: Van Nostrand.

Gafarian, A. V., and Ancker, C. J., Jr. (1966). "Mean Value Estimation from Digital Computer Simulation." *Oper. Res.* **14**, 25–44.

Galliher, H. P., and Wheeler, R. C. (1958). "Nonstationary Queuing Probabilities for Landing Congestion of Aircraft." *Oper. Res.* **6**, 264–275.

Gaver, D. P., Jr. (1959). "Imbedded Markov Chain Analysis of a Waiting Line Process in Continuous Time." *Ann. Math. Statist.* **30**, 698–720.

Gaver, D. P., Jr. (1966). "Observing Stochastic Processes and Approximate Transform Inversion." *Oper. Res.* **14**, 444–459.

Gaver, D. P., Jr. (1968). "Diffusion Approximations and Models for Certain Congestion Problems." *J. Appl. Probl.* **5**, 607–623.

Gebhard, R. F. (1967). "A Queueing Process With Bilevel Hysteretic Service-Rate Control." *Nav. Res. Log. Quart.* **14**, 55–68.

Gordon, G. (1969). *Systems Simulation.* Englewood Cliffs, N. J.: Prentice-Hall.

Greenberg, I. (1973). "Distribution-Free Analysis of $M/G/1$ and $G/M/1$ Queues." *Oper. Res.* **21**, 629–635.

Gross, D., and Harris, C. M. (1971). "On One-for-One Ordering Inventory Policies with State-Dependent Leadtimes." *Oper. Res.* **19**, 735–760.

Haight, F. A. (1957). "Queueing with Balking, I." *Biometrika* **46**, 360–369.

Haight, F. A. (1959). "Queueing with Reneging." *Metrika* **2**, 186–197.

Harris, C. M. (1967). "Queues with State-Dependent Stochastic Service Rates." *Oper. Res.* **15**, 117–130.

Heyman, Daniel P. (1968). "Optimal Operating Policies for $M/G/1$ Queuing Systems." *Oper. Res.* **16**, 362–382.

Hildebrand, F. B. (1949). *Advanced Calculus For Engineers.* Englewood Cliffs, N. J.: Prentice-Hall.

Hildebrand, F. B. (1952). *Methods of Applied Mathematics.* Englewood Cliffs, N. J.: Prentice-Hall.

Hillier, F. S., and Lieberman, G. J. (1967). *Introduction to Operations Reseach,* Chap. 10 and 11. San Francisco, Calif.: Holden-Day.

Hillier, F. S., and Lo, F. D. (1972). "Tables for Multiple-Server Queueing Systems Involving Erlang Distributions." Technical Report No. 149. Departments of Operations Research and Statistics, Stanford University, Stanford, Calif.

Hunt, G. C. (1956). "Sequential Arrays of Waiting Lines." *Oper. Res.* **4**, 674–683.

Hunter, J. S., and Naylor, T. H. (1970). "Experimental Designs for Computer Simulation Experiments." *Manage. Sci.* **16**, 422–434.

IBM Corporation (1970). *General Purpose Simulation System/360 Users Manual.* White Plains, N. Y.: Technical Publications Department.

Iglehart, D. L. (1972). "Extreme Values in the $GI/G/1$ Queue." *Ann. Math. Statist.* **49**, 627–635.

Jaiswal, N. K. (1968). *Priority Queues.* New York: Academic Press.

Jewell, W. S. (1967). "A Simple Proof of $L = \lambda W$." *Oper. Res.* **15**, 1109–1116.

Karlin, S. (1966). *A First Course in Stochastic Processes.* New York: Academic Press.

Karlin, S., and McGregor, J. L. (1957). "The Classification of Birth and Death Processes." *Trans. Amer. Math. Soc.* **86**, 366–400.

Karr, H. W., Luther, E. L., Markowitz, H. M., and Russell, E. C. (1965). "Sim-Optimization Research Phase I." CACI65-P2.0-1. Consolidated Analysis Centers, Inc., Los Angeles, California.

Kendall, D. G. (1951). "Some Problems in the Theory of Queues." *J. Roy. Statist. Soc. Ser. B.* **13**, 151–185.

Kendall, D. G. (1953). "Stochastic Processes Occurring in the Theory of Queues and Their Analysis by the Method of Imbedded Markov Chains." *Ann. Math. Statist.* **24**, 338–354.

Kendall, D. G. (1964). "Some Recent Work and Further Problems in the Theory of Queues." *Theory Probab. Appl.*, **9**, 1–15.

Kesten, H., and Runnenburg, J. T. (1957). "Priority Waiting Line Problems I, II." *Koninki. Ned. Akad. Wetenschap., Proc. Ser. A.* **60**, 312–336.

Khintchine, A. Y. (1932). "Mathematisches über die Erwortung vor einemöffentlichen Schalter." *Mat. Sbornik.* **39**, 73–84.

Kingman, J. F. C. (1962). "Some Inequalities for the Queue $GI/G/1$." *Biometrika* **49**, 315–324.

Kiviat, P. J. (1966). "Development of New Digital Simulation Languages." *J. Ind. Eng.* **17**, 604–608.

Kiviat, P. J., Villanueva, R., and Markowitz, H. M. (1968). *The SIMSCRIPT II Programming Language.* Englewood Cliffs, N. J.: Prentice-Hall.

Kleijnen, J. P., and Naylor, T. H. (1969). "The Use of Multiple Ranking-Procedures to Analyze Simulation of Business and Economic Systems." *Amer. Statist. Assoc. 1969 Proc. Bus. Econ. Statist. Sec.*, 605–615.

Koenigsberg, E. (1966). "On Jockeying in Queues." *Manage. Sci.* **12**, 412–436.

Kolmogorov, A. N. (1931). "Sur la problème d'attente." *Mat. Sbornik.* **8**, 101–106.

Kosten, L. (1948–1949). "On the Validity of the Erlang and Engset Loss Formulae." *Het P.T.T. Bedriff* **2**, 22–45.

Krasnow, H. S., and ,Merikallio, R. A. (1964). "Past, Present and Future of General Simulation Languages." *Manage. Sci.* **11**, 236–267.

Ledermann, W., and Reuter, G. E. (1954). "Spectral Theory for the Differential Equations of Simple Birth and Death Process." *Phil. Trans. Roy. Soc. London Ser. A.* **246**, 321–369.

Lee, A. M. (1966). *Applied Queueing Theory.* Montreal, Canada: St. Martin's Press.

Lilliefors, H. W. (1966). "Some Confidence Intervals for Queues." *Oper. Res.* **14**, 723–727.

Lilliefors, H. W. (1969a). "On the Kolmogorov-Smirnov Test for the Exponential Distribution with Mean Unknown." *J. Amer. Statist. Assoc.* **64**, 387–389.

Lilliefors, H. W. (1969b). "Some Goodness of Fit Tests for the Exponential Distribution with Comparisons of Power." Department of Statistics, George Washington University, Washington, D.C.

Lindley, D. V. (1952). "The Theory of Queues with a Single Server." *Proc. Camb. Phil. Soc.* **48**, 277–289.

Little, J. D. C. (1961). "A Proof for the Queuing Formula $L = \lambda W$." *Oper. Res.* **9**, 383–387.

Marshall, K. T. (1968). Some Inequalities in Queuing." *Oper. Res.* **16**, 651–665.

Mayhugh, J. O., and McCormick, R. E. (1968). "Steady-State Solution of the Queue $M/E_k/r$." *Manage. Sci.* **14**, 692–712.

Meier, R. C., Newell, W. T., and Pazer, H. L. (1969). *Simulation in Business and Economics.* Englewood Cliffs, N. J.: Prentice-Hall.

Miller, B. I. (1969). "A Queueing Reward System with Several Customer Classes." *Manage. Sci.* 16, 234–245.

Moder, J. J., and Phillips, C. R., Jr. (1962). "Queuing with Fixed and Variable Channels." *Oper. Res.* 10, 218–231.

Molina, E. C. (1927). "Application of the Theory of Probability to Telephone Trunking Problems." *Bell Syst. Tech. J.* 6, 461–494.

Morse, P. M. (1958). *Queues, Inventories and Maintenance.* New York: Wiley.

Naylor, T. H., Balintfy, J. L., Burdick, D. S., and Chu, K. (1966). *Computer Simulation Techniques.* New York: Wiley.

Naylor, T. H., and Wonnacott, T. H. (1970). "A Comment on the Analysis of Data Generated by Simulation Experiments." *Manage. Sci.* 17, 233–235.

Neuts, M. F. (1965). "The Busy Period of a Queue with Batch Service." *Oper. Res.* 13, 815–819.

Neuts, M. F. (1966). "An Alternative Proof of a Theorem of Takács on the $GI/M/1$ Queue." *Oper. Res.* 14, 313–316.

Neuts, M. F. (1973). "The Single Server Queue in Discrete Time-Numerical Analysis, I." *Nav. Res. Log. Quart.* 20, 297–304.

Newell, G. F. (1972). *Applications of Queueing Theory.* London: Chapman and Hall Ltd.

Olmsted, J. M. H. (1959). *Real Variables.* New York: Appleton-Century-Crofts.

Palm, C. (1938). "Analysis of the Erlang Traffic Formulae for Busy-Signal Arrangements." *Ericsson Tech.* 6, 39–58.

Panico, J. (1969). *Queuing Theory: A Study of Waiting Lines for Business, Economics, and Science.* Englewood Cliffs, N. J.: Prentice-Hall.

Parzen, E. (1960). *Modern Probability and Its Applications.* New York: Wiley.

Parzen, E. (1962). *Stochastic Processes.* San Francisco, Calif.: Holden-Day.

Peck, L. G., and Hazelwood, R. N. (1958). *Finite Queueing Tables.* New York. Wiley.

Phipps, T. E., Jr. (1956). "Machine Repair As a Priority Waiting-Line Problem." *Oper. Res.* 4, 76–85. (Comments by W. R. Van Voorhis, *ibid.* 4, 86.)

Pollaczek, F. (1932). "Lösung eines Geometrischen Wahrscheinlichkeitsproblems." *Math. Z.* 35, 230–278.

Pollaczek, F. (1934). "Uber das Warteproblem." *Math. Z.* 38, 492–537.

Prabhu, N. U. (1965a). *Stochastic Processes.* New York: Macmillan.

Prabhu, N. U. (1965b). *Queues and Inventories.* New York: Wiley.

Prabhu, N. U. (1967a). "Transient Behavior of a Tandem Queue." *Manage. Sci.* 13, 631–639.

Prabhu, N. U. (1967b). "Queueing Systems Not in Steady State." Technical Report. Department of Operations Research, Cornell University, Ithaca, N. Y..

Prabhu, N. U., and Bhat, U. N. (1963a). "Further Results for the Queue with Poisson Arrivals." *Oper. Res.* 11, 380–386.

Prabhu, N. U., and Bhat, U. N. (1963b). "Some First Passage Problems and Their Application to Queues." *Sankhya Ser. A.* 25, 281–292.

Pyke, R. (1961). "Markov Renewal Processes." *Ann. Math. Statist.* 32, 1231–1242.

Rainville, E. D., and Bedient, P. E. (1969). *A Short Course in Differential Equations.* New York: Macmillan.

Rand Corporation (1955). *A Million Random Digits with 100,000 Normal Deviates*. Glencoe, Ill.: Free Press.

Rao, S. S. (1968). "Queueing with Balking and Reneging in $M/G/1$ Systems." *Metrika* **12**, 173–188.

Reich, E. (1957). "Waiting Times When Queues Are in Tandem." *Ann. Math. Statist.* **28**, 768–773.

Romani, J. (1957). "Un Modelo de la Teoria de Colas con Número Variable de Canales." *Trabajos Estadestica* **8**, 175–189.

Rosenshine, M. (1967). "Queues with State-Dependent Service Times." *Transp. Res.* **1**, 97–104.

Rosenshine, M. (1968). "Operations Research in the Solution of Air Traffic Control Problems." *J. Ind. Eng.* **19**, 122–128.

Ross, S. (1970). *Applied Probability Models with Optimization Examples*. San Francisco, Calif.: Holden-Day.

Saaty, T. L. (1961). *Elements of Queueing Theory with Applications*. New York: McGraw Hill.

Saaty, T. L. (1966). "Seven More Years of Queues: A Lament and a Bibliography." *Nav. Res. Log. Quart.* **13**, 447–476.

Schrage, L. E., and Miller, L. W. (1966). "The Queue $M/G/1$ with the Shortest Remaining Processing Time Discipline." *Oper. Res.* **14**, 670–684.

Smith, D. E. (1973). "Requirements of an 'Optimizer' for Computer Simulations." *Nav. Res. Log. Quart.* **20**, 161–179.

Smith, W. L., and Wilkinson, W. E. (eds.) (1965). *Proceedings of the Symposium on Congestion Theory*. Chapel Hill, N. C.: University of North Carolina Press.

Sobel, M. J. (1969). "Optimal Average Cost Policy for a Queue with Start-Up and Shut-Down Costs." *Oper. Res.* **17**, 145–162.

Stidham, S., Jr. (1970). "On the Optimality of Single-Server Queuing Systems." *Oper. Res.* **18**, 708–732.

Stidham, S., Jr. (1972). "$L = \lambda W$: A Discounted Analogue and a New Proof." *Oper. Res.*, 1115–1126.

Syski, R. (1960). *Introduction to Congestion Theory in Telephone Systems*. London: Oliver and Boyd.

Takács, L. (1955). "Investigations of Waiting Time Problems by Reduction to Markov Processes." *Acta Math. Acad. Sci. Hung.* **6**, 101–129.

Takács, L. (1962). *Introduction to the Theory of Queues*. Oxford, England: Oxford University Press.

Takács, L. (1967). *Combinatorial Methods in the Theory of Stochastic Processes*. New York: Wiley.

Takács, L. (1969). "On Erlang's Formula." *Ann. Math. Stat.* **40**, 71–78.

Thorndyke, F. (1926). "Application of Poisson's Probability Summation." *Bell Syst. Tech. J.* **5**, 604–624.

Tocher, K. D. (1963). *The Art of Simulation*. London: English University Press.

Vaulot, A. E. (1927). "Extension des Formules d'Erlang au Cas où les Durées des Conversations Suivent une Loi Quelconque." *Rev. Gén. d'Electricité* **22**, 1164–1171.

Wilde, D. J. (1964). *Optimum Seeking Methods*. Englewood Cliffs, N. J.: Prentice-Hall.

Wolff, R. W. (1965). "Problems of Statistical Inference for Birth and Death Queuing Models." *Oper. Res.* **13**, 343–357.

Yadin, M., and Naor, P. (1967). "On Queueing Systems with Variable Service Capacities."
Nav. Res. Log. Quart. **14**, 43–54.

Additional Readings

Books

Bagchi, T. P., and Templeton, J. G. C. (1972). *Numerical Methods in Markov Chains and Bulk Queues.* Berlin: Springer-Verlag.

Beckman, P. (1968). *Introduction to Elementary Queuing Theory.* Boulder, Colo.: Golem Press.

Brockmeyer, E., Halstrom, H. L., and Jensen, A. (1948). *The Life and Works of A. K. Erlang.* Trans. of the Danish Acad. Sci., No. 2.

Cox, D. R., and Smith, W. L. (1961). *Queues.* London: Methuen.

Cruon, R. (ed.) (1967). *Queuing Theory, Recent Developments and Applications.* New York: Elsevier.

Descloux, A. (1962). *Delay Tables for Finite- and Infinite-Source Systems.* New York: McGraw-Hill.

Feller, W. (1959). *An Introduction to Probability Theory and Its Applications,* Vol. 1, 2nd Ed. New York: Wiley.

Ghosal, A. (1970). *Some Aspects of Queueing and Storage Systems.* Berlin: Springer Verlag.

Gnedenko, B. V., and Kovalenko, I. N. (1968). *Introduction to Queueing Theory.* Israel Program for Scientific Translations, Jerusalem, Israel.

Haight, F. (1963). *Mathematical Theories of Traffic Flow.* New York: Academic Press.

Kaufmann, A., and Cruon, R. (1961). *Les Phénomènes d'Attente.* Paris, France: Duval.

Khintchine, A. Y. (1960). *Mathematical Methods in the Theory of Queueing.* London: Griffin.

Kleinrock, L. (1964). *Communication Nets.* New York: McGraw-Hill.

Klimov, G. P. (1966). *Stochastic Queueing Systems.* Moscow: Izdat, Nauke, U.S.S.R. (in Russian).

LeGall, P. (1962). *Les Systèmes avec où sans attente et les processes stochastiques.* Tome 1. Paris, France: Dunod.

Riordan, J. (1962). *Stochastic Service Systems.* New York: Wiley.

Ruiz-Pala, E., Avila-Beloso, C., and Hines, W. W. (1967). *Waiting-Line Models.* New York: Reinhold.

Teghem, J., Loris-Teghem, J., and Lambotte, J. P. (1969). *Modèles d'Attente M/G/1 et GI/M/1 à Arrivées et Services en Groupes.* Berlin: Springer-Verlag.

Articles

APPLICATIONS

Chang, W. (1966). "A Queuing Model for a Simple Case of Time Sharing." *IBM Syst. J.* **5**, 115–125.

Edie, L. C. (1954). "Traffic Delays at Toll Booths." *J. Oper. Res. Soc. Amer.* **2**, 107–138.

Edie, L. C. (1956). "Optimization of Traffic Delays at Toll Booths." in *Operation Research for Management*. Vol. 2, J. McCloskey and J. M. Coppinger, (eds.) Johns Hopkins Press, Chap. 3.

Krishnamoorthi, B., and Wood, R. C. (1966). "Time Shared Computer Operations with Interarrival and Service Times Exponential." *J. Assoc. Comput. Mach.* **13**, 317–338.

BULK MODELS

Ancker, C. J., Jr., and Gafarian, A. V. (1961). "Queuing with Multiple Poisson Inputs and Exponential Service Times." *Oper. Res.* **9**, 321–327.

Harris, C. M. (1966). "Queues with Multiple Poisson Input." *J. Ind. Eng.* **17**, 454–460.

Kabak, I. W. (1968). "Blocking and Delays in $M^{(n)}/M/c$ Bulk Queuing Systems." *Oper. Res.* **16**, 830–840.

Miller, R. G., Jr. (1959). "A Contribution to the Theory of Bulk Queues." *J. Roy. Statist Soc. Ser. B.* **21**, 320–337.

DESIGN AND CONTROL

Brosh, I., and Naor, P. (1963). "On Optimal Disciplines in Priority Queueing." *Bull. Inst. Statist. Math.* **40**, 593–607.

Hillier, F. S. (1963). "Economic Models for Industrial Waiting Line Problems." *Manage. Sci.* **10**, 119–130.

Hillier, F. S. (1964). "The Application of Waiting-Line Theory to Industrial Problems." *J. Indust. Eng.* **15**, 3–8.

Kumin, H. J. (1968). "The Design of Markovian Congestion Systems." Technical Memorandum No. 115. Operations Research Department, Case Western Reserve University, Cleveland, Ohio.

FINITE SOURCE QUEUES

Benson, F., and Cox, D. R. (1951). "The Productivity of Machines Requiring Attention at Random Invervals." *J. Roy. Statist. Soc. Ser. B.* **13**, 65–82.

Benson, F., and Cox, D. R. (1952). "Further Notes on the Productivity of Machines Requiring Attention at Random Intervals." *J. Roy. Statist Soc. Ser. B.* **14**, 200–219.

Naor, P. (1956). "On Machine Interference." *J. Roy. Statist Soc. Ser. B.* **18**, 280–287.

GENERAL RESULTS

Bhatia, A., and Garg, A. (1963). "Basic Structure of Queueing Problems." *J. Ind. Eng.* **14**, 13–17.

Cohen, J. W. (1963). "Applications of Derived Markov Chains in Queueing Theory." *Appl. Sci. Res. B* **10**, 269–303.

Harris, T. J. (1967). "Duality of Finite Markovian Queues." *Oper. Res.* **15**, 575–576.

Kiefer, J., and Wolfowitz, J. (1956). "On the Characteristics of the General Queueing Process with Applications to Random Walk." *Ann. Math. Statist.* **27**, 147–161.

Kingman, J. F. C. (1970). "Inequalities in the Theory of Queues." *J. Roy. Statist. Soc. Ser. B.* **32**, 102–110.

Kleinrock, L. (1965). "A Conservation Law for a Wide Class of Queueing Disciplines." *Nav. Res. Log. Quart.* **12**, 181–192.

Marshall, K. T., and Wolff, R. W. (1971). "Customer Average Queue Lengths and Waiting Times." *J. Appl. Probl.* **8**, 535–542.

Page, E. S. (1965). "On Monte Carlo Methods in Congestion Problems." *Oper. Res.* **13**, 291–306.

Saunders, L. R. (1961). "Probability Functions for Waiting Times in Single-Channel Queues, with Emphasis on Simple Approximations." *Oper. Res.*, **9**, 351–362.

Shelton, J. R. (1960). "Solution Methods for Waiting Line Problems." *J. Ind. Eng.* **11**, 293–303.

Spitzer, F. (1956). "A Combinatorial Lemma and Its Applications to Probability Theory." *Trans. Amer. Math. Soc.* **82**, 323–339.

Winsten, C. B. (1959). "Geometric Distributions in the Theory of Queues." *J. Roy. Statist. Soc. Ser. B.* **21**, 1–22.

IMPATIENCE

Ancker, C. J., Jr. and Gafarian, A. V. (1962). "Queueing with Impatient Customers Who Leave at Random." *J. Indust. Eng.* **13**, 84–90.

Ancker, C. J., Jr., and Gafarian, A. V. (1963). "Some Queuing Problems with Balking and Reneging, I." *Oper. Res.* **11**, 88–100.

Ancker, C. J., Jr., and Gafarian, A. V. (1963). "Some Queuing Problems with Balking and Reneging, II." *Oper. Res.* **11**, 928–937.

Finch, P. D. (1960). "Deterministic Customer Impatience in the Queueing System $GI/M/1$." *Biometrika* **47**, 45–52.

Haight, F. A. (1960). "Queueing with Balking, II." *Biometrika* **47**, 285–296.

MODELS—MULTIPLE CHANNELS

Mayhugh, J. O. (1970). "A Note on the Queue $M/E_k/r$." *Manage. Sci.* **16**, 512–513.

Riordan, J. (1953). "Delay Curves for Calls Served at Random." *Bell Syst. Tech. J.* **32**, 100–119.

MODELS—SINGLE CHANNEL

Burke, P. J. (1959). "Equilibrium Delay Distribution for One Channel with Constant Holding Time, Poisson Input and Random Service." *Bell Syst. Tech. J.* **38**, 1021–1031.

Çinlar, E., and Disney, R. L.(1967). "Stream of Overflows from a Finite Queue." *Oper. Res.*, **15**, 131–134.

Crabill, T. B. (1968). "Sufficient Conditions for Positive Recurrence and Recurrence of Specially Structured Markov Chains." *Oper. Res.*, **16**, 858–867.

Greenberg, I. (1967). "The Behavior of a Simple Queue at Various Times and Epochs." *SIAM Rev.* **9**, 234–248.

Harris, C. M. (1967). "Queues with Stochastic Service Rates." *Nav. Res. Log. Quart.* **14**, 219–230.

Harris, C. M. (1968). "The Pareto Distribution as a Queue Service Discipline." *Oper. Res.* **16**, 307–313.

Keilson, J., and Kooharian, A. (1960). "On Time-Dependent Queuing Processes." *Ann. Math. Statist.* **31**, 104–112.

Oliver, R. M. (1964). "An Alternate Derivation of the Pollaczek-Khintchine Formula." *Oper. Res.* **12**, 158–159.

PRIORITIES

Avi-Itzhak, B. (1963). "Preemptive Repeat Priority Queues as a Special Case of the Multi-purpose Server Problem, I." *Oper. Res.* **11**, 597–609.

Avi-Itzhak, B. (1963). "Preemptive Repeat Priority Queues as a Special Case of the Multi-purpose Server Problem, II." *Oper. Res.* **11**, 610–619.

Chang, W. (1965). "Preemptive Priority Queues." *Oper. Res.* **13**, 820–827.

Greenberger, M. (1966). "The Priority Problem and Computer Time Sharing." *Manage. Sci.* **12**, 888–906.

Jackson, J. R. (1961). "Queues with Dynamic Priority Discipline." *Manage. Sci.* **8** 18–34.

Jaiswal, N. K. (1961). "Preemptive Resume Priority Queue." *Oper. Res.* **9**, 732–742.

Jaiswal, N. K., and Pestalozzi, G. (1965). "On a Problem of Optimum Priority Classification." *J. Soc. Ind. Appl. Math.* **13**, 580–591.

Welch, P. D. (1964). "On Preemptive Resume Priority Queues." *Ann. Math. Statist.* **35**, 600–612.

SERIES AND CYCLIC

Burke, P. J. (1956). "The Output of a Queuing System." *Oper. Res.* **4**, 699–704.

Burke, P. J. (1968). "The Output Process of a Stationary $M/M/s$ Queueing System." *Ann. Math. Statist.* **39**, 1144–1152.

Evans, R. V. (1964). "Queueing When Jobs Require Several Services Which Need Not Be Sequenced." *Manage. Sci.* **10**, 298–315.

Evans, R. V. (1967). "Geometric Distribution in Some Two-Dimensional Queuing Systems." *Oper. Res.,* **15**, 830–846.

Evans, R. V. (1967). "Capacity of Queuing Networks." *Oper. Res.* **15**, 530–536.

Finch, P. D. (1959). "The Output Process of the Queueing System $M/G/1$." *J. Roy. Statist. Soc. Ser. B.* **21**, 375–380.

Jackson, J. R. (1963). "Jobshop-Like Queueing Systems." *Manage. Sci.* **10**, 131–142.

STATE DEPENDENCE

Conolly, B. W. (1968). "The Waiting Time Process for a Certain Correlated Queue." *Oper. Res.* **16**, 1006–1015.

Gross, D., Harris, C. M., and Lechner, J. (1971). "Stochastic Inventory Models with Bulk Demand and State-Dependent Leadtimes." *J. Appl. Probl.* **8**, 521–534.

Gross, D., and Harris, C. M. (1972). "Continuous-Review (s, S) Inventory Models with State-Dependent Leadtimes." *Manage. Sci.* **19**, 567–574.

Harris, C. M. (1967). "A Queueing System with Multiple Service Time Distributions." *Nav. Res. Log. Quart.* **14**, 231–239.

Harris, C. M. (1970). "Some Results for Bulk-Arrival Queues with State-Dependent Service Times." *Manage. Sci.* **16**, 313–326.

Harris, C. M. (1971). "On Queues with State-Dependent Erlang Service." *Nav. Res. Log. Quart.* **18**, 103–110.

Appendix 1
DICTIONARY OF SYMBOLS

This appendix contains definitions of common symbols used frequently and consistently throughout the text. Symbols which are used only occasionally in isolated sections of the text are not always included here.

The symbols are listed in alphabetical order. Greek symbols are filed according to their English equivalents; for example, λ (lambda) is found under the L's. Within an alphabetical category, English symbols precede Greek. Listed at the end are nonliteral symbols such as primes and asterisks.

$A/B/X/Y/Z$	Notation for describing queueing models where
	A indicates interarrival pattern
	B indicates service pattern
	X indicates number of channels
	Y indicates system capacity limit
	Z indicates queue discipline
$A(t)$	Cumulative distribution function of interarrival times
$a(t)$	Probability distribution of interarrival times
$B(t)$	Cumulative distribution function of service times
b_n	(1) Probability of n services during an interarrival time; (2) discouragement function in queueing models with balking
$b(t)$	Probability distribution of service times
C	(1) Arbitrary constant; (2) cost per unit time, often used in functional notation as $C(\cdot)$
CDF	Cumulative distribution function
C-K	Chapman-Kolmogorov
C_s	Marginal cost of a server per unit time
C_w	Cost of customer wait per unit time
$C(t)$	CDF of the interdeparture process
$C(\cdot)$	Cost per unit time function
$C(z)$	Probability generating function of c_n
c	Number of parallel channels (servers)
c_n	Probability that in a bulk-arrival queueing model the batch size is n

$c(t)$	Probability distribution of the interdeparture process
D	(1) Deterministic interarrival or service times; (2) linear difference operator, $Dx_n = x_{n+1}$; (3) linear differential operator, $Dy(x) = dy/dx$
$\bar{d}$	Mean observed interdeparture time of a queueing system
Δt	Infinitesimal interval of time
Δy	First finite difference; that is, $\Delta y = y_{n+1} - y_n$
E_k	Erlang type k distributed interarrival or service times
E_n	Possible system state $(n = 0, 1, 2, \ldots)$
$E[\cdot]$	Expected value
η_{ij}	Mean time spent in state i before going to j
ϵ	Denotes set membership
FIFO	First-in, first-out or first-come, first-served queue discipline
$F_n(t)$	Joint probability that n are in the system at time t after the last departure and t is less than the interdeparture time
$F_{ij}(t)$	Conditional probability that, given a process begins in state i and next goes to state j, the transition time is $\leqslant t$ (a conditional CDF)
f_{ij}	Probability that state j of a process is ever reached from state i
$f_{ij}^{(n)}$	Probability that the first passage of a process from state i to state j occurs in exactly n steps
G	General distribution for service times
GCD	Greatest common divisor
GD	General queue discipline
GI	General independent distribution for interarrival times
$G(t)$	Cumulative distribution function of the busy period for $M/G/1$ and $GI/M/1$ models
$G(z)$	Generating function associated with Erlang service steady-state probabilities $\{p_{n,i}\}$
$G_i(t)$	Conditional probability that, given a process starts in state i, the time to the next transition is $\leqslant t$ (a conditional CDF)
$H(z, y)$	(1) Probability generating function for $\{p_{n,i}\}$; (2) joint generating function for a two-priority queueing model
$H_{ij}(t)$	Cumulative distribution function of time until first transition of a process into state j beginning at state i

$H_r(y,z)$	Generating function associated with $P_{mr}(z)$ for a two-priority queueing model
$h(u)$	Failure or hazard rate of a probability distribution
I	Phase of service customer is in for Erlang service models (a random variable)
IID	Independent and identically distributed
I_u	Expected useful clerk idle time
$I_n(\cdot)$	Modified Bessel function of the first kind
$\tilde{I}(t)$	Probability of a clerk being idle for a time $>t$ (a complementary CDF)
$\tilde{I}_n(t)$	Conditional probability that one of the $c-n$ idle clerks remains idle for a time $>t$ (a conditional complementary CDF)
$i(t)$	Probability distribution of idle time
$J_n(\cdot)$	Regular Bessel function
K	System capacity limit (truncation point of system size)
K_q	Greatest queue length at which an arrival would balk (a random variable)
$K(z)$	Probability generating function of $\{k_n\}$
$K_i(z)$	Probability generating function of $\{k_{n,i}\}$
k_n	Probability of n arrivals during a service time
$k_{n,i}$	Probability of n arrivals during a serivce time, given i in the system when service began
L	Expected system size
LIFO	Last-in, first-out queue discipline
LST	Laplace-Stieltjes transform
LT	Laplace transform
$L^{(D)}$	Expected system size at departure points
$L^{(P)}$	Expected number of phases in the system of an Erlang queueing model
$L^{(n)}$	(1) Expected number of customers of type n in system; (2) expected system size at station n in a series or cyclic queue
$L_{(k)}$	The kth factorial moment of system size
L_q	Expected queue size
L_q'	Expected queue size of nonempty queues
$L_q^{(D)}$	Expected queue size at departure points
$L_q^{(P)}$	Expected number of phases in the queue of an Erlang queueing model
$L_q^{(n)}$	(1) Expected queue size for customers of type n; (2) expected queue size in front of station n in a series or cyclic queue

$L_{q(k)}^{(D)}$	The kth factorial moment of the departure-point queue size
$L(\cdot)$	Likelihood function
$\mathcal{L}(\cdot)$	Log-likelihood
$\mathcal{L}\{\cdot\}$	Laplace transform
λ	Mean arrival rate (independent of system size)
λ_n	(1) Mean arrival rate when there are n in the system; (2) mean arrival rate of customers of type n
M	(1) Poisson arrival or service process (or equivalently exponential interarrival or service times); (2) finite population size
MGF	Moment generating function
$M_X(t)$	Moment generating function of the random variable X
m_i	Mean time a process spends in state i during a visit
m_{ij}	Mean first passage time of a process from state i to state j
m_{jj}	Mean recurrence time of a process to state j
μ	Mean service rate (independent of system size)
$\mu^{(B)}$	Mean service rate for a bulk queueing model
μ_n	(1) Mean service rate when there are n in the system; (2) mean service rate of server number n; (3) mean service rate for customers of type n
N	Steady-state number in the system (a random variable)
N_q	Steady-state number in the queue (a random variable)
$N(t)$	Number in the system at time t (a random variable)
$N_q(t)$	Number in the queue at time t (a random variable)
n_a, n_{ae}, n_b	Number of observed arrivals to a system, to an empty system, and to a busy system, respectively ($n_a = n_{ae} + n_b$)
$n(t)$	Number in the system at time t, deterministic queueing model
$o(\Delta t)$	Order Δt; that is, $\lim_{\Delta t \to 0} o(\Delta t)/\Delta t = 0$
PDE	Partial differential equation
PRI	Priority queue discipline
$P(z), P(z,t)$	Probability generating function of $\{p_n\}$ and $\{p_n(t)\}$, respectively
$P_{mr}(z)$	Probability generating function of priority steady-state probabilities $\{p_{mnr}\}$
p_n	Steady-state probability of n in the system

$p_n^{(B)}$	Steady-state probability of n in a bulk queueing system
$p_n^{(P)}$	Steady-state probability of n phases in an Erlang queueing system
$p_n(t)$	Probability of n in the system at time t
p_{ij}	Single-step transition probability of going from state i to state j
$p_{n,i}$	Steady-state probabilities for Erlang models of n in the system and the customer in service (if service is Erlangian) or next to arrive (if arrivals are Erlangian) in phase i
$p_{ij}^{(n)}$	Transition probability of going from state i to state j in n steps
$p_{n,i}(t)$	Probability that, in an Erlang queueing model at time t, n are in the system and the customer in service (if the service is Erlangian) or next to arrive (if arrivals are Erlangian) is in phase i
$p_{mnr}(t)$	Probability at time t of m units of priority 1, n units of priority 2 in the system, and a unit of priority r in service ($r = 1$ or 2)
$p_{i,j}(u,s)$	Transition probability of moving from state i to state j in time beginning at u and ending at s
$p_{n_1,n_2,\ldots,n_k}(t)$	Probability of n_1 customers at station 1, n_2 at station $2,\ldots,n_k$ at station k in a series queue at time t
$p_{n_1,n_2,\ldots,n_k},$ $p(n_1,n_2,\ldots,n_k)$	Steady-state probability of $p_{n_1,n_2,\ldots,n_k}(t)$
$\Pi(z)$	Probability generating function of $\{\pi_n\}$
π_n	Steady-state probability of n in the system at a departure point
$Q_{ij}(t)$	Joint conditional probability that, given a process begins in state i, the next transition will be to state j in an amount of time $\leq t$ (a conditional CDF)
q_n	Steady-state probability that an arriving customer finds n in the system
q_{ik}	Steady-state probability in a series queue of a customer going to station k after being served at station i
RV	Random variable
r	Defined as λ/μ for multichannel models; defined as $\lambda/k\mu$ for Erlang service models

r_i	(1) The ith root of a polynomial equation (if there is only one root r_0 is used); (2) sample covariance of lag i
r_n	The nth uniform-$(0, 1)$ random number
$r(n)$	Reneging function
$\text{Re}(\cdot)$	Real portion of a complex number
ρ	Traffic intensity $(=\lambda/\mu$ for single-channel models and $=\lambda/c\mu$ for multichannel models)
S	Steady-state service time (a random variable)
SMP	Semi-Markov process
SIRO	Served-in-random-order queue discipline
$S^{(n)}$	Service time of the nth arriving customer (a random variable)
$S_k^{(n)}$	Service time of the nth arriving customer of type k
$S_k(S_k')$	Time it takes to serve $n_k(n_k')$ waiting customers of type k (a random variable)
S_0	Time required to finish customer in service (remaining time of service; a random variable)
s_n	Probability that n servers are busy $(c-n$ idle) in a multichannel system
s_X	Sample standard deviation of the random variable X
σ_S^2	Variance of the service-time distribution
σ_k	Sum of traffic intensities in a priority queueing model; that is, $\sigma_k = \sum_{i=1}^{k} \rho_i$, $\rho_i = \lambda_i/\mu_i$
T	(1) Time spent in system (a random variable), with expected value W; (2) the steady-state interarrival time (a random variable); (3) steady-state interdeparture time (a random variable)
$T^{(n)}$	Interarrival time between the nth and $(n+1)$st customers (a random variable)
T_A	Instant of arrival
T_i	Length of time a stochastic process spends in state i (a random variable)
T_S	Instant of service completion
T_{busy}	Length of a busy period for $M/M/1$ (a random variable)
T_q	Time spent in queue (a random variable), with expected value W_q
$T_{b,i}$	Length of i channel busy period for $M/M/c$ (a random variable)

t_b, t_e, t Observed time a system is busy, observed time a system is empty, total observed time, respectively $(t = t_b + t_e)$

t_i (1) Time to first balk, deterministic queueing model; (2) time to steady state, deterministic queueing model

τ Expected clerk idle time in periods longer than τ_{min}

τ_i Time at which the ith arrival to a Poisson process occurred

τ_{max} Maximum productive behind-counter clerk-work time

τ_{min} Minimum time a clerk requires to accomplish behind-counter productive work

U Steady-state difference between service time and interarrival time; that is, $U = S - T$ (a random variable)

$U^{(n)}$ Service time of nth customer minus interarrival time between customer $n + 1$ and n; that is $U^{(n)} = S^{(n)} - T^{(n)}$ (a random variable)

$U(t)$ (1) Cumulative distribution function of $U = S - T$; (2) cumulative distribution function of the time back to the most recent transition

$U^{(n)}(t)$ Cumulative distribution function of $U^{(n)}$

$U_i(t)$ Cumulative distribution function of the time back to the most recent transition, given the process starts in state i

$Var[\cdot]$ Variance

$V(t)$ Virtual waiting time

v_j Steady-state probability of a semi-Markov process being in state j given it starts in state i

W Expected waiting time in system

$W^{(n)}$ Waiting time including service at station n of a series or cyclic queue

W_k Regular kth moment of waiting time in system

W_q Expected waiting time in queue

$W_{q,k}$ Regular kth moment of waiting time in queue

$W(t)$ Cumulative distribution function of waiting time in system

$W_q^{(H)}$ Expected time in queue for a system in heavy traffic

$W_q^{(n)}$ (1) Waiting time in queue for the nth arriving customer (a random variable); (2) expected wait in

	queue for customers of priority class n; (3) expected time in queue at station n of a series or cyclic queue
$W_q(t)$	Cumulative distribution function of the waiting time in queue
$\tilde{W}_q(t\|j)$	Probability ($M/M/c$ model) that the delay undergone by an arbitrary arrival who joined when $c+j$ were in the system is more than t
$w(t)$	Probability distribution of time spent in system
$w_q(t)$	Probability distribution of time spent in queue
$X(t)$	Stochastic process with state space X and parameter t
x, x_i	Observed interval of a queueing system, observed interval of type i (busy, empty, etc.) of a queueing system, respectively
$[x]$	Greatest integer value $\leqslant x$
$\doteq$	Approximately equal to
$\sim$	Asymptotic to
*	Superscript asterisk:
	(1) Denotes the LST; (2) used for various other purposes as specifically defined in text
—	Bar above symbol:
	Denotes the Laplace transform
$\binom{n}{c}$	Binomial coefficient, that is, $n!/[(n-c)!c!]$
[ ]	Superscript bracket:
	Denotes batch queueing model
( )	Superscript parentheses:
	(1) Denotes order of convolution; (2) denotes the order of differentiation
′	Superscript prime:
	(1) Denotes differentiation; (2) used as conditional; for example, p'_n is a conditional probability distribution of n in the system given system not empty; (3) used for various other purposes as specifically defined in text
~	Tilde above symbol:
	Denotes complementary CDF

Appendix 2

DIFFERENTIAL
AND DIFFERENCE
EQUATIONS

Differential and difference equations play a key role in the solution of most queueing models. In this appendix we review some of the fundamentals concerning these types of equations.[1]

A2.1 ORDINARY DIFFERENTIAL EQUATIONS

A differential equation is an equation involving a function and its derivatives. An example of such an equation might be

$$3\frac{d^2y}{dx^2} + 14x\frac{dy}{dx} - x^3y = 6e^x, \qquad (A2.1)$$

where y is a function of x, that is, $y=y(x)$. The problem is to determine the most general $y(x)$ which satisfies Equation (A2.1). Prior to discussing methods of solution to such equations, we first discuss the nomenclature involved with categorizing differential equations.

A2.1.1 Classification

A differential equation is called *ordinary* if it involves only total (as opposed to partial) derivatives. Differential equations are further categorized by *order* and *degree*. Thus a differential equation of the form

$$a_0(x)\frac{d^ny}{dx^n} + a_1(x)\frac{d^{n-1}y}{dx^{n-1}} + \cdots + a_{n-1}(x)\frac{dy}{dx} + a_n(x)y = f(x) \quad (A2.2)$$

---

[1]For a more detailed treatment, the reader is referred to Hildebrand (1949).

is called a *linear* ordinary differential equation of *order n*. The order refers to the highest derivative in the equation, while the degree (linear in this case) refers to the exponent on the dependent variable y and its derivatives. When the coefficients $\{a_n(x)\}$ are independent of x, the equation is said to have *constant coefficients*. Had the right-hand side of equation (A2.2) been zero, the equation would be referred to as homogeneous. Thus the equation

$$a_0 \frac{d^n y}{dx^n} + a_1 \frac{d^{n-1} y}{dx^{n-1}} + \cdots + a_{n-1} \frac{dy}{dx} + a_n y = 0$$

is a linear, homogeneous differential equation of order n with constant coefficients. The descriptor "ordinary" is understood and generally omitted unless one is dealing simultaneously with ordinary and partial differential equations.

A2.1.2   Solutions

Discussion in this appendix is restricted to solutions of linear ordinary differential equations. Solution techniques for nonlinear differential equations are extremely complex and furthermore, the types of differential equations which arise from our interest in queueing analyses are usually linear.

Consider the following linear differential equation of second order with constant coefficients, namely,

$$y'' + 3y' + 2y = 6e^x, \tag{A2.3}$$

where the prime notation is now used to denote differentiation. One solution to (A2.3) is

$$y = e^x, \tag{A2.4}$$

which can be verified by substitution. This is referred to as a *particular* solution to (A2.3). Another solution to (A2.4) is

$$y = C_1 e^{-x} + e^x, \tag{A2.5}$$

where C_1 is any constant. This solution can also be verified by substitution. Notice that it contains the particular solution of Equation (A2.4) and in a sense is a more general solution. We desire the most general solution to any differential equation, which we refer to simply as the *general solution*. It turns out that the general solution of Equation (A2.3) is given by

$$y = C_1 e^{-x} + C_2 e^{-2x} + e^x. \tag{A2.6}$$

Any particular solution can be obtained by specifying the arbitrary constants C_1 and C_2. For example, the particular solution given by (A2.4) results from (A2.6) when $C_1 = C_2 = 0$. The number of arbitrary constants appearing in a general solution of a linear ordinary differential equation can be shown to be equal to the order n. Since (A2.3) is of order two, two constants appear in the general solution given by (A2.6).

Another way of looking at the solution given by (A2.6) is to first consider solutions to a homogeneous equation obtained from (A2.3) by setting the right-hand side to zero. The homogeneous equation then becomes

$$y'' + 3y' + 2y = 0. \qquad (A2.7)$$

We note that $C_1 e^{-x}$ and $C_2 e^{-2x}$ are both solutions to (A2.7). Also, e^x is a solution to the original nonhomogeneous equation (A2.3), so that the general solution consists of a linear combination of all solutions to the homogeneous equation (the general solution to the homogeneous equation) plus a particular solution to the nonhomogeneous equation. It can be proved that for a linear ordinary differential equation of order n there are n solutions to the homogeneous equation, so that the general solution is comprised of a linear combination of the n solutions (thus yielding n arbitrary constants) plus a particular solution to the nonhomogeneous equation. See, for example, Agnew (1942) or Rainville and Bedient (1969).

To determine the constants of a general solution, that is, which particular solution is desired, one must utilize *boundary conditions*. A *boundary condition* is a condition on the function $y(x)$ for a specific x, and results from the model which the differential equation represents. For the equation given by (A2.3), suppose one knows from the physical situation which generated (A2.3) that both the function and its derivative must be zero when x is zero, that is,

$$y(0) = y'(0) = 0.$$

Using these conditions in (A2.6) yields two equations in two unknowns, namely,

$$\begin{cases} 0 = C_1 + C_2 + 1 \\ 0 = -C_1 - 2C_2 + 1 \end{cases},$$

which result in $C_1 = -3$ and $C_2 = 2$, giving the particular solution

$$y = -3e^{-x} + 2e^{-2x} + e^x.$$

We see then, for an nth order equation, n boundary conditions are required to obtain a particular solution from the general solution. Thus the fundamental approach presented here in solving differential equations is to first find the general solution and then, using the boundary conditions, find the particular solution desired. Hence emphasis in this appendix is on finding general solutions. In the following appendix, we discuss the technique of Laplace transforms which can yield a particular solution directly without first obtaining the general solution.

A2.1.3   Separation of Variables

The easiest type of differential equation to solve is one for which *separation of variables* is possible. The general solution can then be obtained by integrating both sides. For example, consider the equation

$$y\frac{dy}{dx} = 3x^2 + 2e^x.$$

We can write

$$ydy = (3x^2 + 2e^x)\,dx.$$

Integrating both sides and combining the arbitrary constants arising from indefinite integration yields

$$\frac{y^2}{2} = x^3 + 2e^x + C.$$

If in general, we have an equation of the form [even for $g(y)$, $u(y)$ nonlinear]

$$f(x)g(y)\frac{dy}{dx} = h(x)u(y),$$

we can separate variables to obtain

$$\frac{g(y)}{u(y)}dy = \frac{h(x)}{f(x)}dx,$$

and the general solution is

$$\int \frac{g(y)}{u(y)}dy = \int \frac{h(x)}{f(x)}dx + C. \qquad (\text{A2.8})$$

Although the examples thus far have been linear differential equations of the first order, it may also be possible to separate variables in higher order linear equations. For example, the solution for

$$\frac{d^2y}{dx^2} = f(x)$$

can be obtained by integrating twice to yield

$$y = \int \left[\int f(x)\,dx \right] dx + C_1 x + C_2$$

since

$$\frac{d^2y}{dx^2} = \frac{d(dy/dx)}{dx}$$

and integrating the first time gives a solution

$$\frac{dy}{dx} = \int f(x)\,dx + C_1.$$

Example A2.1

Find the general solution of

$$y'' = 6x^2.$$

Integrating once gives

$$y' = 2x^3 + C_1$$

and integrating a second time yields

$$y = \tfrac{1}{2}x^4 + C_1 x + C_2.$$

A2.1.4   Linear Differential Equations of First Order

The linear differential equation of the first order can be written in general terms as

$$\frac{dy}{dx} + a(x)y = f(x). \tag{A2.9}$$

If we can determine a function $g(x)$ so that when both sides of (A2.9) are multiplied by it, the equation can be put in the form

$$\frac{d(gy)}{dx} = gf, \qquad (A2.10)$$

then the solution can be determined by separating variables, that is, the solution becomes

$$gy = \int gf\,dx + C$$

or

$$y = \frac{1}{g}\int gf\,dx + \frac{C}{g}. \qquad (A2.11)$$

Such a function as g is referred to as an *integrating factor*. We can, for linear first-order differential equations, find g as follows. Using the product rule of differentiation, and dividing through by g, Equation (A2.10) can be written as

$$\frac{dy}{dx} + \frac{y}{g}\frac{dg}{dx} = f. \qquad (A2.12)$$

For Equation (A2.12) to be equivalent to Equation (A2.9) we must have

$$\frac{1}{g}\frac{dg}{dx} = a(x).$$

Integrating both sides yields

$$\ln g = \int a(x)\,dx + C_1$$

or

$$g = e^{C_1}e^{\int a(x)\,dx}.$$

Since we are seeking only a particular g which will yield equivalency for Equations (A2.9) and (A2.12), we are free to set the constant C_1 to any value we desire. It is most convenient to set $C_1 = 0$. Hence a suitable integrating factor is

$$g = e^{\int a(x)\,dx}. \qquad (A2.13)$$

Using (A2.13) in (A2.11) yields the final solution for y, namely,

$$y = e^{-\int a(x)\,dx} \int e^{\int a(x)\,dx} f(x)\,dx + Ce^{-\int a(x)\,dx}. \qquad (A2.14)$$

Equation (A2.14) is identical to Equation (1.17) which was first utilized in Section 1.8 of Chapter 1. Unfortunately no such similar method is possible for obtaining solutions to higher order linear differential equations. We will consider, however, some higher order equations of specific types.

A2.1.5 Linear Differential Equations with Constant Coefficients

The simplest linear equation of higher order is one where the coefficients are independent of x, namely,

$$a_0 \frac{d^n y}{dx^n} + a_1 \frac{d^{n-1} y}{dx^{n-1}} + \cdots + a_{n-1} \frac{dy}{dx} + a_n y = f(x). \qquad (A2.15)$$

The approach here is to first find the n solutions to the homogeneous equation

$$a_0 \frac{d^n y}{dx^n} + a_1 \frac{d^{n-1} y}{dx^{n-1}} + \cdots + a_{n-1} \frac{dy}{dx} + a_n y = 0, \qquad (A2.16)$$

and then find a particular solution for the nonhomogeneous equation.

The form of Equation (A2.16) suggests that the homogeneous solutions are of the form e^{rx}, since the nth derivative is a multiple of the function itself, that is,

$$\frac{d^n e^{rx}}{dx} = r^n e^{rx}. \qquad (A2.17)$$

Now if e^{rx} is a solution to (A2.16), then we have

$$\left(a_0 r^n + a_1 r^{n-1} + \cdots + a_{n-1} r + a_n \right) e^{rx} = 0$$

which implies for a nontrivial solution ($y = e^{rx} \neq 0$) that

$$a_0 r^n + a_1 r^{n-1} + \cdots + a_{n-1} r + a_n = 0. \qquad (A2.18)$$

Equation (A2.18) is called the *characteristic* or *operator* equation. The characteristic equation can also be obtained directly by looking at the derivative as an operator, say D, so that

$$\begin{cases} Dy = \dfrac{dy}{dx} \\[2mm] D^2y = D(Dy) = \dfrac{d^2y}{dx^2} \\[2mm] \quad \vdots \\[2mm] D^ny = D(D^{n-1}y) = \dfrac{d^ny}{dx^n} \end{cases} \quad ;$$

hence Equation (A2.16) can be written

$$\left(a_0 D^n + a_1 D^{n-1} + \cdots + a_{n-1}D + a_n\right)y = 0,$$

where the characteristic equation is in terms of D instead of r.

Denoting the n roots of the characteristic equation by $r_1, r_2, \ldots, r_n$, we can write

$$(r - r_1)(r - r_2) \cdots (r - r_n)y = 0,$$

and hence theoretically the roots can be found by factorization.[2] If the n roots are distinct, we then have n solutions $e^{r_i x}$ $(i = 1, 2, \ldots, n)$ of the homogeneous equation (A2.16). The most general solution of (A2.16) is then

$$y = C_1 e^{r_1 x} + C_2 e^{r_2 x} + \cdots + C_n e^{r_n x}.$$

If the roots are not all distinct, we have less than n solutions. To find the missing solutions we can proceed as follows. Suppose that r_1 is a double root of the characteristic equation. Then we have

$$(r - r_1)^2(r - r_2) \cdots (r - r_{n-1})e^{rx} = 0. \tag{A2.19}$$

Thus observing that

$$\frac{\partial (r - r_1)^2}{\partial r} = 2(r - r_1),$$

---

[2]Depending on the characteristic equation which results, factorization could be impossible and numerical methods might be required.

we find that the partial derivative with respect to r evaluated at $r = r_1$ also vanishes, so that if $e^{r_1 x}$ is a solution, so too is $\partial e^{rx}/\partial r|_{r=r_1} = xe^{r_1 x}$. To verify that $xe^{r_1 x}$ is a solution consider solutions of the form xe^{rx}. Putting this in for y in Equation (A2.16) yields

$$a_0 \frac{d^n xe^{rx}}{dx^n} + a_1 \frac{d^{n-1} xe^{rx}}{dx^{n-1}} + \cdots + a_{n-1} \frac{dxe^{rx}}{dx} + a_n xe^{rx} = 0.$$

Since

$$xe^{rx} = \frac{\partial e^{rx}}{\partial r},$$

we can write

$$a_0 \frac{\partial^n (\partial e^{rx}/\partial r)}{\partial x^n} + \cdots + a_{n-1} \frac{\partial (\partial e^{rx}/\partial r)}{\partial x} + a_n \frac{\partial e^{rx}}{\partial r} = 0,$$

and changing the order of differentiation gives

$$a_0 \frac{\partial (\partial^n e^{rx}/\partial x^n)}{\partial r} + \cdots + a_{n-1} \frac{\partial (\partial e^{rx}/\partial x)}{\partial r} + a_n \frac{\partial e^{rx}}{\partial r} = 0.$$

Hence we can write

$$\frac{\partial}{\partial r} \left(a_0 \frac{\partial^n e^{rx}}{\partial x^n} + \cdots + a_{n-1} \frac{\partial e^{rx}}{\partial x} + a_n e^{rx} \right) = 0$$

or

$$\frac{\partial}{\partial r} \left[(a_0 r^n + a_1 r^{n-1} + \cdots + a_{n-1} r + a_n) e^{rx} \right] = 0.$$

But we have said that the characteristic equation factors into $n-1$ roots as given in (A2.19) so that

$$\frac{\partial}{\partial r} \left\{ \left[(r - r_1)^2 (r - r_2) \cdots (r - r_{n-1}) \right] e^{rx} \right\} = 0.$$

This equation does hold for $r = r_1$, since the partial with respect to r vanishes at that point.

Thus the two solutions for a double root r_1 are

$$C_1 e^{r_1 x} + C_2 xe^{r_1 x}.$$

This can be generalized to roots of multiplicity k; that is, if r_1 has multiplicity k, the solution associated with r_1 is

$$C_1 e^{r_1 x} + C_2 xe^{r_1 x} + C_3 x^2 e^{r_1 x} + \cdots + C_{k+1} x^k e^{r_1 x}.$$

When we have multiple roots, if factorization is not possible and we must resort to numerical methods, we might only be able to find (say) $n-k$ distinct roots to the characteristic equation. To find which root (or roots) have multiplicity we can simply take partial derivatives of the characteristic equation and check for which root (or roots) they vanish. The roots for which only the first partial derivative of the characteristic equation vanishes have multiplicity two. If a root causes the first, second,,..., kth partial derivatives to vanish, it is of multiplicity $k+1$.

Example A2.2

Find the general solution for

$$\frac{d^3y}{dx^3} - 4\frac{dy}{dx} = 0.$$

The characteristic equation is

$$D^3 - 4D = 0$$

which factors into

$$D(D+2)(D-2) = 0;$$

hence the roots are $r_1 = 0, r_2 = -2, r_3 = +2$. The general solution is then

$$y = C_1 + C_2e^{-2x} + C_3e^{2x}.$$

Example A2.3

Find the general solution for

$$\frac{d^3y}{dx^3} - 4\frac{d^2y}{dx^2} + 5\frac{dy}{dx} - 2y = 0.$$

The characteristic equation is

$$D^3 - 4D^2 + 5D - 2 = 0$$

which factors into

$$(D-2)(D-1)^2 = 0.$$

Thus the roots are $r_1 = 2, r_2 = 1, r_3 = 1$, and we have

$$y = C_1e^{2x} + C_2e^x + C_3xe^x.$$

Had we not been able to factor the characteristic equation but had determined that 2 and 1 were all the distinct roots, we know that since the characteristic equation is a cubic, one root must be double. To find which root it is, we take the partial derivative of the characteristic equation which gives

$$3D^2 - 8D + 5,$$

and evaluating at $D = 2$ and 1, respectively, yields

$$3(2)^2 - 8(2) + 5 = 1$$

and

$$3(1)^2 - 8(1) + 5 = 0,$$

so that the root 1 is the double root.

It remains now to discuss the determination of a particular solution for the nonhomogeneous linear differential equation with constant coefficients. There are four methods for finding a particular solution to the nonhomogeneous equation which are (1) undetermined coefficients, (2) variation of parameters, (3) differential operators, and (4) Laplace transforms. Laplace transforms are presented in some detail in Appendix 3. We briefly discuss the other three methods here.

A2.1.6 Undetermined Coefficients

If the right-hand side of the differential equation given in (A2.15) is of the form x^m (m an integer), $\sin bx$, $\cos bx$, e^{bx}, and/or products of two or more such functions, we can employ the method of *undetermined coefficients* to find a particular solution. We first define a family of a function $f(x)$ to be the set of linearly independent functions consisting of $f(x)$ and its derivatives. The functions specified above are funtions with a finite number of derivatives for which the function and its derivatives are linearly independent. Table A2.1[3] lists the families of the aforementioned functions. The family of a function consisting of a product of n terms of this type consists of all possible products of the family members of each of the n terms. For example, the family of $x^2 \cos x$ would be $x^2 \cos x$, $x \cos x$, $\cos x$, $x^2 \sin x$, $x \sin x$, $\sin x$.

[3]F. B. Hildebrand, *Advanced Calculus for Engineers* (ç) 1948, 1949, p. 12ff, reprinted by permission of Prentice-Hall, Inc., Englewood Cliffs, N.J.

TABLE A2.1 FUNCTIONS AND THEIR FAMILIES

Function	Family
x^m	$x^m, x^{m-1}, x^{m-2}, \ldots, x^2, x, 1$
$\sin bx$	$\sin bx, \cos bx$
$\cos bx$	$\cos bx, \sin bx$
e^{bx}	e^{bx}

The method works as follows in three steps:

(i) Assuming $f(x)$ is a linear combination of functions or products of functions given in Table A2.1, construct the family for each, eliminating families which are included in other families.

(ii) If any family has a member which is also a solution to the homogeneous equation, replace that family by a new one, obtained by multiplying the original family by x (or the lowest power of x necessary) so that the new family has no members which are also solutions to the homogeneous equation.

(iii) The particular solution is assumed to be a linear combination of all members of the constructed families, the constants of the linear combination to be found by the satisfaction of the differential equation when this particular solution is substituted into it.

Example A2.4

Find the general solution for

$$y''' - y' = 2x + 1 - 4\cos x + 2e^x.$$

The general homogeneous solution can be found from previous methods to be

$$y = C_1 + C_2 e^x + C_3 e^{-x}.$$

The families for the right-hand side function are respectively

$$\{x, 1\}, \ \{1\}, \ \{\cos x, \sin x\}, \ \{e^x\}.$$

Since $\{1\}$ is included in $\{x, 1\}$, we omit this. Further since 1 and e^x are in the homogeneous solution, their families are replaced by $\{x^2, x\}$ and

$\{xe^x\}$, respectively. Then the resulting terms to be used are

$$\{x^2, x, \cos x, \sin x, xe^x\}$$

and the particular solution is of the form

$$y_p = Ax^2 + Bx + C\cos x + D\sin x + Exe^x.$$

Substituting y_p into the differential equation yields

$$C\sin x - D\cos x + E(xe^x + 3e^x)$$

$$- [2Ax + B - C\sin x + D\cos x + E(xe^x + e^x)] = 2x + 1 - 4\cos x + 2e^x,$$

or simplifying we get

$$-2Ax - B + 2C\sin x - 2D\cos x + 2Ee^x = 2x + 1 - 4\cos x + 2e^x.$$

Matching coefficients of like terms yields

$$A = -1, B = -1, C = 0, D = 2, E = 1.$$

Hence the particular solution is

$$y_p = -x^2 - x + 2\sin x + xe^x$$

and the general solution becomes

$$y = C_1 + C_2 e^x + C_3 e^{-x} - x^2 - x + 2\sin x + xe^x.$$

A2.1.7 Variation of Parameters

This method determines the general solution for a linear nonhomogeneous equation, and is even applicable for cases with nonconstant coefficients. We assume that the equation is in a form where the coefficient of the highest order derivative is unity; that is, we divide Equation (A2.2) by $a_0(x)$; to obtain

$$\frac{d^n y}{dx^n} + b_1(x)\frac{d^{n-1}y}{dx^{n-1}} + \cdots + b_{n-1}(x)\frac{dy}{dx} + b_n(x)y = g(x). \quad \text{(A2.20)}$$

Let the n solutions to the homogeneous equation be denoted by $y_1(x)$, $y_2(x), \ldots, y_n(x)$, so that the general homogeneous solution is

$$y = C_1 y_1 + C_2 y_2 + \cdots + C_n y_n.$$

It is then possible to determine n equations on the first derivative with respect to x of n unknown functions, say $z_1(x), z_2(x), \ldots, z_n(x)$. It turns out that a particular solution can be expressed in terms of the z's as

$$y_p = z_1(x)y_1(x) + z_2(x)y_2(x) + \cdots + z_n(x)y_n(x). \qquad (A2.21)$$

The n equations in the n unknowns $z_k'(x)$, $k = 1, 2, \ldots, n$, can be solved whereupon the $\{z_k\}$ can be determined by integration. If the n constants of integration resulting from the determination of the $\{z_k\}$ from $\{z_k'\}$ are left as constants, the general solution to the nonhomogeneous equation results. Any fixed setting of the n constants yields, of course, a particular solution.

The equations involving the $\{z_k'(x)\}$ are as follows.[4]

$$\left\{ \begin{aligned}
z_1'y_1 + z_2'y_2 + \cdots + z_n'y_n &= 0 \\
z_1'y_1' + z_2'y_2' + \cdots + z_n'y_n' &= 0 \\
&\;\vdots \\
z_1'y_1^{(n-2)} + z_2'y_2^{(n-2)} + \cdots + z_n'y_n^{(n-2)} &= 0 \\
z_1'y_1^{(n-1)} + z_2'y_2^{(n-1)} + \cdots + z_n'y_n^{(n-1)} &= g(x)
\end{aligned} \right. \qquad (A2.22)$$

Example A2.5

Consider the same equation as that for Example A2.4. The homogeneous solution is

$$y = C_1 + C_2e^x + C_3e^{-x};$$

hence

$$\left\{ \begin{aligned}
y_1 &= 1 \\
y_2 &= e^x \\
y_3 &= e^{-x}
\end{aligned} \right.$$

Utilizing (A2.22) we have

$$\left\{ \begin{aligned}
z_1' + z_2'e^x + z_3'e^{-x} &= 0 \\
z_2'e^x - z_3'e^{-x} &= 0 \\
z_2'e^x + z_3'e^{-x} &= g
\end{aligned} \right.$$

[4]Hildebrand, op. cit., p. 28.

where g is the right-hand side of the nonhomogeneous equation, namely,

$$g = 2x + 1 - 4\cos x + 2e^x.$$

The solution to the equation above can be readily found using Cramer's rule to be

$$\begin{cases} z_1' = -g \\[2mm] z_2' = \dfrac{g}{2} e^{-x} \\[2mm] z_3' = \dfrac{g}{2} e^x \end{cases}.$$

Thus

$$z_1 = -\int (2x + 1 - 4\cos x + 2e^x)\, dx$$

$$= -x^2 - x + 4\sin x - 2e^x + C_1,$$

$$z_2 = \frac{1}{2} \int (2xe^{-x} + e^{-x} - 4e^{-x}\cos x + 2)\, dx$$

$$= e^{-x}(-x - 1) - \frac{e^{-x}}{2} - e^{-x}(-\cos x + \sin x) + x + C_2$$

$$= -xe^{-x} - \frac{3e^{-x}}{2} + e^{-x}\cos x - e^{-x}\sin x + x + C_2,$$

$$z_3 = \frac{1}{2} \int (2xe^x + e^x - 4e^x\cos x + 2e^{2x})\, dx$$

$$= e^x(x + 1) + \frac{e^x}{2} - e^x(\cos x + \sin x) + \frac{1}{2}e^{2x} + C_3$$

$$= xe^x + \frac{3e^x}{2} - e^x\cos x - e^x\sin x + \frac{1}{2}e^{2x} + C_3.$$

Now the general solution can be obtained from (A2.21) as

$$y = -x^2 - x + 4\sin x - 2e^x + C_1 - x - \frac{3}{2} + \cos x - \sin x$$

$$+ xe^x + C_2 e^x + x + \frac{3}{2} - \cos x - \sin x + \frac{1}{2}e^x + C_3 e^{-x}$$

$$= -x^2 - x + 2\sin x + xe^x + C_1 + \hat{C}_2 e^x + C_3 e^{-x},$$

where $\hat{C}_2 = C_2 - \frac{3}{2}$. This agrees with the answer of Example A2.4.

A2.1.8 Differential Operators

We illustrate the use of differential operators on the same equations used in the previous two examples. The equation

$$y''' - y' = 2x + 1 - 4\cos x + 2e^x$$

can be written in operator notation as

$$D^3 y - Dy = g,$$

where g, as before, is the right-hand side. This can be factored as

$$D(D+1)(D-1)y = g.$$

We let

$$y_1 = (D+1)(D-1)y; \qquad\qquad\qquad (A2.23)$$

hence we have

$$Dy_1 = g$$

or

$$\frac{dy_1}{dx} = g.$$

Solving directly by integration gives

$$y_1 = \int g\,dx + C_1$$

$$= \int (2x + 1 - 4\cos x + 2e^x)\,dx + C_1$$

$$= x^2 + x - 4\sin x + 2e^x + C_1.$$

Substituting y_1 into (A2.23) yields the differential equation

$$(D+1)(D-1)y = x^2 + x - 4\sin x + 2e^x + C_1.$$

We next let

$$y_2 = (D-1)y \qquad\qquad\qquad (A2.24)$$

and get

$$(D+1)y_2 = x^2 + x - 4\sin x + 2e^x + C_1$$

or

$$\frac{dy_2}{dx} + y_2 = x^2 + x - 4\sin x + 2e^x + C_1. \qquad\qquad (A2.25)$$

Equation (A2.25) is now a first-order equation which can be solved by the solution previously derived in Section A2.1.4 and given by Equation (A2.14), which yields

$$y_2 = e^{-x} \int e^x (x^2 + x - 4\sin x + 2e^x + C_1) dx + C_2 e^{-x}$$

$$= x^2 - x + 1 - 2\sin x + 2\cos x + e^x + C_1 + C_2 e^{-x}.$$

Now substituting y_2 into (A2.24) yields another first-order equation

$$(D-1)y = x^2 - x + 1 - 2\sin x + 2\cos x + e^x + C_1 + C_2 e^{-x}.$$

Again using the solution for first-order equations we get

$$y = e^x \int e^{-x} (x^2 - x + 1 - 2\sin x + 2\cos x + e^x + C_1 + C_2 e^{-x}) dx + C_3 e^x$$

$$= -x^2 - x - 1 + 2\sin x + xe^x - C_1 - \frac{C_2}{2} e^{-x} + C_3 e^x,$$

which upon redefining the arbitrary constants is

$$y = -x^2 - x - 1 + 2\sin x + xe^x + C_1 + C_2 e^x + C_3 e^{-x}$$

and agrees with our previous solutions. This method of using operators applies only for equations with constant coefficients.

For further use of operators, the reader is referred to Rainville and Bedient (1969), the chapter entitled, "Inverse Differential Operators." Essentially, any equation with constant coefficients can be written in operator notation as

$$f(D)y = g(x).$$

If a function $f^{-1}(D)$ can be found where

$$f(D)f^{-1}(D) = 1,$$

the solution to the equation is

$$y = f^{-1}(D)g(x).$$

The material referenced above deals with determining such inverse differential operators.

A2.1.9 Reduction of Order

Leaving the topic of particular solutions and returning now to the topic of solutions in general, if one homogeneous solution of a linear differential equation of order n is known, the remainder of the solution can be determined by solving a new linear differential equation of order $n-1$ in much the same way one can reduce the degree of an algebraic equation when one root is known. Consider the following second-order equation

$$y'' + a_1 y' + a_2 y = g, \tag{A2.26}$$

where, of course, y and g are functions of x and the coefficients a_1 and a_2 may be also. Suppose one solution to the homogeneous equation can be found from inspection and we denote it by $y_1(x)$, that is,

$$y_1'' + a_1 y_1' + a_2 y_1 = 0. \tag{A2.27}$$

Then if we let

$$y = y_1 v,$$

y will be a solution to (A2.26) if

$$(y_1 v)'' + a_1 (y_1 v)' + a_2 y = g$$

or

$$y_1 v'' + 2y_1' v' + y_1'' v + a_1 (y_1 v' + v y_1') + a_2 y_1 v = g.$$

Simplifying we obtain

$$y_1 v'' + (2y_1' + a_1 y_1) v' + (y_1'' + a_1' y_1' + a_2 y_1) v = g. \tag{A2.28}$$

But since y_1 is a homogeneous solution, using (A2.27) in (A2.28) gives

$$y_1 v'' + (2y_1' + a_1 y_1) v' = g.$$

Letting $u = v'$ we get the following first-order equation involving u, namely,

$$y_1 u' + (2y_1' + a_1 y_1) u = g \tag{A2.29}$$

which can be solved by (A2.14) of Section A2.1.4. Finally, we can get v from $v' = u$ by integration and the general solution $y = y_1 v$ results.

Example A2.6

Consider the equation

$$y'' - y = x.$$

From inspection we can see that one solution to the homogeneous equation is

$$y_1 = e^x.$$

Thus using this in (A2.29) we have the first-order equation

$$e^x u' + 2e^x u = x$$

or

$$u' + 2u = xe^{-x},$$

which by Equation (A2.14) yields

$$u = e^{-x}(x - 1) + Ce^{-2x}.$$

Integrating u we obtain v as

$$v = e^{-x}(-x - 1) + e^{-x} + C_1 e^{-2x} + C_2,$$

and finally

$$y = y_1 v = -x + C_1 e^{-x} + C_2 e^x.$$

A2.1.10 Systems of Simultaneous Linear Differential Equations with Constant Coefficients

Consider the following system of two equations in two unknowns.

$$\begin{cases} \dfrac{d^2 y_1}{dx^2} - y_1 - 2y_2 = g_1(x) \\[3mm] \dfrac{d^2 y_2}{dx^2} - 2y_2 - 3y_1 = g_2(x) \end{cases} \tag{A2.30}$$

These can be rewritten using operator notation as

$$\begin{cases} (D^2 - 1)y_1 - 2y_2 = g_1 \\[2mm] -3y_1 + (D^2 - 2)y_2 = g_2 \end{cases}$$

Using Cramer's rule the solution yields the following two differential equations of a single variable

$$\begin{vmatrix} (D^2-1) & -2 \\ -3 & (D^2-2) \end{vmatrix} y_1 = \begin{vmatrix} g_1 & -2 \\ g_2 & (D^2-2) \end{vmatrix}$$

$$\begin{vmatrix} (D^2-1) & -2 \\ -3 & (D^2-2) \end{vmatrix} y_2 = \begin{vmatrix} (D^2-1) & g_1 \\ -3 & g_2 \end{vmatrix},$$

or rewriting

$$\begin{cases} (D^4-3D^2-4)y_1 = (D^2-2)g_1 + 2g_2 \\ (D^4-3D^2-4)y_2 = (D^2-1)g_2 + 3g_1 \end{cases}$$

Since g_1 and g_2 are known functions of x, the differentiation implied by the operator can be performed and the right-hand side represented by two known functions, $h_1(x)$ and $h_2(x)$, yielding

$$\begin{cases} (D^4-3D^2-4)y_1 = h_1(x) \\ (D^4-3D^2-4)y_2 = h_2(x) \end{cases}$$

Both equations have identical characteristic equations (which can always be easily obtained from the determinant of the left-hand side of the system of equations) so the general solution to the homogeneous equations are of the same form. Denoting the four roots to the characteristic equation by r_1, r_2, r_3, and r_4, we have the homogeneous solutions

$$\begin{cases} y_1 = C_1 e^{r_1 x} + C_2 e^{r_2 x} + C_3 e^{r_3 x} + C_4 e^{r_4 x} \\ y_2 = C_5 e^{r_1 x} + C_6 e^{r_2 x} + C_7 e^{r_3 x} + C_8 e^{r_4 x} \end{cases} \tag{A2.31}$$

While there are eight constants, they are not all independent and their relationships can be obtained by substitution of (A2.31) in either equation of (A2.30) with the right-hand side equal to zero, yielding

$$C_5 = \frac{r_1^2-1}{2}C_1, \qquad C_6 = \frac{r_2^2-1}{2}C_2, \qquad C_7 = \frac{r_3^2-1}{2}C_3, \qquad C_8 = \frac{r_4^2-1}{2}C_4.$$

Equations (A2.31) are the homogeneous solutions to the Equations (A2.30). To obtain particular solutions, one can use the method of undetermined coefficients.

Example A2.7

Solve the following for y_1 and y_2:

$$\begin{cases} y_1' - 2y_1 + 2y_2' = 2 - 4e^{2x} \\ 2y_1' - 3y_1 + 3y_2' - y_2 = 0 \end{cases} \tag{A2.32}$$

Considering first the homogeneous solutions, we rewrite the equations in operator notation as

$$\begin{cases} (D-2)y_1 + 2Dy_2 = 0 \\ (2D-3)y_1 + (3D-1)y_2 = 0 \end{cases} , \tag{A2.33}$$

and the characteristic equation is then

$$\begin{vmatrix} (D-2) & 2D \\ (2D-3) & (3D-1) \end{vmatrix} = 0,$$

which upon expanding yields

$$-D^2 - D + 2 = 0.$$

Factoring gives the two roots as 1 and -2 so that the homogeneous solutions are

$$\begin{cases} y_1 = C_1 e^x + C_2 e^{-2x} \\ y_2 = C_3 e^x + C_4 e^{-2x} \end{cases}$$

To determine the relationship among the constants, we substitute the above in either equation of (A2.32) to get

$$C_3 = \tfrac{1}{2}C_1, \qquad C_4 = -C_2,$$

and thus

$$\begin{cases} y_1 = C_1 e^x + C_2 e^{-2x} \\ \\ y_2 = \dfrac{C_1}{2} e^x - C_2 e^{-2x} \end{cases}$$

To obtain the particular solution, we use the method of undetermined coefficients. The family to be considered is $\{1, e^{2x}\}$ and we proceed as follows:

$$\begin{cases} y_{1,p} = A + Be^{2x} \\ y_{2,p} = C + De^{2x} \end{cases}.$$

Substituting into Equation (A2.32) gives

$$\begin{cases} 2Be^{2x} - 2A - 2Be^{2x} + 4De^{2x} = 2 - 4e^{2x} \\ 4Be^{2x} - 3A - 3Be^{2x} + 6De^{2x} - C - De^{2x} = 0 \end{cases}$$

or upon simplifying we have

$$\begin{cases} -2A + 4De^{2x} = 2 - 4e^{2x} \\ -3A - C + (B + 5D)e^{2x} = 0 \end{cases}.$$

Now equating coefficients of like terms yields

$$-2A = 2, \quad 4D = -4, \quad (-3A - C) = 0, \quad (B + 5D) = 0,$$

which finally gives

$$A = -1, \quad C = 3, \quad B = 5,$$

and the general solutions are

$$\begin{cases} y_1 = C_1 e^x + C_2 e^{-2x} - 1 + 5e^{2x} \\ y_2 = \dfrac{C_1}{2} e^x - C_2 e^{-2x} + 3 - e^{2x} \end{cases}.$$

The procedure, of course, generalizes to more than systems of size two. If we have n simultaneous equations, the characteristic equation is obtained from evaluating an $n \times n$ determinant.

2.1.11   Summary

In attacking the problem of solving ordinary linear differential equations, the first approach should be to determine whether the variables are separable. If they are, the general solution can be obtained directly by integration as discussed in Section A2.1.3.

If separation of variables is not possible, but the equation is of the first order, the solution can be obtained from Equation (A2.14) as derived in Section A2.1.4.

For higher order equations with constant coefficients, the general solution to the homogeneous equation can be obtained by finding the roots of the characteristic equation (Section A2.1.5) and then finding a particular solution via undetermined coefficients (Section A2.1.6) or variation of parameters (Section A2.1.7). Use of operators (Section A2.1.8) can also be employed to determine general solutions for nonhomogeneous linear equations with constant coefficients.

If one (or more) solutions to the homogeneous equation are known, the order of the equation can be reduced (Section A2.1.9) thereby yielding equations of lower order which may be more readily solved.

Finally, in Section A2.1.10 solutions of systems of simultaneous linear differential equations with constant coefficients are discussed.

A2.2 DIFFERENCE EQUATIONS

Consider a function of an independent variable, say x, where x is now a discrete variable, that is, it can take on only integer values. Then the function exists only at discrete points (integer values of x) and we denote this type of function by y_x instead of $y(x)$.

The first finite difference of y_x is defined as

$$\Delta y \equiv y_{x+1} - y_x,$$

the second finite difference as

$$\Delta^2 y = \Delta(\Delta y) = y_{x+2} - y_{x+1} - (y_{x+1} - y_x)$$

$$= y_{x+2} - 2y_{x+1} + y_x,$$

and the nth finite difference as

$$\Delta^n y = \Delta(\Delta^{n-1} y).$$

We define an operator, say D, to be

$$\left\{ \begin{array}{l} Dy_x = y_{x+1} \\ D^2 y_x = D(Dy_x) = y_{x+2} \\ \quad\vdots \\ D^n y_x = D(D^{n-1} y_x) = y_{x+n} \end{array} \right.$$

One can easily see the relationship between Δ and D as

$$D^n = (\Delta + 1)^n.$$

A2.2.1 Linear Difference Equations with Constant Coefficients

An equation involving y_x of the type

$$y_{x+n} + a_1 y_{x+n-1} + \cdots + a_{n-1} y_{x+1} + a_n y_x = g_x \qquad (A2.34)$$

is called a linear difference equation of order n with constant coefficients. We shall not treat here the case where the coefficients are also dependent on x [for such cases see Hildebrand (1952)].

One can see many similarities between difference equations and differential equations and, indeed, the solution techniques are often quite similar. The technique for solving Equation (A2.34) is very much like that used for linear differential equations with constant coefficients. In fact, it can be shown that a general solution of (A2.34) consists of a linear combination of all solutions to the homogeneous equation (g_x replaced by zero) plus a particular solution to (A2.34). Also, for an nth degree equation, there are n arbitrary constants associated with the homogeneous solution, which in any particular case can be found from n boundary conditions.

To find the solution to the homogeneous equation we proceed in a manner similar to Section A2.1.5. We first rewrite (A2.34) using operator notation to get

$$(D^n + a_1 D^{n-1} + \cdots + a_{n-1} D + a_n) y_x = 0.$$

The homogeneous solutions are of the form r^x (as opposed to e^{rx} for differential equations) where r is a root to the characteristic equation

$$D^n + a_1 D^{n-1} + \cdots + a_{n-1} D + a_n = 0.$$

To see this, we let $y_x = r^x$ in (A2.34) and get

$$r^{x+n} + a_1 r^{x+n-1} + \cdots + a_{n-1} r^{x+1} + a_n r^x = 0,$$

whereupon factoring out r^x we have

$$r^x (r^n + a_1 r^{n-1} + \cdots + a_{n-1} r + a_n) = 0.$$

But since r is a root to the characteristic equation (which resulted above when r^x was factored out) we see that the left-hand side does indeed equal zero.

Since the characteristic equation has n roots, the general solution to the homogeneous equation is

$$y_x = C_1 r_1^x + C_2 r_2^x + \cdots + C_n r_n^x.$$

Multiple roots can be handled in a manner analogous to differential equations in that for a root of multiplicity k, the first $k-1$ derivatives of the characteristic equation with respect to D must vanish and the k solutions are of the form $r^x, xr^x, x(x-1)r^x, \ldots, x(x-1)\cdots(x-k+1)r^x$, since in taking the ith derivative of r^x one obtains $x(x-1)\cdots(x-i+1)r^x r^{-i}$ and the r^{-i} can be absorbed in the arbitrary constant.

To find a particular solution to Equation (A2.34), the method of undetermined coefficients can be employed. Here, however, a family consists of the function and all the operators on the function to the highest degree of the operator on the left-hand side. We illustrate the procedure on the following example.

Example A2.8

Consider the difference equation

$$y_{x+2} + 6y_{x+1} + 9y_x = 16x^2.$$

The homegeneous equation in operator notation is

$$(D^2 + 6D + 9)y_x = 0,$$

and the solution to the characteristic equation has two roots at -3. Hence the solution is

$$y_x = C_1(-3)^x + C_2 x(-3)^x.$$

To find the particular solution, the family of x^2 would be $\{x^2, (x+1)^2, (x+2)^2\}$ which gives terms of $\{x^2, x, 1\}$. Therefore

$$y_{x,p} = Ax^2 + Bx + C,$$

and substituting this into the original equation gives

$$A(x+2)^2 + B(x+2) + C + 6[A(x+1)^2 + B(x+1) + C]$$

$$+9[Ax^2 + Bx + C] = 16x^2,$$

or

$$16Ax^2 + (16A + 16B)x + 10A + 8B + 16C = 16x^2.$$

Equating like coefficients yields the conditions

$$16A = 16, \qquad 16A + 16B = 0, \qquad 10A + 8B + 16C = 0,$$

or finally

$$A = 1, \qquad B = -1, \qquad C = -\frac{1}{8}.$$

Thus the particular solution is

$$y_{x,p} = x^2 - x - \frac{1}{8},$$

and the general solution becomes

$$y_x = C_1(-3)^x + C_2 x(-3)^x + x^2 - x - \frac{1}{8}.$$

A2.2.2   Systems of Simultaneous Linear Difference Equations with Constant Coefficients

The solution to systems of difference equations is analogous to the procedure used in Section A2.1.10 for differential equations. One first writes the equation in operator notation, finds the characteristic equation using the determinant of the left-hand side "coefficients," solves for the roots and obtains the homogeneous solutions as linear combinations of r_i^x (instead of $e^{r_i x}$ as for differential equations). The number of constants are then reduced as before by substituting the homogeneous solutions into the homogeneous equations. Then a particular solution can be found (if the equations are nonhomogeneous) by the method of undetermined coefficients.

Example A 2.9

Consider the following system of difference equations to be solved for y and z.

$$\begin{cases} y_{x+1} - 3y_x + z_{x+1} - 3z_x = 2 \\ 2y_{x+1} - 5y_x + 3z_{x+1} - 3z_x = 6(4)^x \end{cases} \qquad (A2.35)$$

We first obtain the homogeneous solutions by solving the characteristic equation obtained after writing the system in operator notation. The characteristic equation is

$$\begin{vmatrix} (D-3) & (D-3) \\ (2D-5) & (3D-3) \end{vmatrix} = 0,$$

which upon calculating the determinant yields

$$D^2 - D - 6 = 0.$$

The roots can be found by factoring to be 3 and -2. Thus the homogeneous solutions are

$$\begin{cases} y_x = C_1(3)^x + C_2(-2)^x \\ z_x = C_3(3)^x + C_4(-2)^x \end{cases}$$

To reduce the number of arbitrary constants we substitute the above in the original equations of (A2.35), with the right-hand side set to zero. This yields the relations

$$C_3 = -\frac{1}{6} C_1, \qquad C_4 = -C_2,$$

and hence the homogeneous solutions become

$$\begin{cases} y_x = C_1(3)^x + C_2(-2)^x \\ z_x = -\frac{C_1}{6}(3)^x - C_2(-2)^x \end{cases}$$

To obtain the particular solution, we employ undetermined coefficients. The family of the first right-hand side of (A2.35) is $\{1\}$ and the family of the second is merely $\{4^x\}$ since $D4^x$ is $4(4)^x$. Thus we have

$$\begin{cases} y_{x,p} = A + B(4)^x \\[2mm] z_{x,p} = C + D(4)^x \end{cases}$$

Substituting into Equations (A2.35) we get

$$\begin{cases} A + 4B(4)^x - 3A - 3B(4)^x + C + 4D(4)^x - 3C - 3D(4)^x = 2 \\[2mm] 2A + 8B(4)^x - 5A - 5B(4)^x + 3C + 12D(4)^x - 3C - 3D(4)^x = 6(4)^x \end{cases}$$

Upon simplification we obtain

$$\begin{cases} -2A - 2C + (B+D)(4)^x = 2 \\[2mm] -3A + (3B+9D)(4)^x = 6(4)^x \end{cases}$$

Equating like coefficients yields

$$-2A - 2C = 2, \qquad B + D = 0, \qquad -3A = 0, \qquad 3B + 9D = 6,$$

which gives

$$A = 0, \qquad B = -1, \qquad C = -1, \qquad D = 1.$$

The general solution is then

$$\begin{cases} y_x = C_1(3)^x + C_2(-2)^x - (4)^x \\[2mm] z_x = -\dfrac{C_1}{6}(3)^x - C_2(-2)^x - 1 + (4)^x \end{cases}$$

This method of finding particular solutions for systems of equations through the use of undetermined coefficients does not always work. For example, one can verify that undetermined coefficients do not yield a particular solution to the following set of equations:

$$\begin{cases} y_{x+1} - 2y_x + 2z_x = 2 \\[2mm] 2y_{x+1} - 3y_x + 3z_{x+1} - z_x = 6(4)^x \end{cases}$$

For such cases, other methods are necessary: however, further detailed treatment of finding particular solutions is not necessary since differential and difference equations encountered in queueing theory are, for the most part, homogeneous.

Appendix 3
TRANSFORMS AND GENERATING FUNCTIONS

In this appendix we describe briefly the concepts of transforms and generating functions, and present some major properties and results concerning Laplace transforms and probability generating functions which are useful in queueing theory.

A3.1 THE LAPLACE TRANSFORM

A *transform* is merely a mapping of a function from one space to another. While it may be very difficult to solve certain equations directly for a particular function of interest, it is often easier to solve a corresponding equation in terms of a transform of the function, and then *invert* the transform to obtain the function. One particular transform that is quite useful for solving some types of differential equations as well as certain integral equations (our particular interest lies in those differential and integral equations resulting from queueing analyses) is the Laplace transform. The Laplace transform (hereafter abbreviated by LT) of a function $f(t)$ is defined as[1]

$$\mathcal{L}\{f(t)\} \equiv \int_0^\infty e^{-st}f(t)\,dt \qquad [\text{Re}(s) > s_0] \qquad (A3.1)$$

Since the LT maps a function of t into a new function of s, it will also be denoted by $\bar{f}(s)$. If $\bar{f}(s)$ exists, its inverse is unique $[=f(t)]$. Using this one-to-one correspondence between the LT and its inverse we can study the function $f(t)$ by studying its LT $\bar{f}(s)$. We illustrate below an example of using the LT to solve a differential equation encountered in Section 1.8.

---

[1]The parameter s is not restricted to be a real number. If the integral converges it will do so for all s where the real part is greater than some fixed value, say s_0.

500

Example A3.1

Equation (1.16) was the differential equation

$$\frac{dy(x)}{dx} + \Phi(x)y(x) = \Psi(x). \tag{1.16}$$

Equation (1.17) gave the solution to (1.16) as

$$y(x) = Ce^{-\int \Phi(x)dx} + e^{-\int \Phi(x)dx} \int e^{\int \Phi(x)dx} \Psi(x)\, dx. \tag{1.17}$$

Consider now the specific case where we desire to find $y(x)$ from the equation

$$\frac{dy(x)}{dx} - y(x) = e^{ax} \tag{A3.2}$$

and the boundary condition

$$y(0) = -1.$$

Using the formula given by Equation (1.17) and the boundary condition yields

$$y(x) = \frac{e^{ax} - ae^{x}}{a - 1}. \tag{A3.3}$$

To illustrate the use of Laplace transforms, we take the LT of both sides of (A3.2), making use of the additive property of LTs (property 1) given in Table A3.1, and obtain

$$\mathcal{L}\left\{ \frac{dy(x)}{dx} \right\} - \mathcal{L}\{ y(x) \} = \mathcal{L}\{ e^{ax} \}.$$

Now using property 2 of Table A3.1 involving the LT of a derivative we have

$$s\mathcal{L}\{ y(x) \} - y(0) - \mathcal{L}\{ y(x) \} = \mathcal{L}\{ e^{ax} \},$$

and using the boundary condition we get

$$s\mathcal{L}\{ y(x) \} + 1 - \mathcal{L}\{ y(x) \} = \mathcal{L}\{ e^{ax} \}.$$

The LT of the right-hand side can be readily evaluated by direct integration and is also given in Table A3.2. Utilizing this and changing to the

TABLE A3.1 PROPERTIES OF LAPLACE TRANSFORMS

1. $\mathcal{L}\{af(t)+bg(t)\}$	$= a\bar{f}(s)+b\bar{g}(s)$	(Additive)
2a. $\mathcal{L}\{f'(t)\}$	$= s\bar{f}(s)-f(0)$	(Derivative)
2b. $\mathcal{L}\{f''(t)\}$	$= s^2\bar{f}(s)-sf(0)-f'(0)$	
2c. $\mathcal{L}\{f^{(n)}(t)\}$	$= s^n\bar{f}(s)-\sum_{i=1}^{n} s^{n-i}f^{(i-1)}(0)$	
3a. $\mathcal{L}\left\{\int_0^t f(u)\,du\right\}$	$= \dfrac{1}{s}\bar{f}(s)$	(Integral)
3b. $\mathcal{L}\left\{\int_0^t\int_0^u f(v)\,dv\,du\right\}$	$= \dfrac{1}{s^2}\bar{f}(s)$	
4. $\mathcal{L}\left\{\int_0^t f(t-u)g(u)\,du\right\}$	$= \bar{f}(s)\bar{g}(s)$	(Convolution)
5. $\mathcal{L}\{e^{at}f(t)\}$	$= \bar{f}(s-a)$	
6a. $\mathcal{L}\{tf(t)\}$	$= -\bar{f}'(s)$	
6b. $\mathcal{L}\{t^n f(t)\}$	$= (-1)^n\bar{f}^{(n)}(s)$	
7. $\mathcal{L}\left\{\dfrac{1}{t}f(t)\right\}$	$= \int_s^\infty \bar{f}(y)\,dy$	
8. $\mathcal{L}\left\{\begin{array}{ll} f(t-a) & (t>a) \\ 0 & (t<a) \end{array}\right\}$	$= e^{-as}\bar{f}(s)$	

alternate notation for LTs we obtain

$$s\bar{y}(s)+1-\bar{y}(s)=\frac{1}{s-a}.$$

Solving for $\bar{y}(s)$ yields

$$\bar{y}(s)=\frac{1-s+a}{(s-a)(s-1)}. \tag{A3.4}$$

It is now necessary to find the function $y(x)$ whose LT is the right-hand side of (A3.4); that is, we must invert (A3.4) to find $y(x)$. This can often be tedious and is generally the most difficult part of working with LTs. Hopefully, we can recognize the answer as being the LT of a known function given in a table (such as Table A3.2) or operate on the answer to put it in the form of the sum of LTs of known functions. It turns out that

TABLE A3.2   TABLE OF LAPLACE TRANSFORMS

	$\bar{f}(s)$	$f(t)$
1.	$\bar{f}(s)$	$f(t)$
2.	$\dfrac{1}{s}$	1
3.	$\dfrac{1}{s^2}$	t
4.	$\dfrac{1}{s^n}$, $n=2,3,\ldots$	$\dfrac{t^{n-1}}{(n-1)!}$
5.	$\dfrac{n!}{s^{n+1}}$, $n=2,3,\ldots$	t^n
6.	$\dfrac{1}{s^a}$, $a>0$	$\dfrac{t^{a-1}}{\Gamma(a)}$
7.	$\dfrac{1}{\sqrt{s}}$	$\dfrac{1}{\sqrt{\pi t}}$
8.	$\dfrac{1}{s-a}$	e^{at}
9.	$\dfrac{1}{(s-a)(s-b)}$, $a\neq b$	$\dfrac{1}{a-b}(e^{at}-e^{bt})$
10.	$\dfrac{s}{(s-a)(s-b)}$, $a\neq b$	$\dfrac{1}{a-b}(ae^{at}-be^{bt})$
11.	$\dfrac{1}{(s-a)^2}$	te^{at}
12.	$\dfrac{1}{(s-a)^n}$, $n=3,4,\ldots$	$\dfrac{t^{n-1}e^{at}}{(n-1)!}$
13.	$\dfrac{a}{s^2+a^2}$	$\sin at$

14. $\dfrac{s}{s^2+a^2}$ $\cos at$

15. $\dfrac{a}{s^2-a^2}$ $\sinh at$

16. $\dfrac{s}{s^2-a^2}$ $\cosh at$

17. $\dfrac{2as}{\left(s^2+a^2\right)^2}$ $t\sin at$

18. $\dfrac{s^2-a^2}{\left(s^2+a^2\right)^2}$ $t\cos at$

19. $\dfrac{2a^3}{\left(s^2+a^2\right)^2}$ $\sin at - at\cos at$

20. $\dfrac{2as^2}{\left(s^2+a^2\right)^2}$ $\sin at + at\cos at$

21. $\dfrac{2as}{\left(s^2-a^2\right)^2}$ $t\sinh at$

22. $\dfrac{s^2+a^2}{\left(s^2-a^2\right)^2}$ $t\cosh at$

23. $\dfrac{b^2-a^2}{\left(s^2+a^2\right)\left(s^2+b^2\right)}, \quad a^2\neq b^2$ $\dfrac{1}{a}\sin at - \dfrac{1}{b}\sin bt$

24. $\dfrac{s\left(b^2-a^2\right)}{\left(s^2+a^2\right)\left(s^2+b^2\right)}, \quad a^2\neq b^2$ $\cos at - \cos bt$

25. $\dfrac{b}{\left(s-a\right)^2+b^2}$ $e^{at}\sin bt$

26. $\dfrac{\left(s-a\right)}{\left(s-a\right)^2+b^2}$ $e^{at}\cos bt$

27.	$\ln\left(1+\dfrac{1}{s}\right)$	$\dfrac{1-e^{-t}}{t}$
28.	$\ln\dfrac{s+a}{s-a}$	$\dfrac{2\sinh at}{t}$
29.	$\ln\left(1+\dfrac{a^2}{s^2}\right)$	$\dfrac{2(1-\cos at)}{t}$
30.	$\ln\left(1-\dfrac{a^2}{s^2}\right)$	$\dfrac{2(1-\cosh at)}{t}$
31.	$\dfrac{e^{-as}}{s}$	$f(t)=\begin{cases} 0, & 0<t<a \\ 1, & a<t \end{cases}$
32.	$\dfrac{e^{-as}}{s^2}$	$f(t)=\begin{cases} 0, & 0<t<a \\ t-a, & a<t \end{cases}$
33.	$\dfrac{e^{-a/s}}{s}$	$J_0(2\sqrt{at}\,)$
34.	$\dfrac{e^{-a/s}}{s^{n+1}},\qquad n=1,2,\dots$	$\left(\dfrac{t}{x}\right)^{n/2}J_n(2\sqrt{at}\,)$
35.	$\dfrac{1}{\sqrt{s^2+a^2}}$	$J_0(at)$
36.	$\dfrac{\left(\sqrt{s^2+a^2}-s\right)^n}{\sqrt{s^2+a^2}},\qquad n=1,2,\dots$	$x^nJ_1(at)$
37.	$\dfrac{\left(s-\sqrt{s^2-a^2}\,\right)^n}{\sqrt{s^2-a^2}},\quad n=1,2,\dots$	$x^nI_n(at)$
38.	$\dfrac{a^n}{\left(s+\sqrt{s^2-a^2}\,\right)^n},\qquad n=1,2,\dots$	$\dfrac{n}{t}I_n(at)$

the right-hand side of (A3.4) is of the form of functions 9 and 10 of Table A3.2, which is

$$\frac{1-s+a}{(s-a)(s-1)} = \frac{1}{(s-a)(s-1)} - \frac{s}{(s-a)(s-1)} + \frac{a}{(s-a)(s-1)}$$

and we get

$$y(x) = \frac{e^{ax}-e^x}{a-1} - \frac{ae^{ax}-e^x}{a-1} + \frac{a(e^{ax}-e^x)}{a-1}$$

$$= \frac{e^{ax}-ae^x}{a-1}.$$

To illustrate a technique which must often be employed in inverting LTs, suppose function 9 and 10 had not been in the table. Function 8 could have been employed by expanding the right-hand side of (A3.4) through the method of partial fractions[2] to obtain

$$\frac{1-s+a}{(s-a)(s-1)} = \frac{1}{(a-1)(s-1)} - \frac{a}{(a-1)(s-1)}.$$

Now using the additive property of LTs and function 8 yields

$$y(x) = \frac{e^{ax}}{a-1} - \frac{ae^{ax}}{a-1} = \frac{e^{ax}-ae^x}{a-1}.$$

While for this example one might not be convinced that using LTs was easier, reconsider the situation had the formula of Equation (1.17) not been available. Also, the use of LTs becomes even more powerful for higher order differential equations.

A3.2 LAPLACE-STIELTJES TRANSFORM

The Laplace-Stieltjes Transform (LST) of a function $f(t)$ is defined as

$$f^*(s) \equiv \int_0^\infty e^{-st} df(t) .$$

We can relate $f^*(s)$ to $\bar{f}(s)$ [when $\bar{f}(s)$ exists] by integrating by parts as follows:

$$f^*(s) = \int_0^\infty e^{-st} df(t) = e^{-st} f(t) \Big|_0^\infty + \int se^{-st} f(t) dt .$$

[2]Refer to any basic calculus test.

Assuming $f(t)$ is such that as t goes to ∞, it does not go to ∞ as fast as e^t, that is,

$$\lim_{t \to \infty} e^{-st} f(t) = 0,$$

we have

$$f^*(s) = s\bar{f}(s) - f(0). \tag{A3.5}$$

If we consider the LST of a continuous CDF where the random variable is nonnegative, then $F(\infty) = 1$ and $F(0) = 0$ so that

$$F^*(s) = s\bar{F}(s). \tag{A3.6}$$

Example A3.2

Suppose we desire to calculate the LST for the empirical distribution of service time given in Example 5.2 of Section 5.1.2; namely, $B^*(s)$ where

t	$B(t)$
9	$\frac{2}{3}$
12	1

Now,

$$B^*(s) = \int_0^\infty e^{-st} dB(t)$$

$$= \sum_t e^{-st} [B(t^+) - B(t^-)]$$

$$= \frac{2}{3} e^{-9s} + \frac{1}{3} e^{-12s}.$$

Example A3.3

It is desired to find the LST of the exponential CDF

$$F(t) = 1 - e^{-\theta t}.$$

We can calculate this in either of two ways: (1) by actually taking $dF(t)$ and hence calculating the LT of $F'(t)$ or (2) by using the relationship given in (A3.6). Hence

(i) $F^*(s) = \mathcal{L}\{F'(t)\} = \mathcal{L}\{\theta e^{-\theta t}\} = \dfrac{\theta}{s + \theta}$

or

(ii) $\quad F^*(s) = s\mathcal{L}\{1 - e^{-\theta t}\} = s[\mathcal{L}\{1\} - \mathcal{L}\{e^{-\theta t}\}]$

$$= s\left[\frac{1}{s} - \frac{1}{s + \theta}\right]$$

$$= \frac{\theta}{s + \theta}.$$

A3.3　GENERATING FUNCTIONS

Consider a function $G(z)$ which has a power series expansion,

$$G(z) = \sum_{n=0}^{\infty} g_n z^n = g_0 + g_1 z + g_2 z^2 + g_3 z^3 + \cdots. \tag{A3.7}$$

If the series converges for some range of z, $G(z)$ is called the generating function (GF) of the sequence $g_0, g_1, g_2, g_3, \ldots$.[3] Generating functions are helpful in solving certain difference equations in a manner analogous to the way Laplace transforms were helpful in solving certain differential equations; that is, it is often easier to solve a corresponding equation for a generating function, determine its series expansion and then "pick off" the coefficients $\{g_n\}$ rather than to solve the original equation(s) directly for the $\{g_n\}$.

Example A3.4

We desire to find f_n where

$$f_{n+2} - 2f_{n+1} + f_n = a \qquad (n = 1, 2, 3, \ldots)$$

with the boundary conditions $f_0 = f_1 = 0$. We first multiply all terms by z^n and then sum from 0 to ∞ and get

$$\sum_{n=0}^{\infty} f_{n+2} z^n - 2 \sum_{n=0}^{\infty} f_{n+1} z^n + \sum_{n=0}^{\infty} f_n z^n = a \sum_{n=0}^{\infty} z^n$$

$$z^{-2} \sum_{n=0}^{\infty} f_{n+2} z^{n+2} - 2z^{-1} \sum_{n=0}^{\infty} f_{n+1} z^{n+1} + F(z) = \frac{a}{1-z}$$

$$z^{-2}[F(z) - f_1 - f_0] - 2z^{-1}[F(z) - f_0] + F(z) = \frac{a}{1-z}.$$

[3]The generating function as defined above is closely related to the z-transform often used in engineering work and defined as $\sum_{n=0}^{\infty} g_n z^{-n}$.

Using the boundary conditions we have then that

$$z^{-2}F(z) - 2z^{-1}F(z) + F(z) = \frac{a}{1-z}$$

or

$$F(z) = \frac{az^2}{(1-z)^3}. \qquad (A3.8)$$

It remains now to expand the right-hand side of (A3.8) in a power series in order to "pick off" the coefficients which are the $\{f_n\}$. We know the geometric series is given by

$$\sum_{n=0}^{\infty} z^n = \frac{1}{1-z} \qquad (|z| < 1).$$

Also, taking the derivative of both sides twice yields

$$\sum_{n=2}^{\infty} n(n-1)z^{n-2} = \frac{2}{(1-z)^3}.$$

Hence

$$F(z) = \sum_{n=2}^{\infty} \frac{az^2}{2} n(n-1)z^{n-2}$$

$$= \sum_{n=2}^{\infty} \frac{an(n-1)}{2} z^n.$$

Thus the coefficient of z^n is $an(n-1)/2$; hence

$$f_n = \frac{an(n-1)}{2}.$$

Had it not been possible to expand the closed-form solution for $F(z)$ given by Equation (A3.8) in a power series by recognizing or manipulating known series, one might have to resort to expansion by using a Maclaurin's series which would require successive differentiation, that is,

$$F(z) = F(0) + F'(0)z + \frac{F''(0)z^2}{2} + \cdots + \frac{F^{(n)}(0)z^n}{n!} + \cdots.$$

This can be quite cumbersome in specific cases, especially if it is desired to obtain a general formula for f_n which requires a general expression for $F^{(n)}(0)$.

A3.4 PROBABILITY GENERATING FUNCTION

If f_n is a probability distribution of a discrete nonnegative random variable X so that

$$\Pr\{X=n\}=f_n \qquad (n=0,1,2,\ldots),$$

then

$$F(z)=\sum_{n=0}^{\infty}f_n z^n$$

is called a probability generating function, since it generates the probabilities $\{f_n\}$. Furthermore,

$$F'(1)=\sum_{n=0}^{\infty}nf_n$$

is the expected value of the random variable, and

$$F(1)=\sum_{n=0}^{\infty}f_n=1.$$

Since in queueing theory analyses, we are mostly working with equations involving system state probabilities, the probability generating function is quite often useful in determining these probabilities via the generating function methodology described in the preceding section. Even if it is not possible in some cases to obtain the series expansion for the generating function, as long as the generating function itself is obtainable in closed form, it is possible to get moments of the probability distribution by calculating $F^{(n)}(1)$, that is, the nth derivative evaluated at $z=1$, which gives the nth factorial moment of f_n. Since it is possible to relate factorial moments to regular moments, it turns out that merely knowing $F(z)$ can be quite useful, even if its power series expansion is unobtainable.

Example A3.5

Suppose the probability generating function is found to be

$$F(z)=e^{-\lambda}e^{\lambda z},$$

and we desire the mean and variance. Then

$$\mu=F'(1)=e^{-\lambda}\lambda e^{\lambda z}\big|_{z=1}=\lambda.$$

To find the variance we note that the second factorial moment

$$\sum_{n=0}^{\infty} n(n-1)f_n = \sum_{n=0}^{\infty} n^2 f_n - \mu.$$

But the variance σ^2 can be given as

$$\sigma^2 = \sum_{n=0}^{\infty} n^2 f_n - \mu^2$$

so that

$$\sigma^2 = \sum_{n=0}^{\infty} n(n-1)f_n + \mu - \mu^2$$

$$= F''(1) + \mu - \mu^2$$

$$= e^{-\lambda}\lambda^2 e^{\lambda z}\big|_{z=1} + \lambda - \lambda^2$$

$$= \lambda^2 + \lambda - \lambda^2 = \lambda.$$

We, of course, can obtain the entire probability distribution here since it is an easy matter to obtain the power series expansion for $F(z)$, namely,

$$F(z) = e^{-\lambda}e^{\lambda z} = e^{-\lambda} \sum_{n=0}^{\infty} \frac{(\lambda z)^n}{n!} = \sum_{n=0}^{\infty} \frac{e^{-\lambda}(\lambda z)^n}{n!}.$$

Hence the coefficient of z^n is $e^{-\lambda}\lambda^n/n!$ and

$$f_n = \frac{e^{-\lambda}\lambda^n}{n!},$$

which is the Poisson distribution.

A3.5   MOMENT GENERATING FUNCTION

Given a random variable X, the moment generating function (MGF) is defined as

$$M_X(t) \equiv E[e^{tX}], \tag{A3.9}$$

where E is the expected-value operator. For nonnegative continuous random variables we have

$$M_X(t) = \int_0^{\infty} e^{tx}f(x)\,dx$$

which is almost identical to the Laplace transform of $f(x)$.

Moment generating functions have properties very similar to LTs. Of particular interest is the convolution property. Since the sum of two random variables can be expressed as a convolution, we have the useful result that the MGF of a random variable which is the sum of two other random variables is merely the product of the respective MGFs; that is, if

$$Z = X + Y$$

then

$$M_Z(t) = M_X(t)M_Y(t).$$

The fact that the inverse of the MGF is unique often allows one to determine sums of random variables by inverting the MGF resulting from a product.

Example A3.6

We desire to find the density function of the sum of two independent and identically distributed exponential random variables. Letting X_i be distributed according to

$$f(x) = \theta e^{-\theta x}$$

we desire the density of

$$Z = X_1 + X_2.$$

The MGF for an exponential random variable can be readily calculated (or found from Table A5.3) to be

$$M_X(t) = \frac{\theta}{\theta - t}.$$

Hence the MGF of the sum is

$$M_Z(t) = \left(\frac{\theta}{\theta - t}\right)^2$$

$$= \left(\frac{2\theta/2}{2\theta/2 - t}\right)^2$$

which is the MGF (see Table A5.3) for an Erlang type 2 random variable with mean $2/\theta$.

If a random variable is discrete, the MGF becomes [from Equation (A3.9)]

$$M_X(t) = \sum_x e^{tx} f_x.$$

The fact that the MGF does indeed generate moments can be seen by expanding Equation (A3.9) to get

$$M_X(t) = E\left[1 + tX + \frac{t^2 X^2}{2!} + \frac{t^3 X^3}{3!} + \cdots\right].$$

The coefficient of t^n is $(1/n!)E[X^n]$, where $E[X^n]$ is the nth moment about the origin. Thus if $M_X(t)$ can be gotten in closed form and is differentiable, $E[X^n]$ can be obtained by

$$E[X^n] = \frac{d^{(n)} M_X(t)}{dt^n}\bigg|_{t=0}.$$

Also, if $M_X(t)$ can be expanded in a power series, $E[X^n]$ can be picked off by taking $n!$ times the coefficient of t^n.

Appendix 4

STOCHASTIC PROCESSES AND MARKOV CHAINS

In this appendix we summarize some important results concerning stochastic processes and Markov chains.

A4.1 STOCHASTIC PROCESSES

A stochastic process is the mathematical abstraction of an empirical process whose development is governed by probabilistic laws. From the point of view of the mathematical theory of probability, a stochastic process is best defined as a family of random variables, $\{X(t),\ t\epsilon T\}$[1], defined over some index set or parameter space T. T is sometimes also called the time range and $X(t)$ denotes the observation at time t. Depending upon the nature of the time range, the process is classified as a discrete-parameter or continuous-parameter process as follows:

(i) If T is an infinite sequence, for example,

$$T = \{0, \pm 1, \pm 2, \ldots\}$$

or

$$T = \{0, 1, 2, \ldots\},$$

then the stochastic process $\{X(t),\ t\epsilon T\}$ is said to be a discrete-parameter process defined on the index set T;

(ii) If T is an interval or an algebraic combination of intervals, for example,

$$T = \{t: -\infty < T < +\infty\}$$

---

[1]The symbol ϵ denotes set membership, that is, t is a member of T.

or

$$T = \{ t : 0 \leqslant t < +\infty \},$$

then the stochastic process $\{X(t),\ t\epsilon T\}$ is called a contintinuous-parameter process defined on the index set T.

A4.2   MARKOV PROCESS

A discrete-parameter stochastic process $\{X(t),\ t = 0, 1, 2, \ldots\}$ or a continuous-parameter stochastic process $\{X(t),\ t > 0\}$ is said to be a Markov process if, for any set of n time points $t_1 < t_2 < \cdots < t_n$ in the index set or time range of the process, the conditional distribution of $X(t_n)$, given the values of $X(t_1), X(t_2), X(t_3), \ldots, X(t_{n-1})$, depends only on $X(t_{n-1})$, the immediately preceding value; more precisely, for any real numbers $x_1, x_2, \ldots, x_n$,

$$\Pr\{ X(t_n) \leqslant x_n | X(t_1) = x_1, \ldots, X(t_{n-1}) = x_{n-1} \}$$

$$= \Pr\{ X(t_n) \leqslant x_n | X(t_{n-1}) = x_{n-1} \}.$$

In nonmathematical language one says that, given the "present" condition of the process, the "future" is independent of the "past."

Markov processes are classified according to:

(i) The nature of the index set of the process (whether discrete parameter or continuous parameter); and
(ii) The nature of the state space of the process.

A real number x is said to be a state of a stochastic process $\{X(t), t\epsilon T\}$ if there exists a time point t in T such that the $\Pr\{x - h < X(t) < x + h\}$ is positive for every $h > 0$. The set of possible states constitutes the state space of the process. If the state space is discrete, the Markov process is called a Markov chain.

A discrete-parameter Markov process with discrete state space is called a discrete-parameter Markov chain. A Markov chain is finite if the state space is finite; otherwise it is a denumerable or infinite Markov chain. Since the system is observed at a discrete set of time points, let the successive observations be denoted by $X_0, X_1, X_2, \ldots, X_n, \ldots$. It is assumed that X_n is a random variable whose value represents the state of the system at the nth time point. The sequence $\{X_n\}$ is called a chain if it is assumed that there are only a finite or countably infinite number of states in which the system may be found at any point within the given time range. The sequence $\{X_n\}$ is thus a Markov chain if each random variable X_n is discrete and the following holds: for any integer $m > 2$ and any set of m points $n_1 < n_2 < \cdots < n_m$, the conditional distribution of X_{n_m}, given values of $X_{n_1}, X_{n_2}, \ldots, X_{n_{m-1}}$, depends only on $X_{n_{m-1}}$, the immediately preceding value; that is,

$$\Pr\left\{ X_{n_m} = x_{n_m} \mid X_1 = x_{n_1}, \ldots, X_{n_{m-1}} = x_{n_{m-1}} \right\}$$

$$= \Pr\left\{ X_{n_m} = x_{n_m} \mid X_{n_{m-1}} = x_{n_{m-1}} \right\}.$$

A continuous-parameter Markov process with discrete state space is called a continuous-parameter Markov chain, while for continuous-state space and discrete-parameter space, the process is called a discrete-parameter Markov process. If both the state space and parameter space are continuous, it is called a continuous-parameter Markov process.

An important generalization of the Markov chain which is very useful in queueing is the semi-Markov process (SMP). The state transitions in an SMP form a discrete Markov chain, but the times between successive transitions are essentially random variables dependent upon both the *to* and *from* state.

A4.3 MARKOV CHAINS

Consider then a sequence of random variables, $\{X_n, n = 0, 1, 2, \ldots \mid X_n = 0, 1, 2, \ldots\}$, which forms a Markov chain with discrete parameter space; that is, for all n,

$$\Pr\left\{ X_n = j \mid X_0 = i_0, X_1 = i_1, \ldots, X_{n-1} = i_{n-1} \right\} = \Pr\left\{ X_n = j \mid X_{n-1} = i_{n-1} \right\}.$$

If the value of the random variable X_n is j, then the system is said to be in state j after n steps or transitions. The conditional probabilities $\Pr\{X_n = j \mid X_{n-1} = i\}$ are called the *single-step transition probabilities* or just *transition probabilities*. If these probabilities are independent of n, then the chain is said to be *homogeneous* and the probabilities $\Pr\{X_n = j \mid X_{n-1} = i\}$ can be written as p_{ij}. The matrix formed by placing p_{ij} in the (i,j) location is known as the *transition matrix* or *chain matrix*. For homogeneous chains, the m-step transition probabilities

$$\Pr\left\{ X_{n+m} = j \mid X_n = i \right\} \equiv p_{ij}^{(m)}$$

are also independent of n. The unconditional probability of state j at the nth trial will be written as

$$\Pr\left\{ X_n = j \right\} = p_j^{(n)},$$

so that the initial distribution is given by $p_j^{(0)}$.

From the basic laws of probability, one can easily show that the matrix formed by the elements $\{p_{ij}^{(m)}\}$, say $P^{(m)}$, can be found by simply multiplying $P^{(m-k)}$ by $P^{(k)}$ for any value of k, $0 < k < m$. This is the matrix equivalent of the well-known Chapman-Kolmogorov equations for this Markov process, namely,

$$p_{ij}^{(m)} = \sum_r p_{ir}^{(m-k)} p_{rj}^{(k)} \qquad (0 < k < m).$$

Two states, i and j, are said to *communicate* ($i \leftrightarrow j$) if i is accessible from j ($j \rightarrow i$) and j is accessible from i ($i \rightarrow j$). A chain is said to be *irreducible* if all of its states communicate, that is, if there exists an n such that $p_{ij}^{(n)} > 0$ for all pairs (i,j).

The period of a return state k of a chain is defined as the greatest common divisor (GCD) of the set of integers $\{n\}$ for which $p_{kk}^{(n)} > 0$. A state is said to be *aperiodic* if this GCD is 1, that is, if it has period 1. A chain is said to be aperiodic if each of its states is aperiodic.

Define $f_{jj}^{(n)}$ as the probability that a chain starting at state j returns for the first time to j in n transitions. Hence the probability that the chain ever returns to j is

$$f_{jj} = \sum_{n=1}^{\infty} f_{jj}^{(n)}.$$

If $f_{jj} = 1$, then j is said to be a *recurrent state*; if $f_{jj} < 1$, j is said to be a *transient state*. When $f_{jj} = 1$,

$$m_{jj} = \sum_{n=1}^{\infty} n f_{jj}^{(n)}$$

is the *mean recurrence time*. If $m_{jj} < \infty$, then j is known as a *positive recurrent state*, while if $m_{jj} = \infty$, we say that j is a *null recurrent state*.

Define $f_{ij}^{(n)}$, $i \neq j$, as the probability that the first passage from state i to state j occurs in exactly n steps. Then the probability that state j is ever reached from i is

$$f_{ij} = \sum_{n=1}^{\infty} f_{ij}^{(n)}.$$

The expected value of the sequence $\{f_{ij}^{(n)}, n = 1, 2, \ldots\}$, of first passage probabilities for a fixed pair (i,j), $i \neq j$, is denoted by m_{ij} and is called the *mean first passage time*; that is,

$$m_{ij} = \sum_{n=1}^{\infty} n f_{ij}^{(n)} \qquad (i \neq j).$$

If $i=j$, then m_{ij} becomes the mean recurrence time of state i.

We are now in a position to state some of the more important theorems pertaining to Markov chains.

THEOREM A4.1:   *Let a chain C be irreducible. Then C is either recurrent or transient; that is, either all the states are recurrent or all are nonrecurrent* [see Parzen (1962)]

THEOREM A4.2.   *Let C be irreducible, and let k be a fixed state in C. Then C is recurrent if, and only if, for every state j, $j \neq k$, $f_{jk} = 1$* [see Parzen (op. cit.)].

THEOREM A4.3.   *A chain is recurrent if there exists a solution, $\{y_i\}$, of the inequalities*

$$\sum_{j=0}^{\infty} p_{ij} y_j \leqslant y_i \qquad (i \neq 0)$$

such that $y_i \to \infty$ as $i \to \infty$ [see Foster (1953)].

THEOREM A4.4.   *An irreducible Markov chain is transient if, and only if, there exists a bounded nonconstant solution of the equations*

$$\sum_{j=0}^{\infty} p_{ij} y_j = y_i \qquad (i \neq 0) \qquad [\text{Foster (op.cit.)}].$$

THEOREM A4.5.   *Let k be a fixed state in an irreducible, recurrent chain. Then the set of mean first passage times, $\{m_{jk}, j \neq k\}$, uniquely satisfies the system of equations*

$$m_{jk} = 1 + \sum_{i \neq k} p_{ji} m_{ik} \qquad (j \neq k),$$

and the mean recurrence times satisfy

$$m_{kk} = 1 + \sum_{i \neq k} p_{ki} m_{ik} \qquad [\text{Parzen (op.cit.)}].$$

THEOREM A4.6.   *If C is an irreducible, recurrent chain, then either all its states are positive or all are null.*

We now add the following set of definitions before a final set of theorems:

(i) An aperiodic chain which is irreducible and positive recurrent is said to be *ergodic*.

(ii) A probability distribution $\{\pi_j, j \in C\}$ is called a *stationary distribution* if

$$\pi_j = \sum_{i=0}^{\infty} \pi_i p_{ij} \qquad (j \in C).$$

(iii) A Markov chain is said to possess a *long-run* or *limiting distribution* if there exists a probability distribution $\{\pi_j, j \in C\}$ having the property that

$$\lim_{n \to \infty} p_{ij}^{(n)} = \pi_j \qquad (\text{for all } i,j).$$

THEOREM A4.7. *In an irreducible and aperiodic chain, the limiting probabilities,*

$$\lim_{n \to \infty} \Pr\{ X_n = j \} = \pi_j \qquad (\text{all } j),$$

always exist and are independent of the distribution of initial states. If the states are all transient or null recurrent, then $\pi_j = 0$ for all j, and there exists no stationary distribution. If, however, all states are positive recurrent (i.e., ergodic), then $\pi_j > 0$, for all j, and $\{\pi_j\}$ is a probability distribution, with $\pi_j = 1/m_{jj}$. The limiting distribution is the unique solution of the stationary system of equations

$$\pi_j = \sum_{i=0}^{\infty} \pi_i p_{ij}$$

and

$$\sum_{i=0}^{\infty} \pi_i = 1.$$

Note that, among other things, an ergodic chain possesses identical stationary and limiting distributions, commonly known as *steady-state probabilities.*

THEOREM A4.8. *An irreducible, aperiodic chain is ergodic if there exists a nonnegative solution of the system*

$$\sum_{j=0}^{\infty} p_{ij} x_j \leqslant x_i - 1 \qquad (i \neq 0)$$

such that

$$\sum_{j=0}^{\infty} p_{0j} x_j < \infty \qquad [\text{Foster (op.cit.)}].$$

520

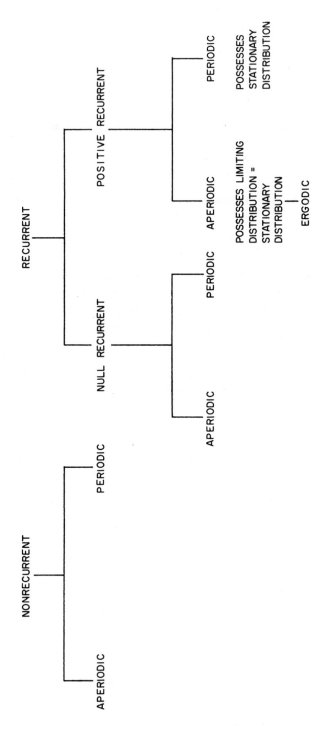

Fig. A4.1 Classification of irreducible infinite chains.

THEOREM A4.9. *An aperiodic, irreducible system is ergodic if, and only if, there exists a nonnull solution of the equations*

$$\sum_{j=0}^{\infty} x_j p_{ji} = x_i$$

such that

$$\sum_{j=0}^{\infty} |x_j| < \infty.$$

Figure A4.1 shows in diagramatic form how irreducible chains can be classified with respect to periodicity and ergodicity.

Appendix 5

TABLES

TABLE A5.1   PSEUDORANDOM NUMBERS

0.05954251	0.35724688	0.60759890	0.43037140	0.11383766
0.80968320	0.83356035	0.71421254	0.78323233	0.27148068
0.57979262	0.03542919	0.99444067	0.64778125	0.93672144
0.79029751	0.31129181	0.75507295	0.72881150	0.57721210
0.90396905	0.22890544	0.23771131	0.36611867	0.05731013
0.04879197	0.77696061	0.22263598	0.34317052	0.05529903
0.24325907	0.96186364	0.58184946	0.83432412	0.76929927
0.10687858	0.71757793	0.34356034	0.60316026	0.52691877
0.73306966	0.65614879	0.33926558	0.13025415	0.72813427
0.19651842	0.62590182	0.98674500	0.28735316	0.84341335
0.47430158	0.25508869	0.26181769	0.27510786	0.29428720
0.28975248	0.08993012	0.93180823	0.78147817	0.30259502
0.78226650	0.97024333	0.78106093	0.95417523	0.69550312
0.58544147	0.25312018	0.24974775	0.22040415	0.07469541
0.46453428	0.11494690	0.50887275	0.01871454	0.53243136
0.02615793	0.36506438	0.95496488	0.44420981	0.07057512
0.42556214	0.91819644	0.67911947	0.81094849	0.75361574
0.22315800	0.55640650	0.33001673	0.97244179	0.86449945
0.43502080	0.82962906	0.06258726	0.90886164	0.88988471
0.15955305	0.94835520	0.25415385	0.98972595	0.65097106
0.99829304	0.13101816	0.80147123	0.62966418	0.56474340
0.72148252	0.24620426	0.98388267	0.68745732	0.26980007
0.43168426	0.16190481	0.08627015	0.06047676	0.58642900
0.97428346	0.56783950	0.63848579	0.72035849	0.57577884
0.97144663	0.64667046	0.13700235	0.00198018	0.77885914
0.65533376	0.92226970	0.63561404	0.51325703	0.35901570
0.53478038	0.97754085	0.05222104	0.51545823	0.62276042
0.09743780	0.97978294	0.00175762	0.19249809	0.13917053
0.10253990	0.36270392	0.25336444	0.25585151	0.25482881
0.22630894	0.06439447	0.34958601	0.51796579	0.96152067
0.10743207	0.99090576	0.97854638	0.95312572	0.91183710
0.89289045	0.15080893	0.86883914	0.85575473	0.31497562
0.18806040	0.29358196	0.06894779	0.77144909	0.00816469
0.10594547	0.56219065	0.41963494	0.45809317	0.97184432
0.70822716	0.50276387	0.64253843	0.33035588	0.19928920
0.22253215	0.54158962	0.24674773	0.60617971	0.41634810
0.04247101	0.50769258	0.66391635	0.41426492	0.51034236
0.33367014	0.40893888	0.45060217	0.02316305	0.08355772
0.29287910	0.00525482	0.39561665	0.32640707	0.39789200
0.44968820	0.11710054	0.65540934	0.87855124	0.37262273
0.32877541	0.61904740	0.75530529	0.96040463	0.96467972
0.14443588	0.18449783	0.80706418	0.18190432	0.82784796
0.32994902	0.52906215	0.20483172	0.46743095	0.96110034
0.55972326	0.70843637	0.21310818	0.90272200	0.49835765
0.86564803	0.70866895	0.46118164	0.38906884	0.18377757
0.60104597	0.95227730	0.30425012	0.25500524	0.79177976
0.45563102	0.60776854	0.54593146	0.80567169	0.92064679
0.27283525	0.35119009	0.65162337	0.74902916	0.62956452
0.03612430	0.55066407	0.97886622	0.91722012	0.69352448
0.90616572	0.19527328	0.01614823	0.33942974	0.89124477

IBM 370/145 George Washington University Computing Center Subroutine RANDU Starting Seed: 1951.

A5.2.1 Exponential Functions[a]

x	e^x	e^{-x}	x	e^x	e^{-x}	x	e^x	e^{-x}	x	e^x	e^{-x}
.00	1.0000	1.00000	**0.30**	1.3499	0.74082	**0.60**	1.8221	0.54881	**0.90**	2.4596	0.40657
0.01	1.0101	0.99005	0.31	1.3634	0.73345	0.61	1.8404	0.54335	0.91	2.4843	0.40252
0.02	1.0202	0.98020	0.32	1.3771	0.72615	0.62	1.8589	0.53794	0.92	2.5093	0.39852
0.03	1.0305	0.97045	0.33	1.3910	0.71892	0.63	1.8776	0.53259	0.93	2.5345	0.39455
0.04	1.0408	0.96079	0.34	1.4049	0.71177	0.64	1.8965	0.52729	0.94	2.5600	0.39063
0.05	1.0513	0.95123	0.35	1.4191	0.70469	0.65	1.9155	0.52205	0.95	2.5857	0.38674
0.06	1.0618	0.94176	0.36	1.4333	0.69768	0.66	1.9348	0.51685	0.96	2.6117*	0.38289
0.07	1.0725	0.93239	0.37	1.4477	0.69073	0.67	1.9542	0.51171	0.97	2.6379	0.37908
0.08	1.0833	0.92312	0.38	1.4623	0.68386	0.68	1.9739	0.50662	0.98	2.6645	0.37531
0.09	1.0942	0.91393	0.39	1.4770	0.67706	0.69	1.9937	0.50158	0.99	2.6912	0.37158
0.10	1.1052	0.90484	**0.40**	1.4918	0.67032	**0.70**	2.0138	0.49659	**1.00**	2.7183	0.36788
0.11	1.1163	0.89583	0.41	1.5068	0.66365	0.71	2.0340	0.49164	1.01	2.7456	0.36422
0.12	1.1275	0.88692	0.42	1.5220	0.65705	0.72	2.0544	0.48675	1.02	2.7732	0.36059
0.13	1.1388	0.87810	0.43	1.5373	0.65051	0.73	2.0751	0.48191	1.03	2.8011	0.35701
0.14	1.1503	0.86936	0.44	1.5527	0.64404	0.74	2.0959	0.47711	1.04	2.8292	0.35345
0.15	1.1618	0.86071	0.45	1.5685	0.63763	0.75	2.1170	0.47237	1.05	2.8577	0.34994
0.16	1.1735	0.85214	0.46	1.5841	0.63128	0.76	2.1383	0.46767	1.06	2.8864	0.34646
0.17	1.1853	0.84366	0.47	1.6000	0.62500	0.77	2.1598	0.46301	1.07	2.9154	0.34301
0.18	1.1972	0.83527	0.48	1.6161	0.61878	0.78	2.1815	0.45841	1.08	2.9447	0.33960
0.19	1.2092	0.82696	0.49	1.6323	0.61263	0.79	2.2034	0.45384	1.09	2.9743	0.33622
0.20	1.2214	0.81873	**0.50**	1.6487	0.60653	**0.80**	2.2255	0.44933	**1.10**	3.0042	0.33287
0.21	1.2337	0.81058	0.51	1.6653	0.60050	0.81	2.2479	0.44486	1.11	3.0344	0.32956
0.22	1.2461	0.80252	0.52	1.6820	0.59452	0.82	2.2705	0.44043	1.12	3.0649	0.32628
0.23	1.2586	0.79453	0.53	1.6989	0.58860	0.83	2.2933	0.43605	1.13	3.0957	0.32303
0.24	1.2712	0.78663	0.54	1.7160	0.58275	0.84	2.3164	0.43171	1.14	3.1268	0.31982
0.25	1.2840	0.77880	0.55	1.7333	0.57695	0.85	2.3396	0.42741	1.15	3.1582	0.31664
0.26	1.2969	0.77105	0.56	1.7507	0.57121	0.86	2.3632	0.42316	1.16	3.1899	0.31349
0.27	1.3100	0.76338	0.57	1.7683	0.56553	0.87	2.3869	0.41895	1.17	3.2220	0.31037
0.28	1.3231	0.75578	0.58	1.7860	0.55990	0.88	2.4109	0.41478	1.18	3.2544	0.30728
0.29	1.3364	0.74826	0.59	1.8040	0.55433	0.89	2.4361	0.41066	1.19	3.2871	0.30422
0.30	1.3499	0.74082	**0.60**	1.8221	0.54881	**0.90**	2.4596	0.40657	**1.20**	3.3201	0.30119

Reprinted with permission from Joseph B. Rosenbach, Edwin A. Whitman, and David Moskovitz (eds.), *Mathematical Tables*. New York: Ginn and Co., 1943.

x	e^x	e^{-x}	x	e^x	e^{-x}	x	e^x	e^{-x}	x	e^x	e
1.20	3.3201	0.30119	**1.50**	4.4817	0.22313	**1.80**	6.0496	0.16530	**2.10**	8.1662	0.1
1.21	3.3535	0.29820	1.51	4.5267	0.22091	1.81	6.1104	0.16365	2.11	8.2482	0.1
1.22	3.3872	0.29523	1.52	4.5722	0.21871	1.82	6.1719	0.16203	2.12	8.3311	0.1
1.23	3.4212	0.29229	1.53	4.6182	0.21654	1.83	6.2339	0.16041	2.13	8.4149	0.1
1.24	3.4556	0.28938	1.54	4.6646	0.21438	1.84	6.2965	0.15882	2.14	8.4994	0.1
1.25	3.4903	0.28650	1.55	4.7115	0.21225	1.85	6.3598	0.15724	2.15	8.5849	0.1
1.26	3.5254	0.28365	1.56	4.7588	0.21014	1.86	6.4237	0.15567	2.16	8.6711	0.1
1.27	3.5609	0.28083	1.57	4.8066	0.20805	1.87	6.4883	0.15412	2.17	8.7583	0.1
1.28	3.5966	0.27804	1.58	4.8550	0.20598	1.88	6.5535	0.15259	2.18	8.8463	0.1
1.29	3.6328	0.27527	1.59	4.9037	3.20393	1.89	6.6194	0.15107	2.19	8.9352	0.1
1.30	3.6693	0.27253	**1.60**	4.9530	0.20190	**1.90**	6.6859	0.14957	**2.20**	9.0250	0.1
1.31	3.7062	0.26982	1.61	5.0028	0.19989	1.91	6.7531	0.14808	2.21	9.1157	0.1
1.32	3.7434	0.26714	1.62	5.0531	0.19790	1.92	6.8210	0.14661	2.22	9.2073	0.1
1.33	3.7810	0.26448	1.63	5.1039	0.19593	1.93	6.8895	0.14515	2.23	9.2999	0.1
1.34	3.8190	0.26185	1.64	5.1552	0.19398	1.94	6.9588	0.14370	2.24	9.3933	0.1
1.35	3.8574	0.25924	1.65	5.2070	0.19205	1.95	7.0287	0.14227	2.25	9.4877	0.1
1.36	3.8962	0.25666	1.66	5.2593	0.19014	1.96	7.0993	0.14086	2.26	9.5831	0.1
1.37	3.9354	0.25411	1.67	5.3122	0.18825	1.97	7.1707	0.13946	2.27	9.6794	0.1
1.38	3.9749	0.25158	1.68	5.3656	0.18637	1.98	7.2427	0.13807	2.28	9.7767	0.1
1.39	4.0149	0.24908	1.69	5.4195	0.18452	1.99	7.3155	0.13670	2.29	9.8749	0.1
1.40	4.0552	0.24660	**1.70**	5.4739	0.18268	**2.00**	7.3891	0.13534	**2.30**	9.9742	0.1
1.41	4.0960	0.24414	1.71	5.5290	0.18087	2.01	7.4633	0.13399	2.31	10.074	0.0
1.42	4.1371	0.24171	1.72	5.5845	0.17907	2.02	7.5383	0.13266	2.32	10.176	0.0
1.43	4.1787	0.23931	1.73	5.6407	0.17728	2.03	7.6141	0.13134	2.33	10.278	0.0
1.44	4.2207	0.23693	1.74	5.6973	0.17552	2.04	7.6906	0.13003	2.34	10.381	0.0
1.45	4.2631	0.23457	1.75	5.7546	0.17377	2.05	7.7679	0.12873	2.35	10.486	0.0
1.46	4.3060	0.23224	1.76	5.8124	0.17204	2.06	7.8460	0.12745	2.36	10.591	0.0
1.47	4.3492	0.22993	1.77	5.8709	0.17033	2.07	7.9248	0.12619	2.37	10.697	0.0
1.48	4.3929	0.22764	1.78	5.9299	0.16864	2.08	8.0045	0.12493	2.38	10.805	0.0
1.49	4.4371	0.22537	1.79	5.9895	0.16696	2.09	8.0849	0.12369	2.39	10.913	0.0
1.50	4.4817	0.22313	**1.80**	6.0496	0.16530	**2.10**	8.1662	0.12246	**2.40**	11.023	0.0

TABLE A5.2.1 Exponential Functions (*Continued*)

x	e^x	e^{-x}	x	e^x	e^{-x}	x	e^x	e^{-x}	x	e^x	e^{-x}
2.40	11.023	0.09072	**2.70**	14.880	0.06721	**3.00**	20.086	0.04979	**3.30**	27.113	0.03688
2.41	11.134	0.08982	2.71	15.029	0.06654	3.01	20.287	0.04929	3.31	27.385	0.03652
2.42	11.246	0.08892	2.72	15.180	0.06587	3.02	20.491	0.04880	3.32	27.660	0.03615
2.43	11.359	0.08804	2.73	15.333	0.06522	3.03	20.697	0.04832	3.33	27.938	0.03579
2.44	11.473	0.08716	2.74	15.487	0.06457	3.04	20.905	0.04783	3.34	28.219	0.03544
2.45	11.588	0.08629	2.75	15.643	0.06393	3.05	21.115	0.04736	3.35	28.503	0.03508
2.46	11.705	0.08543	2.76	15.800	0.06329	3.06	21.328	0.04689	3.36	28.789	0.03474
2.47	11.822	0.08458	2.77	15.959	0.06266	3.07	21.542	0.04642	3.37	29.079	0.03439
2.48	11.941	0.08374	2.78	16.119	0.06204	3.08	21.758	0.04596	3.38	29.371	0.03405
2.49	12.061	0.08291	2.79	16.281	0.06142	3.09	21.977	0.04550	3.39	29.666	0.03371
2.50	12.182	0.08208	**2.80**	16.445	0.06081	**3.10**	22.198	0.04505	**3.40**	29.964	0.03337
2.51	12.305	0.08127	2.81	16.610	0.06020	3.11	22.421	0.04460	3.41	30.265	0.03304
2.52	12.429	0.08046	2.82	16.777	0.05961	3.12	22.646	0.04416	3.42	30.569	0.03271
2.53	12.554	0.07966	2.83	16.945	0.05901	3.13	22.874	0.04372	3.43	30.877	0.03239
2.54	12.680	0.07887	2.84	17.116	0.05843	3.14	23.104	0.04328	3.44	31.187	0.03206
2.55	12.807	0.07808	2.85	17.288	0.05784	3.15	23.336	0.04285	3.45	31.500	0.03175
2.56	12.936	0.07730	2.86	17.462	0.05727	3.16	23.571	0.04243	3.46	31.817	0.03143
2.57	13.066	0.07654	2.87	17.637	0.05670	3.17	23.807	0.04200	3.47	32.137	0.03112
2.58	13.197	0.07577	2.88	17.814	0.05613	3.18	24.047	0.04159	3.48	32.460	0.03081
2.59	13.330	0.07502	2.89	17.993	0.05558	3.19	24.288	0.04117	3.49	32.786	0.03050
2.60	13.464	0.07427	**2.90**	18.174	0.05502	**3.20**	24.533	0.04076	**3.50**	33.115	0.03020
2.61	13.599	0.07353	2.91	18.357	0.05448	3.21	24.779	0.04036	3.51	33.448	0.02990
2.62	13.736	0.07280	2.92	18.541	0.05393	3.22	25.028	0.03996	3.52	33.784	0.02960
2.63	13.874	0.07208	2.93	18.728	0.05340	3.23	25.280	0.03956	3.53	34.124	0.02930
2.64	14.013	0.07136	2.94	18.916	0.05287	3.24	25.534	0.03916	3.54	34.467	0.02901
2.65	14.154	0.07065	2.95	19.106	0.05234	3.25	25.790	0.03877	3.55	34.813	0.02872
2.66	14.296	0.06995	2.96	19.298	0.05182	3.26	26.050	0.03839	3.56	35.163	0.02844
2.67	14.440	0.06925	2.97	19.492	0.05130	3.27	26.311	0.03801	3.57	35.517	0.02816
2.68	14.585	0.06856	2.98	19.688	0.05079	3.28	26.576	0.03763	3.58	35.874	0.02788
2.69	14.732	0.06788	2.99	19.886	0.05029	3.29	26.843	0.03725	3.59	36.234	0.02760
2.70	14.880	0.06721	**3.00**	20.086	0.04979	**3.30**	27.113	0.03688	**3.60**	36.598	0.02732

x	e^x	e^{-x}	x	e^x	e^{-x}	x	e^x	e^{-x}	x	e^x	e^{-x}
3.60	36.598	0.02732	**3.90**	49.402	0.02024	**4.20**	66.686	0.01500	**4.50**	90.017	0.01111
3.61	36.966	0.02705	3.91	49.899	0.02004	4.21	67.357	0.01485	4.51	90.922	0.01110
3.62	37.338	0.02678	3.92	50.400	0.01984	4.22	98.033	0.01470	4.52	91.836	0.01108
3.63	37.713	0.02652	3.93	50.907	0.01964	4.23	68.717	0.01455	4.53	92.759	0.01107
3.64	38.092	0.02625	3.94	51.419	0.01945	4.24	69.408	0.01441	4.54	93.691	0.01106
3.65	38.475	0.02599	3.95	51.935	0.01925	4.25	70.105	0.01426	4.55	94.632	0.01105
3.66	38.861	0.02573	3.96	52.457	0.01906	4.26	70.810	0.01412	4.56	95.583	0.01104
						4.27	71.522	0.01398			
3.67	39.252	0.02548	3.97	52.985	0.01887	4.28	72.240	0.01384	4.57	96.544	0.01103
3.68	39.646	0.02522	3.98	53.517	0.01869	4.29	72.966	0.01370	4.58	97.514	0.01102
3.69	40.045	0.02497							4.59	98.494	0.01101
3.70	40.447	0.02472	**4.00**	54.598	0.01832	**4.30**	73.700	0.01357	**4.60**	99.484	0.01100
3.71	40.854	0.02448	4.01	55.147	0.01813	4.31	74.440	0.01343	4.61	100.48	0.00995
3.72	41.264	0.02423	4.02	55.701	0.01795	4.32	75.189	0.01330	4.62	101.49	0.00985
3.73	41.679	0.02399	4.03	56.261	0.01777	4.33	75.944	0.01317	4.63	102.51	0.00975
3.74	42.098	0.02375	4.04	56.826	0.01760	4.34	76.708	0.01304	4.64	103.54	0.00966
3.75	42.521	0.02352	4.05	57.397	0.01742	4.35	77.478	0.01291	4.65	104.58	0.00956
3.76	42.948	0.02328	4.06	57.974	0.01725	4.36	78.257	0.01278	4.66	105.64	0.00947
3.77	43.380	0.02305	4.07	58.557	0.01708	4.37	79.044	0.01265	4.67	106.70	0.00937
3.78	43.816	0.02282	4.08	59.145	0.01691	4.38	79.838	0.01253	4.68	107.77	0.00928
3.79	44.256	0.02260	4.09	59.740	0.01674	4.39	80.640	0.01240	4.69	108.85	0.00919
3.80	44.701	0.02237	**4.10**	60.340	0.01657	**4.40**	81.451	0.01228	**4.70**	109.95	0.00910
3.81	45.150	0.02215	4.11	60.947	0.01641	4.41	82.269	0.01216	4.71	111.05	0.00900
3.82	45.604	0.02193	4.12	61.559	0.01624	4.42	83.096	0.01203	4.72	112.17	0.00892
3.83	46.063	0.02171	4.13	62.178	0.01608	4.43	83.931	0.01191	4.73	113.30	0.00883
3.84	46.525	0.02149	4.14	62.803	0.01592	4.44	84.775	0.01180			
3.85	46.993	0.02128	4.15	63.434	0.01576	4.45	85.627	0.01168	4.74	114.43	0.00874
3.86	47.465	0.02107	4.16	64.072	0.01561	4.46	86.488	0.01156	4.75	115.58	0.00865
									4.76	116.75	0.00857
3.87	47.942	0.02086	4.17	64.715	0.01545	4.47	87.357	0.01145	4.77	117.92	0.00848
3.88	48.424	0.02065	4.18	65.366	0.01530	4.48	88.235	0.01133	4.78	119.10	0.00840
3.89	48.911	0.02045	4.19	66.023	0.01515	4.49	89.121	0.01122	4.79	120.30	0.00831
3.90	49.402	0.02024	**4.20**	66.686	0.01500	**4.50**	90.017	0.01111	**4.80**	121.51	0.00823

TABLE A5.2.1 Exponential Functions (*Continued*)

x	e^x	e^{-x}	x	e^x	e^{-x}	x	e^x	e^{-x}	x	e^x	e^{-x}
4.80	121.51	0.00823	**5.10**	164.02	0.00610	**5.40**	221.41	0.00452	**5.70**	298.87	0.00335
4.81	122.73	0.00815	5.11	165.67	0.00604	5.41	223.63	0.00447	5.71	301.87	0.00331
4.82	123.97	0.00807	5.12	167.34	0.00598	5.42	225.88	0.00443	5.72	304.90	0.00328
4.83	125.21	0.00799	5.13	169.02	0.00592	5.43	228.15	0.00438	5.73	307.97	0.00325
4.84	126.47	0.00791	5.14	170.72	0.00586	5.44	230.44	0.00434	5.74	311.06	0.00321
4.85	127.74	0.00783	5.15	172.43	0.00580	5.45	232.76	0.00430	5.75	314.19	0.00318
4.86	129.02	0.00775	5.16	174.16	0.00574	5.46	235.10	0.00425	5.76	317.35	0.00315
4.87	130.32	0.00767	5.17	175.91	0.00568	5.47	237.46	0.00421	5.77	320.54	0.00312
4.88	131.63	0.00760	5.18	177.68	0.00563	5.48	239.85	0.00417	5.78	323.76	0.00309
4.89	132.95	0.00752	5.19	179.47	0.00557	5.49	242.26	0.00413	5.79	327.01	0.00306
4.90	134.29	0.00745	**5.20**	181.27	0.00552	**5.50**	244.69	0.00409	**5.80**	330.30	0.00303
4.91	135.64	0.00737	5.21	183.09	0.00546	5.51	247.15	0.00405	5.81	333.62	0.00300
4.92	137.00	0.00730	5.22	184.93	0.00541	5.52	249.64	0.00401	5.82	336.97	0.00297
4.93	138.38	0.00723	5.23	186.79	0.00535	5.53	252.14	0.00397	5.83	340.36	0.00294
4.94	139.77	0.00715	5.24	188.67	0.00530	5.54	254.68	0.00393	5.84	343.78	0.00291
4.95	141.17	0.00708	5.25	190.57	0.00525	5.55	257.24	0.00389	5.85	347.23	0.00288
4.96	142.59	0.00701	5.26	192.48	0.00520	5.56	259.82	0.00385	5.86	350.72	0.00285
4.97	144.03	0.00694	5.27	194.42	0.00514	5.57	262.43	0.00381	5.87	354.25	0.00282
4.98	145.47	0.00687	5.28	196.37	0.00509	5.58	265.07	0.00377	5.88	357.81	0.00279
4.99	146.94	0.00681	5.29	198.34	0.00504	5.59	267.74	0.00374	5.89	361.41	0.00277
5.00	148.41	0.00674	**5.30**	200.34	0.00499	**5.60**	270.43	0.00370	**5.90**	365.04	0.00274
5.01	149.90	0.00667	5.31	202.35	0.00494	5.61	273.14	0.00366	5.91	368.71	0.00271
5.02	151.41	0.00660	5.32	204.38	0.00489	5.62	275.89	0.00362	5.92	372.41	0.00269
5.03	152.93	0.00654	5.33	206.44	0.00484	5.63	278.66	0.00359	5.93	376.15	0.00266
5.04	154.47	0.00647	5.34	208.51	0.00480	5.64	281.46	0.00355	5.94	379.93	0.00263
5.05	156.02	0.00641	5.35	210.61	0.00475	5.65	284.29	0.00352	5.95	383.75	0.00261
5.06	157.59	0.00635	5.36	212.72	0.00470	5.66	287.15	0.00348	5.96	387.61	0.00258
5.07	159.17	0.00628	5.37	214.86	0.00465	5.67	290.03	0.00345	5.97	391.51	0.00255
5.08	160.77	0.00622	5.38	217.02	0.00461	5.68	292.95	0.00341	5.98	395.44	0.00253
5.09	162.39	0.00616	5.39	219.20	0.00456	5.69	295.89	0.00338	5.99	399.41	0.00250
5.10	164.02	0.00610	**5.40**	221.41	0.00452	**5.70**	298.87	0.00335	**6.00**	403.43	0.00248

TABLE A5.2.1 Exponential Functions (*Continued*)

x	e^x	e^{-x}	x	e^x	e^{-x}	x	e^x	e^{-x}	x	e^x	e^{-x}
6.00	403.43	0.00248									
6.25	518.01	0.00193									
6.50	665.14	0.00150									
6.75	854.06	0.00117									
7.00	1096.6	0.00091									
7.50	1808.0	0.00055									
8.00	2981.0	0.00034									
8.50	4914.8	0.00020									
9.00	8103.1	0.00012									
9.50	13360.	0.00007									
10.00	22026.	0.00005									

TABLE A5.2.2   Natural Logarithms[a]

N	0	1	2	3	4	5	6	7	8	9
1.0	0.0 0000	0995	1980	2956	3922	4879	5827	6766	7696	8618
1.1	0.0 9531	*0436	*1333	*2222	*3103	*3976	*4842	*5700	*6551	*7395
1.2	0.1 8232	9062	9885	*0701	*1511	*2314	*3111	*3902	*4686	*5464
1.3	0.2 6236	7003	7763	8518	9267	*0010	*0748	*1481	*2208	*2930
1.4	0.3 3647	4359	5066	5767	6464	7156	7844	8526	9204	9878
1.5	0.4 0547	1211	1871	2527	3178	3825	4469	5108	5742	6373
1.6	0.4 7000	7623	8243	8858	9470	*0078	*0682	*1282	*1879	*2475
1.7	0.5 3063	3649	4232	4812	5389	5962	6531	7098	7661	8222
1.8	0.5 8779	9333	9884	*0432	*0977	*1519	*2058	*2594	*3127	*3658
1.9	0.6 4185	4710	5233	5752	6269	6783	7294	7803	8310	8813
2.0	0.6 9315	9813	*0310	*0804	*1295	*1784	*2271	*2755	*3237	*3716
2.1	0.7 4194	4669	5142	5612	6081	6547	7011	7473	7932	8390
2.2	0.7 8846	9299	9751	*0200	*0648	*1093	*1536	*1978	*2418	*2855
2.3	0.8 3291	3725	4157	4587	5015	5442	5866	6289	6710	7129
2.4	0.8 7547	7963	8377	8789	9200	9609	*0016	*0422	*0826	*1228
2.5	0.9 1629	2028	2426	2822	3216	3609	4001	4391	4779	5166
2.6	5551	5935	6317	6698	7078	7456	7833	8208	8582	8954
2.7	0.9 9325	9695	*0063	*0430	*0796	*1160	*1523	*1885	*2245	*2604
2.8	1.0 2962	3318	3674	4028	4380	4732	5082	5431	5779	6126
2.9	6471	6815	7158	7500	7841	8181	8519	8856	9192	9527
3.0	1.0 9861	*0194	*0526	*0856	*1186	*1514	*1841	*2168	*2493	*2817
3.1	1.1 3140	3462	3783	4103	4422	4740	5057	5373	5688	6002
3.2	6315	6627	6938	7248	7557	7865	8173	8479	8784	9089
3.3	1.1 9392	9695	9996	*0297	*0597	*0896	*1194	*1491	*1788	*2083
3.4	1.2 2378	2671	2964	3256	3547	3837	4127	4415	4703	4990
3.5	5276	5562	5846	6130	6413	6695	6976	7257	7536	7815
3.6	1.2 8093	8371	8647	8923	9198	9473	9746	*0019	*0291	*0563
3.7	1.3 0833	1103	1372	1641	1909	2176	2442	2708	2972	3237
3.8	3500	3763	4025	4286	4547	4807	5067	5325	5584	5841
3.9	6098	6354	6609	6864	7118	7372	7624	7877	8128	8379
4.0	1.3 8629	8879	9128	9377	9624	9872	*0118	*0364	*0610	*0854

[a]Reprinted with permission from Joseph B. Rosenbach, Edwin A. Whitman, and David Moskovitz (eds.), *Mathematical Tables*. New York: Ginn and Co., 1943.

529

TABLE A5.2.2 Natural Logarithms (*Continued*)

N	0	1	2	3	4	5	6	7	8	9
4.0	1.3 8629	8879	9128	9377	9624	9872	*0118	*0364	*0610	*0854
4.1	1.4 1099	1342	1585	1828	2070	2311	2552	2792	3031	3270
4.2	3508	3746	3984	4220	4456	4692	4927	5161	5395	5629
4.3	5862	6094	6326	6557	6787	7018	7247	7476	7705	7933
4.4	1.4 8160	8387	8614	8840	9065	9290	9515	9739	9962	*0185
4.5	1.5 0408	0630	0851	1072	1293	1513	1732	1951	2170	2388
4.6	2606	2823	3039	3256	3471	3687	3902	4116	4330	4543
4.7	4756	4969	5181	5393	5604	5814	6025	6235	6444	6653
4.8	6862	7070	7277	7485	7691	7898	8104	8309	8515	8719
4.9	1.5 8924	9127	9331	9534	9737	9939	*0141	*0342	*0543	*0744
5.0	1.6 0944	1144	1343	1542	1741	1939	2137	2334	2531	2728
5.1	2924	3120	3315	3511	3705	3900	4094	4287	4481	4673
5.2	4866	5058	5250	5441	5632	5823	6013	6203	6393	6582
5.3	6771	6959	7147	7335	7523	7710	7896	8083	8269	8455
5.4	1.6 8640	8825	9010	9194	9378	9562	9745	9928	*0111	*0293
5.5	1.7 0475	0656	0838	1019	1199	1380	1560	1740	1919	2098
5.6	2277	2455	2633	2811	2988	3166	3342	3519	3695	3871
5.7	4047	4222	4397	4572	4746	4920	5094	5267	5440	5613
5.8	5786	5958	6130	6302	6473	6644	6815	6985	7156	7326
5.9	7495	7665	7834	8002	8171	8339	8507	8675	8842	9009
6.0	1.7 9176	9342	9509	9675	9840	*0006	*0171	*0336	*0500	*0665
6.1	1.8 0829	0993	1156	1319	1482	1645	1808	1970	2132	2294
6.2	2455	2616	2777	2938	3098	3258	3418	3578	3737	3896
6.3	4055	4214	4372	4530	4688	4845	5003	5160	5317	5473
6.4	5630	5786	5942	6097	6253	6408	6563	6718	6872	7026
6.5	7180	7334	7487	7641	7794	7947	8099	8251	8403	8555
6.6	1.8 8707	8858	9010	9160	9311	9462	9612	9762	9912	*0061
6.7	1.9 0211	0360	0509	0658	0806	0954	1102	1250	1398	1545
6.8	1692	1839	1986	2132	2279	2425	2571	2716	2862	3007
6.9	3152	3297	3442	3586	3730	3874	4018	4162	4305	4448
7.0	1.9 4591	4734	4876	5019	5161	5303	5445	5586	5727	5869

TABLE A5.2.2 Natural Logarithms (*Continued*)

N	0	1	2	3	4	5	6	7	8	9
7.0	1.94591	4734	4876	5019	5161	5303	5445	5586	5727	5869
7.1	6009	6150	6291	6431	6571	6711	6851	6991	7130	7269
7.2	7408	7547	7685	7824	7962	8100	8238	8376	8513	8650
7.3	1.98787	8924	9061	9198	9334	9470	9606	9742	9877	*0013
7.4	2.00148	0283	0418	0553	0687	0821	0956	1089	1223	1357
7.5	1490	1624	1757	1890	2022	2155	2287	2419	2551	2683
7.6	2815	2946	3078	3209	3340	3471	3601	3732	3862	3992
7.7	4122	4252	4381	4511	4640	4769	4898	5027	5156	5284
7.8	5412	5540	5668	5796	5924	6051	6179	6306	6433	6560
7.9	6686	6813	6939	7065	7191	7317	7443	7568	7694	7819
8.0	2.07944	8069	8194	8318	8443	8567	8691	8815	8939	9063
8.1	2.09186	9310	9433	9556	9679	9802	9924	*0047	*0169	*0291
8.2	2.10413	0535	0657	0779	0900	1021	1142	1263	1384	1505
8.3	1626	1746	1866	1986	2106	2226	2346	2465	2585	2704
8.4	2823	2942	3061	3180	3298	3417	3535	3653	3771	3889
8.5	4007	4124	4242	4359	4476	4593	4710	4827	4943	5060
8.6	5176	5292	5409	5524	5640	5756	5871	5987	6102	6217
8.7	6332	6447	6562	6677	6791	6905	7020	7134	7248	7361
8.8	7475	7589	7702	7816	7929	8042	8155	8267	8380	8493
8.9	8605	8717	8830	8942	9054	9165	9277	9389	9500	9611
9.0	2.19722	9834	9944	*0055	*0166	*0276	*0387	*0497	*0607	*0717
9.1	2.20827	0937	1047	1157	1266	1375	1485	1594	1703	1812
9.2	1920	2029	2138	2246	2354	2462	2570	2678	2786	2894
9.3	3001	3109	3216	3324	3431	3538	3645	3751	3858	3965
9.4	4071	4177	4284	4390	4496	4601	4707	4813	4918	5024
9.5	5129	5234	5339	5444	5549	5654	5759	5863	5968	6072
9.6	6176	6280	6384	6488	6592	6696	6799	6903	7006	7109
9.7	7213	7316	7419	7521	7624	7727	7829	7932	8034	8136
9.8	8238	8340	8442	8544	8646	8747	8849	8950	9051	9152
9.9	2.29253	9354	9455	9556	9657	9757	9858	9958	*0058	*0158
10.0	2.30259	0358	0458	0558	0658	0757	0857	0956	1055	1154

TABLE A5.2.2 Natural Logarithms (*Continued*)

To find $\log_e P$ when $0 < P < 1$ or when $P > 10$, write P in the form $10^k \cdot N$ where $1 \leqslant N < 10$ and k is a positive or a negative integer; then use $\log_e P = \log_e(10^k \cdot N) = \log_e 10^k + \log_e N$

k	10^k	$\text{Log}_e 10^k$	k	10^k	$\text{Log}_e 10^k$
1	10	2.30259	−1	0.1	7.69741 − 10
2	100	4.60517	−2	0.01	5.39483 − 10
3	1 000	6.90776	−3	0.001	3.09224 − 10
4	10 000	9.21034	−4	0.0001	0.78966 − 10
5	100 000	11.51293	−5	0.00001	8.48707 − 20
6	1 000 000	13.81551	−6	0.000001	6.18449 − 20

N	$\text{Log } N$	$\text{Log } \frac{1}{10} N$	$\text{Log } \frac{1}{100} N$	N	$\text{Log } N$	$\text{Log } \frac{1}{10} N$	$\text{Log } \frac{1}{100} N$
0	———	———	———	20	2.99573	0.69315	8.39056 − 10
1	0.00000	7.69741*	5.39483 − 10	21	3.04452	0.74194	8.43935 − 10
2	0.69315	8.39056*	6.08798 − 10	22	3.09104	0.78846	8.48587 − 10
3	1.09861	8.79603*	6.49344 − 10	23	3.13549	0.83291	8.53032 − 10
4	1.38629	9.08371*	6.78112 − 10	24	3.17805	0.87547	8.57288 − 10
5	1.60944	9.30685*	7.00427 − 10	25	3.21888	0.91629	8.61371 − 10
6	1.79176	9.48917*	7.18659 − 10	26	3.25810	0.95551	8.65293 − 10
7	1.94591	9.64333*	7.34074 − 10	27	3.29584	0.99325	8.69067 − 10
8	2.07944	9.77686*	7.47427 − 10	28	3.33220	1.02962	8.72703 − 10
9	2.19722	9.89464*	7.59205 − 10	29	3.36730	1.06471	8.76213 − 10
10	2.30259	0.00000	7.69741 − 10	**30**	3.40120	1.09861	8.79603 − 10
11	2.39790	0.09531	7.79273 − 10	31	3.43399	1.13140	8.82882 − 10
12	2.48491	0.18232	7.87974 − 10	32	3.46574	1.16315	8.86057 − 10
13	2.56495	0.26236	7.95978 − 10	33	3.49651	1.19392	8.89134 − 10
14	2.63906	0.33647	8.03389 − 10	34	3.52636	1.22378	8.92119 − 10
15	2.70805	0.40547	8.10288 − 10	35	3.55535	1.25276	8.95018 − 10
16	2.77259	0.47000	8.16742 − 10	36	3.58352	1.28093	8.97835 − 10
17	2.83321	0.53063	8.22804 − 10	37	3.61092	1.30833	9.00575 − 10
18	2.89037	0.58779	8.28520 − 10	38	3.63759	1.33500	9.03242 − 10
19	2.94444	0.64185	8.33927 − 10	39	3.66356	1.36098	9.05839 − 10
20	2.99573	0.69315	8.39056 − 10	**40**	3.68888	1.38629	9.08371 − 10

TABLE A5.2.2   Natural Logarithms (*Continued*)

N	Log N	Log $\frac{1}{10}N$	Log $\frac{1}{100}N$	N	Log N	Log $\frac{1}{10}N$	Log $\frac{1}{100}N$
40	3.68888	1.38629	9.08371 − 10	**70**	4.24850	1.94591	9.64333 − 10
41	3.71357	1.41099	9.10840 − 10	71	4.26268	1.96009	9.65751 − 10
42	3.73767	1.43508	9.13250 − 10	72	4.27667	1.97408	9.67150 − 10
43	3.76120	1.45862	9.15603 − 10	73	4.29046	1.98787	9.68529 − 10
44	3.78419	1.48160	9.17902 − 10	74	4.30407	2.00148	9.69889 − 10
45	3.80666	1.50408	9.20149 − 10	75	4.31749	2.01490	9.71232 − 10
46	3.82864	1.52606	9.22347 − 10	76	4.33073	2.02815	9.72556 − 10
47	3.85015	1.54756	9.24498 − 10	77	4.34381	2.04122	9.73864 − 10
48	3.87120	1.56862	9.26603 − 10	78	4.35671	2.05412	9.75154 − 10
49	3.89182	1.58924	9.28665 − 10	79	4.36945	2.06686	9.76428 − 10
50	3.91202	1.60944	9.30685 − 10	**80**	4.38203	2.07944	9.77686 − 10
51	3.93183	1.62924	9.32666 − 10	81	4.39445	2.09186	9.78928 − 10
52	3.95124	1.64866	9.34607 − 10	82	4.40672	2.10413	9.80155 − 10
53	3.97029	1.66771	9.36512 − 10	83	4.41884	2.11626	9.81367 − 10
54	3.98898	1.68640	9.38381 − 10	84	4.43082	2.12823	9.82565 − 10
55	4.00733	1.70475	9.40216 − 10	85	4.44265	2.14007	9.83748 − 10
56	4.02535	1.72277	9.42018 − 10	86	4.45435	2.15176	9.84918 − 10
57	4.04305	1.74047	9.43788 − 10	87	4.46591	2.16332	9.86074 − 10
58	4.06044	1.75786	9.45527 − 10	88	4.47734	2.17475	9.87217 − 10
59	4.07754	1.77495	9.47237 − 10	89	4.48864	2.18605	9.88347 − 10
60	4.09434	1.79176	9.48917 − 10	**90**	4.49981	2.19722	9.89464 − 10
61	4.11087	1.80829	9.50570 − 10	91	4.51086	2.20827	9.90569 − 10
62	4.12713	1.82455	9.52196 − 10	92	4.52179	2.21920	9.91662 − 10
63	4.14313	1.84055	9.53796 − 10	93	4.53260	2.23001	9.92743 − 10
64	4.15888	1.85630	9.55371 − 10	94	4.54329	2.24071	9.93812 − 10
65	4.17439	1.87180	9.56922 − 10	95	4.55388	2.25129	9.94871 − 10
66	4.18965	1.88707	9.58448 − 10	96	4.56435	2.26176	9.95918 − 10
67	4.20469	1.90211	9.59952 − 10	97	4.57471	2.27213	9.96954 − 10
68	4.21951	1.91692	9.61434 − 10	98	4.58497	2.28238	9.97980 − 10
69	4.23411	1.93152	9.62894 − 10	99	4.59512	2.29253	9.98995 − 10
70	4.24850	1.94591	9.64333 − 10	**100**	4.60517	2.30259	10.00000 − 10

*Subtract 10 from each entry which is followed by an asterisk.

TABLE A5.3 SOME IMPORTANT PROBABILITY DISTRIBUTIONS, MOMENTS, AND GENERATING FUNCTIONS

Name	Probability Function $p(x)$, discrete; $f(x)$, continuous	Parameters	Mean $E[X]$	Variance $E\{X-E[X]\}^2$	Moment Generating Function $E[e^{tX}]$	Characteristic Function $E[e^{itX}]$	Probability Generating Function (Discrete) $\sum_{m=0}^{\infty} p(m)z^m$
Bernoulli	$p(x) = \begin{cases} p & (x=1) \\ 1-p & (x=0) \end{cases}$	$0 \le p \le 1$	p	$p(1-p)$	$pe^t + 1 - p$	$pe^{it} + 1 - p$	$1 - p + pz$
Binomial	$p(x) = \binom{n}{x} p^x (1-p)^{n-x}$ $(x=0,1,\dots,n)$	$n = 1,2,\dots$ $0 \le p \le 1$	np	$np(1-p)$	$(pe^t + 1 - p)^n$	$(pe^{it} + 1 - p)^n$	$(pz + 1 - p)^n$
Poisson	$p(x) = \dfrac{e^{-\lambda}\lambda^x}{x!}$ $(x=0,1,2,\dots)$	$\lambda > 0$	λ	λ	$e^{\lambda(e^t - 1)}$	$e^{\lambda(e^{it} - 1)}$	$e^{-\lambda(1-z)}$
Geometric	$p(x) = p(1-p)^x$ $(x=0,1,2,\dots)$	$0 \le p \le 1$	$\dfrac{1-p}{p}$	$\dfrac{1-p}{p^2}$	$\dfrac{p}{1-(1-p)e^t}$	$\dfrac{p}{1-(1-p)e^{it}}$	$\dfrac{p}{1-(1-p)z}$
Negative binomial	$p(x) = \binom{k+x-1}{x} p^k (1-p)^x$ $(x=0,1,2,\dots)$	$k > 0$ $0 \le p \le 1$	$\dfrac{k(1-p)}{p}$	$\dfrac{k(1-p)}{p^2}$	$\left(\dfrac{p}{1-(1-p)e^t}\right)^k$	$\left(\dfrac{p}{1-(1-p)e^{it}}\right)^k$	$\left(\dfrac{p}{1-(1-p)z}\right)^k$
Uniform	$f(x) = \dfrac{1}{b-a}$ $(a < x \le b)$	$-\infty < a < b < \infty$	$\dfrac{a+b}{2}$	$\dfrac{(b-a)^2}{12}$	$\dfrac{e^{tb} - e^{ta}}{t(b-a)}$	$\dfrac{e^{itb} - e^{ita}}{it(b-a)}$	

	$f(x)$		μ	σ^2	$e^{t\mu+(1/2)t^2\sigma^2}$	$e^{it\mu-(1/2)t^2\sigma^2}$
Normal	$f(x)=\dfrac{1}{\sigma\sqrt{2\pi}}e^{-(1/2)[(x-\mu)/\sigma]^2}$ $(-\infty<x<\infty)$	$-\infty<\mu<\infty$ $\sigma>0$	μ	σ^2		
Exponential	$f(x)=\theta e^{-\theta x}$ $(x>0)$	$\theta>0$	$\dfrac{1}{\theta}$	$\dfrac{1}{\theta^2}$	$\dfrac{\theta}{\theta-t}$	$\dfrac{\theta}{\theta-it}$
Gamma	$f(x)=\dfrac{1}{\Gamma(\alpha)\beta^\alpha}x^{\alpha-1}e^{-x/\beta}$ $(x>0)$	$\alpha,\beta>0$	$\alpha\beta$	$\alpha\beta^2$	$\left(\dfrac{1/\beta}{1/\beta-t}\right)^\alpha$	$\left(\dfrac{1/\beta}{1/\beta-it}\right)^\alpha$
Erlang-k	$f(x)=\dfrac{(\theta k)^k}{(k-1)!}x^{k-1}e^{-k\theta x}$ $(x>0)$	$\theta>0$ $k=1,2,\dots$	$\dfrac{1}{\theta}$	$\dfrac{1}{k\theta^2}$	$\left(\dfrac{k\theta}{k\theta-t}\right)^k$	$\left(\dfrac{k\theta}{k\theta-it}\right)^k$
Chi-square	$f(x)=\dfrac{1}{2^{n/2}\Gamma(n/2)}x^{(n/2)-1}e^{-x/2}$ $(x>0)$	$n=1,2,\dots$	n	$2n$	$\left(\dfrac{1/2}{1/2-t}\right)^{n/2}$	$\left(\dfrac{1/2}{1/2-it}\right)^{n/2}$
Beta	$f(x)=\dfrac{\Gamma(\alpha+\beta)}{\Gamma(\alpha)\Gamma(\beta)}x^{\alpha-1}(1-x)^{\beta-1}$ $(0<x<1)$	$\alpha,\beta>0$	$\dfrac{\alpha}{\alpha+\beta}$	$\dfrac{\alpha\beta}{(\alpha+\beta)^2(\alpha+\beta+1)}$	$E[e^{itX}]=\dfrac{\Gamma(\alpha+\beta)}{\Gamma(\alpha)}\displaystyle\sum_{j=0}^{\infty}\dfrac{\Gamma(\alpha+j)(it)^j}{\Gamma(\alpha+\beta+j)\Gamma(j+1)}$ (to get $E[e^{tX}]$ replace (it) by t)	

TABLE A5.4 POISSON PROBABILITIES,[a] $e^{-\lambda}\lambda^x/x!$

$\lambda \backslash x$	0	1	2	3	4	5	6	7	8	9	10	11	12
.1	.9048	.0905	.0045	.0002									
.2	.8187	.1637	.0164	.0011	.0001								
.3	.7408	.2222	.0333	.0033	.0002								
.4	.6703	.2681	.0536	.0072	.0007	.0001							
.5	.6065	.3033	.0758	.0126	.0016	.0002							
.6	.5488	.3293	.0988	.0198	.0030	.0004	.0000						
.7	.4966	.3476	.1217	.0284	.0050	.0007	.0001	.0000					
.8	.4493	.3595	.1438	.0383	.0077	.0012	.0002	.0000					
.9	.4066	.3659	.1647	.0494	.0111	.0020	.0003	.0000					
1.0	.3679	.3679	.1839	.0613	.0153	.0031	.0005	.0001	.0000				
1.1	.3329	.3662	.2014	.0738	.0203	.0045	.0008	.0001	.0000				
1.2	.3012	.3614	.2169	.0867	.0260	.0062	.0012	.0002	.0000				
1.3	.2725	.3543	.2303	.0998	.0324	.0084	.0018	.0003	.0001	.0000			
1.4	.2466	.3452	.2417	.1128	.0395	.0111	.0026	.0005	.0001	.0000			
1.5	.2231	.3347	.2510	.1255	.0471	.0141	.0035	.0008	.0001	.0000			
1.6	.2019	.3230	.2584	.1378	.0551	.0176	.0047	.0011	.0002	.0000			
1.7	.1827	.3106	.2640	.1496	.0636	.0216	.0061	.0015	.0003	.0001			
1.8	.1653	.2975	.2678	.1607	.0723	.0260	.0078	.0020	.0005	.0001	.0000		
1.9	.1496	.2842	.2700	.1710	.0812	.0309	.0098	.0027	.0006	.0001	.0000		
2.0	.1353	.2707	.2707	.1804	.0902	.0361	.0120	.0034	.0009	.0002	.0000		

λ \ x	0	1	2	3	4	5	6	7	8	9	10	11	12
2.2	.1108	.2438	.2681	.1966	.1082	.0476	.0174	.0055	.0015	.0004	.0001	.0000	.0000
2.4	.0907	.2177	.2613	.2090	.1254	.0602	.0241	.0083	.0025	.0007	.0002	.0000	.0000
2.6	.0743	.1931	.2510	.2176	.1414	.0735	.0319	.0118	.0038	.0011	.0003	.0001	.0001
2.8	.0608	.1703	.2384	.2225	.1557	.0872	.0407	.0163	.0057	.0018	.0005	.0001	.0000
3.0	.0498	.1494	.2240	.2240	.1680	.1008	.0504	.0216	.0081	.0027	.0008	.0002	.0001
3.2	.0408	.1304	.2087	.2226	.1781	.1140	.0608	.0278	.0111	.0040	.0013	.0004	.0001
3.4	.0334	.1135	.1929	.2186	.1858	.1264	.0716	.0348	.0148	.0056	.0019	.0006	.0002
3.6	.0273	.0984	.1771	.2125	.1912	.1377	.0826	.0425	.0191	.0076	.0028	.0009	.0003
3.8	.0224	.0850	.1615	.2046	.1944	.1477	.0936	.0508	.0241	.0102	.0039	.0013	.0004
4.0	.0183	.0733	.1465	.1954	.1954	.1563	.1042	.0595	.0298	.0132	.0053	.0019	.0006
5.0	.0067	.0337	.0842	.1404	.1755	.1755	.1462	.1044	.0653	.0363	.0181	.0082	.0034
6.0	.0025	.0149	.0446	.0892	.1339	.1606	.1606	.1377	.1033	.0688	.0413	.0225	.0113
7.0	.0009	.0064	.0223	.0521	.0912	.1277	.1490	.1490	.1304	.1014	.0710	.0452	.0264
8.0	.0003	.0027	.0107	.0286	.0573	.0916	.1221	.1396	.1396	.1241	.0993	.0722	.0481
9.0	.0001	.0011	.0050	.0150	.0337	.0607	.0911	.1171	.1318	.1318	.1186	.0970	.0728
10.0	.0000	.0005	.0023	.0076	.0189	.0378	.0631	.0901	.1126	.1251	.1251	.1137	.0948

x λ	13	14	15	16	17	18	19	20	21	22	23	24
5.0	.0013	.0005	.0002									
6.0	.0052	.0022	.0009	.0003								
7.0	.0142	.0071	.0033	.0014	.0006	.0002	.0001					
8.0	.0295	.0169	.0090	.0045	.0021	.0009	.0004	.0002	.0001			
9.0	.0504	.0324	.0194	.0109	.0058	.0029	.0014	.0006	.0003	.0001		
10.0	.0729	.0521	.0347	.0217	.0128	.0071	.0037	.0019	.0009	.0004	.0002	.0001

aReprinted with permission from Emanuel Parzen, *Modern Probability Theory and Its Applications*. New York: Wiley, 1960, pp. 444–445.

m \ α	.25	.1	.05	.025	.01	.005
1	1.000	3.078	6.314	12.706	31.821	63.657
2	.816	1.886	2.920	4.303	6.965	9.925
3	.765	1.638	2.353	3.182	4.541	5.841
4	.741	1.533	2.132	2.776	3.747	4.604
5	.727	1.476	2.015	2.571	3.365	4.032
6	.718	1.440	1.943	2.447	3.143	3.707
7	.711	1.415	1.895	2.365	2.998	3.499
8	.706	1.397	1.860	2.306	2.896	3.355
9	.703	1.383	1.833	2.262	2.821	3.250
10	.700	1.372	1.812	2.228	2.764	3.169
11	.697	1.363	1.796	2.201	2.718	3.106
12	.695	1.356	1.782	2.179	2.681	3.055
13	.694	1.350	1.771	2.160	2.650	3.012
14	.692	1.345	1.761	2.145	2.624	2.977
15	.691	1.341	1.753	2.131	2.602	2.947
16	.690	1.337	1.746	2.120	2.583	2.921
17	.689	1.333	1.740	2.110	2.567	2.898
18	.688	1.330	1.734	2.101	2.552	2.878
19	.688	1.328	1.729	2.093	2.539	2.861
20	.687	1.325	1.725	2.086	2.528	2.845
21	.686	1.323	1.721	2.080	2.518	2.831
22	.686	1.321	1.717	2.074	2.508	2.819
23	.685	1.319	1.714	2.069	2.500	2.807
24	.685	1.318	1.711	2.064	2.492	2.797
25	.684	1.316	1.708	2.060	2.485	2.787
26	.684	1.315	1.706	2.056	2.479	2.779
27	.684	1.314	1.703	2.052	2.473	2.771
28	.683	1.313	1.701	2.048	2.467	2.763
29	.683	1.311	1.699	2.045	2.462	2.756
30	.683	1.310	1.697	2.042	2.457	2.750
40	.681	1.303	1.684	2.021	2.423	2.704
60	.679	1.296	1.671	2.000	2.390	2.660
120	.677	1.289	1.658	1.980	2.358	2.617
∞	.674	1.282	1.645	1.960	2.326	2.576

[a]Tables A5.5, A5.6, and A5.7 are reprinted with permission from *Introductory Engineering Statistics* by I. Guttman, S. S. Wilks and J. S. Hunter, pp. 501, 499 and 507. New York: Wiley, 1971.

TABLE A5.6 PERCENTAGE POINTS OF THE χ_m^2 DISTRIBUTION

m \ α	.995	.990	.975	.950	.050	.025	.010	.005
1	392704×10^{-10}	157088×10^{-9}	982069×10^{-9}	393214×10^{-8}	3.84146	5.02389	6.63490	7.87944
2	.0100251	.0201007	.0506356	.102587	5.99147	7.37776	9.21034	10.5966
3	.0717212	.114832	.215795	.351846	7.81473	9.34840	11.3449	12.8381
4	.206990	.297110	.484419	.710721	9.48773	11.1433	13.2767	14.8602
5	.411740	.554300	.831211	1.145476	11.0705	12.8325	15.0863	16.7496
6	.675727	.872085	1.237347	1.63539	12.5916	14.4494	16.8119	18.5476
7	.989265	1.239043	1.68987	2.16735	14.0671	16.0128	18.4753	20.2777
8	1.344419	1.646482	2.17973	2.73264	15.5073	17.5346	20.0902	21.9550
9	1.734926	2.087912	2.70039	3.32511	16.9190	19.0228	21.6660	23.5893
10	2.15585	2.55821	3.24697	3.94030	18.3070	20.4831	23.2093	25.1882
11	2.60321	3.05347	3.81575	4.57481	19.6751	21.9200	24.7250	26.7569
12	3.07382	3.57056	4.40379	5.22603	21.0261	23.3367	26.2170	28.2995
13	3.56503	4.10691	5.00874	5.89186	22.3621	24.7356	27.6883	29.8194
14	4.07468	4.66043	5.62872	6.57063	23.6848	26.1190	29.1413	31.3193
15	4.60094	5.22935	6.26214	7.26094	24.9958	27.4884	30.5779	32.8013
16	5.14224	5.81221	6.90766	7.96164	26.2962	28.8454	31.9999	34.2672
17	5.69724	6.40776	7.56418	8.67176	27.5871	30.1910	33.4087	35.7185
18	6.26481	7.01491	8.23075	9.39046	28.8693	31.5264	34.8053	37.1564
19	6.84398	7.63273	8.90655	10.1170	30.1435	32.8523	36.1908	38.5822

TABLE A5.6 PERCENTAGE POINTS OF THE χ_m^2 DISTRIBUTION (*Continued*)

α / m	.995	.990	.975	.950	.050	.025	.010	.005
20	7.43386	8.26040	9.59083	10.8508	31.4104	34.1696	37.5662	39.9968
21	8.03366	8.89720	10.28293	11.5913	32.6705	35.4789	38.9321	41.4010
22	8.64272	9.54249	10.9823	12.3380	33.9244	36.7807	40.2894	42.7956
23	9.26042	10.19567	11.6885	13.0905	35.1725	38.0757	41.6384	44.1813
24	9.88623	10.8564	12.4011	13.8484	36.4151	39.3641	42.9798	45.5585
25	10.5197	11.5240	13.1197	14.6114	37.6525	40.6465	44.3141	46.9278
26	11.1603	12.1981	13.8439	15.3791	38.8852	41.9232	45.6417	48.2899
27	11.8076	12.8786	14.5733	16.1513	40.1133	43.1944	46.9630	49.6449
28	12.4613	13.5648	15.3079	16.9279	41.3372	44.4607	48.2782	50.9933
29	13.1211	14.2565	16.0471	17.7083	42.5569	45.7222	49.5879	52.3356
30	13.7867	14.9535	16.7908	18.4926	43.7729	46.9792	50.8922	53.6720
40	20.7065	22.1643	24.4331	26.5093	55.7585	59.3417	63.6907	66.7659
50	27.9907	29.7067	32.3574	34.7642	67.5048	71.4202	76.1539	79.4900
60	35.5346	37.4848	40.4817	43.1879	79.0819	83.2976	88.3794	91.9517
70	43.2752	45.4418	48.7576	51.7393	90.5312	95.0231	100.425	104.215
80	51.1720	53.5400	57.1532	60.3915	101.879	106.629	112.329	116.321
90	59.1963	61.7541	65.6466	69.1260	113.145	118.136	124.116	128.299
100	67.3276	70.0648	74.2219	77.9295	124.342	129.561	135.807	140.169

TABLE A5.7   PERCENTAGE POINTS OF THE F_{m_1, m_2} DISTRIBUTION

$$\alpha = .05$$

m_1 m_2	1	2	3	4	5	6	7	8	9
1	161.45	199.50	215.71	224.58	230.16	233.99	236.77	238.88	240.54
2	18.513	19.000	19.164	19.247	19.296	19.330	19.353	19.371	19.385
3	10.128	9.5521	9.2766	9.1172	9.0135	8.9406	8.8868	8.8452	8.8123
4	7.7086	6.9443	6.5914	6.3883	6.2560	6.1631	6.0942	6.0410	5.9988
5	6.6079	5.7861	5.4095	5.1922	5.0503	4.9503	4.8759	4.8183	4.7725
6	5.9874	5.1433	4.7571	4.5337	4.3874	4.2839	4.2066	4.1468	4.0990
7	5.5914	4.7374	4.3468	4.1203	3.9715	3.8660	3.7870	3.7257	3.6767
8	5.3177	4.4590	4.0662	3.8378	3.6875	3.5806	3.5005	3.4381	3.3881
9	5.1174	4.2565	3.8626	3.6331	3.4817	3.3738	3.2927	3.2296	3.1789
10	4.9646	4.1028	3.7083	3.4780	3.3258	3.2172	3.1355	3.0717	3.0204
11	4.8443	3.9823	3.5874	3.3567	3.2039	3.0946	3.0123	2.9480	2.8962
12	4.7472	3.8853	3.4903	3.2592	3.1059	2.9961	2.9134	2.8486	2.7964
13	4.6672	3.8056	3.4105	3.1791	3.0254	2.9153	2.8321	2.7669	2.7144
14	4.6001	3.7389	3.3439	3.1122	2.9582	2.8477	2.7642	2.6987	2.6458
15	4.5431	3.6823	3.2874	3.0556	2.9013	2.7905	2.7066	2.6408	2.5876
16	4.4940	3.6337	3.2389	3.0069	2.8524	2.7413	2.6572	2.5911	2.5377
17	4.4513	3.5915	3.1968	2.9647	2.8100	2.6987	2.6143	2.5480	2.4943
18	4.4139	3.5546	3.1599	2.9277	2.7729	2.6613	2.5767	2.5102	2.4563
19	4.3808	3.5219	3.1274	2.8951	2.7401	2.6283	2.5435	2.4768	2.4227
20	4.3513	3.4928	3.0984	2.8661	2.7109	2.5990	2.5140	2.4471	2.3928
21	4.3248	3.4668	3.0725	2.8401	2.6848	2.5727	2.4876	2.4205	2.3661
22	4.3009	3.4434	3.0491	2.8167	2.6613	2.5491	2.4638	2.3965	2.3419
23	4.2793	3.4221	3.0280	2.7955	2.6400	2.5277	2.4422	2.3748	2.3201
24	4.2597	3.4028	3.0088	2.7763	2.6207	2.5082	2.4226	2.3551	2.3002
25	4.2417	3.3852	2.9912	2.7587	2.6030	2.4904	2.4047	2.3371	2.2821
26	4.2252	3.3690	2.9751	2.7426	2.5868	2.4741	2.3883	2.3205	2.2655
27	4.2100	3.3541	2.9604	2.7278	2.5719	2.4591	2.3732	2.3053	2.2501
28	4.1960	3.3404	2.9467	2.7141	2.5581	2.4453	2.3593	2.2913	2.2360
29	4.1830	3.3277	2.9340	2.7014	2.5454	2.4324	2.3463	2.2782	2.2229
30	4.1709	3.3158	2.9223	2.6896	2.5336	2.4205	2.3343	2.2662	2.2107
40	4.0848	3.2317	2.8387	2.6060	2.4495	2.3359	2.2490	2.1802	2.1240
60	4.0012	3.1504	2.7581	2.5252	2.3683	2.2540	2.1665	2.0970	2.0401
120	3.9201	3.0718	2.6802	2.4472	2.2900	2.1750	2.0867	2.0164	1.9588
∞	3.8415	2.9957	2.6049	2.3719	2.2141	2.0986	2.0096	1.9384	1.8799

TABLE A5.7 PERCENTAGE POINTS OF THE F_{m_1, m_2} DISTRIBUTION (*Continued*)

$\alpha = .05$

m_2 \ m_1	10	12	15	20	24	30	40	60	120	∞
1	241.88	243.91	245.95	248.01	249.05	250.09	251.14	252.20	253.25	254.32
2	19.396	19.413	19.429	19.446	19.454	19.462	19.471	19.479	19.487	19.496
3	8.7855	8.7446	8.7029	8.6602	8.6385	8.6166	8.5944	8.5720	8.5494	8.5265
4	5.9644	5.9117	5.8578	5.8025	5.7744	5.7459	5.7170	5.6878	5.6581	5.6281
5	4.7351	4.6777	4.6188	4.5581	4.5272	4.4957	4.4638	4.4314	4.3984	4.3650
6	4.0600	3.9999	3.9381	3.8742	3.8415	3.8082	3.7743	3.7398	3.7047	3.6688
7	3.6365	3.5747	3.5108	3.4445	3.4105	3.3758	3.3404	3.3043	3.2674	3.2298
8	3.3472	3.2840	3.2184	3.1503	3.1152	3.0794	3.0428	3.0053	2.9669	2.9276
9	3.1373	3.0729	3.0061	2.9365	2.9005	2.8637	2.8259	2.7872	2.7475	2.7067
10	2.9782	2.9130	2.8450	2.7740	2.7372	2.6996	2.6609	2.6211	2.5801	2.5379
11	2.8536	2.7876	2.7186	2.6464	2.6090	2.5705	2.5309	2.4901	2.4480	2.4045
12	2.7534	2.6866	2.6169	2.5436	2.5055	2.4663	2.4259	2.3842	2.3410	2.2962
13	2.6710	2.6037	2.5331	2.4589	2.4202	2.3803	2.3392	2.2966	2.2524	2.2064
14	2.6021	2.5342	2.4630	2.3879	2.3487	2.3082	2.2664	2.2230	2.1778	2.1307
15	2.5437	2.4753	2.4035	2.3275	2.2878	2.2468	2.2043	2.1601	2.1141	2.0658
16	2.4935	2.4247	2.3522	2.2756	2.2354	2.1938	2.1507	2.1058	2.0589	2.0096
17	2.4499	2.3807	2.3077	2.2304	2.1898	2.1477	2.1040	2.0584	2.0107	1.9604
18	2.4117	2.3421	2.2686	2.1906	2.1497	2.1071	2.0629	2.0166	1.9681	1.9168
19	2.3779	2.3080	2.2341	2.1555	2.1141	2.0712	2.0264	1.9796	1.9302	1.8780
20	2.3479	2.2776	2.2033	2.1242	2.0825	2.0391	1.9938	1.9464	1.8963	1.8432
21	2.3210	2.2504	2.1757	2.0960	2.0540	2.0102	1.9645	1.9165	1.8657	1.8117
22	2.2967	2.2258	2.1508	2.0707	2.0283	1.9842	1.9380	1.8895	1.8380	1.7831
23	2.2747	2.2036	2.1282	2.0476	2.0050	1.9605	1.9139	1.8649	1.8128	1.7570
24	2.2547	2.1834	2.1077	2.0267	1.9838	1.9390	1.8920	1.8424	1.7897	1.7331
25	2.2365	2.1649	2.0889	2.0075	1.9643	1.9192	1.8718	1.8217	1.7684	1.7110
26	2.2197	2.1479	2.0716	1.9898	1.9464	1.9010	1.8533	1.8027	1.7488	1.6906
27	2.2043	2.1323	2.0558	1.9736	1.9299	1.8842	1.8361	1.7851	1.7307	1.6717
28	2.1900	2.1179	2.0411	1.9586	1.9147	1.8687	1.8203	1.7689	1.7138	1.6541
29	2.1768	2.1045	2.0275	1.9446	1.9005	1.8543	1.8055	1.7537	1.6981	1.6377
30	2.1646	2.0921	2.0148	1.9317	1.8874	1.8409	1.7918	1.7396	1.6835	1.6223
40	2.0772	2.0035	1.9245	1.8389	1.7929	1.7444	1.6928	1.6373	1.5766	1.5089
60	1.9926	1.9174	1.8364	1.7480	1.7001	1.6491	1.5943	1.5343	1.4673	1.3893
120	1.9105	1.8337	1.7505	1.6587	1.6084	1.5543	1.4952	1.4290	1.3519	1.2539
∞	1.8307	1.7522	1.6664	1.5705	1.5173	1.4591	1.3940	1.3180	1.2214	1.0000

$$\alpha = .025$$

m_1 m_2	1	2	3	4	5	6	7	8	9
1	647.79	799.50	864.16	899.58	921.85	937.11	948.22	956.66	963.28
2	38.506	39.000	39.165	39.248	39.298	39.331	39.355	39.373	39.387
3	17.443	16.044	15.439	15.101	14.885	14.735	14.624	14.540	14.473
4	12.218	10.649	9.9792	9.6045	9.3645	9.1973	9.0741	8.9796	8.9047
5	10.007	8.4336	7.7636	7.3879	7.1464	6.9777	6.8531	6.7572	6.6810
6	8.8131	7.2598	6.5988	6.2272	5.9876	5.8197	5.6955	5.5996	5.5234
7	8.0727	6.5415	5.8898	5.5226	5.2852	5.1186	4.9949	4.8994	4.8232
8	7.5709	6.0595	5.4160	5.0526	4.8173	4.6517	4.5286	4.4332	4.3572
9	7.2093	5.7147	5.0781	4.7181	4.4844	4.3197	4.1971	4.1020	4.0260
10	6.9367	5.4564	4.8256	4.4683	4.2361	4.0721	3.9498	3.8549	3.7790
11	6.7241	5.2559	4.6300	4.2751	4.0440	3.8807	3.7586	3.6638	3.5879
12	6.5538	5.0959	4.4742	4.1212	3.8911	3.7283	3.6065	3.5118	3.4358
13	6.4143	4.9653	4.3472	3.9959	3.7667	3.6043	3.4827	3.3880	3.3120
14	6.2979	4.8567	4.2417	3.8919	3.6634	3.5014	3.3799	3.2853	3.2093
15	6.1995	4.7650	4.1528	3.8043	3.5764	3.4147	3.2934	3.1987	3.1227
16	6.1151	4.6867	4.0768	3.7294	3.5021	3.3406	3.2194	3.1248	3.0488
17	6.0420	4.6189	4.0112	3.6648	3.4379	3.2767	3.1556	3.0610	2.9849
18	5.9781	4.5597	3.9539	3.6083	3.3820	3.2209	3.0999	3.0053	2.9291
19	5.9216	4.5075	3.9034	3.5587	3.3327	3.1718	3.0509	2.9563	2.8800
20	5.8715	4.4613	3.8587	3.5147	3.2891	3.1283	3.0074	2.9128	2.8365
21	5.8266	4.4199	3.8188	3.4754	3.2501	3.0895	2.9686	2.8740	2.7977
22	5.7863	4.3828	3.7829	3.4401	3.2151	3.0546	2.9338	2.8392	2.7628
23	5.7498	4.3492	3.7505	3.4083	3.1835	3.0232	2.9024	2.8077	2.7313
24	5.7167	4.3187	3.7211	3.3794	3.1548	2.9946	2.8738	2.7791	2.7027
25	5.6864	4.2909	3.6943	3.3530	3.1287	2.9685	2.8478	2.7531	2.6766
26	5.6586	4.2655	3.6697	3.3289	3.1048	2.9447	2.8240	2.7293	2.6528
27	5.6331	4.2421	3.6472	3.3067	3.0828	2.9228	2.8021	2.7074	2.6309
28	5.6096	4.2205	3.6264	3.2863	3.0625	2.9027	2.7820	2.6872	2.6106
29	5.5878	4.2006	3.6072	3.2674	3.0438	2.8840	2.7633	2.6686	2.5919
30	5.5675	4.1821	3.5894	3.2499	3.0265	2.8667	2.7460	2.6513	2.5746
40	5.4239	4.0510	3.4633	3.1261	2.9037	2.7444	2.6238	2.5289	2.4519
60	5.2857	3.9253	3.3425	3.0077	2.7863	2.6274	2.5068	2.4117	2.3344
120	5.1524	3.8046	3.2270	2.8943	2.6740	2.5154	2.3948	2.2994	2.2217
∞	5.0239	3.6889	3.1161	2.7858	2.5665	2.4082	2.2875	2.1918	2.1136

TABLE A5.7 PERCENTAGE POINTS OF THE F_{m_1, m_2} DISTRIBUTION (*Continued*)

$\alpha = .025$

m_2 \ m_1	10	12	15	20	24	30	40	60	120	∞
1	968.63	976.71	984.87	993.10	997.25	1001.4	1005.6	1009.8	1014.0	1018.3
2	39.398	39.415	39.431	39.448	39.456	39.465	39.473	39.481	39.490	39.498
3	14.419	14.337	14.253	14.167	14.124	14.081	14.037	13.992	13.947	13.902
4	8.8439	8.7512	8.6565	8.5599	8.5109	8.4613	8.4111	8.3604	8.3092	8.2573
5	6.6192	6.5246	6.4277	6.3285	6.2780	6.2269	6.1751	6.1225	6.0693	6.0153
6	5.4613	5.3662	5.2687	5.1684	5.1172	5.0652	5.0125	5.9589	4.9045	4.8491
7	4.7611	4.6658	4.5678	4.4667	4.4150	4.3624	4.3089	4.2544	4.1989	4.1423
8	4.2951	4.1997	4.1012	3.9995	3.9472	3.8940	3.8398	3.7844	3.7279	3.6702
9	3.9639	3.8682	3.7694	3.6669	3.6142	3.5604	3.5055	3.4493	3.3918	3.3329
10	3.7168	3.6209	3.5217	3.4186	3.3654	3.3110	3.2554	3.1984	3.1399	3.0798
11	3.5257	3.4296	3.3299	3.2261	3.1725	3.1176	3.0613	3.0035	2.9441	2.8828
12	3.3736	3.2773	3.1772	3.0728	3.0187	2.9633	2.9063	2.8478	2.7874	2.7249
13	3.2497	3.1532	3.0527	2.9477	2.8932	2.8373	2.7797	2.7204	2.6590	2.5955
14	3.1469	3.0501	2.9493	2.8437	2.7888	2.7324	2.6742	2.6142	2.5519	2.4872
15	3.0602	2.9633	2.8621	2.7559	2.7006	2.6437	2.5850	2.5242	2.4611	2.3953
16	2.9862	2.8890	2.7875	2.6808	2.6252	2.5678	2.5085	2.4471	2.3831	2.3163
17	2.9222	2.8249	2.7230	2.6158	2.5598	2.5021	2.4422	2.3801	2.3153	2.2474
18	2.8664	2.7689	2.6667	2.5590	2.5027	2.4445	2.3842	2.3214	2.2558	2.1869
19	2.8173	2.7196	2.6171	2.5089	2.4523	2.3937	2.3329	2.2695	2.2032	2.1333
20	2.7737	2.6758	2.5731	2.4645	2.4076	2.3486	2.2873	2.2234	2.1562	2.0853
21	2.7348	2.6368	2.5338	2.4247	2.3675	2.3082	2.2465	2.1819	2.1141	2.0422
22	2.6998	2.6017	2.4984	2.3890	2.3315	2.2718	2.2097	2.1446	2.0760	2.0032
23	2.6682	2.5699	2.4665	2.3567	2.2989	2.2389	2.1763	2.1107	2.0415	1.9677
24	2.6396	2.5412	2.4374	2.3273	2.2693	2.2090	2.1460	2.0799	2.0099	1.9353
25	2.6135	2.5149	2.4110	2.3005	2.2422	2.1816	2.1183	2.0517	1.9811	1.9055
26	2.5895	2.4909	2.3867	2.2759	2.2174	2.1565	2.0928	2.0257	1.9545	1.8781
27	2.5676	2.4688	2.3644	2.2533	2.1946	2.1334	2.0693	2.0018	1.9299	1.8527
28	2.5473	2.4484	2.3438	2.2324	2.1735	2.1121	2.0477	1.9796	1.9072	1.8291
29	2.5286	2.4295	2.3248	2.2131	2.1540	2.0923	2.0276	1.9591	1.8861	1.8072
30	2.5112	2.4120	2.3072	2.1952	2.1359	2.0739	2.0089	1.9400	1.8664	1.7867
40	2.3882	2.2882	2.1819	2.0677	2.0069	1.9429	1.8752	1.8028	1.7242	1.6371
60	2.2702	2.1692	2.0613	1.9445	1.8817	1.8152	1.7440	1.6668	1.5810	1.4822
120	2.1570	2.0548	1.9450	1.8249	1.7597	1.6899	1.6141	1.5299	1.4327	1.3104
∞	2.0483	1.9447	1.8326	1.7085	1.6402	1.5660	1.4835	1.3883	1.2684	1.0000

544

$$\alpha = .01$$

m_2 \ m_1	1	2	3	4	5	6	7	8	9
1	4052.2	4999.5	5403.3	5624.6	5763.7	5859.0	5928.3	5981.6	6022.5
2	98.503	99.000	99.166	99.249	99.299	99.332	99.356	99.374	99.388
3	34.116	30.817	29.457	28.710	28.237	27.911	27.672	27.489	27.345
4	21.198	18.000	16.694	15.977	15.522	15.207	14.976	14.799	14.659
5	16.258	13.274	12.060	11.392	10.967	10.672	10.456	10.289	10.158
6	13.745	10.925	9.7795	9.1483	8.7459	8.4661	8.2600	8.1016	7.9761
7	12.246	9.5466	8.4513	7.8467	7.4604	7.1914	6.9928	6.8401	6.7188
8	11.259	8.6491	7.5910	7.0060	6.6318	6.3707	6.1776	6.0289	5.9106
9	10.561	8.0215	6.9919	6.4221	6.0569	5.8018	5.6129	5.4671	5.3511
10	10.044	7.5594	6.5523	5.9943	5.6363	5.3858	5.2001	5.0567	4.9424
11	9.6460	7.2057	6.2167	5.6683	5.3160	5.0692	4.8861	4.7445	4.6315
12	9.3302	6.9266	5.9526	5.4119	5.0643	4.8206	4.6395	4.4994	4.3875
13	9.0738	6.7010	5.7394	5.2053	4.8616	4.6204	4.4410	4.3021	4.1911
14	8.8616	6.5149	5.5639	5.0354	4.6950	4.4558	4.2779.	4.1399	4.0297
15	8.6831	6.3589	5.4170	4.8932	4.5556	4.3183	4.1415	4.0045	3.8948
16	8.5310	6.2262	5.2922	4.7726	4.4374	4.2016	4.0259	3.8896	3.7804
17	8.3997	6.1121	5.1850	4.6690	4.3359	4.1015	3.9267	3.7910	3.6822
18	8.2854	6.0129	5.0919	4.5790	4.2479	4.0146	3.8406	3.7054	3.5971
19	8.1850	5.9259	5.0103	4.5003	4.1708	3.9386	3.7653	3.6305	3.5225
20	8.0960	5.8489	4.9382	4.4307	4.1027	3.8714	3.6987	3.5644	3.4567
21	8.0166	5.7804	4.8740	4.3688	4.0421	3.8117	3.6396	3.5056	3.3981
22	7.9454	5.7190	4.8166	4.3134	3.9880	3.7583	3.5867	3.4530	3.3458
23	7.8811	5.6637	4.7649	4.2635	3.9392	3.7102	3.5390	3.4057	3.2986
24	7.8229	5.6136	4.7181	4.2184	3.8951	3.6667	3.4959	3.3629	3.2560
25	7.7698	5.5680	4.6755	4.1774	3.8550	3.6272	3.4568	3.3239	3.2172
26	7.7213	5.5263	4.6366	4.1400	3.8183	3.5911	3.4210	3.2884	3.1818
27	7.6767	5.4881	4.6009	4.1056	3.7848	3.5580	3.3882	3.2558	3.1494
28	7.6356	5.4529	4.5681	4.0740	3.7539	3.5276	3.3581	3.2259	3.1195
29	7.5976	5.4205	4.5378	4.0449	3.7254	3.4995	3.3302	3.1982	3.0920
30	7.5625	5.3904	4.5097	4.0179	3.6990	3.4735	3.3045	3.1726	3.0665
40	7.3141	5.1785	4.3126	3.8283	3.5138	3.2910	3.1238	2.9930	2.8876
60	7.0771	4.9774	4.1259	3.6491	3.3389	3.1187	2.9530	2.8233	2.7185
120	6.8510	4.7865	3.9493	3.4796	3.1735	2.9559	2.7918	2.6629	2.5586
∞	6.6349	4.6052	3.7816	3.3192	3.0173	2.8020	2.6393	2.5113	2.4073

$$\alpha = .01$$

m_2 \ m_1	10	12	15	20	24	30	40	60	120	∞
1	6055.8	6106.3	6157.3	6208.7	6234.6	6260.7	6286.8	6313.0	6339.4	6366.0
2	99.399	99.416	99.432	99.449	99.458	99.466	99.474	99.483	99.491	99.501
3	27.229	27.052	26.872	26.690	26.598	26.505	26.411	26.316	26.221	26.125
4	14.546	14.374	14.198	14.020	13.929	13.838	13.745	13.652	13.558	13.463
5	10.051	9.8883	9.7222	9.5527	9.4665	9.3793	9.2912	9.2020	9.1118	9.0204
6	7.8741	7.7183	7.5590	7.3958	7.3127	7.2285	7.1432	7.0568	6.9690	6.8801
7	6.6201	6.4691	6.3143	6.1554	6.0743	5.9921	5.9084	5.8236	5.7372	5.6495
8	5.8143	5.6668	5.5151	5.3591	5.2793	5.1981	5.1156	5.0316	4.9460	4.8588
9	5.2565	5.1114	4.9621	4.8080	4.7290	4.6486	4.5667	4.4831	4.3978	4.3105
10	4.8492	4.7059	4.5582	4.4054	4.3269	4.2469	4.1653	4.0819	3.9965	3.9090
11	4.5393	4.3974	4.2509	4.0990	4.0209	3.9411	3.8596	3.7761	3.6904	3.6025
12	4.2961	4.1553	4.0096	3.8584	3.7805	3.7008	3.6192	3.5355	3.4494	3.3608
13	4.1003	3.9603	3.8154	3.6646	3.5868	3.5070	3.4253	3.3413	3.2548	3.1654
14	3.9394	3.8001	3.6557	3.5052	3.4274	3.3476	3.2656	3.1813	3.0942	3.0040
15	3.8049	3.6662	3.5222	3.3719	3.2940	3.2141	3.1319	3.0471	2.9595	2.8684
16	3.6909	3.5527	3.4089	3.2588	3.1808	3.1007	3.0182	2.9330	2.8447	2.7528
17	3.5931	3.4552	3.3117	3.1615	3.0835	3.0032	2.9205	2.8348	2.7459	2.6530
18	3.5082	3.3706	3.2273	3.0771	2.9990	2.9185	2.8354	2.7493	2.6597	2.5660
19	3.4338	3.2965	3.1533	3.0031	2.9249	2.8442	2.7608	2.6742	2.5839	2.4893
20	3.3682	3.2311	3.0880	2.9377	2.8594	2.7785	2.6947	2.6077	2.5168	2.4212
21	3.3098	3.1729	3.0299	2.8796	2.8011	2.7200	2.6359	2.5484	2.4568	2.3603
22	3.2576	3.1209	2.9780	2.8274	2.7488	2.6675	2.5831	2.4951	2.4029	2.3055
23	3.2106	3.0740	2.9311	2.7805	2.7017	2.6202	2.5355	2.4471	2.3542	2.2559
24	3.1681	3.0316	2.8887	2.7380	2.6591	2.5773	2.4923	2.4035	2.3099	2.2107
25	3.1294	2.9931	2.8502	2.6993	2.6203	2.5383	2.4530	2.3637	2.2695	2.1694
26	3.0941	2.9579	2.8150	2.6640	2.5848	2.5026	2.4170	2.3273	2.2325	2.1315
27	3.0618	2.9256	2.7827	2.6316	2.5522	2.4699	2.3840	2.2938	2.1984	2.0965
28	3.0320	2.8959	2.7530	2.6017	2.5223	2.4397	2.3535	2.2629	2.1670	2.0642
29	3.0045	2.8685	2.7256	2.5742	2.4946	2.4118	2.3253	2.2344	2.1378	2.0342
30	2.9791	2.8431	2.7002	2.5487	2.4689	2.3860	2.2992	2.2079	2.1107	2.0062
40	2.8005	2.6648	2.5216	2.3689	2.2880	2.2034	2.1142	2.0194	1.9172	1.8047
60	2.6318	2.4961	2.3523	2.1978	2.1154	2.0285	1.9360	1.8363	1.7263	1.6006
120	2.4721	2.3363	2.1915	2.0346	1.9500	1.8600	1.7628	1.6557	1.5330	1.3805
∞	2.3209	2.1848	2.0385	1.8783	1.7908	1.6964	1.5923	1.4730	1.3246	1.0000

Index

Prior to presenting the standard index, we first give Table I.1 which lists the models treated, the types of results developed, and the section of the book in which each model is discussed.

Types of results are categorized by (1) analytical or closed form expressions, such as $p_n = (1 - \rho)\rho^n$ or $L = \rho/(1 - \rho)$, (2) numerical results where expressions must be calculated iteratively or have terms involving infinite summations, (3) expressions in terms of generating functions or transforms where the actual form of results to follow from these depends on specific distributions, and (4) no results obtained.

TABLE I.1
SUMMARY OF MODELS TREATED AND TYPES OF RESULTS – STEADY STATE

Notation

a = analytical results
n = numerical results
g = results in form of generating function or Laplace transform
0 = no results
$-$ = not applicable

Model (Notation Explained in Table 1.1, Section 1.4)	Types of Results for:			Section
	$\{p_n\}$	Expected Value Measures (L, L_q, W, W_q)	Waiting Time Distribution	
FIFO				
$D/D/1$ $(D/D/c)$		Graphical solutions obtainable		1.7
$M/M/1$	a	a	a	2.1–2.4
$M/M/1/K$	a	a	a	2.5
$M/M/c$	a	a	a	3.2
$M/M/c/K$	a	a	a^1	3.3
$M/M/c/c$	a	a	$-$	3.4
$M/M/\infty$	a	a	$-$	3.5.1
Finite source $M/M/c$	a	a	a	3.6
State-Dep. Serv. $M/M/1$	a, n^2	a, n^2	0	3.7
Impatience $M/M/c$	a, n^2	a, n^2	a, n^2	3.8
$M[X]/M/1$	g	a	0	4.1
$M[X]/M/c$		Results indicated		4.1
$M/M[K]/1$	a^3	a^3	0	4.2
$M/M[K]/c$		Results indicated		4.2
$M/E_k/1$ $(M/D/1)$	$g, (a)$	a	0	4.3.1, (5.1.5)
$M/D/c$	g	a	0	6.2
$E_k/M/1$ $(D/M/1)$	a^3	n	n	4.3.2
$E_k/E_l/c$ $(E_k/E_l/1)$		Some numerical results possible—reference given		4.3.2, (6.1)
$M/G/1$	n	a	n	5.1.1, 5.1.2, 5.1.6
$M/G/1/K$	g	g	0	5.1.8

[1] Indicated but not presented.
[2] Depends on particular model.
[3] Analytical results follow after a root to a nonlinear equation is found; finding the root may require numerical analysis.